# Springer Complexity

Springer Complexity is an interdisciplinary program publishing the best research and academic-level teaching on both fundamental and applied aspects of complex systems – cutting across all traditional disciplines of the natural and life sciences, engineering, economics, medicine, neuroscience, social and computer science.

Complex Systems are systems that comprise many interacting parts with the ability to generate a new quality of macroscopic collective behavior the manifestations of which are the spontaneous formation of distinctive temporal, spatial or functional structures. Models of such systems can be successfully mapped onto quite diverse "real-life" situations like the climate, the coherent emission of light from lasers, chemical reaction-diffusion systems, biological cellular networks, the dynamics of stock markets and of the internet, earthquake statistics and prediction, freeway traffic, the human brain, or the formation of opinions in social systems, to name just some of the popular applications.

Although their scope and methodologies overlap somewhat, one can distinguish the following main concepts and tools: self-organization, nonlinear dynamics, synergetics, turbulence, dynamical systems, catastrophes, instabilities, stochastic processes, chaos, graphs and networks, cellular automata, adaptive systems, genetic algorithms and computational intelligence.

The two major book publication platforms of the Springer Complexity program are the monograph series "Understanding Complex Systems" focusing on the various applications of complexity, and the "Springer Series in Synergetics", which is devoted to the quantitative theoretical and methodological foundations. In addition to the books in these two core series, the program also incorporates individual titles ranging from textbooks to major reference works.

# Understanding Complex Systems

**Founding Editor: J.A. Scott Kelso**

Future scientific and technological developments in many fields will necessarily depend upon coming to grips with complex systems. Such systems are complex in both their composition – typically many different kinds of components interacting simultaneously and nonlinearly with each other and their environments on multiple levels – and in the rich diversity of behavior of which they are capable.

The Springer Series in Understanding Complex Systems series (UCS) promotes new strategies and paradigms for understanding and realizing applications of complex systems research in a wide variety of fields and endeavors. UCS is explicitly transdisciplinary. It has three main goals: First, to elaborate the concepts, methods and tools of complex systems at all levels of description and in all scientific fields, especially newly emerging areas within the life, social, behavioral, economic, neuro- and cognitive sciences (and derivatives thereof); second, to encourage novel applications of these ideas in various fields of engineering and computation such as robotics, nano-technology and informatics; third, to provide a single forum within which commonalities and differences in the workings of complex systems may be discerned, hence leading to deeper insight and understanding.

UCS will publish monographs, lecture notes and selected edited contributions aimed at communicating new findings to a large multidisciplinary audience.

Peter beim Graben · Changsong Zhou ·
Marco Thiel · Jürgen Kurths (Eds.)

# Lectures in Supercomputational Neuroscience

## Dynamics in Complex Brain Networks

With 179 Figures and 18 Tables

 Springer

Dr. Peter beim Graben
University of Reading
School of Psychology and Clinical Language
Sciences
Whiteknights, PO Box 217
Reading RG6 6AH
United Kingdom

Prof. Dr. Changsong Zhou
Hong Kong Baptist University
Department of Physics
224 Waterloo Road
Kowloon Tong, Hong Kong
China

Dr. Marco Thiel
University of Aberdeen
School of Engineering and Physical Sciences
Aberdeen AB24 3UE
United Kingdom

Prof. Dr. Jürgen Kurths
Universität Potsdam
Institut für Physik
LS Theoretische Physik
Am Neuen Palais 10
14469 Potsdam, Germany

Library of Congress Control Number: 2007932774

ISSN 1860-0832
ISBN 978-3-540-73158-0 Springer Berlin Heidelberg New York

Springer is a part of Springer Science+Business Media
springer.com
© Springer-Verlag Berlin Heidelberg 2008

Typesetting: by the author and Integra, India using a Springer LATEX macro package
Cover design: WMX Design, Heidelberg

Printed on acid-free paper        SPIN: 12066483        5 4 3 2 1 0

# Preface

Computational neuroscience has become a very active field of research in the last decades. Improved experimental facilities, new mathematical techniques and especially the exponential increase of computational power have lead to stunning new insights into the functioning of the brain. Scientists begin to endeavor simulating the brain from the bottom level of single neurons to the top-level of cognitive behavior. The "Blue Brain Project" (http://bluebrainproject.epfl.ch/) for example, is a hallmark for this approach.

Many scientists are attracted to this highly interdisciplinary field of research, in which only the combined efforts of neuroscientists, biologists, psychologists, physicists and mathematicians, computer scientists, engineers and other specialists, e.g. from anthropology, linguistics, or medicine, seem to be able to shift the limits of our knowledge. However, one of the most common problems of interdisciplinary work is to find a "common language", i.e., an effective way to discuss problems with colleagues with a different scientific background. Therefore, an introduction into this field has to familiarize the reader with aspects from various relevant fields in an intelligible way.

This book is an introduction to the field of computational neuroscience from a physicist's perspective, regarded as *neurophysics*, with in depth contributions of systems neuroscientists. It is based upon the lectures delivered during the *5th Helmholtz Summer School on Supercomputational Physics*:

"Complex Networks in Brain Dynamics"

held in September 2005 at the University of Potsdam.

The book-title *Lectures in Supercomputational Neuroscience: Dynamics in Complex Brain Networks* is motivated by the methods and outcomes of the Summer School: A conceptual model for complex networks of neurons is introduced, which incorporates many important features of the "real" brain, such as different types of neurons, various brain areas, inhibitory and excitatory coupling and plasticity of the network. The model is then implemented in an MPI (message-passing interface)-based parallel computer code, running at appropriate supercomputers, that is introduced and discussed in detail in this book. But beyond the mere presentation of the C-program, the text will enable the reader to modify and adapt the algorithm for his/her own research.

The first part of the book (**Neurophysiology**) gives an introduction to the physiology of the brain on different levels, ranging from the rather large areas of the brain down to individual neurons. Various models for individual neurons are discussed as well as models for the "communication" among these neuronal oscillators. An outlook on cognition and learning is also given in this first part.

The second part (**Complex Networks**) outlines the dynamics of ensembles of neurons forming different types of networks. Recently developed new approaches based on complex networks with special emphasis on the relationships between structure and function of complex systems are presented. The topology of such a network, i.e., how the neurons are coupled, plays an important role for the behavior of the ensemble. Even though the network of the $10^{10}$ neurons in a human brain is much too complex to be modeled with our present knowledge, the conceptual models presented here are a promising starting point and allow gaining insight into the principles of complex networks in brain dynamics. This part covers all aspects from the basics of networks, their topology and how to quantify them, to the structure and function of complex cortical networks up to collective behavior of large networks such as clustered synchronization. New techniques for the analysis of data of complex networks are also introduced. They allow not only to study large populations of neurons but also to study (neural) oscillators with more than one time scale, e.g. spiking and bursting neurons.

The third part (**Cognition and Higher Perception**) presents results about how structural units of the brain (columns) can be described and how networks of neurons can be used to model cognition and perception as measured by the electroencephalogram. It is shown how networks of simple neuronal models can be used to model, e.g., reaction times from psychological experiments.

The forth part (**Implementations**) discusses the implementation of a model of a network of networks of neurons in an MPI-based C-code. The code is modular in the sense that the model(s) for the neurons, the topology of the network, the coupling and many further parameters can easily be changed and adapted. The main point is to outline how in principle many different features can be implemented in a computer code, rather than presenting a cutting-edge algorithm. The computer code is available for download (http://www.agnld.uni-potsdam.de). In this part we also discuss that computational neuroscience is not simply about parallelizing normal computer code. A very important component of it is the implementation of specially adapted algorithms. An example of such a computer code will be given in this chapter.

In the fifth and final part (**Applications**), three groups of students of the Summer School, discuss the results they obtained running the code on supercomputers. After studying the parameter space for large networks of Morris-Lecar neurons, they use a map of cortical connections from a cat's brain, which was obtained based on experimental studies. In their simulations they consider multiple spatio-temporal scales and study the patterns for synchronized firing of neurons in different brain areas. Results of simulations for different network topologies and neuronal models are also summarized here. These chapters will be helpful to those who are planning to apply the parallel code for their own research, as they give a very practical account of how to actually perform simulations. They point at crucial problems and show how to overcome pitfalls when simulating based on the MPI code.

We hope that this book will help graduate students and researchers to access the field of computational neuroscience and to develop and improve high-end, parallel computer codes for the simulation of large networks of neurons.

Last, but not least, we wish to thank all lecturers and the coordinators of the 5th Helmholtz Summer School on Supercomputational Physics; Mamen Romano, Lucia Zemanová, and Gorka Zamora-López for their assistance; the Land Brandenburg for main funding, EU, NoE, and EU-Network BioSim (contract No. LSHB–CT–2004–005137) for further support; the University of Potsdam for making access to its supercomputer cluster available and also for logistics. Finally, we thank James Ong for his careful proof-reading of the complete book.

Nonlinear Dynamics Group                              Peter beim Graben
University of Potsdam                                    Changsong Zhou
                                                          Marco Thiel
                                                        Jürgen Kurths

# Contents

# Part I

## Neurophysiology

# 1

# Foundations of Neurophysics

Peter beim Graben[1,2]

[1] School of Psychology and Clinical Language Sciences,
University of Reading, United Kingdom
p.r.beimgraben@reading.ac.uk
[2] Institute of Physics, Nonlinear Dynamics Group, Universität Potsdam,
Germany

**Summary.** This chapter presents an introductory course to the biophysics of neurons, comprising a discussion of ion channels, active and passive membranes, action potentials and postsynaptic potentials. It reviews several conductance-based and reduced neuron models, neural networks and neural field theories. Finally, the basic principles of the neuroelectrodynamics of mass potentials, i.e. dendritic fields, local field potentials, and the electroencephalogram are elucidated and their putative functional role as a mean field is discussed.

## 1.1 Introduction

Metaphorically, the brain is often compared with a digital computer [1, 2] that runs *software* algorithms in order to perform cognitive computations. In spite of its usefulness as a working hypothesis in the cognitive [3–6] and computational [7–18] neurosciences, this metaphor does obviously not apply to the *hardware* level. Digital computers consist of circuit boards equipped with chips, transistors, resistors, capacitances, power supplies, and other electronic components wired together. Digital computation is essentially based on controlled switching processes in semiconductors which are nonlinear physical systems. On the other hand, brains consist to 80% of water contained in cells and also surrounding cells. How can this physical *wet-ware* substrate support computational dynamics? This question should be addressed in the present chapter. Starting from the physiological facts about neurons, their cell membranes, electrolytes, and ions [19–21], I shall outline the biophysical principles of neural computation [12, 13, 15, 18, 22–25] in parallel to those of computation in electronic circuits. Thus, the interesting physiological properties will be described by electric "equivalent circuits" providing a construction kit of building blocks that allow the modeling of membranes, single neurons, and eventually neural networks. This field of research is broadly covered by *computational neuroscience*. However, since this discipline also deals with more abstract approximations of real neurons (see Sect. 1.4.3) and with artificial neural networks, I prefer to speak about *neurophysics*, i.e. the biophysics of real neurons.

The chapter is organized as a journey along a characteristic neuron where the stages are Sects. 1.2–1.4. Looking at Fig. 8.1 in Chap. 8, the reader recognizes the *cell bodies*, or *somata*, of three cortical neurons as the triangular knobs. Here, our journey will start by describing the microscopically observable membrane potentials. Membranes separating electrolytes with different ion concentrations exhibit a characteristic resting potential. In a corresponding equivalent circuit, this voltage can be thought of being supplied by a battery. Moreover, passive membranes act as a capacitance while their semipermeability with respect to particular kinds of ions leads to an approximately ohmic resistance. This property is due to the existence of leaky ion channels embedded in the cell membrane. At the neuron's axon hillock (trigger zone), situated at the base of the soma, the composition of the cell membrane changes. Here and along the axon, voltage-gated sodium and potassium channels appear in addition to the leakage channels, both making the membrane active and excitable. As we shall see, the equivalent circuit of the membrane allows for the derivation of the famous *Hodgkin-Huxley equations* of the *action potentials* which are the basic of neural *conductance models*. Traveling along the axon, we reach the presynaptic terminals, where the Hodgkin-Huxley equations have to be supplemented by additional terms describing the dynamics of voltage-gated calcium channels. Calcium flowing into the terminal causes the release of transmitter vesicles that pour their content of neurotransmitter into the synaptic cleft of a chemical synapse. Then, at the postsynapse, transmitter molecules dock onto receptor molecules, which indirectly open other ion channels. The kinetics of these reactions give rise to the *impulse response functions* of the postsynaptic membranes. Because these membranes behave almost passively, a linear differential equation describes the emergence of *postsynaptic potentials* by the convolution product of the postsynaptic pulse response with the *spike train*, i.e. the sequence of action potentials. Postsynaptic potentials propagate along the *dendrites* and the soma of the neuron and superimpose linearly to a resulting signal that eventually arrives at the axon hillock, where our journey ends.

In Sect. 1.5, we shall change our perspective from the microscopic to the macroscopic. Here, the emergence of mass potentials such as the local field potential (LFP) and the electroencephalogram (EEG) will be discussed.

## 1.2 Passive Membranes

Neurons are cells specialized for the purpose of fast transfer and computation of information in an organism. Like almost every other cell, they posses a cell body containing a nucleus and other organelles and they are surrounded by a membrane separating their interior from the extracellular space. In order to collect information from their environment, the soma of a characteristic neuron branches out into a *dendritic tree* while another thin process, the *axon*, provides an output connection to other neurons [19–21]. The cell plasma in the interior as well as the liquid in the extracellular space are electrolytes, i.e. solutions of different kinds of ions such as sodium ($Na^+$), potassium ($K^+$), calcium ($Ca^{2+}$), chloride ($Cl^-$), and large organic ions. However, the concentrations of these ions (denoted by $[Na^+], [K^+], [Ca^{2+}]$, etc.) can differ drastically from one side of the cell membrane to the other (see Fig. 2.1 of Chap. 2). Therefore, the membrane is subjected to two competing forces: the osmotic force aiming at a compensation of these concentration gradients on the one hand, and the Coulomb force aiming at a compensation of the electric potential gradient. Biochemically, cell membranes are lipid bi-layers swimming like fat blobs in the plasma soup [19, 20], which makes them perfect electric isolators. Putting such a dielectric between two opposite electric charges yields a capacitance of capacity

$$C_m = \frac{Q}{U}, \tag{1.1}$$

where $Q$ is the total charge stored in the capacitance and $U$ is the voltage needed for that storage. Hence, a membrane patch of a fixed area $A$ that separates different ion concentrations can be represented by a single capacitance $C_m = 1 \, \mu F \, cm^{-1} \times A$ in an equivalent "circuit" shown in Fig. 1.1 [19, 20].

Generally, we interpret such equivalent circuits in the following way: The upper clamp refers to the extracellular space whereas the clamp at the bottom measures the potential within the cell. Due to its higher conductance, the extracellular space is usually assumed to be equipotential, which can be designated as $U = 0 \, mV$ without loss of generality.

### 1.2.1 Ion Channels

If neuron membranes were simply lipid bi-layers, there would be nothing more to say. Of course, they are not. All the dynamical richness and computational complexity of neurons is due to the presence of particular proteins, called *ion channels*, embedded in the cell membranes. These molecules form tubes traversing the membrane that are permeable to certain kinds of ions [19–25]. The "zoo" of ion channels is comparable with that of elementary particles. There are channels whose pores are always open (*leakage channels*) but permeable only for sodium or potassium or chloride. Others possess *gates* situated in their pores which are controlled by the membrane potential, or the presence of certain substances or even both. We shall refer to the first kind of channels as to *voltage-gated channels*, and to the second kind as to *ligand-gated*

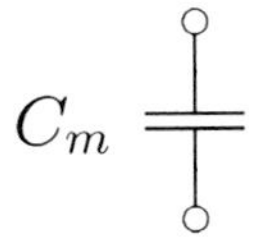

$$C_m \; \rightvert\kern-0.4em\equiv$$

**Fig. 1.1.** Equivalent "circuit" for the capacitance $C_m$ of a membrane patch

*channels.* Furthermore, the permeability of a channel can depend on the direction of the ionic current such that it behaves as a rectifier whose equivalent "circuit" would be a diode [19, 20]. Eventually, the permeability could be a function of the concentration of particular reagents either in the cell plasma or in the extracellular space, which holds not only for ligand-gated channels. Such substances are used for classifying ion channels. Generally, there are two types of substances. Those from the first class facilitate the functioning of a channel and are therefore called *agonists*. The members of the second class are named *antagonists* as they impede channel function.

Omitting these complications for a while, we assume that a single ion channel of kind $k$ behaves as an ohmic resistor with conductance

$$\gamma_k = \frac{1}{\rho_k} , \tag{1.2}$$

where $\rho_k$ is the resistivity of the channel. A typical value (for the gramicidin-A channel) is $\gamma_{GRAMA} \approx 12\,\text{pS}$. Figure 1.2 displays the corresponding equivalent "circuit".

In the remainder of this chapter, we will always consider membrane patches of a fixed area $A$. In such a patch, many ion channels are embedded, forming the parallel circuit shown in Fig. 1.3(a).

According to Kirchhoff's First Law, the total conductance of the parallel circuit is

$$g_k = N_k \gamma_k \tag{1.3}$$

when $N_k$ channels are embedded in the patch, or, equivalently, expressed by the channel concentration $[k] = N_k/A$,

$$g_k = [k] A \gamma_k .$$

### 1.2.2 Resting Potentials

By embedding leakage channels into the cell membrane, it becomes *semipermeable*, i.e. permeable for certain kinds of ions while impenetrable for others. If there is a concentration gradient of a permeable ion across a semipermeable membrane, a diffusion current $I_{\text{diff}}$ through the membrane patch $A$ is created, whose density obeys Fick's Law

$$j_{\text{diff}} = -D q \frac{\text{d}[I]}{\text{d}x} , \tag{1.4}$$

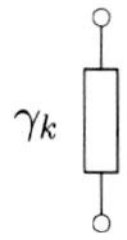

**Fig. 1.2.** Equivalent "circuit" for a single ohmic ion channel with conductance $\gamma_k$

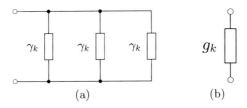

(a)                    (b)

**Fig. 1.3.** Equivalent circuits (**a**) for ion channels of one kind $k$ connected in parallel; (**b**) Substituted by a single resistor of conductance $g_k = 3\gamma_k$

where $\mathrm{d}[\mathrm{I}]/\mathrm{d}x$ denotes the concentration gradient for ion I, $q$ its charge, and $D = k_\mathrm{B}T/\mu$ is the diffusion constant given by Einstein's relation [26] ($k_\mathrm{B}$ is Boltzmann's constant, $T$ is the temperature and $\mu$ is the viscosity of the electrolyte) [22–25]. This diffusion current can be described by an equivalent "circuit" given by a current source $I_\mathrm{diff}$ (Fig. 1.4).

The separation of charges by the diffusion current leads to an increasing potential gradient $\mathrm{d}U/\mathrm{d}x$ across the membrane. Therefore, a compensating ohmic current

$$j_\mathrm{ohm} = -\sigma \frac{\mathrm{d}U}{\mathrm{d}x} \tag{1.5}$$

flows back through the leakage channels ($\sigma = q^2[\mathrm{I}]/\mu$ is the conductance of the electrolyte expressed by the ion concentration and its charge). Then the total current $j = j_\mathrm{diff} + j_\mathrm{ohm}$ (visualized by the circuit in Fig. 1.5) is described by the *Nernst-Planck equation*

$$j = -Dq\frac{\mathrm{d}[\mathrm{I}]}{\mathrm{d}x} - [\mathrm{I}]\frac{q^2}{\mu}\frac{\mathrm{d}U}{\mathrm{d}x} . \tag{1.6}$$

**The Nernst Equation**

The general quasi-stationary solution of (1.6), the Goldman-Hodgkin-Katz equation ((2.4) in Chap. 2), clearly exhibits a nonlinear dependence of the ionic current on the membrane voltage [22–25]. However, for only small deviations from the stationary solution — given by the *Nernst equation*

$$E_\mathrm{I} = \frac{k_\mathrm{B}T}{q}\ln\frac{[\mathrm{I}]_\mathrm{out}}{[\mathrm{I}]_\mathrm{int}} , \tag{1.7}$$

$I$

**Fig. 1.4.** Equivalent "circuit" either for the diffusion currents through the cell membrane or for the active ion pumps

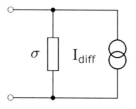

**Fig. 1.5.** Equivalent circuit for the derivation of the Nernst-Planck equation (1.6)

where $[I]_{out}$ is the ion concentration in the extracellular space and $[I]_{int}$ within the cell — the current can be regarded as being ohmic.

For room temperature, the factor $k_B T/q \approx 25\,\text{mV}$. With the concentrations from Fig. 2.1, Chap. 2, this leads to the characteristic resting potentials; e.g. $U_{K+} = -101\,\text{mV}$, and $U_{Na+} = +56\,\text{mV}$.

Each sort of ion possesses its own Nernst equilibrium potential. We express this fact by a battery in an equivalent "circuit" shown in Fig. 1.6.

Now, we are able to combine different ion channels $k$ all selective for one sort of ions I with their corresponding power supplies. This is achieved by a serial circuit as shown in Fig. 1.7. This equivalent circuit will be our basic building block for all other subsequent membrane models.

If the clamp voltage of this circuit has the value $U$, we have to distribute this voltage according to Kirchhoff's Second Law as

$$U = \frac{I_k}{g_k} + E_I \,,$$

leading to the fundamental equation

$$I_k = g_k(U - E_I)\,. \tag{1.8}$$

### The Goldman Equation

As an example, we assume that three types of ion channels are embedded in the membrane patch, one pervious for sodium with the conductance $g_{Na+}$, another pervious for potassium with the conductance $g_{K+}$, and the third pervious for chloride with the conductance $g_{Cl-}$, respectively. Figure 1.8 displays the corresponding equivalent circuit.

Interpreting the top of the circuit as the extracellular space and the bottom as the interior of the neuron, we see that the resting potential for potassium

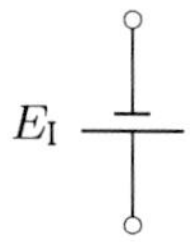

$$E_I \quad \begin{array}{c} \circ \\ \vert \\ \overline{\phantom{aa}} \\ \vert \\ \circ \end{array}$$

**Fig. 1.6.** Equivalent circuit for the Nernst equilibrium potential (1.7)

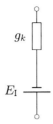

**Fig. 1.7.** Equivalent circuit for a population of ion channels of kind $k$ selective for the ion sort I embedded in a membrane with resting potential $E_I$

and chloride is negative (denoted by the short tongue of the battery symbol) while the sodium equilibrium potential is positive in comparison to the extracellular space.

According to Kirchhoff's First Law, the total current through the circuit is

$$I = I_{Na^+} + I_{K^+} + I_{Cl^-} \,. \tag{1.9}$$

To obtain the stationary equilibrium, we have to set $I = 0$. Using the fundamental equation (1.8), we get the equation

$$0 = g_{Na^+}(U - E_{Na^+}) + g_{K^+}(U - E_{K^+}) + g_{Cl^-}(U - E_{Cl^-}) \,,$$

whose resolution entails the equilibrium potential

$$U = \frac{g_{Na^+} E_{Na^+} + g_{K^+} E_{K^+} + g_{Cl^-} E_{Cl^-}}{g_{Na^+} + g_{K^+} + g_{Cl^-}} \,. \tag{1.10}$$

Equation (1.10) is closely related to the Goldman equation that can be derived from the Goldman-Hodgkin-Katz equation [24]. It describes the net effect of all leakage channels. Therefore, the circuit in Fig. 1.8 can be replaced by the simplification found in Fig. 1.9.

Accordingly, the leakage current is again given by (1.8)

$$I_l = g_l(U - E_l) \,. \tag{1.11}$$

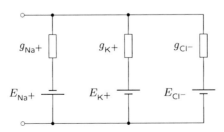

**Fig. 1.8.** Equivalent circuit for three populations of ion channels permeable for sodium, potassium and chloride with their respective Nernst potentials

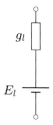

**Fig. 1.9.** Equivalent circuit for the total leakage current and its corresponding leakage potential $E_l$

Characteristic values are $g_l = 13\,\mu S$ for the leakage conductance and $E_l = -69\,\mathrm{mV}$ for the leakage potential as the solution of (1.10) [19, 20].

While the Nernst potential for one kind of ions denotes a stationary state, the Goldman equilibrium potential results from a continuous in- and outflow of ions that would cease when all concentration gradients had been balanced. To stabilize the leakage potential the cell exploits active *ion pumps* modeled by a current source as displayed in Fig. 1.4. These ion pumps are proteins embedded in the cell membrane that transfer ions against their diffusion gradients by consuming energy. Maintaining resting potentials is one of the energetically most expensive processes in the nervous system [27]. This consumption of energy is, though rather indirectly, measurable by neuroimaging techniques such as positron emission tomography (PET) or functional magnetic resonance imaging (fMRI) [19, 20, 28, 29].

## 1.3 Active Membranes

The resting potentials we have discussed so far are very sensitive to changes in the conductances of the ion channels. While these are almost constant for the leakage channels, there are other types of channels whose conductances are functions of certain parameters such as the membrane potential or the occurrence of particular reagents. These channels make membranes active and dynamic. The former are called voltage-gated whereas the latter are referred to as ligand-gated. Basically, these channels occur in two dynamical states: their pore may be open ($O$) or closed ($C$). The conductance of closed channels is zero, while that of an open channel assumes a particular value $\gamma_k$. Therefore, a single gated channel can be represented by a serial circuit of a resistor with conductance $\gamma_k$ and a switch $S$, as depicted in Fig. 1.10.

Let $N_k$ be the number of gated channels of brand $k$ embedded in our membrane patch of area $A$, and let $O_k$ and $C_k$ the number of momentarily open and closed channels of this kind, respectively. As argued in Sect. 1.2.1, the total conductance of all open channels is given by Kirchhoff's First Law as

**Fig. 1.10.** Equivalent circuit for a single gated channel with open-conductance $\gamma_k$

$$g_k = O_k\,\gamma_k\,,\tag{1.12}$$

while

$$\bar{g}_k = N_k\,\gamma_k\tag{1.13}$$

is now the maximal conductance of these channels.

### 1.3.1 Action Potentials

Signals propagate mainly passively along the dendritic and somatic membranes until they reach the axon hillock, or trigger zone of the neuron. Here, the composition of the membrane changes significantly and voltage-gated sodium and potassium channels supplement the all-pervasive leakage channels. Above, we have modeled these channels by switches connected serially with ohmic resistors. Now, the crucial question arises: Who opens the switches?

Here, for the first time, a stochastic account is required. Ion channels are macro-molecules and hence quantum objects. Furthermore, these objects are weakly interacting with their environments. Therefore the cell membrane and the electrolytes surrounding it provide a *heat bath* making a thermodynamical treatment necessary. From a statistical point of view, an individual channel has a probability of being open, $p_k$, such that the number of open channels is the expectation value

$$O_k = p_k\,N_k\,.\tag{1.14}$$

Inserting (1.14) into (1.12) yields the conductance

$$g_k = p_k\,N_k\,\gamma_k = p_k\,\bar{g}_k\,.\tag{1.15}$$

The problem of determining the probability $p_k$ is usually tackled by modeling Markov chains [24, 25]. The simplest approach is a two-state Markov process shown in Fig. 1.11, where $C$ and $O$ denote the closed and the open state, respectively, while $\alpha, \beta$ are transition rates.

The state probabilities of the Markov chain in Fig. 1.11 obey a *master equation* [30, 31]

$$\frac{\mathrm{d}p_k}{\mathrm{d}t} = \alpha_k\,(1 - p_k(t)) - \beta_k\,p_k(t)\,,\tag{1.16}$$

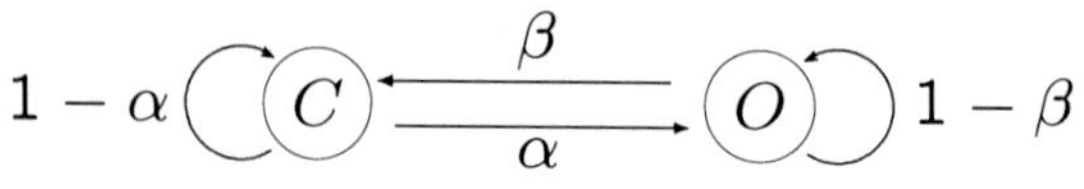

$$1 - \alpha \,\,\bigcirc\!\!\!\!\!\bigcirc\,\, C \xrightarrow[\;\;\alpha\;\;]{\;\;\beta\;\;} O \,\,\bigcirc\!\!\!\!\!\bigcirc\,\, 1 - \beta$$

**Fig. 1.11.** Two-state Markov model of a voltage-gated ion channel

whose transition rates are given by the thermodynamic Boltzmann weights

$$\alpha_k = e^{\frac{W(C \to O)}{k_{\mathrm{B}} T}}, \tag{1.17}$$

where $W(C \to O)$ is the necessary amount of energy that has to be supplied by the heat bath to open the channel pore.

Channel proteins consist of amino acids that are to some extent electrically polarized [19–21]. The gate blocking the pore is assumed to be a subunit with charge $Q$. Call $W_0(C \to O)$ the work that is necessary to move $Q$ through the electric field generated by the other amino acids to open the channel pore. Superimposing this field with the membrane potential $U$ yields the total transition energy

$$W(C \to O) = W_0(C \to O) + QU. \tag{1.18}$$

If $QU < 0$, $W(C \to O)$ is diminished and the transition $C \to O$ is facilitated [12], thereby increasing the rate $\alpha_k$ according to

$$\alpha_k(U) = e^{\frac{W_0(C \to O) + QU}{k_{\mathrm{B}} T}}. \tag{1.19}$$

The equations (1.15, 1.16, 1.19) describe the functioning of voltage-gated ion channels [12,13,15,23–25]. Yet, voltage-gated resistors are also well-known in electric engineering: *transistors* are *transient resistors*. Though not usual in the literature, I would like to use the transistor symbol to denote voltage-gated ion channels here (Fig. 1.12). In contrast to batteries, resistors and capacitors, which are *passive* building blocks of electronic engineering, transistors are *active* components thus justifying our choice for active membranes.

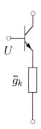

**Fig. 1.12.** Equivalent circuit for a population of voltage-gated ion channels. The maximal conductance $\bar{g}_k$ is reached when the transistor is in saturation

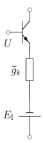

**Fig. 1.13.** Equivalent circuit for a population of voltage-gated ion channels of kind $k$ selective for the ion sort I embedded in a membrane with resting potential $E_I$

Corresponding to Fig. 1.7, the equivalent circuit for a population of voltage-gated channels of kind $k$ permeable for ions I supplied by their respective resting potential $E_I$ is provided in Fig. 1.13.

**The Hodgkin-Huxley Equations**

Now we are prepared to derive the Nobel-prize-winning Hodgkin-Huxley equations for the action potential [32] (see also [12–15, 23–25]). Looking again at Fig. 1.8, one easily recognizes that an increase of the sodium conductance leads to a more positive membrane potential, or, to a *depolarization*, while an increasing conductance either of potassium or of chloride entails a further negativity, or *hyperpolarization* of the membrane potential. These effects are in fact achieved by voltage-gated sodium and potassium channels which we refer here to as $AN$ and $AK$, respectively. Embedding these into the cell membrane yields the equivalent circuit shown in Fig. 1.14.[3]

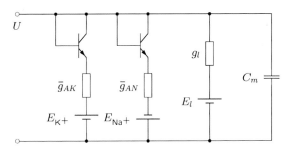

**Fig. 1.14.** Equivalent circuit for the Hodgkin-Huxley equations (1.25, 1.27–1.29)

---

[3] I apologize to all electrical engineers for taking their notation rather symbolically. Certainly, this circuit has neither protection resistors nor voltage stabilizers and should not be reproduced. Sorry for that!

The first and second branches represent the voltage-gated potassium and sodium channels, respectively. The third is taken from the stationary descriptions of the leakage potential (Sect. 1.2.2) while the capacitance is now necessary to account for the dynamics of the membrane potential. According to Kirchhoff's First Law, the total current through the circuit adds up to an injected current $I_m$,

$$I_m = I_{AK} + I_{AN} + I_l + I_C. \tag{1.20}$$

The partial currents are

$$I_{AK} = p_{AK}\,\bar{g}_{AK}(U - E_{K^+}) \tag{1.21}$$

$$I_{AN} = p_{AN}\,\bar{g}_{AN}(U - E_{Na^+}) \tag{1.22}$$

$$I_l = g_l(U - E_l) \tag{1.23}$$

$$I_C = C_m\frac{dU}{dt}, \tag{1.24}$$

where (1.21, 1.22) are produced from (1.15) and (1.8), (1.23) is actually (1.11) and (1.24) is the temporal derivative of (1.1). Taken together, the membrane potential $U(t)$ obeys the differential equation

$$C_m\frac{dU}{dt} + p_{AK}\,\bar{g}_{AK}(U - E_{K^+}) + p_{AN}\,\bar{g}_{AN}(U - E_{Na^+}) + g_l(U - E_l) = I_m. \tag{1.25}$$

Equation (1.25) has to be supplemented by two master equations: (1.16) for the open probabilities $p_{AK}, p_{AN}$ and the rate equations (1.19) for $\alpha_{AK}, \alpha_{AN}$.

Unfortunately, this approach is inconsistent with the experimental findings of Hodgkin and Huxley [32]. They reported two other relations

$$p_{AK} = n^4; \qquad p_{AN} = m^3 h, \tag{1.26}$$

where $n, m$ and $h$ now obey three master equations

$$\frac{dn}{dt} = \alpha_n(1 - n) - \beta_n n \tag{1.27}$$

$$\frac{dm}{dt} = \alpha_m(1 - m) - \beta_m m \tag{1.28}$$

$$\frac{dh}{dt} = \alpha_h(1 - h) - \beta_h h. \tag{1.29}$$

The equations (1.25, 1.27–1.29) are called Hodgkin-Huxley equations [12–15, 23–25, 32]. They constitute a four-dimensional nonlinear dynamical system controlled by the parameter $I_m$. Figure 1.15 displays numerical solutions for three different values of $I_m$.

Figure 1.15 illustrates only two of a multitude of dynamical patters of the Hodgkin-Huxley system. Firstly, it exhibits a threshold behavior that is due to a Hopf bifurcation [18]. For subthreshold currents (solid line: $I_m = 7.09\,\mu A$), one observes a damped oscillation corresponding to a stable fixed point in

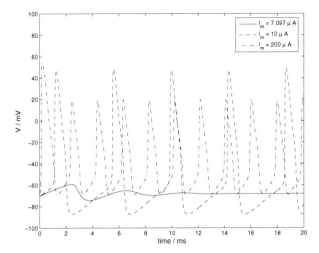

**Fig. 1.15.** Numeric solutions of the Hodgkin-Huxley equations (1.25, 1.27–1.29) according to the Rinzel-Wilson model (Sect. 1.4.3) for three different values of the control parameter $I_m$. *Solid: subthreshold current* $I_m = 7.09\,\mu A$; *dashed: super-threshold current* $I_m = 10\,\mu A$; *dashed-dotted: even higher current* $I_m = 200\,\mu A$

the phase space. If the control parameter $I_m$ exceeds a certain threshold $\theta$, this fixed point destabilizes and a limit cycle emerges (dashed line: $I_m = 10\,\mu A$). Secondly, further heightening of $I_m$ leads to limit cycles of increased frequencies (dashed-dotted line: $I_m = 200\,\mu A$). This *regular spiking* dynamics explains the law of all-or-nothing as well as the encoding principle by frequency modulation in the nervous system [19–21].

In order to interpret the Hodgkin-Huxley equations (1.25, 1.27–1.29) biologically, we have to consider (1.26) first. It tells that our simple two-state Markov chain (Fig. 1.11) is not appropriate. Instead, the description of the active potassium channel requires a four-state Markov chain comprising three distinct closed and one open state [24, 25]. However, (1.26) allows for another instructive interpretation: According to a fundamental theorem of probability theory, the joint probability of disjunct events equals the product of the individual probabilities upon their stochastic independence. Since $p_{AK} = n^4$, we can assume the existence of four independently moving *gating charges* within the channel molecule. Correspondingly, for the sodium channel we expect three independent gating charges and one inhibiting subunit since $p_{AN} = m^3\,h$. This is supported by patch clamp measurements where the channel's pores were blocked by the Fugu's fish tetradotoxin [19–21]. Although the blocked channel could not pass any ions, about three brief currents were observed. We can imagine these charges as key cylinders that have to be brought into the right positions to unlock a cylinder lock (thus opening the channel).

The emergence of an action potential results from different kinetics of the ion channels. If the cell membrane is slightly depolarized by the current $I_m$, the opening rate $\alpha_n$ for the sodium channels increases, thus entailing a further depolarization of the membrane. The positive feed-back loop started in this way leads to a torrent of inflowing sodium until the peak of the action potential is reached. Then, the membrane potential is positive in comparison to the extracellular space, causing voltage-gated potassium channels to open. Due to its negative equilibrium potential, potassium leaves the cell thereby hyperpolarizing the interior. Contrastingly, the hyperpolarization of the membrane reduces the open probability of the sodium channels, which become increasingly closed. Another positive feed-back loop enhances the hyperpolarization thereby overshooting the resting potential. While the potassium channels change very slowly back to their closed state, the sodium channels become additionally inactivated by a stopper subunit of the channel molecule whose kinetics is governed by the $h$ term. This inhibition process is responsible for the refractory time prohibiting the occurrence of another action potential within this period.

### 1.3.2 Presynaptic Potentials

A spike train, generated in the way described by the Hodgkin-Huxley equations, travels along the axon and, after several branches, reaches the presynaptic terminals. Here, the composition of the membrane changes again. Voltage-gated calcium channels are present in addition to the voltage-gated potassium and sodium channels, and can be described by another branch in Fig. 1.14. The class of voltage-gated calcium channels is quite extensive and they operate generally far from the linear (ohmic) domain of the Goldman-Hodgkin-Katz equation [13, 15, 24, 25]. However, according to Johnston & Wu [24], an ohmic treatment of presynaptic $Ca^{2+}$ channels is feasible such that their current is given by

$$I_{AC} = l^5 \, \bar{g}_{AC} \left( U - E_{Ca^{2+}} \right), \tag{1.30}$$

where $l$ obeys another master equation

$$\frac{dl}{dt} = \alpha_l \left( 1 - l \right) - \beta_l \, l. \tag{1.31}$$

In the absence of an injected current ($I_m = 0$), the presynaptic potential $U(t)$ is then governed by the differential equation

$$C_m \frac{dU}{dt} + I_{AK} + I_{AN} + I_{AC} + I_l = 0. \tag{1.32}$$

Neglecting calcium leakage, the current (1.30) leads to an enhancement of the intracellular concentration $[Ca^{2+}]_{int}$ that is described by a continuity equation [12]

$$\frac{d[Ca^{2+}]_{int}}{dt} = -\frac{I_{AC}}{qN_A V}. \tag{1.33}$$

Here, $q = 2e$ is the charge of the calcium ion ($e$ denoting the elementary charge). Avogadro's constant $N_A$ scales the ion concentration to moles contained in the volume $V$. The accumulation of calcium in the cell plasma gives rise to a cascade of metabolic reactions. Calcium does not only serve as an electric signal; it also acts as an important messenger and chemical reagent, enabling or disenabling the functioning of enzymes.

The movement of neurotransmitter into the synaptic cleft comprises two sub-processes taking place in the presynaptic terminal: Firstly, transmitter must be allocated, and secondly, it must be released. The allocation of transmitter depends on the intracellular calcium concentration (1.33), while it is stochastically released by increased calcium currents (1.30) as a consequence of an arriving action potential with a probability $p$.

In the resting state, transmitter vesicles are anchored at the cytoskeleton by proteins called *synapsin*, which act like a wheel clamp. The probability to loosen these joints increases with the concentration $[Ca^{2+}]_{int}$. Liberated vesicles wander to one of a finite number $Z$ of *active zones* where vesicles can fuse with the terminal membrane thereby releasing their content into the synaptic cleft by the process of *exocytosis* [19–21]. Allocation means that $Y \leq Z$ active zones are provided with vesicles, where

$$Y = \kappa([Ca^{2+}]_{int})\, Z \qquad (1.34)$$

is the average number of occupied active zones, and $\kappa([Ca^{2+}]_{int})$ is a monotonic function of the calcium concentration that must be determined from the reaction kinetics between calcium and synapsin mediated by kinases. The release of transmitter is then described by a *Bernoulli process* started by an arriving action potential. The probability that $k$ of the $Y$ occupied active zones release a vesicle is given by the *binomial distribution*

$$p(k, Y) = \binom{Y}{k} p^k (1 - p)^{Y-k}. \qquad (1.35)$$

For the sake of mathematical convenience, we shall replace the binomial distribution by a normal distribution

$$\rho(k, Y) = \frac{1}{\sqrt{2\pi y(1-p)}} \exp\left[-\frac{(k-y)^2}{2y(1-p)}\right], \qquad (1.36)$$

where $y = Yp$ is the average number of transmitter releasing active zones. Assuming that a vesicle contains on average $n_T = 5000$ transmitter molecules [19, 20], we can estimate the mean number of transmitter molecules that are released by an action potential as

$$T = n_T Y p = n_T Z p\, \kappa([Ca^{2+}]_{int}). \qquad (1.37)$$

Correspondingly, the expected number of transmitter molecules released by $k$ vesicles is given by

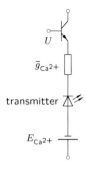

**Fig. 1.16.** Equivalent circuit for the calcium-controlled transmitter release (indicated by the arrows of the LED)

$$T(k) = \frac{n_T Y}{\sqrt{2\pi y(1-p)}} \exp\left[-\frac{(k-y)^2}{2y(1-p)}\right]. \qquad (1.38)$$

Finally, we need an equivalent circuit symbol for the transmitter release. Electronics suggests the use of the LED symbol (light-emitting diode). Connected all together, the calcium controlled transmitter release might be represented by the branch shown in Fig. 1.16.

### 1.3.3 Postsynaptic Potentials

After being poured out into the synaptic cleft of a chemical synapse, transmitter molecules diffuse to the opposite postsynaptic membrane, unless they have not been decomposed by enzymic reactions. There, they dock onto *receptor molecules*, which fall into two classes: *ionotropic receptors* are actually transmitter-gated ion channels, whereas *metabotropic receptors* are proteins that, once activated by transmitter molecules, start metabolic processes from second messenger release up to gene expression. At particular pathways, they control the opening of other ion channels gated by intracellular reaction products. The directly transmitter-gated channels are fast and effective, while the intracellularly gated channels react very slowly [19–21, 33]. In this section, I shall treat two distinct examples from each receptor class.

### Excitatory Postsynaptic Potentials

One important transmitter-gated ion channel is (among others, such as the AMPA, GABA$_A$, and NMDA receptors) the nACh receptor that has nicotine as an antagonist. It becomes open if three or four molecules of the neurotransmitter acetylcholine (ACh) dock at its surface rising into the synaptic cleft. These molecules cause shifts of the electric polarization within the molecule which opens the gate in the pore. This process can be modeled by a Markov chain similarly to the exposition in Sect. 1.3.1. However, another treatment is also feasible, using *chemical reaction networks* [30, 33].

The open nACh channel is conductive for sodium as well as for potassium ions, such that its reversal (resting) potential is provided by the Goldman equation (1.10). Yet the sodium conductance is slightly larger than that for potassium yielding a net current of inflowing sodium ions. Since this current is depolarizing, the nACh channels constitute *excitatory synapses*. Therefore, they generate *excitatory postsynaptic potentials* (EPSP). On the other hand, hyperpolarizing channels, such as the $GABA_A$ channel, constitute *inhibitory synapses* generating *inhibitory postsynaptic potentials* (IPSP).

Let us once more consider a membrane patch of area $A$ containing $N_{nACh}$ receptors. Again, let $O_{nACh}$ be the number of momentarily opened and $C_{nACh}$ the number of closed channels. According to (1.12), the conductance of all open channels connected in parallel is then $g_{nACh} = O_{nACh} \gamma_{nACh}$. Opening of the channels can now be described by the *chemical reaction equation*

$$C + 3T \rightleftarrows O \,, \tag{1.39}$$

where $C$ denotes the closed and $O$ the opened molecules. $T$ stands for the transmitter ACh. Because in each single reaction, three molecules $T$ react with one molecule $C$ to produce one molecule $O$, the corresponding *kinetic equation* [30, 31, 33] comprises a cubic nonlinearity,

$$\frac{dO}{dt} = \nu_1 C T^3 - \nu_2 O \,, \tag{1.40}$$

where $\nu_1$ denotes the production and $\nu_2$ the decomposition rate of open channels in (1.39). These reaction rates depend on the temperature of the heat bath and probably on metabolic circumstances such as phosphorylation. This equation has to be supplemented by a *reaction-diffusion equation* for the neurotransmitter reservoir in the synaptic cleft

$$\frac{dT}{dt} = \nu_2 O - \nu_3 TE - \sigma T \,, \tag{1.41}$$

where $\nu_2 O$ is the intake of transmitter due to decaying receptor-transmitter complexes, which is the same as the loss of open channels in (1.40), $\nu_3 TE$ is the decline due to reactions between the transmitter with enzyme $E$, and $\sigma T$ denotes the diffusion out of the synaptic cleft. Its initial condition $T(t = 0)$ is supplied by (1.38). Taken together, the equations (1.40, 1.41) describe the *reaction-diffusion kinetics* of the ligand-gated ion channel nACh.

Expressing the electric conductivity (1.12) through the maximal conductivity $\bar{g}_{nACh}$,

$$g_k = \frac{O_{nACh}}{N_{nACh}} \bar{g}_{nACh} \,, \tag{1.42}$$

suggests a new equivalent circuit symbol for ligand-gated channels. The conductance is controlled by the number of transmitter molecules, i.e. the number of particular particles in the environment. This corresponds to the phototransistor in electronic engineering which is controlled by the number of photons

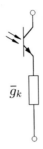

**Fig. 1.17.** Equivalent circuit for a population of ligand-gated ion channels of kind $k$

collected by its base. Hence, I would like to suggest the circuit shown in Fig. 1.17 as an equivalent to the nACh receptor.

In order to compute the postsynaptic potential, the circuit in Fig. 1.17 has to be connected in parallel with the leakage conductance and the membrane capacitance as in Fig. 1.18.

The EPSP for the nACh receptor then obeys the equations

$$C_m \frac{dU}{dt} + \frac{O_{\text{nACh}}}{N_{\text{nACh}}} \bar{g}_{\text{nACh}}(U - E_{\text{nACh}}) + g_l(U - E_l) = 0 \qquad (1.43)$$

together with (1.40, 1.41), and initial condition (1.38).

However, instead of solving these differential equations, most postsynaptic potentials can be easily described by alpha functions as synaptic gain functions [12–16, 18]

$$U^{\text{PSP}}(t) = E_{\text{PSP}}\, \alpha^2 t\, e^{-\alpha t}\, \Theta(t)\,, \qquad (1.44)$$

where $\alpha$ is the characteristic time constant of the postsynaptic potential (PSP) and

$$\Theta(t) = \begin{cases} 0 & \text{for} \quad t \le 0 \\ 1 & \text{for} \quad t > 0 \end{cases} \qquad (1.45)$$

is Heaviside's jump function.

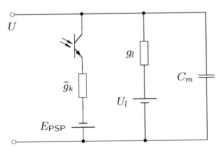

**Fig. 1.18.** Equivalent circuit for the postsynaptic potential generated by ligand-gated channels of kind $k$ with reversal potential PSP

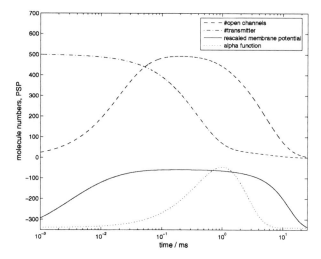

**Fig. 1.19.** Numeric solutions of the kinetic equations (1.40, 1.41, 1.43) for the nACh receptor. *Dashed-dotted*: the number of ACh molecules (max = 500); *dashed*: the number of open nACh channels (max = 500); *solid*: the EPSP $U(t)$; *dotted*: an alpha function (1.44). The time axis is logarithmically scaled; the functions are rescaled and (if necessary) shifted for better visibility

Figure 1.19 displays the numerical solution of equations (1.40, 1.41, 1.43) for arbitrarily chosen parameters together with a "fitted" alpha function for comparison. Obviously, the correspondence is not that large.

### Inhibitory Postsynaptic Potentials

Synapses are excitatory if they open sodium or calcium channels with more positive reversal potentials compared to the resting state. Their neurotransmitters are generally acetylcholine (ACh) or the amino acid glutamate. Contrastingly, most inhibitory synapses employ the amino acids glycine or GABA (gamma-amino-butyric-acid) to open potassium or chloride channels with more negative reversal potentials. While the $GABA_A$ receptor is transmitter-gated such as the nACh receptor discussed in the previous section, the $GABA_B$- and mACh receptors (having the toadstool toxin *muscarine* as an antagonist) activate intracellular *G proteins* which subsequently open G protein-gated potassium channels [19–21]. The activation of G protein-gated potassium channels comprises the following chemical reactions [12]:

$$R_0 + T \rightleftarrows R^* \rightleftarrows D \qquad (1.46)$$
$$R^* + G_0 \rightleftarrows RG^* \rightarrow R^* + G^*$$
$$G^* \rightarrow G_0$$
$$C + n\,G^* \rightleftarrows O,$$

where $R_0$ is the metabotropic GABA$_B$ receptor in its resting state, $T$ the transmitter GABA, $R^*$ the transmitter-activated receptor on the one hand, and $D$ the same transmitter-receptor complex in its inactivated state on the other hand; furthermore, $G_0$ is the G protein in its resting state, $(RG)^*$ a short-lived activated receptor-G protein complex and $G^*$ the activated G protein; finally, $C$ is the G protein-gated potassium channel in its closed state and $O$ in the open state. The channel possesses $n$ docking sites for G protein molecules. Translating (1.46) into kinetic equations and adding (1.41) yields

$$\frac{dR_0}{dt} = -\nu_1 R_0 T + \nu_2 R^* \tag{1.47}$$

$$\frac{dT}{dt} = -\nu_1 R_0 T + \nu_2 R^* - \nu_{11} T E - \sigma T \tag{1.48}$$

$$\frac{dR^*}{dt} = \nu_1 R_0 T - \nu_2 R^* + \nu_3 D - \nu_4 R^* \tag{1.49}$$
$$-\nu_5 R^* G_0 + \nu_6 (RG)^* + \nu_8 (RG)^*$$

$$\frac{dD}{dt} = -\nu_3 D + \nu_4 R^* \tag{1.50}$$

$$\frac{dG_0}{dt} = -\nu_5 R^* G_0 + \nu_6 (RG)^* + \nu_7 G^* \tag{1.51}$$

$$\frac{d(RG)^*}{dt} = \nu_5 R^* G_0 - \nu_6 (RG)^* - \nu_8 (RG)^* \tag{1.52}$$

$$\frac{dG^*}{dt} = -\nu_7 G^* + \nu_8 (RG)^* + \nu_{10} O \tag{1.53}$$

$$\frac{dO}{dt} = \nu_9 C {G^*}^n - \nu_{10} O \tag{1.54}$$

for the metabolic dynamics. Equations (1.47–1.54) together with (1.43) describe the inhibitory GABA-ergic potential

$$C_m \frac{dU}{dt} + \frac{O}{O + C} \bar{g}_{GP} (U - E_{K^+}) + g_l (U - E_l) = 0 \,, \tag{1.55}$$

where $\bar{g}_{GP}$ denotes the maximal conductance and $E_{K^+}$ the reversal potential of the G protein-gated potassium channels.

## Temporal Integration

Each action potential arriving at the presynaptic terminal causes the (binomially distributed) release of one or more vesicles that pour their total amount of transmitter molecules into the synaptic cleft. Here, transmitter molecules react either with ionotropic or with metabotropic receptors which open — more or less directly — ion channels such that a postsynaptic current

$$I^{PSC}(t) = \frac{O_k(t)}{N_k} \bar{g}_k (U - E_k) \,, \tag{1.56}$$

either excitatory or inhibitory, flows through the "phototransistor" branch of Fig. 1.17. This current gives rise to the EPSP or IPSP according to (1.43, 1.55). Since these potentials were determined for the transmitter released by one action potential, we can consider them as *impulse response functions*. Inserting $I^{PSC}$ into (1.43, 1.55) and shifting the resting potential to $E_l = 0$ yields the inhomogeneous linear differential equation

$$\tau \frac{dU}{dt} + U = -\frac{I^{PSC}}{C_m} \, , \tag{1.57}$$

with $\tau = C_m/g_l$ as the characteristic *time constant* of the membrane patch. If we describe the current $I^{PSC}$ by a pulse of height $I_0$ at time $t_0$,

$$I^{PSC}(t) = I_0 \delta(t - t_0) \, , \tag{1.58}$$

the solution of (1.57) is given by a Green's function $U^{PSP} = G(t, t')$ [13,18,22]. By virtue of this Green's function, we can easily compute the postsynaptic potential evoked by an arbitrary spike train

$$I^{PSC}(t) = I_0 \sum_k \delta(t - t_k) \tag{1.59}$$

as the *convolution product*

$$U^{PSP}(t) = \int G(t, t') \, I^{PSC}(t) \, dt' = G * I^{PSC} \, . \tag{1.60}$$

Inserting (1.59) into (1.60) gives

$$U^{PSP}(t) = I_0 \sum_k G(t, t_k) \, . \tag{1.61}$$

If the action potentials are generated by the presynaptic neuron in the regular spiking mode with frequency $f$ (see Sect. 1.3.1), the event times are given by

$$t_k = \frac{k}{f} \, . \tag{1.62}$$

Eventually, (1.61, 1.62) lead to

$$U^{PSP}(t) = I_0 \sum_k G\left(t, \frac{k}{f}\right) \, . \tag{1.63}$$

Figure 1.20 displays two postsynaptic potentials obtained by the convolution of the Green's function

$$G(t, t') = \Theta(t - t') \cdot \frac{I_0}{C_m} \exp\left(-\frac{t - t'}{\tau}\right)$$

with regular spike trains.

By means of the convolution mechanism, an analogue continuously varying membrane potential is regained from a frequency-modulated spike train. This process is called *temporal integration*.

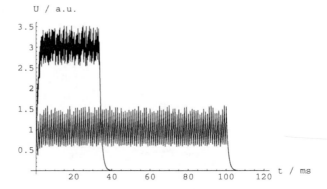

**Fig. 1.20.** Temporal integration of postsynaptic pulse responses for a lower (*lower curve*) and a higher regular spiking frequency (*upper curve*)

## 1.4 Neuron Models

In the preceding sections, we completed our construction kit for neurophysical engineering. In the remaining ones, we are going to apply these building blocks. There are three main threads of neural modeling. The first one builds point models, where all kinds of ion channels are connected in parallel. Secondly, compartment models additionally take into account the cable properties of cell membranes that are responsible for *spatial integration* processes. However, all these models are computationally very expensive. Therefore several simplifications and abstractions have been proposed to cope with these problems especially for the modeling of *neural networks*. Nevertheless, for the simulation of relatively small networks of point or compartment models, powerful software tools such as GENESIS [34], or NEURON [35] have been developed.

### 1.4.1 Point Models

In the point model account, all membrane patches of a nerve cell are assumed to be equipotential, disregarding its spacial extension [12, 15, 36, 37]. In our equivalent circuits, this assumption is reflected by connecting all different ion channels in parallel. Figure 1.21 shows such a point model.

In order to simulate the neuron in Fig. 1.21, all discussed differential equations are to be solved simultaneously. Neural networks then consist of many circuits of this form that are "optically" coupled, i.e. by the transmitter releasing and receiving devices at both ends of one circuit. The efficacy of the coupling between two neurons $i$ and $j$ is expressed by the *synaptic weight* $w_{ij}$. Physiologically, these weights depend on the maximal synaptic conductances $\bar{g}_{PSP}$.

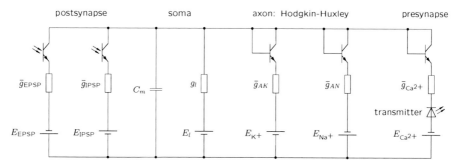

**Fig. 1.21.** Equivalent circuit for a neural point model

## 1.4.2 Compartment Models

Point models have one serious disadvantage: They completely disregard the spatial extension of the cell and the fact that different membrane patches, e.g. at the soma, the axon, the terminals, or the dendrites, exhibit different potentials (cf. Chap. 2). The gradients between these sites then lead to ion currents through the cell plasma thus contributing to the mechanisms of *spatial integration*. Moreover, currents moving back through the extracellular space give rise to the *somato-dendritic field potentials* (DFP). These fields sum to the local field potentials (LFP) at a mesoscopic and to electrocorticogram (ECoG) and electroencephalogram (EEG) at macroscopic scales [38–42] (cf. Chaps. 8 and 7 in this volume).

Correctly, the spatiotemporal dynamics of neuronal membranes must be treated by the *cable equation* [12, 13, 15, 18, 22, 24, 25]. This is a second-order partial differential equation for the membrane potential $U(\boldsymbol{r}, t)$. For the sake of numerical simulations, its discretized form leads to compartment models where individual membrane patches are described by the equivalent circuits discussed in the previous sections [12, 13, 15, 16, 18, 40, 41]. As an example, I shall create an equivalent circuit for a three-compartment model for the cortical pyramidal cells that is able to describe somato-dendritic field potentials.

Pyramidal cells are characterized by their axial symmetry. They consist of an *apical dendritic tree* comprising only excitatory synapses and a *basal dendritic tree* where mainly inhibitory synapses are situated. Both types of synapses are significantly separated in space thus forming a dipole of current sources (the inhibitory synapses) and sinks (the excitatory synapses) [38, 39, 41, 42]. The extracellular current flows from the sources to the sinks through a non-negligible resistance $R_{\mathrm{out}}$ which entails the somato-dendritic field. Therefore, we divide the pyramidal cell into three compartments: the first represents the apical dendrites, the second the basal dendrites, and the third takes firing into account. Figure 1.22 depicts its equivalent circuit. The internal resistance $R_{\mathrm{int}}$ accounts for the *length constant* of the neuron and contributes also to the synaptic weights (see Chap. 7).

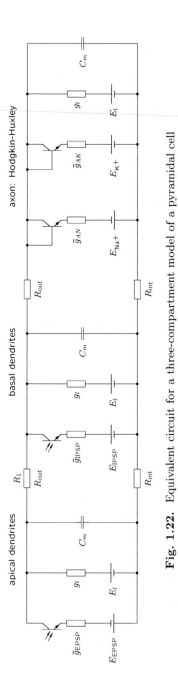

**Fig. 1.22.** Equivalent circuit for a three-compartment model of a pyramidal cell

In this model, the extracellular current $I_{\text{out}}$ flowing from the inhibitory to the excitatory compartment through $R_1$ entails the dendritic field potential

$$U^{\text{DFP}} = \frac{I_{\text{out}}}{R_1}. \tag{1.64}$$

Note that $I_{\text{out}}$ is large for a large difference between the EPSP and the IPSP, i.e. when both types of synapses are synchronously activated. In this case, however, the remaining current $I - I_{\text{out}}$ flowing through the axonal compartment can be too small to elicit action potentials. Therefore, DFP and spiking are inversely related with each other [41].

### 1.4.3 Reduced Models

The computational complexity of conductance models prevents numerical simulations of large neural networks. Therefore, simplifications and approximations have been devised and employed by several authors [10–18, 43–50].

In the following, we shall consider networks composed from $n$ model neurons. Their membrane potentials $U_i$ ($1 \leq i \leq n$) span the *observable state space*, such that $\boldsymbol{U} \in \mathbb{R}^n$; note that the proper phase space of the neural network might be of higher dimension. The observables $U_i$ depend on the total postsynaptic current

$$I_i^{\text{PSC}} = -\sum_{j=1}^{n} w_{ij} I_j - I_i^{\text{ext}}, \tag{1.65}$$

where $w_{ij}$ is the *synaptic weight* of the connection from unit $j$ to unit $i$, dependent on the synaptic gain $\bar{g}_{ij}$ that evolves during learning, thus reflecting *synaptic plasticity* (see Chap. 2), and the intracellular resistances (see Chap. 7). The capacitance in (1.57) has been deliberately neglected and $I_i^{\text{ext}}$ denotes the externally controlled input to neuron $i$.

### The McCulloch-Pitts Model

The coarsest simplification by McCulloch and Pitts [16, 18, 51, 52] replaces the involved Hodgkin-Huxley system (1.25, 1.27–1.29) by a threshold device with only two states: $X_i \in \{0, 1\}$ where 0 denotes the inactivated, silent, and 1 denotes the activated, firing state. The dynamics of a network of $n$ McCulloch-Pitts units is governed by the equations

$$X_i(t+1) = \Theta(I_i^{\text{PSC}} - \theta_i), \tag{1.66}$$

where $t$ is the discretized time, $\theta_i$ the activation threshold for unit $i$, and $I_i = X_i$ have been identified.

## Integrate-and-fire Models

The next step to make threshold units biologically more plausible is by taking the passive membrane properties as described by (1.57) into account. This leads to the class of (leaky) integrate-and-fire models [12, 13, 15, 16, 18, 46, 50]:

$$\tau_i \frac{dU_i}{dt} + U_i = I_i^{\mathrm{PSC}} \tag{1.67}$$

$$X_i(t_k) = \Theta(U_i(t_k) - \theta_i)$$

$$U_i(t_k) \leftarrow E.$$

Here, $U_i(t)$ describes the membrane potential of unit $i$, $X_i$ and $\theta_i$ model the action potentials and the firing thresholds as in (1.66), and $t_k$ are the firing times where the membrane potential is reset to its resting value $E$ (indicated by the arrow).

## Rate Models

In Sects. 1.3.1 and 1.3.3, we saw that the principles of frequency modulation are exploited for neural en- and decoding — at least for regular spiking dynamics. Therefore, it seems to be appropriate to replace the exact time-course of a spike train by its frequency, *firing rate*, or firing probability [53, 54]. The latter approach leads to the problem of determining the value

$$R_i(t) = \mathrm{Prob}(U_i(t) \geq \theta_i) = \int \mathrm{d}^{n-1}u \int_{\theta_i}^{\infty} \rho(\boldsymbol{u}, t) \, \mathrm{d}u_i, \tag{1.68}$$

where we have to regard the membrane potentials $\boldsymbol{U}(t)$ as a multivariate stochastic variable in the observable space with expectation values $\bar{U}_i(t)$ and probability density function $\rho(\boldsymbol{u}, t)$. The first integral in (1.68) provides the marginal distribution in the $i$th observable subspace. The stochasticity assumption is justified by our treatment of the presynaptic potential in Sect. 1.3.2. Because every action potential starts a Bernoulli process which describes how many vesicles are to be released, this element of stochasticity propagates along the synapse. As we have characterized the distribution of the number of released vesicles by (1.36), the postsynaptic currents are normally distributed about their voltage-dependent means $\bar{I}_i(\bar{\boldsymbol{U}})$,

$$I_i^{\mathrm{PSC}} = -\sum_j w_{ij} \left( \bar{I}_j(\bar{\boldsymbol{U}}(t)) + \eta_j(t) \right), \tag{1.69}$$

where $\eta_j(t)$ are independent normally distributed stochastic processes with

$$\langle \eta_j(t) \rangle = 0 \tag{1.70}$$

$$\langle \eta_j(t), \eta_k(t') \rangle = Q_{jk} \, \delta(t - t').$$

Therefore, (1.57) has to be read as a stochastic differential (Langevin) [30,31] equation

$$\frac{dU_i}{dt} = K_i(\boldsymbol{U}) - \sum_j \alpha_i w_{ij} \eta_j(t)) \,, \tag{1.71}$$

where

$$K_i(\boldsymbol{U}) = -\alpha_i U_i - \sum_j \alpha_i w_{ij} \bar{I}_j(\bar{U}) \tag{1.72}$$

are the deterministic drifting forces and $\sum_j \alpha_i w_{ij} \eta_j$ are stochastic fluctuations, obeying

$$\left\langle \sum_j \alpha_i w_{ij} \eta_j(t), \sum_l \alpha_k w_{kl} \eta_l(t') \right\rangle = R_{ik} \delta(t - t') \,, \tag{1.73}$$

with

$$R_{ik} = \sum_{jl} \alpha_i w_{ij} \alpha_k w_{kl} Q_{jl} \,; \tag{1.74}$$

here, we substituted $\alpha_i = \tau_i^{-1}$.

The probability distribution density $\rho(\boldsymbol{u}, t)$ is then obtained by solving the Fokker-Planck equation [30,31] associated to (1.71),

$$\frac{\partial \rho}{\partial t} = \sum_i \frac{\partial}{\partial u_i}[K_i(\boldsymbol{u})\rho] + \frac{1}{2} \sum_{ik} R_{ik} \frac{\partial^2 \rho}{\partial u_i \partial u_k} \,. \tag{1.75}$$

In order to solve (1.75), we assume that the currents $\bar{I}_j$ do not explicitly depend on the mean membrane potential, and that they change rather slowly in comparison to the density $\rho$ (the "adiabatic ansatz"). Then, (1.75) is linear and hence solved by the Gaussians

$$\rho(u,t) = \frac{1}{\sqrt{2\pi\sigma_U^2(t)}} \exp\left[-\frac{(u - \bar{U}(t))^2}{2\sigma_U^2(t)}\right] \tag{1.76}$$

as its stationary marginal distributions, where $\bar{U}(t)$ and $\sigma_U^2(t)$ have to be determined from $\bar{I}(t)$ and $R_{ik}$. Integrating (1.68) with respect to (1.76) yields the spike rate

$$R_i = f(\bar{U}_i) = \frac{1}{2} \operatorname{erfc}\left(\frac{\theta_i - \bar{U}_i}{\sqrt{2}\sigma_U}\right) \,, \tag{1.77}$$

with "erfc" denoting the complementary error function. In such a way, the stochastic threshold dynamics are translated into the typical *sigmoidal activation functions* $f(x)$ employed in computational neuroscience [7–9,11–13,15,16,18].

Gathering (1.67, 1.77), a leaky integrator model [46] is obtained as

$$\tau_i \frac{dU_i}{dt} + U_i = \sum_j w_{ij} f(U_j) \,. \tag{1.78}$$

An alternative derivation of (1.78) can be found in Chap. 7 by disregarding the postsynaptic impulse response functions $G(t, t')$. If these are taken into account, instead an integro-differential equation

$$\tau_i \frac{\mathrm{d}U_i}{\mathrm{d}t} + U_i = \sum_j w_{ij} \int_{-\infty}^{t} G(t - t') f(U_j(t')) \, \mathrm{d}t' \tag{1.79}$$

applies.

## The Rinzel-Wilson Model

The models to be discussed next are approximations for the full Hodgkin-Huxley equations (1.25, 1.27–1.29). Following Rinzel, Wilson and Trappenberg [14,16], the Hodgkin-Huxley equations exhibit two separated time-scales: at the fast scale, the opening of the sodium channels characterized by $m(t)$ happens nearly instantaneously such that $m(t)$ can be replaced by its stationary value $m_\infty$. On the other hand, the opening rate for the potassium channels $n$ and the inactivation rate $h$ for the sodium channels exhibit an almost linear relationship $h = 1 - n$. The corresponding substitutions then lead to a two-dimensional system for each neuron $i$.

$$I_i = C_m \frac{\mathrm{d}U_i}{\mathrm{d}t} + n_i^4 \bar{g}_{AK}(U_i - E_{K+}) + \tag{1.80}$$
$$+ m_\infty^3 (1 - n_i) \bar{g}_{AN}(U_i - E_{Na+}) + g_l(U_i - E_l)$$
$$\frac{\mathrm{d}n_i}{\mathrm{d}t} = \alpha_n (1 - n_i) - \beta_n \, n_i$$

which was applied for the plot in Fig. 1.15.

## The FitzHugh-Nagumo Model

The same observation as in above led FitzHugh and Nagumo to their approximation of the Hodgkin-Huxley equations [13, 14, 18, 43, 50]. Here, a general linear relation $h = a - bn$ is used in combination with a coordinate transformation and rescaling to arrive at the *Bonhoeffer-Van-der-Pol-*, or likewise, *FitzHugh-Nagumo equations*,

$$\frac{\mathrm{d}U_i}{\mathrm{d}t} = U_i - \frac{1}{3}U_i^3 - W_i + I_i \tag{1.81}$$
$$\frac{\mathrm{d}W_i}{\mathrm{d}t} = \phi(U_i + a - bW_i),$$

with parameters $\phi, a, b$.

## The Morris-Lecar Model

Originally, the *Morris-Lecar model* was devised to describe the spiking dynamics of potassium- and calcium-controlled muscle fibers [12–14,18,50]. After introducing dimensionless variables and rescaled parameters, they read

$$\frac{dU_i}{dt} = -m_\infty \bar{g}_{AC}(U_i - 1) - W_i\,\bar{g}_{AK}(U_i - E_{K^+}) - g_l(U_i - E_l) + I_i \quad (1.82)$$

$$\frac{dW_i}{dt} = \alpha_W\,(1 - W_i) - \beta_W\,W_i\,.$$

The Morris-Lecar model has been extensively employed during the Summer School, see Chaps. 9, 11, 12, 14.

## The Hindmarsh-Rose Model

The FitzHugh-Nagumo and Morris-Lecar models have the disadvantage that they do not have a bursting regime in their parameter space [50]. In order to overcome this obstacle, a third dimension for the phase space is necessary. The *Hindmarsh-Rose equations*, which exhibit this third dimension, are [14,47]

$$\frac{dU_i}{dt} = V_i - U_i^3 + 3U_i^2 + I_i - W_i \quad (1.83)$$

$$\frac{dV_i}{dt} = 1 - 5U_i^2 - V_i$$

$$\frac{dW_i}{dt} = r[s(U_i - U_0) - W_i]\,,$$

with parameters $r, s, U_0$.

For applications of the Hindmarsh-Rose model in this book, see Chap. 6.

## The Izhikevich Model

Making use of arguments from bifurcation theory, Izhikevich [49] approximated the Hodgkin-Huxley equations by the two-dimensional flow

$$\frac{dU_i}{dt} = 0.04U_i^2 + 5U_i + 140 - U_i + I_i \quad (1.84)$$

$$\frac{dV_i}{dt} = a(bV_i - U_i)\,, \quad (1.85)$$

disrupted by an auxiliary after-spike resetting

$$\text{if} \quad U_i \geq 30\,\text{mV}, \quad \text{then} \begin{cases} U_i & \leftarrow & E \\ V_i & \leftarrow & V_i + c \end{cases}$$

with parameters $a, b, c, E$, where $E$ denotes the resting potential.

A comprehensive comparison of different spiking neuron models with respect to their biological plausibility and computational complexity can be found in [50].

Also the Izhikevich model has been used during the Summer School. These results are presented in Chap. 13.

### 1.4.4 Neural Field Theories

For very large neural networks, a continuum approximation by spatial coarse-graining suggests itself [55–70]. Starting from the rate equation (1.79), the sum over the nodes connected with unit $i$ has to be replaced by an integral transformation of a neural field quantity $U(x,t)$, where the continuous parameter $x$ now indicates the position $i$ in the network. Correspondingly, the synaptic weights $w_{ij}$ turn into a kernel function $w(x,y)$. In addition, for large networks, the propagation velocity $c$ of neural activation has to be taken into account. Therefore, (1.79) assumes the retarded form

$$\tau(x)\,\frac{\partial U(x,t)}{\partial t}+U(x,t)=\int\limits_{-\infty}^{t}\mathrm{d}t'\int\mathrm{d}x'w(x,x')G(t-t')f\left[U\left(x',t'-\frac{|x-x'|}{c}\right)\right],$$

(1.86)

which can be transformed into a wave equation under additional assumptions. For further details, consult Chap. 8 and the references above.

## 1.5 Mass Potentials

Neural field theories [55–70] as well as population models of cortical modules [39, 71–81] (see also Chap. 7) describe mass potentials such as LFP or EEG as spatial sums of the EPSPs and IPSPs of cortical pyramidal cells. In these accounts, the somato-dendritic field potential (DFP) of an infinitesimally small volume element of cortical tissue, or of a single neuron, respectively, is described [79] by

$$U^{\mathrm{DFP}} = U^{\mathrm{EPSP}} + U^{\mathrm{IPSP}}$$

(1.87)

when $U^{\mathrm{EPSP}} > 0, U^{\mathrm{IPSP}} < 0$.[4] Unfortunately, this description is at variance with the physiological origin of the DFP. Looking at the equivalent circuit of the three-compartment model in Fig. 1.22, one easily recognizes that simultaneously active excitatory and inhibitory synapses give rise to a large voltage drop along the resistor $R_1$ separating both kind of synapses in space. Therefore, a large extracellular current yields a large DFP according to (1.64). On the other hand, the sum in (1.87) becomes comparatively small since EPSP and IPSP almost compensate each other. Therefore, the geometry and anatomy of pyramidal cells and cortex have to be taken into account. To this end, I shall mainly review the presentation of Nunez and Srinivasan [42, 82] in the following.

---

[4] The signs in (1.87) are physiologically plausible (cf. (1.43, 1.55)), whereas Jansen et al. [75, 76], Wendling et al. [77, 78], and David and Friston [79] assume that EPSP and IPSP both have positive signs such that their estimate for the DFP reads $U^{\mathrm{DFP}} = U^{\mathrm{EPSP}} - U^{\mathrm{IPSP}}$ (cf. Chap. 5).

## 1.5.1 Dendritic Field Potentials

If the reader takes a look at Fig. 8.1 in Chap. 8, she or he sees three triangular knobs that are the cell bodies of three pyramidal cells. Starting from their bases, axons proceed downwards like the roots of plants. In the other direction, they send strongly branched trees of dendrites towards the surface of the cortex. Pyramidal cells exhibit roughly an axonal symmetry and they are very densely packed in parallel, forming a fibrous tissue. Excitatory and inhibitory synapses are spatially separated along the dendritic tree: Excitatory synapses are mainly situated at the apical (i.e. the top-most) dendrites, while inhibitory synapses are arranged at the soma and the basal dendrites of the cells. This arrangement is functionally significant as the inhibitory synapses very effectively suppress the generation of action potentials by establishing short-cuts [41].

From the viewpoint of the extracellular space, inhibitory synapses act as current sources while excitatory synapses are current sinks. The extracellular space itself can be regarded as an electrolyte with (volume-) conductance $\sigma(\boldsymbol{x})$, where $\boldsymbol{x}$ indicates the dependence on the spatial position. From Maxwell's equations for the electromagnetic field, a continuity equation

$$-\nabla \cdot (\sigma \nabla \phi) + \frac{\partial \rho}{\partial t} = 0 \qquad (1.88)$$

can be derived for the "wet-ware" [42, 82]. Here, $\phi(\boldsymbol{x})$ denotes the electric potential and $\rho(\boldsymbol{x}, t)$ the charge density, and $\boldsymbol{j} = -\sigma \nabla \phi$ is the current density according to Ohm's Law (1.5). Assuming that the conductivity $\sigma(\boldsymbol{x})$ is piecewise constant in the vicinity of a pyramidal cell, $\sigma$ can be removed from the scope of the first gradient, yielding

$$-\sigma \Delta \phi + \frac{\partial \rho}{\partial t} = 0 \,. \qquad (1.89)$$

Next, we have to describe the change of the current density. Setting

$$\frac{\partial \rho(\boldsymbol{x})}{\partial t} = \sum_i I_i \delta(\boldsymbol{x} - \boldsymbol{x_i}) \qquad (1.90)$$

describes the postsynaptic transmembrane currents in the desired way as point sources and sinks located at $\boldsymbol{x_i}$. When we insert (1.90) into (1.89), we finally arrive at a Poisson equation

$$\sigma \Delta \phi = \sum_i I_i \delta(\boldsymbol{x} - \boldsymbol{x_i}) \qquad (1.91)$$

in complete analogy to electrostatics.

Equation (1.91) can be easily solved by choosing appropriate boundary conditions that exclude the interiors and the membranes of the cells from the domain of integration.[5] Integrating (1.91) over the extracellular space gives

$$\phi(\boldsymbol{x}) = \frac{1}{4\pi\sigma} \sum_i \frac{I_i}{r_i} , \tag{1.92}$$

where $\boldsymbol{x}$ denotes the observation site and $r_i = |\boldsymbol{x} - \boldsymbol{x}_i|$ abbreviates the distance between the point sources and sinks $I_i$ and $\boldsymbol{x}$.

If the distance of the observation site $\boldsymbol{x}$ is large in comparison to the respective distances of the current sources and sinks from each other, the potential $\phi(\boldsymbol{x})$ can be approximated by the first few terms of a multipole expansion,

$$\phi(\boldsymbol{x}) = \frac{1}{4\pi\sigma} \left( \frac{1}{x} \sum_i I_i + \frac{1}{x^3} \sum_i I_i \boldsymbol{x}_i \cdot \boldsymbol{x} + \dots \right) . \tag{1.93}$$

Now, $x$ denotes the distance of the observation point from the center of mass of the current cloud $I_i$. Due to the conservation of charge, the monopole term vanishes, whereas the higher order multipoles strongly decline with $x \to \infty$. Therefore, only the dipole term accounts for the DFP,

$$\phi^{\mathrm{DFP}}(\boldsymbol{x}) = \frac{1}{4\pi\sigma} \frac{1}{x^3} \sum_i I_i \boldsymbol{x}_i \cdot \boldsymbol{x} = \frac{1}{4\pi\sigma} \frac{\boldsymbol{p} \cdot \boldsymbol{x}}{x^3} , \tag{1.94}$$

where the dipole moment of the currents

$$\boldsymbol{p} = I(\boldsymbol{x}_1 - \boldsymbol{x}_2) = I\boldsymbol{d} \tag{1.95}$$

can be introduced when a current source $+I$ and a sink $-I$ are separated by the distance $d$. The unit vector $\boldsymbol{d}/d$ points from the source to the sink.

Equation (1.95) now suggests a solution for the problem with (1.87). The DFP is proportional to the dipole moment which depends monotonically on the absolute value $I$. Assuming that the excitatory and the inhibitory branch in the equivalent circuit in Fig. 1.22 are symmetric, the dipole moment, which is determined by $I_{\mathrm{out}}$, depends on the difference between the (positive) EPSP and the (negative) IPSP,

$$U^{\mathrm{DFP}} = U^{\mathrm{EPSP}} - U^{\mathrm{IPSP}} , \tag{1.96}$$

such that a simple change of the sign corrects (1.87).

---

[5] My presentation here deviates from that given by Nunez and Srinivasan [42] who assume quasi-stationary currents $[\nabla \cdot \boldsymbol{j} = 0]$. As a consequence of (1.88), the change of the charge density would also vanish leading to a trivial solution. In order to circumvent this obstacle, Nunez and Srinivasan [42, p. 166] decompose the current into an "ohmic" part $-\sigma \nabla \phi$ and peculiar "impressed currents" $\boldsymbol{J}_S$ corresponding to EPSC and IPSC that cross the cell membranes. However, they concede that "the introduction of this pseudo-current may, at first, appear artificial and mysterious" aiming at the representation of the boundary conditions. This distinction is actually unnecessary when boundary conditions are appropriately chosen [82]. Nevertheless, Nunez' and Srinivasan's argument became very popular in the literature, e.g. in [4,67].

## 1.5.2 Local Field potentials

Since pyramidal cells are aligned in parallel, they form a dipole layer of thickness $d$ when they are synchronized within a cortical module. Subsequently, we will identify such modules with the anatomical columns (cf. Chap. 8) in order to compute the collective DFP, i.e. the local field potential (LFP) generated by a mass of approximately 10,000 pyramidal cells.

The current differential $dI$ is then proportional to the infinitesimal area in cylindrical coordinates

$$dI = jdA, \tag{1.97}$$

where the current density $j$ is assumed to be a constant scalar within one column. Hence, the differential of the potential $d\phi$ at a distance $z$ perpendicular to a cortical column of radius $R$ that is contributed by the current $dI$ is given by

$$d\phi(\boldsymbol{x}) = \frac{1}{4\pi\sigma} \frac{j\,\boldsymbol{d} \cdot (\boldsymbol{x} - \boldsymbol{x}')}{|\boldsymbol{x} - \boldsymbol{x}'|^3} dA, \tag{1.98}$$

where $\boldsymbol{x}'$ varies across the area of the module. Making use of the geometry depicted in Fig. 1.23 yields

$$d\phi(z) = \frac{1}{4\pi\sigma} \frac{jd\sqrt{r^2 + z^2} \cos\vartheta}{(r^2 + z^2)^{3/2}} r\, dr\, d\varphi$$

$$= \frac{jd}{4\pi\sigma} \frac{z\sqrt{r^2 + z^2}}{\sqrt{r^2 + z^2}(r^2 + z^2)^{3/2}} r\, dr\, d\varphi$$

$$= \frac{jd}{4\pi\sigma} \frac{rz}{(r^2 + z^2)^{3/2}} dr d\varphi$$

$$\phi(z) = \frac{jd}{4\pi\sigma} \int_0^{2\pi} d\varphi \int_0^R dr \frac{rz}{(r^2 + z^2)^{3/2}}.$$

Performing the integration then gives the LFP perpendicular to the dipole layer

$$\phi^{\text{LFP}}(z) = \frac{jd}{2\sigma} \left( 1 - \frac{z}{\sqrt{R^2 + z^2}} \right). \tag{1.99}$$

## 1.5.3 Electroencephalograms

Equation (1.99) describes the summed potential resulting from the synchronized synaptic activity of all pyramidal neurons in a column in a distance $z$ above the cortical gray matter. By integrating over a larger domain of cortical tissue, e.g. over a macrocolumn, one obtains an estimator of the electrocorticogram (ECoG) [83]. In order to compute the electroencephalogram (EEG), one has to take the different conductances of skull and scalp into account. Nunez and Srinivasan [42] discuss different scenarios with different

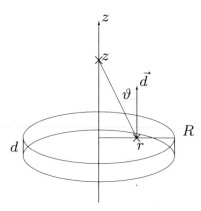

**Fig. 1.23.** Geometry of a cortical column

geometries. In the simplest case, only one interface between a conductor $(G_1)$ with conductance $\sigma_1$ and an isolator $G_2$ (conductance $\sigma_2 = 0$) is considered. According to the respective interpretation, either the skull, or the air above the subject's head is regarded to be the isolator.[6]

In one case, one has to consider the potential generated by a point source (or sink) $I$ at a distance $-h$ from the interface in the semi-space $G_1$ by attaching a mirror source (or sink) $I'$ at a distance $h$ from the interface in order to solve Dirichlet's boundary problem if $\boldsymbol{x} \in G_2, z > 0$ [42]. In the other case, one has to replace the source (or sink) $I$ in the semi-space $G_1$ by another source (or sink) $I + I''$ if $\boldsymbol{x} \in G_1, z < 0$. The geometrical situation is shown in Fig. 1.24.

In the semi-space $G_2$ relevant for the EEG measurement, (1.91) is then solved by

$$\phi^{\mathrm{DFP}}(\boldsymbol{x}) = \frac{1}{2\pi(\sigma_1 + \sigma_2)} \frac{I}{\sqrt{r^2 + (z + h)^2}} \,. \qquad (1.100)$$

When $G_2$ is assumed to be an isolator, we set $\sigma_2 = 0$, $\sigma_1 \equiv \sigma$. Hence the potential in the semi-space $G_2$ is simply twice the potential in a homogeneous medium. Provided that all current sources and sinks are distributed in $G_1$, the superposition principle entails

$$\phi^{\mathrm{DFP}}(\boldsymbol{x}) = \frac{1}{2\pi\sigma} \sum_i \frac{I_i}{\sqrt{r_i^2 + (z + h_i)^2}} \,. \qquad (1.101)$$

From (1.99) follows

---

[6] Occasionally a misunderstanding occurs in the literature where ionic currents and dielectric displacement, i.e. polarization, are confused [84, 85]. Do we really measure sodium or potassium ions that have traversed the skull during an EEG measurement, or is the skull merely a polarizable medium?

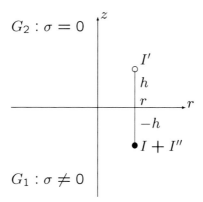

**Fig. 1.24.** Geometry of Dirichlet's boundary problem for the EEG

$$\Phi(z) \equiv \phi^{\mathrm{EEG}} = \frac{jd}{\sigma}\left(1 - \frac{z}{\sqrt{R^2 + z^2}}\right) \tag{1.102}$$

for the EEG generated by a cortical column measurable at the scalp.

### 1.5.4 Mean Fields

In this last subsection, I shall discuss the question of whether mass potentials such as LFP or EEG are mere epiphenomena [42, 86], or whether they play a functional role in the organization of brain functioning. If the latter were the case, they would be described as *order parameters* which couple as *mean fields* onto the microscopic neurodynamics with the ability to enslave its behavior [30, 87].

In order to encounter this problem, one has to estimate the average field strengths and voltage differences that are generated by synchronized activity of all pyramidal cells of a cortical module. These quantities then have to be compared with experimental findings on the susceptibility of nerve cells through electromagnetic fields. As mentioned above, the spiking threshold is around $\theta = -50\,\mathrm{mV}$, i.e. the membrane must be polarized by $\Delta U = 10\,\mathrm{mV} - 20\,\mathrm{mV}$ from its resting value given by the Nernst equation (1.7). This corresponds to an electric field strength $E = 10^6\,\mathrm{V/m}$ for a thickness of 5 nm of the cell membrane [88].

On the other hand, neurophysiological experiments have revealed that much smaller field strengths of about 1 V/m entail significant changes of neural excitability [89–93]. Event-related potentials can be modulated by values around 4 V/m. In the hippocampus, where pyramidal cells are very densely packed, effective field strengths are in the range of 5–7 V/m, whereas 10–15 V/m are needed in the cerebellum [91].

To estimate the field strength generated by a cortical column, we have to solve the Dirichlet boundary problem within the electrolyte $G_1$ as shown in

Fig. 1.24. The potential of a point source (or sink) $I$ at $\boldsymbol{x} \cong (0,0,z)$ is given in the semi-space $G_1$ as

$$\phi(\boldsymbol{x}) = \frac{I}{4\pi\sigma_1} \left( \frac{1}{\sqrt{r^2 + (z+h)^2}} + \frac{\sigma_1 - \sigma_2}{\sigma_1 + \sigma_2} \frac{1}{\sqrt{r^2 + (z-h)^2}} \right). \tag{1.103}$$

Since we assume $G_2$ to be an isolator again ($\sigma_2 = 0$), the superposition principle yields

$$\phi(\boldsymbol{x}) = \frac{1}{4\pi\sigma} \sum_i I_i \left( \frac{1}{\sqrt{r_i^2 + (z+h_i)^2}} + \frac{1}{\sqrt{r_i^2 + (z-h_i)^2}} \right). \tag{1.104}$$

Next, we apply (1.104) to a current dipole made up by a source (or sink) $I$ at $(r, \varphi, l - d/2)$ and a sink (or source) $-I$ at $(r, \varphi, l + d/2)$, where the dipole's center is situated in a distance $l$ below the interface:

$$\phi(\boldsymbol{x}) = \frac{I}{4\pi\sigma} \left( \frac{1}{\sqrt{r^2 + (z+l-d/2)^2}} + \frac{1}{\sqrt{r^2 + (z-l+d/2)^2}} - \right.$$

$$\left. \frac{1}{\sqrt{r^2 + (z+l+d/2)^2}} + \frac{1}{\sqrt{r^2 + (z-l-d/2)^2}} \right).$$

Approximating the quotients by $(1+x)^{-1/2} \approx 1 - x/2$ gives

$$\phi(\boldsymbol{x}) = \frac{Ild}{2\pi\sigma r^3}, \tag{1.105}$$

i.e. the potential depends only on the radial direction in the conductor. Therefore, the field strength is given by the $r$-component of the gradient

$$E = -\frac{\partial}{\partial r}\phi(\boldsymbol{x}) = \frac{3Ild}{2\pi\sigma r^4}. \tag{1.106}$$

In order to replace the column by an equivalent current dipole moment generating the same field, we have to compute the current density through the surface of the column according to (1.102) from the measured scalp EEG. Rearrangement of (1.102) yields

$$j = \frac{\sigma}{d} \frac{\Phi(z)}{1 - \frac{z}{\sqrt{R^2 + z^2}}}. \tag{1.107}$$

Then, the current through the column would be

$$I = j\pi R^2 \tag{1.108}$$

if the tissue were a continuum as presupposed in Sect. 1.5.2. Here, $R \approx 150\,\mu m$ is the radius of a column. By contrast, one has to take into account that a

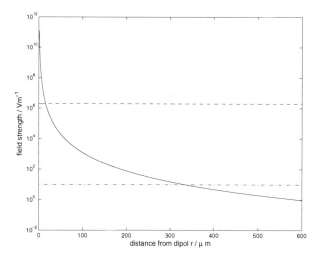

**Fig. 1.25.** Electric field strength estimated from the EEG depending on the distance of an equivalent dipole. The horizontal lines indicate critical field strengths for evoking action potentials (*upper line*), and for detectable physiological impact (*bottom line*)

column contains about 10,000 pyramidal cells. Thus, the current along a single pyramidal cell is

$$I_{\mathrm{Pyr}} = \frac{j\pi R^2}{N} \tag{1.109}$$

with $N = 10,000$. Inserting (1.109) and (1.107) into (1.106) gives with $z = l$

$$E(r) = \frac{3lR^2}{2Nr^4} \frac{\Phi(l)}{1 - \frac{l}{\sqrt{R^2+l^2}}} . \tag{1.110}$$

Figure 1.25 presents a plot of $E(r)$ for the parameter values $R = 150\,\mu\mathrm{m}$, $N = 10,000$, where a distance between the cortex surface and the skull $l = 8\,\mathrm{mm}$ and a peak EEG amplitude of $\Phi(l) = 100\,\mu\mathrm{V}$ have been assumed.

Additionally, Fig. 1.25 displays two lines: the upper line reflects the spiking threshold of a single neuron, $E = 10^6\,\mathrm{V/m}$; the bottom one indicates the limit of physiological efficacy, $E = 10\,\mathrm{V/m}$ [91]. These thresholds correspond to the distances $r_1 = 16.39\,\mu\mathrm{m}$, and $r_2 = 346.77\,\mu\mathrm{m}$ from the equivalent dipole. Because we took $R = 150\,\mu\mathrm{m}$ as the radius of a cortical module, $r_2$ reaches far into the neighboring column. With 10,000 pyramidal cells per module, their average distance amounts to $3\,\mu\mathrm{m}$, such that approximately 120 closely packed pyramidal cells can be excited by the mass potential. Interestingly, the radius $r_2$ coincides nicely with the size of the column. Hence, this rough estimate suggests that cortical columns are functional modules controlled by their own electric mean fields that are very likely not mere epiphenomena.

This is consistent with empirical findings. Adey [89] reported a change of the calcium conductivity of neuron membranes under the impact of sustained high-frequency fields. In the hippocampus of rats, field effects are particularly pronounced due to the extreme packing density of the pyramidal cells and the resulting low conductivity of the extracellular liquid. Suppressing synaptic transmission experimentally by reducing the amount of available extracellular calcium, leads to the emergence of spontaneous *bursting* that can be synchronized by mass potentials [91]. Bracci et al. [90] demonstrated that the synchronization of hippocampal neurons is facilitated by the application of external electric fields. They showed also that the conductance of the extracellular electrolyte is a control parameter which can be tuned in such a way that spontaneous synchronization takes place if the conductance is lowered below a critical value. In this case, the fall of the dendritic field potentials along the extracellular resistors contribute to larger LFP and EEG that in turn enslave the whole population. Most recently, Richardson, Schiff and Gluckman [92, 93] studied the propagation of traveling waves through an excitable neural medium under the influence of external fields. They reported a dependence of the group velocity on the polarity of the applied field, and modeled these effects through a neural field theory analogue of (1.86).

Coming back to our neuron models, the impact of mass potentials can be formally taken into account by introducing a *mean field* coupling. Concerning, for example, the Hodgkin-Huxley equations (1.25, 1.27–1.29), one has to replace the membrane potential $U_i$ of neuron $i$ by a shifted value

$$U_i' = U_i - \sum_j U_j^{\mathrm{DFP}} \,, \tag{1.111}$$

where either the dendritic field potential is given by (1.64), or, in a continuum account, the whole sum is provided by (1.99). This idea has been expressed by Braitenberg and Schüz [94, p. 198], in that a cortical module controls its own activation thresholds.

## 1.6 Discussion

In this chapter, I have sketched the basic principles of neurophysics, i.e. the biophysics of membranes, neurons, neural networks and neural masses. Let me finally make some concluding remarks on neural modeling and descriptions. I hope the reader has recognized that there is no unique *physical model* of *the* neuron, or of *the* neural network of the brain. Even a single neuron can be described by models from different complexity classes. It can be regarded as a continuous system governed by a nonlinear partial differential equation which describes its cable properties. Decomposing the cable into compartments, one obtains either compartment models comprised of lots of coupled ordinary differential equations, or point models that are still described by many coupled ordinary differential equations, one for the kinetics of each population of ion channels. Further simplifying these models, one eventually arrives at the coarsest McCulloch-Pitts neuron [51].

On the other hand, each neuron model dictates the dimensionality of its phase space and, as its projection, its observable space. Observables provide the interface to experimental neuroscience in that they should be *observable*. The best theoretical neuron model is not much good if it contains quantities that are not observable in real experiments. In most cases, observables are membrane potentials either collected from the axons, i.e. action potentials, or measured from the dendro-somatic membranes such as EPSP and IPSP. However, a few electrode tips put into a couple of neurons will not provide sufficiently many observables for describing the behavior of a neural network. At the network level, the abovementioned problems greatly worsen with increasing size and complexity of the network, ending in an unmanageable number of degrees of freedom for continuum models.

In this case, spatial coarse-graining [62] is the method of choice. By averaging activity across regions of appropriate size, one obtains mass potentials such as LFP or EEG. LFP is experimentally observable through the use of multi-electrode arrays placed into the extracellular space. Each sensor collects averaged dendritic field potentials from several thousands of neurons, as well as some spiking activity in its vicinity. On the other hand, ECoG (intracranial EEG) and EEG are gathered by electrodes placed at the cerebral membrane or at the scalp, respectively. Each electrode registers the mean activity of billions of neurons. Using conventional 32, 64, or 128 channel amplifiers thereby collapses the huge microscopic observable space of single neurons and the large mesoscopic observable space of LFP to a macroscopic observable space of 32, 64, or 128 dimensions.

As we have seen in Sect. 1.5.3, such mass potentials are not in the least irrelevant because they serve as order parameters [30,87], both indicating and causing macroscopic ordering of the system. Yet there is another important aspect of mass potentials. They do not only comprise a spatial coarse-graining by definition, but also provide a coarse-graining of the high-dimensional microscopic phase space. Consider a mean field observable

$$F(x) = \sum_i f_i(\boldsymbol{x}) \,, \tag{1.112}$$

where the sum extends over a population of $n$ neurons and $f_i$ denotes a projection of the microscopic state $\boldsymbol{x} \in X$ onto the $i$-th coordinate axis measuring the activation $f_i(\boldsymbol{x}) = x_i$ of the $i$th neuron. Obviously, the outcomes of $F$ may have multiple realizations as the terms in the sum in (1.111) can be arbitrarily arranged. Therefore, two neural activation vectors $\boldsymbol{x}, \boldsymbol{y}$ can lead to the same value $F(\boldsymbol{x}) = F(\boldsymbol{y})$ (e.g. when $f_i(\boldsymbol{x}) - \epsilon = f_j(\boldsymbol{x}) + \epsilon$, $i \neq j$), so that they are indistinguishable by means of $F$. Beim Graben and Atmanspacher [95] call such microstates *epistemically equivalent*. All microstates that are epistemically equivalent to each other form an equivalence class, and, as it is well known from set theory, all equivalence classes partition the phase space $X$. If the equivalence classes of $F$ in $X$ form a finite partition $\mathcal{Q} = \{A_1, \ldots A_I\}$ of $X$, one can assign symbols $a_i$ from an alphabet $\mathbf{A}$ to the cells $A_i$ and obtain a *symbolic dynamics* [96–98]. In this way, experimentally well-defined meso- and macroscopic brain observables, LFP and EEG, form a coarse-grained description of the underlying microscopic neurodynamics.

Atmanspacher and beim Graben [99, 100] discuss this coarse-graining with respect to its stability properties. The microscopic dynamics $x \mapsto \Phi^t(x)$ where the *flow* $\Phi$ solves the microscopic differential equations, is captured by transitions from one symbol to another one $a_i \mapsto a_j$. If these transitions can be described by an ergodic Markov chain, the symbolic dynamics exhibits particular stability properties. If the Markov chain is aperiodic, a distinguished thermal equilibrium state can be constructed for the symbolic description. If, contrarily, the Markov chain is periodic, the system possesses stable fixed point or limit cycle attractors. Atmanspacher and beim Graben argue that in both cases, the concept of *contextual emergence* applies where higher-level descriptions emerge from contingently supplied contexts that are not merely reducible to lower-level descriptions. As an application, Atmanspacher and beim Graben [99, 100] demonstrated the contextual emergence of neural correlates of consciousness [101] from neurodynamics where arbitrary contexts are given by phenomenal families partitioning the space of phenomenal experiences [102]. Other examples were discussed by beim Graben [103] and Dale and Spivey [104] where symbolic cognition emerges from partitioned dynamical systems.

The problem of finding reasonable macroscopic observables for neural networks has been addressed by Amari [52]. He considered random networks of McCulloch-Pitts neurons (cf. Chaps. 3, 5, 7), and defined a proper macrostate as a macroscopic observable such as (1.112) if two conditions hold: Firstly, the temporal evolution of the observable should be compatible with the coarse-graining, and, secondly, the observable should be structurally stable against topological deformations of the network. The second requirement is closely related to ergodicity of the resulting symbolic dynamics [99]. Accordingly, the first demand entails that all initial conditions that are mapped onto the same value of a macrostate are epistemically equivalent. As an example for a good macrostate, at least for his toy-model, Amari [52] provided the mass potential (1.112). Hence, macroscopic descriptions of neural masses provide important insights into the functional organization of neurodynamical systems. They are by far more then mere epiphenomena.

# Acknowledgements

This chapter is based on notes from a lecture course "Introduction to Neurophysics", I taught together with J. Kurths, D. Saddy, and T. Liebscher in the years 2000 and 2002 at the University of Potsdam for the DFG Research Group "Conflicting Rules in Cognitive Systems". Section 1.5 contains parts of my thesis [82] that were previously available only in German. Section 1.6 presents some results from a cooperation with H. Atmanspacher. S. J. Nasuto,

J. J. Wright and C. Zhou helped me improving this chapter. I greatly acknowledge their respective contributions and inspiring discussions.

# References

1. J. von Neumann. *The Computer and the Brain.* Yale University Press, New Haven (CT), 1958. Partly reprinted in J. A. Anderson and E. Rosenfeld (1988), p. 83ff.
2. Z. W. Pylyshyn. *Computation and Cognition: Toward a Foundation for Cognitive Science.* MIT Press, Cambrigde (MA), 1986.
3. J. R. Anderson. *Cognitive Psychology and its Implications.* W. H. Freeman and Company, New York (NY), 4th edition, 1995.
4. M. Kutas and A. Dale. Electrical and magnetic readings of mental functions. In M. Rugg, editor, *Cognitive Neuroscience*, pp. 197–242. Psychology Press, Hove East Sussex, 1997.
5. R. C. O'Reilly and Y. Munakata. *Computational Explorations in Cognitive Neuroscience. Understanding the Mind by Simulating the Brain.* MIT Press, Cambridge (MA), 2000.
6. M. S. Gazzaniga, R. B. Ivry, and G. R. Mangun, editors. *Cognitive Neuroscience. The Biology of the Mind.* W. W. Norton, New York (NY), 2nd edition, 2002.
7. J. A. Anderson and E. Rosenfeld, editors. *Neurocomputing. Foundations of Research*, Vol. 1. MIT Press, Cambridge (MA), 1988.
8. J. A. Anderson, A. Pellionisz, and E. Rosenfeld, editors. *Neurocomputing. Directions for Research*, Vol. 2. MIT Press, Cambridge (MA), 1990.
9. P. S. Churchland and T. J. Sejnowski. *The Computational Brain.* MIT Press, Cambridge (MA), 1994.
10. F. Riecke, D. Warland, R. de Ruyter van Steveninck, and W. Bialek. *Spikes: Exploring the Neural Code.* Computational Neurosciences. MIT Press, Cambridge (MA), 1997.
11. M. A. Arbib, editor. *The Handbook of Brain Theory and Neural Networks.* MIT Press, Cambridge (MA), 1998.
12. C. Koch and I. Segev, editors. *Methods in Neuronal Modelling. From Ions to Networks.* Computational Neuroscience. MIT Press, Cambridge (MA), 1998.
13. C. Koch. *Biophysics of Computation. Information Processing in Single Neurons.* Computational Neuroscience. Oxford University Press, New York (NY), 1999.
14. H. R. Wilson. *Spikes, Decisions and Actions. Dynamical Foundations of Neuroscience.* Oxford University Press, New York (NY), 1999.
15. P. Dayan and L. F. Abbott. *Theoretical Neuroscience.* Computational Neuroscience. MIT Press, Cambridge (MA), 2001.
16. T. P. Trappenberg. *Fundamentals of Computational Neuroscience.* Oxford University Press, Oxford (GB), 2002.
17. R. P. N. Rao, B. A. Olshausen, and M. S. Lewicky, editors. *Probabilistic Models of the Brain: Perception and Neural Function.* MIT Press, Cambridge (MA), 2002.
18. W. Gerstner and W. Kistler. *Spiking Neuron Models. Single Neurons, Populations, Plasticity.* Cambridge University Press, Cambridge (UK), 2002.

19. E. R. Kandel, J. H. Schwartz, and T. M. Jessel, editors. *Principles of Neural Science*. Appleton & Lange, East Norwalk, Connecticut, 1991.

20. E. R. Kandel, J. H. Schwartz, and T. M. Jessel, editors. *Essentials of Neural Science and Behavior*. Appleton & Lange, East Norwalk, Connecticut, 1995.

21. J. G. Nicholls, A. R-Martin, B. G. Wallace, and P. A. Fuchs. *From Neuron to Brain*. Sinauer, Sunderland (MA), 2001.

22. H. C. Tuckwell. *Introduction to Theoretical Neurobiology*, Vol. 1. Cambridge University Press, Cambridge (UK), 1988.

23. H. C. Tuckwell. *Introduction to Theoretical Neurobiology*, Vol. 2. Cambridge University Press, Cambridge (UK), 1988.

24. D. Johnston and S. M.-S. Wu. *Foundations of Cellular Neurophysiology*. MIT Press, Cambridge (MA), 1997.

25. B. Hille. *Ion Channels of Excitable Membranes*. Sinauer, Sunderland, 2001.

26. A. Einstein. Eine neue Bestimmung der Moleküldimensionen. *Annalen der Physik*, 19:289–306, 1906.

27. S. B. Laughlin, R. R. de Ruyter van Steveninck, and J. C. Anderson. The metabolic cost of neural information. *Nature Neuroscience*, 1(1): 36–41, 1998.

28. W. W. Orrison Jr., J. D. Lewine, J. A. Sanders, and M. F. Hartshorne. *Functional Brain Imaging*. Mosby, St. Louis, 1995.

29. N. K. Logothetis, J. Pauls, M. Augath, T. Trinath, and A. Oeltermann. Neurophysiological investigation of the basis of the fMRI signal. *Nature*, 412: 150–157, 2001.

30. H. Haken. *Synergetics. An Introduction*, Vol. 1 of *Springer Series in Synergetics*. Springer, Berlin, 1983.

31. N. G. van Kampen. *Stochastic Processes in Physics and Chemistry*. Elsevier, Amsterdam, 1992.

32. A. L. Hodgkin and A. F. Huxley. A quantitative description of membrane current and its application to conduction and excitation in nerve. *J. Physiol.*, 117: 500–544, 1952.

33. I. Swameye, T. G. Müller, J. Timmer, O. Sandra, and U. Klingmüller. Identification of nucleocytoplasmatic cycling as a remote sensor in cellular signaling by databased modeling. *Proceedings of the National Academy of Sciences of the U.S.A.*, 100(3): 1028–1033, 2003.

34. J. M. Bower and D. Beeman. *The Book of GENESIS. Exploring Realistic Neural Models with the GEneral NEural SImulation System*. Springer, New York (NY), 1998.

35. J. W. Moore and M. L Hines. *Simulations with NEURON*. Duke and Yale University, 1994.

36. A. Destexhe, D. Contreras, and M. Steriade. Cortically-induced coherence of a thalamic-generated oscillation. *Neuroscience*, 92(2): 427–443, 1999.

37. C. Bédard, H. Kröger, and A. Destexhe. Modeling extracellular field potentials and the frequency-filtering properties of extracellular space. *Biophys. J.*, 86(3): 1829–1842, 2004.

38. O. Creutzfeld and J. Houchin. Neuronal basis of EEG-waves. In *Handbook of Electroencephalography and Clinical Neurophysiology*, Vol. 2, Part C, pp. 2C-5–2C-55. Elsevier, Amsterdam, 1974.

39. W. J. Freeman. *Mass Action in the Nervous System*. Academic Press, New York (NY), 1975.

40. D. T. J. Liley, D. M. Alexander, J. J. Wright, and M. D. Aldous. Alpha rhythm emerges from large-scale networks of realistically coupled multicompartmental model cortical neurons. *Network: Comput. Neural Syst.*, 10: 79–92, 1999.

41. A. J. Trevelyan and O. Watkinson. Does inhibition balance excitation in neocortex? *Prog. Biophys. Mol. Biol.,*, 87: 109–143, 2005.

42. P. L. Nunez and R. Srinivasan. *Electric Fields of the Brain: The Neurophysics of EEG.* Oxford University Press, New York, 2006.

43. R. FitzHugh. Impulses and physiological states in theoretical models of nerve membrane. *Biophys. J.*, 1: 445–466, 1961.

44. T. Pavlidis. A new model for simple neural nets and its application in the design of a neural oscillator. *Bull. Math. Biol.*, 27: 215–229, 1965.

45. R. B. Stein, K. V. Leung, M. N. Oğuztöreli, and D. W. Williams. Properties of small neural networks. *Kybernetik*, 14:223–230, 1974.

46. R. B. Stein, K. V. Leung, D. Mangeron, and M. N. Oğuztöreli. Improved neuronal models for studying neural networks. *Kybernetik*, 15: 1–9, 1974.

47. J. L. Hindmarsh and R. M. Rose. A model of neuronal bursting using three coupled first-order differential equations. *Proceedings of the Royal Society London*, B221:87–102, 1984.

48. N. F. Rulkov. Modeling of spiking-bursting neural behavior using two-dimensional map. *Phys. Rev. E*, 65: 041922, 2002.

49. E. M. Izhikevich. Simple model of spiking neurons. *IEEE Trans. Neural Networks*, 14(6): 1569–1572, 2003.

50. E. M. Izhikevich. Which model to use for cortical spiking neurons? *IEEE Trans. Neural Networks*, 15(5): 1063–1070, 2004.

51. W. S. McCulloch and W. Pitts. A logical calculus of ideas immanent in nervous activity. *Bull. Math. Biophys.*, 5:115–133, 1943. Reprinted in J. A. Anderson and E. Rosenfeld (1988) [7], p. 83ff.

52. S. Amari. A method of statistical neurodynamics. *Kybernetik*, 14: 201–215, 1974.

53. D. J. Amit. *Modeling Brain Function. The World of Attractor Neural Networks.* Cambridge University Press, Cambridge (MA), 1989.

54. A. Kuhn, A. Aertsen, and S. Rotter. Neuronal integration of synaptic input in the fluctuation-driven regime. *J. Neurosci.*, 24(10): 2345–2356, 2004.

55. J. S. Griffith. A field theory of neural nets: I. derivation of field equations. *Bull. Math. Biophys.*, 25:111–120, 1963.

56. J. S. Griffith. A field theory of neural nets: II. properties of the field equations. *Bull. Math. Biophys.*, 27: 187–195, 1965.

57. H. R. Wilson and J. D. Cowan. A mathematical theory of the functional dynamics of cortical and thalamic nervous tissue. *Kybernetik*, 13: 55–80, 1973.

58. P. L. Nunez, editor. *Neocortical Dynamics and Human EEG Rhythms.* Oxford University Press, New York (NY), 1995.

59. V. K. Jirsa and H. Haken. Field theory of electromagnetic brain activity. *Phys. Rev. Lett.*, 77(5): 960–963, 1996.

60. V. K. Jirsa and H. Haken. A derivation of a macroscopic field theory of the brain from the quasi-microscopic neural dynamics. *Physica D*, 99: 503–526, 1997.

61. J. J. Wright and D. T. J. Liley. Dynamics of the brain at global and microscopic scales: Neural networks and the EEG. *Behavioral and Brain Sciences*, 19: 285–320, 1996.

62. D. T. J. Liley, P. J. Cadusch, and J. J. Wright. A continuum theory of electro-cortical activity. *Neurocomputing*, 26–27: 795–800, 1999.

63. P. A. Robinson, C. J. Rennie, J. J. Wright, H. Bahramali, E. Gordon, and D. L. Rowe. Prediction of electroencephalic spectra from neurophysiology. *Phys. Rev. E*, 63, 2001. 021903.

64. C. J. Rennie, P. A. Robinson, and J. J. Wright. Effects of local feedback on dispersion of electrical waves in the cerebral cortex. *Phys. Rev. E.*, 59(3): 3320–3329, 1999.

65. P. A. Robinson, C. J. Rennie, J. J. Wright, and P. D. Bourke. Steady states and global dynamics of electrical activity in the cerebral cortex. *Phys. Rev. E.*, 58(3): 3557–3571, 1998.

66. J. J. Wright, C. J. Rennie, G. J. Lees, P. A. Robinson, P. D. Bourke, C. L. Chapman, E. Gordon, and D. L. Rowe. Simulated electrocortical activity at microscopic, mesoscopic, and global scales. *Neuropsychopharmacology*, 28: S80–S93, 2003.

67. V. K. Jirsa. Information processing in brain and behavior displayed in large-scale scalp topographies such as EEG and MEG. *Int. J. Bifurcation and Chaos*, 14(2): 679–692, 2004.

68. J. J. Wright, C. J. Rennie, G. J. Lees, P. A. Robinson, P. D. Bourke, C. L. Chapman, E. Gordon, and D. L. Rowe. Simulated electrocortical activity at microscopic, mesoscopic and global scales. *Int. J. Bifurcation and Chaos*, 14(2): 853–872, 2004.

69. J. J. Wright, P. A. Robinson, C. J. Rennie, E. Gordon, P. D. Burke, C. L. Chapman, N. Hawthorn, G. J. Lees, and D. Alexander. Toward an integrated continuum model of cerebral dynamics: the cerebral rhythms, synchronous oscillation and cortical stability. *Biosystems*, 63: 71–88, 2001.

70. V. K. Jirsa and J. A. S. Kelso. Spatiotemporal pattern formation in neural systems with heterogeneous connection toplogies. *Phys. Rev. E.*, 62(6): 8462–8465, 2000.

71. H. R. Wilson and J. D. Cowan. Excitatory and inhibitory interactions in localized populations of model neurons. *Biophys. J.*, 12: 1–24, 1972.

72. F. H. Lopes da Silva, A. Hoecks, H. Smits, and L. H. Zetterberg. Model of brain rhythmic activity: The alpha-rhythm of the thalamus. *Kybernetik*, 15: 27–37, 1974.

73. F. H. Lopes da Silva, A. van Rotterdam, P. Bartels, E. van Heusden, and W. Burr. Models of neuronal populations: The basic mechanisms of rhythmicity. In M. A. Corner and D. F. Swaab, editors, *Perspectives of Brain Research*, Vol. 45 of *Prog. Brain Res.*, pp. 281–308. 1976.

74. W. J. Freeman. Simulation of chaotic EEG patterns with a dynamic model of the olfactory system. *Biol. Cybern.*, 56: 139–150, 1987.

75. B. H. Jansen, G. Zouridakis, and M. E. Brandt. A neurophysiologically-based mathematical model of flash visual evoked potentials. *Biol. Cybern.*, 68: 275–283, 1993.

76. B. H. Jansen and V. G. Rit. Electroencephalogram and visual evoked potential generation in a mathematical model of coupled cortical columns. *Biol. Cybern.*, 73: 357–366, 1995.

77. F. Wendling, J. J. Bellanger, F. Bartolomei, and P. Chauvel. Relevance of nonlinear lumped-parameter models in the analysis of depth-EEG epileptic signals. *Biol. Cybern.*, 83: 367–378, 2000.

78. F. Wendling, F. Bartolomei, J. J. Bellanger, and P. Chauvel. Epileptic fast activity can be explained by a model of impaired GABAergic dendritic inhibition. *Eur. J. Neurosci.*, 15: 1499–1508, 2002.
79. O. David and K. J. Friston. A neural mass model for MEG/EEG: coupling and neuronal dynamics. *Neuroimage*, 20: 1743–1755, 2003.
80. O. David, D. Cosmelli, and K. J. Friston. Evaluation of different measures of functional connectivity using a neural mass model. *Neuroimage*, 21: 659–673, 2004.
81. O. David, L. Harrison, and K. J. Friston. Modelling event-related respones in the brain. *Neuroimage*, 25: 756–770, 2005.
82. P. beim Graben. *Symbolische Dynamik Ereigniskorrelierter Potentiale in der Sprachverarbeitung*. Berichte aus der Biophysik. Shaker Verlag, Aachen, 2001.
83. C. Baumgartner. Clinical applications of source localisation techniques — the human somatosensory cortex. In F. Angelieri, S. Butler, S. Giaquinto, and J. Majkowski, editors, *Analysis of the Electrical Activity of the Brain*, pp. 271–308. Wiley & Sons, Chichester, 1997.
84. W. Lutzenberger, T. Elbert, B. Rockstroh, and N. Birbaumer. *Das EEG*. Springer, Berlin, 1985.
85. N. Birbaumer and R. F. Schmidt. *Biologische Psychologie*. Springer, Berlin, 1996.
86. S. Zschocke. *Klinische Elektroenzephalographie*. Springer, Berlin, 1995.
87. A. Wunderlin. On the slaving principle. In R. Graham and A. Wunderlin, editors, *Lasers and Synergetics*, pp. 140–147, Springer, Berlin, 1987.
88. J. Dudel, R. Menzel, and R. F. Schmidt, editors. *Neurowissenschaft. Vom Molekül zur Kognition*. Springer, Berlin, 1996.
89. W. R. Adey. Molecular aspects of cell membranes as substrates for interaction with electromagnetic fields. In E. Başar, H. Flohr, H. Haken, and A. J. Mandell, editors, *Synergetics of the Brain*, pp. 201–211, Springer, Berlin, 1983.
90. E. Bracci, M. Vreugdenhil, S. P. Hack, and J. G. R. Jefferys. On the synchronizing mechanism of tetanically induced hippocampal oscillations. *J. Neurosci.*, 19(18): 8104–8113, 1999.
91. J. G. R. Jefferys. Nonsynaptic modulation of neuronal activity in the brain: Electric currents and extracellular ions. *Physiol. Rev.*, 75: 689–723, 1995.
92. K. A. Richardson, S. J. Schiff, and B. J. Gluckman. Electric field control of seizure propagation: From theory to experiment. In S. Boccaletti, B. Gluckman, J. Kurths, L. M. Pecora, R. Meucci, and O. Yordanov, editors, *Proceeding of the 8th Experimental Chaos Conference 2004*, pp. 185–196, American Institute of Physics, Melville (NY), 2004.
93. K. A. Richardson, S. J. Schiff, and B. J. Gluckman. Control of traveling waves in the mammalian cortex. *Phys. Rev. Lett.*, 94: 028103, 2005.
94. V. Braitenberg and A. Schüz. *Cortex: Statistics and Geometry of Neuronal Connectivity*. Springer, Berlin, 1998.
95. P. beim Graben and H. Atmanspacher. Complementarity in classical dynamical systems. *Found. Phys.*, 36(2): 291–306, 2006.
96. D. Lind and B. Marcus. *An Introduction to Symbolic Dynamics and Coding*. Cambridge University Press, Cambridge (UK), 1995.
97. P. beim Graben, J. D. Saddy, M. Schlesewsky, and J. Kurths. Symbolic dynamics of event–related brain potentials. *Phys. Rev. E.*, 62(4): 5518–5541, 2000.

98. P. beim Graben and J. Kurths. Detecting subthreshold events in noisy data by symbolic dynamics. *Phys. Rev. Let.*, 90(10): 100602, 2003.

99. H. Atmanspacher and P. beim Graben. Contextual emergence of mental states from neurodynamics. *Chaos and Complexity Letters*, 2(2/3), 151–168, 2007.

100. H. Atmanspacher. Contextual emergence from physics to cognitive neuroscience. *J. of Consciousness Stud.*, 14(1–2): 18–36, 2007.

101. T. Metzinger, editor. *Neural Correlates of Consciousness*. MIT Press, Cambridge (MA), 2000.

102. D. J. Chalmers. What is a neural correlate of consciousness? In Metzinger [101], Chap. 2, pp. 17–39, 2000.

103. P. beim Graben. Incompatible implementations of physical symbol systems. *Mind and Matter*, 2(2): 29–51, 2004.

104. R. Dale and M. J. Spivey. From apples and oranges to symbolic dynamics: A framework for conciliating notions of cognitive representation. *J. Exp. & Theor. Artific. Intell.*, 17(4): 317–342, 2005.

# Synapses and Neurons:
# Basic Properties and Their Use in Recognizing Environmental Signals

Henry D. I. Abarbanel[1,2,3], Julie S. Haas[1] and Sachin S. Talathi[1,2]

[1] Institution for Nonlinear Science
   haas@ucsd.edu
[2] Department of Physics, University of California San Diego
   talathi@physics.ucsd.edu
[3] Marine Physical Laboratory (Scripps Institute for Oceanography)
   hdia@jacobi.ucsd.edu

## 2.1 Introduction

This chapter incorporates two lectures presented at the 2005 Helmholtz School in Potsdam, Germany in September, 2005. The goal of this chapter is to cover two topics, both briefly:

- some basics of the biophysics of neurons and synapses. Many students had studied this material, and for them it was a lightning fast review. Some students had not studied this material, and for them it was a lightning fast introduction. See the book by Johnston and Wu [1] to make the introduction slower.

- the use of these basics to develop a neural time delay circuit and explore its use in the fundamental nervous system task of recognizing signals sent from the sensory system to a central nervous system as spike trains [2].

In a broad sense, modeling or computation in neuroscience takes place at two levels:

- Top-down: Start with the analysis of those macroscopic aspects of an animal's behavior that are robust, reproducible and important for survival. Try to represent the nervous system as a collection of circuits, perhaps based on biological physics, but perhaps not, needed to perform these function. The top-down approach is a speculative "big picture" view.

- Bottom-up: Start from a description of individual neurons and their synaptic connections; use facts about the details of their dynamical behavior from observed anatomical and electrophysiological data. Using these data, the pattern of connectivity in a circuit is reconstructed. Using the patterns of connectivity (the "wiring diagram") along with the dynamical features of the neurons and synapses, bottom-up models have been able to predict functional properties of neural circuits and their role in animal behavior.

The point of view in these lectures is distinctly "bottom-up." This is a much harder task than "top-down" approaches as it relies on observations to dictate the road ahead and the constraints in navigating that road. In our opinion, biological phenomena, neurobiological as well as others, are complex because the networks are designed by the necessity of performing functions for living things. They are not designed by optimality principles so far as we know.

The point of view we take here is that if one can understand in a well formulated predictive and quantitative mathematical model how biological processes are constructed in nature, then general principles for the design and construction of these can be analyzed. This requires working closely with experiments as guides, making predictions with models that are never totally correct, and refining those models within the framework of the outcome of experiments suggested by the models or formulated on other grounds.

Some will find the effort required rather formidable, but looking back on the historical interplay between experiment and theory that, say, led to the 20th century uncovering of the structure of quantum theory from the application of the Planck heat radiation law to wave equations, one should be ready for the engagement.

The dynamical point of view emphasized in this chapter is expanded in a review article appeared in *Reviews of Modern Physics* [3].

## 2.2 Lightning Fast (Review, Introduction) to Neural and Synaptic Dynamics

### 2.2.1 Neurons-points

The basic biophysical phenomena involved in the operation of neurons and a phenomenological set of equations describing them were identified by Hodgkin, Katz, Huxley, and many others in the mid part of the 20th century. Neurons are cells producing electrical signals through protein channels penetrating their plasma membrane allowing ions to flow into or out of the cells with permeabilities and rates often controlled by the voltage $V(t)$ across the membrane. There are two competing sources of current leading to the voltage difference across the membrane: ions flowing from higher concentrations $C(x,t)$ to lower concentrations giving rise to a current

$$J_{\text{concentration}}(x,t) = -D\frac{\partial C(x,t)}{\partial x}, \tag{2.1}$$

with $D$ the diffusion coefficient; and ions flowing because of the electric field associated with the electrostatic potential difference between the inside and the outside of the cell, giving rise to a competing current

$$J_{\text{electric}}(x,t) = -D\frac{zF}{RT}C(x,t)\frac{\partial V(x,t)}{\partial x}. \tag{2.2}$$

$z$ is the magnitude of the ion charge in units of $|e|$, $F$ is the Faraday constant 96485.34 C/mol, $T$ is the temperature, and $R$ is the universal gas constant 8.3144 J/mol K.

Approximating the gradient of the voltage across the thickness $l$ of the cell membrane by $V(t)/l$, we have for the total current

$$J(t) = -D\left\{\frac{\partial C(x,t)}{\partial x} + \frac{zF}{RT}C(x,t)\frac{V(t)}{l}\right\}. \tag{2.3}$$

Denoting the intracellular concentration of the ion in question as $C_{\text{in}}$ and the extracellular concentration as $C_{\text{out}}$, one may solve for the current $J(t)$, resulting in the Goldman-Hodgkin-Katz (GHK) equation

$$J(t) = -\frac{V(t)zFD}{RTl}\frac{C_{\text{in}} - C_{\text{out}}e^{-\frac{zFV(t)}{RT}}}{1 - e^{-\frac{zFV(t)}{RT}}}. \tag{2.4}$$

The GHK current has a zero at the *reversal potential or Nernst potential* $V_{\text{rev}}$ for each ionic species

$$V_{\text{rev}} = \frac{RT}{zF}\ln\frac{C_{\text{out}}}{C_{\text{in}}}$$
$$= \frac{61.5\,\text{mV}}{z}\ln\frac{C_{\text{out}}}{C_{\text{in}}}. \tag{2.5}$$

Using the values of intracellular and extracellular concentrations for various common ions present in nerve cells, we have the resting potentials listed in Fig. 2.1.

In the study of many neurons and their processes (axons carrying signals to other receiver cells and dendrites receiving signals from other transmitter cells), Hodgkin-Huxley (HH), along with many others, formulated a quite general, phenomenological form for ion currents that are gated, or controlled, by the membrane voltage. This form is

$$I_{\text{voltage-gated}}(t) = g_{\text{ion}}m(t)^p h(t)^q(V(t) - V_{\text{rev}}), \tag{2.6}$$

where $g_{\text{ion}}$ is the maximal conductance of the ion channel, $p$ and $q$ are integers, and $Z(t) = \{m(t), h(t)\}$ are activation and inactivation variables depending on the voltage $V(t)$ through phenomenological first order kinetic equations (master equations)

$$\frac{dZ(t)}{dt} = \alpha_Z(V(t))(1 - Z(t)) - \beta_Z(V(t))Z(t)$$
$$= \frac{Z_0(V(t)) - Z(t)}{\tau(V(t))}, \tag{2.7}$$

| Ion Name | Intracellular Concentration nM | Extracellular Concentration nM | Resting Potential |
|----------|-------------------------------|-------------------------------|-------------------|
| $Na^+$ | 18 | 145 | 56 mV |
| $K^+$ | 135 | 3 | -101 mV |
| $Cl^-$ | 120 | 7 | -76 mV |
| $Ca^{2+}$ | 0.1 | 3000 | 136 mV |

**Fig. 2.1.** Resting potentials for various common ions in nerve cells

where $\alpha_Z(V)$ and $\beta_Z(V)$ are empirically determined functions of voltage whose functional form is dictated by considerations of activations of gating variables. A good discussion of this formalism is found in Chap. 2 by Christopher Fall and Joel Keizer of the Joel Keizer memorial volume *Computational Cell Biology* [4]. The activation variable $m(t)$ normally resides near zero for voltages near $V_{\text{rev}}$ and rises to order unity when the voltage rises toward positive values. The inactivation variables $n(t)$ and $h(t)$ normally reside near unity and decrease towards zero as the voltage rises.

In the original HH model, we have three channels, one for $Na^+$ ions, one for $K^+$ ions, and one for a generalized lossy effect called a "leak" channel. With an added DC current, the HH equations are

$$C\frac{dV(t)}{dt} = -(g_{Na}m(t)^3 h(t)(V(t) - V_{Na})$$
$$+ g_K n(t)^4 (V(t) - V_K) + g_L(V(t) - V_L)) + I_{DC}, \tag{2.8}$$

along with kinetic equations for $m(t), h(t)$, and $n(t)$ (2.7). Using $g_{Na} = 120\,\text{mS cm}^{-2}, g_K = 36\,\text{mS cm}^{-2}, g_L = 0.3\,\text{mS cm}^{-2}, V_{Na} = 55\,\text{mV}, V_K = -72\,\text{mV}, V_L = -49\,\text{mV}$ and $C = 1\,\mu\text{F cm}^{-2}$, we can numerically solve these HH equations for various values of $I_{DC}$. The standard HH equations for the $\alpha_Z(V)$ and the $\beta_Z(V)$ are given in Fig. 2.2.

For low values of $I_{DC}$, the neuron is at rest at various voltages. Above a threshold, action potentials are observed as shown in Fig. 2.3 for $I_{DC} = 0.55\,\mu\text{A}$. The behavior of the $K^+$ activation variable $n(t)$ is shown in Fig. 2.4. The action potential comes from $Na^+$ flowing rapidly into the cell from higher concentrations outside the cell. The $K^+$ activation variable shows the $K^+$

$$\alpha_m(V) = -0.1\,\frac{35+V}{e^{-(35+V)/10}-1} \qquad \beta_m(V) = e^{-(60+V)/18}$$

$$\alpha_n(V) = 0.07\,e^{-(60+V)/20} \qquad \beta_n(V) = \frac{1}{e^{-(30+V)/10}+1}$$

$$\alpha_h(V) = -0.01\,\frac{50+V}{e^{-(50+V)}-1} \qquad \beta_h(V) = 0.125\,e^{-(60+V)/80}$$

**Fig. 2.2.** Empirical forms for the voltage-gating functions in the original HH equations

channel opening subsequently and allowing $K^+$ to flow out of the cell, thus assisting in terminating the action potential (cf. Chap. 1).

The HH model has four degrees of freedom, and with the parameters we have used exhibits a limit cycle or periodic behavior. To represent the phase space variation of a limit cycle solution of a differential equation generically takes three coordinates [5], however, we are fortunate here in that we can see the full limit cycle in two dimensions as shown in Fig. 2.5. In Fig. 2.6 we show the functions $m_0(V), h_0(V)$, and $n_0(V)$ along with the associated voltage dependent times, $\tau(V)$.

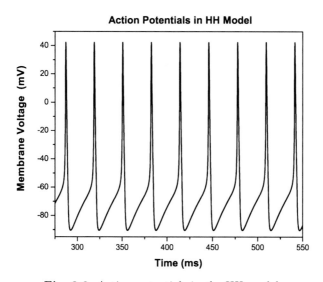

**Fig. 2.3.** Action potentials in the HH model

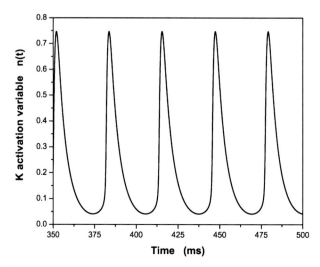

**Fig. 2.4.** The potassium activation variable $n(t)$ in the standard HH model

The HH model is phenomenological in origin, yet has the distinct scientific advantage of identifying measurable quantities in each expression. It comes from a "bottom-up" analysis of cellular dynamics. Reduced models of neural behavior often lose the latter property, and this makes the connection with how biology solves problems less satisfactory. In the "bottom-up" approach we are following, one of the goals is to use "biological parts" to construct neural circuits, not so much for simplicity at times but for the ability to connect

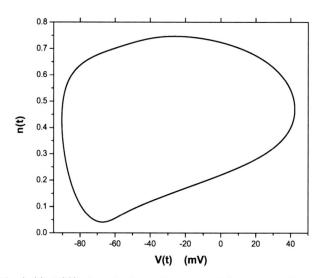

**Fig. 2.5.** The $(n(t), V(t))$-plane during action potential generation by the HH model

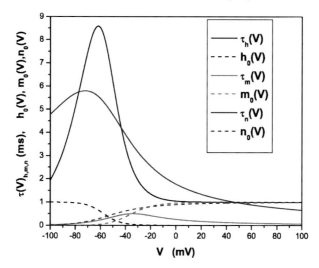

**Fig. 2.6.** Activation $\{m_0(V), n_0(V)\}$ and inactivation $h_0(V)$ variables and the time constants associated with them for the standard HH model. Note that the time scale of $Na^+$ activation $\tau_m(V)$ is much smaller than the times for the others, indicating the rapidity with which the $Na^+$ channels open

with biological networks and biophysical realizations of those networks. If one invents a lovely "top-down" network but has no way to establish whether it is implemented by biology, that may prove to be amusing applied mathematics but it tells us little about how biology, evolution and environment, achieved its functional goals.

Not all real neurons are periodic action potential generators as we have seen of the HH model neurons. Indeed, even the HH model has complex behavior for parameter values outside the range quoted. It is interesting, and perhaps sobering, to have a look at some real neuron data which is quite different, lest one settle into thinking of neurons as periodic oscillators.

In Fig. 2.7, we show a long time series of (scaled) membrane voltage measured from a neuron in a rhythm generator circuit in the digestive system of a California spiny lobster (not the tasty kind, alas). The circuit is comprised of fourteen neurons, see Fig. 2.8. The data is from an isolated LP neuron; it was physically isolated from the rest. The data were taken every 0.2 ms for several minutes. Only a portion of the data is shown.

The membrane voltage shows a nearly periodic oscillation at about 1 Hz, and on top of the peaks of this oscillation occur bursts of spikes with frequencies in the range of 30–50 Hz. While inspection of a time series by eye is not a recommended diagnostic, look carefully at the graphic and you will notice the nonperiodic nature of the signal. The tools for analyzing such a signal are given in [3, 5]. We can achieve some, again visual, insight if we reconstruct a proxy version of the full phase space of the system in a manner which imitates

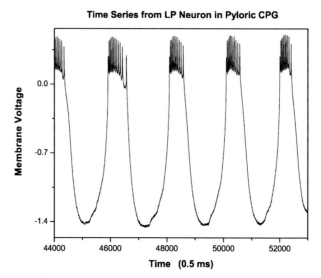

Fig. 2.7. Experimental membrane voltage data from an isolated LP neuron in the Pyloric Central Pattern Generator

the plot of $n(t)$ versus $V(t)$ we presented for the HH model. In Fig. 2.9, we plot the membrane voltage against itself, but time delayed. This is the start of an unfolding of the full phase space of the neuron [5] from the voltage measurements which are a projection of the data. In this figure, we see the region of nearly regular behavior, as well as a structured picture of the spiking activity.

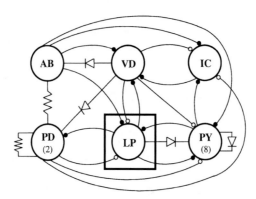

Fig. 2.8. Circuit diagram of the Pyloric Central Pattern Generator. The data in Fig. 2.7 were taken from the LP neuron, in the box, after it was physically isolated from the remainder of the circuit

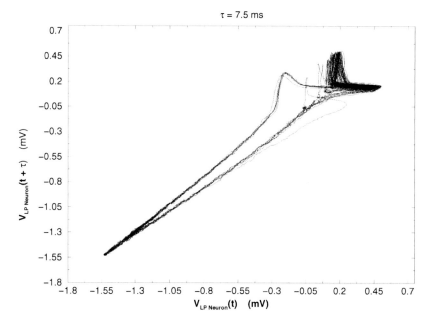

**Fig. 2.9.** Two dimensional reconstructed phase space plot [5] for the LP membrane voltage. This indicates that two dimensions is not sufficient to unfold the projection of the attractor onto the voltage axis; that takes four dimensions

### 2.2.2 Neurons-non-points

Of course, neurons are not points; they have spatial structure, and this can be very important. The point neuron which we have constructed so far is an abstraction which represents the idea that the membrane voltage is constant over the spatial extent of a neuron at any given time. To address spatially varying voltages across a nerve cell, it is common practice to make a many-compartment model comprised of the various sections of a neuron (see Chap. 1). Each section is assumed to be an equipotential, and one connects them together by ohmic couplings.

One might construct an HH model determining the voltage in the soma of a neuron $V_S(t)$ and another HH model for a dendrite compartment with potential $V_d(t)$. In the soma equation, one would then include a coupling term $g_{SD}(V_d(t) - V_S(t))$; similarly, in the dendrite equation. This allows one to more realistically represent the distribution of ion channels across a spatially extended neuron and to introduce interesting time delays as potential variations propagate from one part of the neuron to another. We do not explore this here as there is an extensive literature on compartment neuron models [6].

### 2.2.3 Synapses

The isolated neuron of HH type we have constructed is an interesting dynamical system having so far no functional role in describing biological systems. We must connect these nonlinear oscillators to create a network. There are two major types of connections among neurons:

- Ohmic or gap junction or electrotonic connections are implemented by means of a protein which penetrates the membrane of two adjacent cells and allows the flow of the various intracellular ion species. The current entering an HH neuron when the two connected cells have voltage $V_1(t)$ and $V_2(t)$ is (for the equation for $V_1(t)$) $I_{\text{gapjunction}}(t) = g_{GJ}(V_2(t) - V_1(t))$.
- Chemical synaptic connections, which activate when an action potential signal from a transmitting (presynaptic) cell arrives at the termination of an axon process. This starts processes which release neurotransmitters of various compositions to diffuse across a gap (the synaptic cleft) and dock on receptors penetrating the membrane of the receiving (or postsynaptic) cell. The docking of the neurotransmitter changes the permeability of the channel associated with the receptor and allows ions to flow. When those neurotransmitter molecules undock, the receptor closes down the ion flow. The times for the docking and undocking of the neurotransmitter on receptors vary substantially with the receptor type.

  Each synaptic connection can be represented by a maximal conductance $g$, a function $S(t)$ taking values between zero and unity representing the percentage of the maximum allowed docked neurotransmitter presently on postsynaptic receptors, and an "ohmic term" $(V_{\text{postsynaptic}}(t) - V_{\text{reversal}})$. This makes for a net current $I_{\text{synaptic}}(t) = gS(t)(V_{\text{postsynaptic}}(t) - V_{\text{reversal}})$.

  - Excitatory synaptic connections have reversal potentials near 0 mV and allow mixtures of $Na^+$, $Ca^{2+}$ and $K^+$ ions to flow. They tend to cause the postsynaptic voltage to rise from its resting value upon receipt of a presynaptic action potential, thus exciting the postsynaptic cell toward action potential generation. The rise, often small, in the postsynaptic potential is called an "excitatory postsynaptic potential" or EPSP.

    There are two quite important excitatory synapses we need to know about for our discussion of synaptic plasticity:

    - AMPA receptors, which are the main excitatory connection in many nervous systems and where synaptic strength change (plasticity) occurs. AMPA connections are fast, opening and closing in a few ms.

    - NMDA receptors, which have a very high permeability to $Ca^{2+}$ ions, are blocked by $Mg^{2+}$ ions at low membrane voltages, and allow the entry of $Ca^{2+}$ for order of 100 ms after being opened. In addition to the form of the synaptic current noted above, one must add a multiplicative term

$$B(V) = \frac{1}{1 + 0.288\,e^{-0.062V}}$$

to the NMDA synaptic current to represent the $Mg^{2+}$ block:
$$I_{\text{NMDA}}(t) = g_{\text{NMDA}}\, S_{\text{NMDA}}(t)B(V_{\text{post}})(V_{\text{post}}(t) - V_{\text{reversal}}).$$
– Inhibitory synaptic connections have reversal potentials near $-80\,\text{mV}$ and allow $Cl^-$ ions to flow. They tend to cause the postsynaptic voltage to fall from its resting value upon receipt of a presynaptic action potential, thus inhibiting the postsynaptic cell from action potential generation. The fall, often small, in the postsynaptic potential is called an "inhibitory postsynaptic potential" or IPSP.

A useful equation for $S(t)$ associates two time constants with its temporal evolution: one for docking the neurotransmitter and one for undocking it. The following describes this:

$$\frac{dS(t)}{dt} = \frac{S_0(V_{\text{pre}}(t)) - S(t)}{\tau(S_1 - S_0(V_{\text{pre}}(t)))}, \tag{2.9}$$

where $S_0(x)$ is zero for negative arguments and rises rapidly to unity for $x \geq 0$. A convenient form for this is

$$S_0(x) = 0.5\,(1 + \tanh(120(x - 0.1))). \tag{2.10}$$

This equation tells us that when the presynaptic voltage is below $0\,\text{mV}$, so no action potential is present, $S(t)$ decays to zero with a time constant $\tau S_1$. This is the undocking time. When an action potential arrives at the presynaptic terminal, $S_0$ rises rapidly to unity, driving $S(t)$ toward unity with a time constant $\tau(S_1 - 1)$. This is the docking time. This formulation applies for both inhibition and excitation when the rise time of the synapse is comparable to the width of the action potential, as in the case of fast excitatory AMPA synapses and fast inhibitory $\text{GABA}_A$ synapses. In order to model synapses

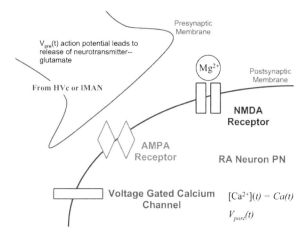

**Fig. 2.10.** Graphic depicting the arrival of an action potential at a synaptic terminal

with slower rise times such as NMDA, we need coupled first order kinetic equations of the form given in (2.10).

Figure 2.10 is a "cartoon" indicating the action at a synapse and postsynaptic neuron.

The ion channels discussed in the context of the HH model are called voltage-gated channels. The synaptic connections just described are called ligand-gated. There are other synaptic connections which do not allow ion flow after receipt of a presynaptic potential but through changes in the properties of the receptor-induced postsynaptic biochemical processes. These metabotropic receptors are discussed in Chap. 1.

## 2.3 Synaptic Plasticity

An important dynamical process in nervous systems is the activity dependent change in synaptic strength associated with both inhibitory and excitatory synaptic connections. Though far from "proven", it is widely believed that changes in these connection strengths among neurons produce the rewiring occurring when the nervous systems learns.

As early as 1973, Lomø and Bliss [7] showed that if one presents a series of spikes (a tetanus) with interspike intervals (ISIs) as small as 10 ms, namely a spiking frequency of 100 Hz, to certain hippocampal cells through an excitatory AMPA synapse, the baseline EPSP before and after the presentation of the tetanus showed increased amplitude that persisted for hours after the presentation. This was called "long term potentiation" or LTP. Experiments by Malinow and Miller [8] that presented lower frequency tetani and controlled the postsynaptic voltage at the same time showed that one could induce long lasting decreases in EPSPs. This is called long term depression or LTD.

Experiments in the 1990s showed that both LTP and LTD could be induced at excitatory AMPA synapses by pairing isolated single presynaptic and postsynaptic spikes. An evoked presynaptic spike arrives at the synaptic terminal at $t_{pre}$ and a postsynaptic spike is induced by a short current injection at time $t_{post}$. As a function of the time difference $\tau = t_{post} - t_{pre}$, one observes both LTP and LTD. This is nicely summarized in the data of Bi and Poo [9] shown in Fig. 2.11.

To provide an explanation of these phenomena, several groups [10–12] have made biophysically based plasticity models founded on the observation [13] that postsynaptic $Ca^{2+}$ concentration is critical to inducing the competing biochemical pathways in the postsynaptic cell.

In our formulation [11], we attribute a dynamical degree of freedom $P(t)$ to kinases which lead to potentiation and another $D(t)$ to phosphatases which lead to depression. We hypothesize first order kinematics for each of these

$$\frac{dP(t)}{dt} = f_P(\Delta[Ca^{2+}](t))(1 - P(t)) - \beta_P P(t)$$

$$\frac{dD(t)}{dt} = f_D(\Delta[Ca^{2+}](t))(1 - D(t)) - \beta_D D(t), \qquad (2.11)$$

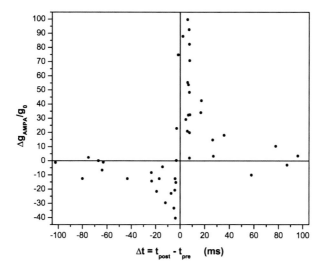

**Fig. 2.11.** Data from Bi and Poo [9] on LTP and LTD induction by presentation of a presynaptic spike at $t_{\mathrm{pre}}$ and stimulation of a postsynaptic spike at $t_{\mathrm{post}}$ as a function of $\tau = t_{\mathrm{post}} - t_{\mathrm{pre}}$

with $f_P(\Delta[\mathrm{Ca}^{2+}](t))$ and $f_D(\Delta[\mathrm{Ca}^{2+}](t))$ functions of the time dependent elevation of intracellular $\mathrm{Ca}^{2+}$ over its equilibrium value $C_0$, and $\beta_P$ and $\beta_D$ being rates for the return of each process to zero. The change in excitatory synaptic strength is hypothesized to be proportional to the nonlinear competition between these processes

$$\frac{\mathrm{d}g_E(t)}{\mathrm{d}t} = g_0(P(t)D(t)^\eta - D(t)P(t)^\eta), \tag{2.12}$$

with $g_0$ being a baseline conductance.

The competition of these two processes is supported by the data from O'Connor, et al. [14] and presented in Fig. 2.12. To develop these data, the potentiation processes mediated by kinases were first blocked by the application of K252a, and then the depression processes mediated by phosphatases were blocked by okadaic acid. One can clearly see from their very nice data the presence of competing dynamical postsynaptic mechanisms.

The time dependence of intracellular calcium $[\mathrm{Ca}^{2+}](t)$ is determined by a rate equation of the form

$$\frac{\mathrm{d}[\mathrm{Ca}^{2+}](t)}{\mathrm{d}t} = \frac{C_0 - [\mathrm{Ca}^{2+}](t)}{\tau_C} + \mathrm{Sources}(t) \tag{2.13}$$

where $\tau_C \approx 20\,\mathrm{ms}$ is the relaxation rate of $[\mathrm{Ca}^{2+}]$ back to $C_0$. The sources include AMPA currents, NMDA currents, and voltage-gated $\mathrm{Ca}^{2+}$ channels [11]. Coupling this to a HH model of the postsynaptic cell allows us to simulate

LTD and LTP are separable

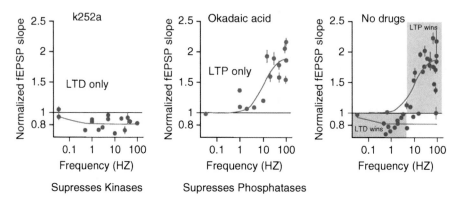

**Fig. 2.12.** Separation of the biochemical postsynaptic processes associated with LTP (kinases) and LTD (phosphatases) (adapted from [14])

the effect of an experimental electrophysiological stimulation protocol — for example, one presynaptic spike at $t_{pre}$ and one postsynaptic spike at $t_{post}$ — to determine the postsynaptic membrane voltage $V_{post}(t)$, the intracellular $Ca^{2+}$ time course $[Ca^{2+}](t)$ and from those quantities deduce the change in $g_E(t)$ from its value $g_0$ before the electrophysiological induction, to the value $g_0(1 + \int_{-\infty}^{\infty} dt((P(t)D(t)^\eta - D(t)P(t)^\eta)))$ after the induction.

In Fig. 2.13, we show the result of a calculation of $\frac{\Delta g(\tau)}{g_0}$ for such a model.

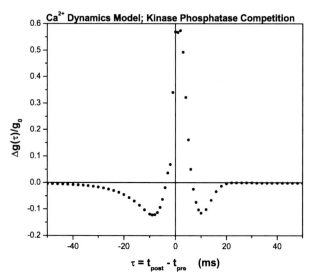

**Fig. 2.13.** $\frac{\Delta g(\tau)}{g_0}$ for $Ca^{2+}$ dynamics model

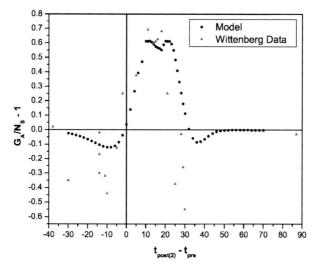

**Fig. 2.14.** Data from Gayle Wittenberg as a function of $\tau = t_{\text{post}(2)} - t_{\text{pre}}$ compared to our model calculation based on $Ca^{2+}$ postsynaptic dynamics

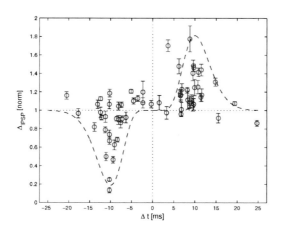

**Fig. 2.15.** Spike timing plasticity at an inhibitory synapse: Summary results of change in postsynaptic IPSP initial slope, expressed as a function of $\Delta t = t_{\text{post}} - t_{\text{pre}}$. No change is represented by normalized IPSP equal to unity (equal to the baseline value before pairing). Each point represents data from one cell. Change is evaluated as the mean IPSP slope over the 20 min. following pairings, normalized to the mean of the slopes for 15 minutes preceding pairings. Cells for which the change in IPSP slope was significant ($p < .01$, ANOVA) are plotted in blue. Empirical fit to the observed data is given by $\Delta_{\text{IPSP}} = 1 + \frac{g_0}{\beta e^{-\beta}} \alpha^\beta \Delta t |\Delta t|^{\beta - 1} e^{-\alpha|\Delta t|}$, with $g_0 = 0.8$, $\alpha = 1$, and $\beta = 10$

Gayle Wittenberg [15] has performed experiments in which she presents cells in the hippocampus with a single presynaptic spike at $t_{\text{pre}}$ and two postsynaptic spikes separated by $\Delta t$. She used the time of the second postsynaptic spike and $t_{\text{pre}}$ to determine $\tau = t_{\text{post}(2)} - t_{\text{pre}}$, and her results are shown in Fig. 2.14. Along with her data is the result of a calculation by our laboratory for the same process based on a $Ca^{2+}$ dynamics model.

Finally, we note, and we will use below, the experimental evidence for spike timing dependent plasticity at an inhibitory synapse. This is shown in Fig. 2.15 and comes from data by [16].

That ends the lightning fast review (or introduction). As suggested, the introductory book by Johnston and Wu [1], or Chap. 1 expands on these topics in rather more detail.

## 2.4 Synchronization and Plasticity

It is widely thought, though hardly proven, that synchronization among populations of neurons can play an important role in their performing important functional activity in biological neural networks. Here we look at the microcircuit of one periodically oscillating HH neuron driving another periodically oscillating HH neuron. The question we ask is whether synaptic plasticity has an interesting effect on the ability of these two neurons to synchronize. The answer is yes, and though interesting, does not answer a biological question yet. Let's look at what we can show, then pose some harder questions for further investigation.

The setup we examine is that of a periodically oscillating HH neuron with a period $T_1$. We control $T_1$ by injection of a selected level of DC current. This is the "transmitter". The receiver, or postsynaptic neuron, is another HH neuron oscillating with a period $T_2^0$ before receiving synaptic input from neuron 1. When the synaptic current has begun, the receiver neuron changes its period to $T_2$. A schematic of the setup is in Fig. 2.16. The coupling is through an excitatory synapse with current $I_{\text{synapse}}(t) = g_E(t)S_E(t, V_{\text{pre}}(t))(V_{\text{post}}(t) - V_{\text{reversal}})$.

We have selected the postsynaptic neuron to be our two-compartment model as described in [17] and set it to produce autonomous oscillations with a period $T_2^0$. This period is a function of the injected DC current into the somatic compartment. We hold this fixed while we inject a synaptic AMPA current

$$I_{\text{synapse}}(t) = g_{\text{AMPA}}(t)S_A(t)(V_{\text{post}}(t) - V_{\text{rev}}), \qquad (2.14)$$

into the postsynaptic somatic compartment. $V_{\text{post}}(t)$ is the membrane voltage of this postsynaptic compartment. $g_{\text{AMPA}}(t)$ is our time-dependent maximal AMPA conductance, and $S_A(t)$ satisfies

$$\frac{dS_A(t)}{dt} = \frac{1}{\tau_A} \frac{S_0(V_{\text{pre}}(t)) - S_A(t)}{S_{1A} - S_0(V_{\text{pre}}(t))} \qquad (2.15)$$

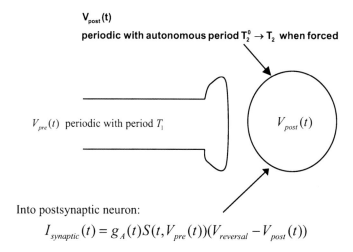

**Fig. 2.16.** Setup for exploring effect of a dynamical synapse on synchronization of two periodic neural HH neurons

as described above. $V_{\text{pre}}(t)$ is the periodic presynaptic voltage which we adjust by selecting the injected DC current into the presynaptic HH neuron. We call the period of this oscillation $T_1$.

When $g_{\text{AMPA}} = 0$, the neurons are disconnected and oscillate autonomously. When $g_{\text{AMPA}}(t) \neq 0$, the synaptic current into the postsynaptic neuron changes its period of oscillation from the autonomous $T_2^0$ to the driven value of $T_2$, which we evaluate for various choices of $T_1$. We expect from general arguments [18] that there will be regimes of synchronization where $\frac{T_1}{T_2}$ equal ratios of integers over the range of frequencies $\frac{1}{T_1}$ presented presynaptically. This will be true both for fixed $g_{\text{AMPA}}$ and when $g_{\text{AMPA}}(t)$ varies as determined by our model.

In Fig. 2.17 we present $\frac{T_1}{T_2}$ as function of the frequency $\frac{1000}{T_1}$ ($T_1$ is given in milliseconds, so this is in units of Hz) for fixed $g_{\text{AMPA}} = 0.1\,\text{mS}\,\text{cm}^{-2}$ and for $g_{\text{AMPA}}(t)$ determined from our model. This amounts to a choice for the baseline value of the AMPA conductance. The fixed $g_{\text{AMPA}}$ results are in filled upright triangles and, as expected, show a regime of one-to-one synchronization over a range of frequencies. One also sees regions of two-to-one and hints of five-to-two and three-to-one synchronization. These are expected from general arguments on the parametric driving of a nonlinear oscillator by periodic forces.

When we allow $g_{\text{AMPA}}$ to change in time according to the model we have discussed, we see (unfilled inverted triangles) a substantial increase in the regime of one-to-one synchronization, the appearance of some instances of

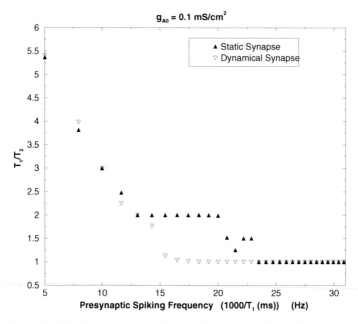

**Fig. 2.17.** $T_1/T_2$, the ratio of the interspike interval $T_1$ of the presynaptic neuron to the interspike interval $T_2$ of the postsynaptic neuron, is plotted as a function of the presynaptic input frequency, $1000/T_1$ Hz, for a synapse starting at a base AMPA conductance of $g_{\mathrm{AMPA}}(t = 0) = 0.1\,\mathrm{mS\,cm^{-2}}$. We see that the one-to-one synchronization window is broadened when the static synapse is replaced by a plastic synapse

three-to-two synchronization, and a much smaller regime with two-to-one synchronization. This suggests that the one-to-one synchronization of oscillating neurons, which is what one usually means by neural synchrony, is substantially enhanced when the synaptic coupling between neurons is allowed to vary by the rules we have described.

## 2.5 Marking Time Biologically

Biological systems have to mark time to keep pace with events and transmit information. There are three distinct ways known to us by which this is accomplished. One is to use the same principle of a delay line in physics: a signal has a propagation velocity $v$ along some cable, and then a signal traversing a length of this cable $L$ takes a time $\frac{L}{v}$. This is manifest in the interaural time differences used by the barn owl, for example, in actively locating prey or passively detecting sources of sound. With a velocity $v \approx 5\,\frac{m}{s}$ and axon lengths of a few mm, time delays as short as tens of microseconds are used [19–22].

Time delays of the order of hours or days are connected with circadian rhythms and are marked using limit cycle oscillators. A detailed model of the biochemical processes thought to underly the $\approx$ 24 h circadian rhythm is found in recent work by Forger and Peskin [23, 24], where a limit cycle oscillator with a period slightly more than 24 h is identified and analyzed.

Environmental signals are passed on in an animal from its sensory systems to central nervous system neurons as sequences of spikes with interspike intervals (ISIs) of a few to hundreds of milliseconds. Since these spikes are essentially identical in waveform, all information contained in these sequences are in the ISIs. Our focus is on these signals.

There are many examples of sensitive stimulus-response properties characterizing how neurons respond to specific stimuli. These include whisker-selective neural response in barrel cortex [25, 26] of rats and motion sensitive cells in the visual cortical areas of primates [27, 28].

One striking example is the selective auditory response of neurons in the songbird telencephalic nucleus HVC [29–32]. Projection neurons within HVC fire sparse bursts of spikes when presented with auditory playback of the bird's own song (BOS) and are quite unresponsive to other auditory inputs. Nucleus NIf, through which auditory signals reach HVC [32–36], also strongly responds to BOS in addition to responding to a broad range of other auditory stimuli. NIf projects to HVC, and the similarity of NIf responses to the auditory input and the subthreshold activity in HVC neurons suggests that NIf could be acting as a nonlinear filter for BOS, preferentially passing that important signal on to HVC. It was these examples from birdsong that led us to address the ISI reading problem we consider here.

How can these neural circuits be sensitive to specific sequences of ISIs? One way is that they act as a nonlinear filter for such sequences. The circuit is trained on a particular sequence which the animal has found important to recognize with great sensitivity. We identify the ISI sequence we wish to recognize as a set of times $S_{ISI} = \{T_0, T_1, T_2, \ldots, T_N\}$ coming from a set of spike times $S_{spikes} = \{t_0, t_1, \ldots, t_{N+1}\}$ with $T_j = t_{j+1} - t_j$. If we have a set of time delay units which is trained to create a signal at $\tau_j$ after it receives a spike at time $t_j$, then we can use the output sequence $\tau_0, \tau_1, \ldots, \tau_N$ coming from the input $S_{spikes}$ as a comparison to the original ISI sequence $S_{ISI}$. The comparison can be achieved by introducing both into a detection circuit which fires only when two spikes occur within a few ms of each other. If the detection unit fires, the correct ISI $T_j$ has matched the time delay $\tau_j$.

How are we to build a time delay circuit with tunable synapses? For this, we turn to the observations of Kimpo et al. [37] in which they identify a pathway in the birdsong system which quite reliably produces a time delay of $50 \pm 10$ ms when a short burst of spikes is introduced into its entry point. The importance of this timing in the birdsong systems has been examined in [38]; not surprisingly, given the tone of these lectures, it is connected with a specific timing required by spike timing plasticity in the stabilization of

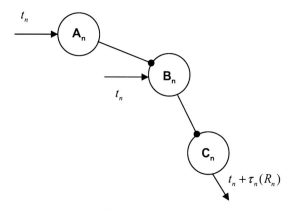

**Fig. 2.18.** Time delay circuit adapted from the birdsong system

the bird's song. How are we to "tune" $\tau_j$? For this, we use the spike time dependent inhibitory plasticity rules observed by Haas [16].

The time delay circuit is displayed in Fig. 2.18. There are three neurons in the circuit, and they are connected by inhibitory synapses. Neuron A receives an excitatory input signal from some source at time $t_n$. It is at rest when the source is quiet, and when activated, it inhibits neuron B. Neuron B receives an excitatory input from the same source at the same time $t_n$. Neuron B oscillates periodically when there is no input from the source. Neuron B inhibits neuron C. Neuron C produces periodic spiking in the absence of inhibition from neuron B. The tunable synaptic strength is that connecting neuron B to neuron C. The dimensionless number $R_n$ is the magnitude of the B→C maximal conductance relative to some baseline conductance.

When the inhibition from neuron B to neuron C is released by the inhibitory signal from neuron A to neuron B, neuron C rebounds and produces an action potential some time later. This is due to the intrinsic stable spiking of neuron C in the absence of any inhibition from neuron B.

This time delay is dependent on the strength of the B→C inhibition, as the stronger that is set the further below threshold neuron C is driven and the further it must rise in membrane voltage to reach the action potential threshold. This means the larger the B→C inhibition, the longer the time delay produced by the circuit. Other parameters in the circuit, such as the cellular membrane time constants, set the scale of the overall time delay.

The direct excitation of neuron B by the signal source is critical. It serves to reset the phase of the neuron B oscillation, as a result of which the spike from neuron C is measured with respect to the input signal and thus makes the timing of the circuit precise relative to the arrival of the initiating spike.

Without this excitation to neuron B, the phase of its oscillation is uncorrelated with the arrival time of a signal from the source, and the time delay of the circuit varies over the period of oscillation of neuron B. This is not a desirable outcome.

We have constructed this circuit using HH conductance based neurons and realistic synaptic connections. The dynamical equations for the three HH neurons shown in Fig. 2.18 are these:

$$C_M \frac{dV_i(t)}{dt} = g_{\text{Na}} m(t, V_i(t))^3 h(t, V_i(t))(V_i(t) - V_{\text{Na}})$$
$$+ g_{\text{K}} n(t, V_i(t))^4 (V_i(t) - V_{\text{K}}) + g_L(V_i(t) - V_L)$$
$$+ g_{ij}^I S_I(t)(V_i(t) - V_{\text{revI}}) + I_i^{syn}(t) + I_i^{DC}, \qquad (2.16)$$

where $(i, j) = [\text{A, B, C}]$. The membrane capacitance is $C_M$, and $V_{\text{Na}}, V_{\text{K}}, V_L$, and $V_{\text{revI}}$ are reversal potentials for the sodium, potassium, leak, and inhibitory synaptic connections, respectively. $m(t), h(t)$, and $n(t)$ are the usual activation and inactivation dynamical variables. $g_{Na}, g_K$ and $g_L$ are the maximal conductances of sodium, potassium and leak channels respectively. $I_i^{DC}$ is the DC current into the A, B or C neuron. These are selected such that neuron A is resting at $-63.74\,\text{mV}$ in the absence of any synaptic input, neuron B is spiking at around 20 Hz, and neuron C would also be spiking at around 20 Hz in the absence of any synaptic inputs.

$I_i^{syn} = g_i^E S_E(t)(V_i(t) - V_{\text{revE}})$ is the synaptic input to the delay circuit at neuron A and B. It receives a spike from the signal source at time $t_0$; $g_i^E = (g_A, g_B, 0)$. The nonzero inhibitory synaptic strengths $g_{ij}^I$ in the delay circuit are $g_{BA} = R^0 g_I$ and $g_{CB} = R g_I$. The dimensionless factors, $R$ and $R^0$, set the strength of A→B and B→C inhibitory connections respectively, relative to baseline strength $g_I$, which is set to $1\,\text{mS cm}^{-2}$ in all the calculations presented here.

$g_A = g_B = 0.5\,\text{mS cm}^{-2}$. $g_I = 1\,\text{mS cm}^{-2}$, $R^0 = 50.0$, and $R$ varies as given in text. $V_{\text{revE}} = 0\,\text{mV}$, and $V_{\text{revI}} = -80\,\text{mV}$. $\tau_E = 1.0\,\text{ms}$, $S_{1E} = 1.5$, $\tau_I = 1.2\,\text{ms}$, $S_{1I} = 4.6$. The DC currents in the neurons are taken as $I_A^{DC} = 0.0\,\mu\text{A cm}^{-2}$, $I_B^{DC} = 1.97\,\mu\text{A cm}^{-2}$ and $I_C^{DC} = 1.96\,\mu\text{A cm}^{-2}$.

$S_E(t)$ represents the fraction of excitatory neurotransmitter docked on the postsynaptic cell receptors as a function of time. It varies between 0 and 1 and has two time constants: one for the docking time of the neurotransmitter and one for its release time. It satisfies the dynamical equation:

$$\frac{dS_E(t)}{dt} = \frac{S_0(V_{\text{pre}}(t)) - S_E(t)}{\tau_E(S_{1E} - S_0(V_{\text{pre}}(t)))}. \qquad (2.17)$$

The docking time constant for the neurotransmitter is $\tau_E(S_{1E} - 1)$, while the undocking time is $\tau_E S_{1E}$. For neurons A and B, the presynaptic voltage is given by the incoming spike or burst of spikes arriving from some source at time $t_0$. For our excitatory synapses, we take $\tau_E = 1\,\text{ms}$ and $S_{1E} = 1.5$, for

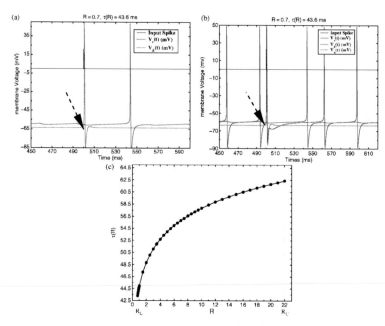

**Fig. 2.19. (a)** For $R = 0.7$, we show the membrane voltages of neuron A (*blue*) and neuron C (*red*) in response to single spike input (*black*) arriving at neuron A and neuron B at time $t_0 = 500$ ms. We see the output spike from neuron C occurring at $t = 543.68$ ms, corresponding to $\tau(R) = 43.6$ ms; **(b)** For $R = 0.7$ we again show the membrane voltages of neuron A (*blue*) and neuron C (*red*), and in addition now display the membrane voltage of neuron B (*green*). A single spike input (*black*) arrives at time $t = 500$ ms. We see that the periodic action potential generation by neuron B is reset by the incoming signal; **(c)** The delay $\tau(R)$ produced by the three neuron time delay unit as a function of $R$, the strength of the inhibitory synaptic connection B→C. All other parameters of the time delay circuit are fixed to values given in the text. For $R < R_L$, the inhibition is too weak to prevent spiking of neuron C. For $R > R_U$, the inhibitory synapse is so strong that neuron C does not produce any action potential, so effectively the delay is infinity. In Figs. 2.19(a) and 2.19(b), the arrows indicate the time of the spike input to units A and B of our delay unit

a docking time of 0.5 ms and an undocking time of 1.5 ms. These times are characteristic of AMPA excitatory synapses.

Similarly, $S_I(t)$ represents the percentage of inhibitory neurotransmitter docked on the postsynaptic cell as a function of time. It satisfies the following equation:

$$\frac{dS_I(t)}{dt} = \frac{S_0(V_{\mathrm{pre}}(t)) - S_I(t)}{\tau_I(S_{1I} - S_0(V_{\mathrm{pre}}(t)))}, \tag{2.18}$$

aa

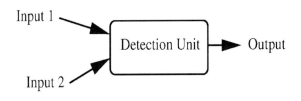

b

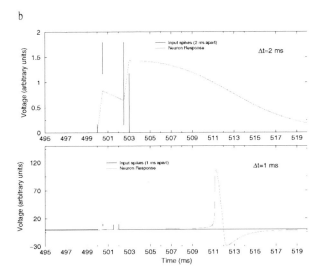

**Fig. 2.20.** (a) Schematic of the detection unit. It receives two input spikes with various time intervals between them. It responds with a spike if the two inputs are within 1 ms of each other; (b) *Top panel* The scaled response of the detection unit when two inputs arrive within 2 ms of each other. We see that the integrated input arriving at this delay does not result in neuron spiking. In the *bottom panel*, we show the scaled neuron response to two input spikes arriving within 1 ms of each other. The detection unit produces a spike output, indicating coincidence detection

where we select $\tau_I = 1.2$ ms and $S_{1I} = 4.6$ for a docking time of 4.32 ms and undocking time of 5.52 ms. The range of time delays produced by the three neuron delay circuit depends sensitively on the docking and undocking times of this synapse.

For these values, we find $\tau(R)$ as shown in Fig. 2.19(c). For $R$ too small, $R < R_L$ in Fig. 2.19(c), the inhibition from B $\rightarrow$ C does not prevent the production of action potentials. For $R$ too large, $R > R_U$ in Fig. 2.19(c), the C neuron is inhibited so strongly that it never spikes. Over the range of $R_L \leq R \leq R_U$, we typically find that $\tau(R)$ has a range of about 20 ms within an overall scale of about 10 ms to 100 ms.

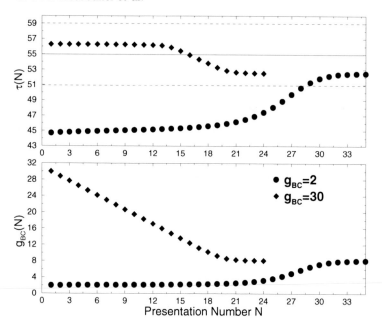

**Fig. 2.21.** Training an IRU to learn an ISI of $T = 55$ ms. The initial values of $g_{BC}(N = 0)$ are set to explore the two scenarios described in the text. $\tau(R)$ (*top panel*) and $g_{BC}$ (*bottom panel*) are plotted as function of the number of presentations of the training sequence N. The resolution limit $\delta = 4$ ms is shown in dotted lines for $\tau(R)$ and $T = 55$ ms is shown as a solid line

The ISI recognition unit is now constructed as follows:

- A circuit we call the "spike separation unit" [2] separates the individual spikes at time $t_0, t_1, t_{N+1}$ in the sequence. These are then presented to a set of time delay units in pairs $\{t_j, t_{j+1}\}$.
- After introducing an excitatory feedback from neuron C to neuron C of each time delay unit, we use the inhibitory plasticity rule in Fig. 2.15 to adjust $R_j$ until $\tau_j \approx t_{j+1} - t_j = T_j$. When this occurs, the training is completed.
- A "detection unit" comprised of a neuron or neurons fires when two spikes within a few ms of each other are received, but not when a single spike is received. The operation of the detection unit is illustrated in Fig. 2.20. When the detection unit fires, the information that the replica ISI sequence $\tau_0(R_0), \tau_1(R_1), \ldots, \tau_N(R_N)$ has matched the desired sequence $S_{ISI}$ is passed on to other functions.

As an example of the training of a delay unit, we show in Fig. 2.21 the training of one time delay unit to a time delay of 55 ms. The spike sequence $t_0, t_1 = t_0 + T_0$ with $T_0 = 55$ ms, is presented $N = 0, 1, 2, \ldots$ times. We present the spike sequence many times $N = 0, 1, 2, \ldots$ to the IRU to train the time delays to accurately reflect the individual ISIs in the sequence. In Fig. 2.21, we show results from training two IRU units tuned to detect an ISI of $T = 55$ ms. The first IRU has $g_{BC}(N = 0) = 2$, corresponding to $\tau(R) \approx 43$ ms, so $T > \tau(R)$. The second has $g_{BC}(N = 0) = 30$ leading to $\tau(R) \approx 60$ ms, so $T < \tau(R)$. Each IRU trains itself on the given ISI input presented $N = 1, 2, \ldots$ times. In the detection unit, we set the time within which the spikes must arrive at 4 ms. This is a resolution which approximates the refractory period of a typical neuron.

As a final comment, we return to the matter which led to a discussion of the sensitivity of a neural circuit, such as that in the sensory-motor junction of the birdsong system [32]. We suggest, based on the construction discussed here, that within NIf (or possibly before NIf in the auditory pathway) there will be neural structures which select for specific ISI sequences. In the first phase of song learning, called the sensory phase, we suggest that a genetically determined circuit with time delays nearly those appropriate to the tutor's song is tuned by repeated presentation of the song. This tunes the nonlinear filter for ISI sequences we just constructed. Once this has happened, the birdsong development moves into its next phase, the sensori-motor phase, where the muscles of the bird's songbox are trained by other plasticity events [38] through auditory feedback, filtered by the ISI recognition structure. Once the bird's own song matches the tutor's song, the training of the songbox is completed.

## Acknowledgements

This work was partially funded by a grant from the National Science Foundation, NSF PHY0097134. HDIA and SST are partially supported by the NSF sponsored Center for Theoretical Biological Physics at UCSD.

## References

1. Johnston, D. and S. Miao-Sin Wu: *Foundations of Cellular Neurophysiology*, (MIT Press, Cambridge, MA, 1994).
2. Abarbanel, H. D. I. and S.S. Talathi: Phys. Rev. Lett. **96**, 148104 (2006).
3. Rabinovich, M. I., P. Varona, A. I. Selverston, and H. D. I. Abarbanel: Rev. Mod. Phys. **78**, 1213–1265 (2006).
4. Fall, C. P., E. S. Marland, J. M. Wagner, and J. J. Tyson, Editors, *Computational Cell Biology*, (Springer-Verlag, New York, 2002).
5. Abarbanel, H. D. I: *Analysis of Observed Chaotic Data*, (Springer-Verlag, New York, 1996).
6. Traub, R. D. and R. Miles: *Neuronal Networks of the Hippocampus*, (Cambridge University Press, 1991).
7. Bliss T. V. and T. I. Lomø: J. Physiol. **232**, 331–356 (1973).
8. Malinow, R. and J. P. Miller: Nature **320**, 529–530 (1986).

9. Bi, G-Q and M-m Poo: Annu. Rev. Neurosci. **24**, 139–166 (2001).

10. Castellani, G. C., E. M. Quinlan, L. N. Cooper, and H. Z. Shouval: Proc. Natl. Acad. Sci. USA. **98**, 12772–12777 (2001).

11. Abarbanel, H. D. I., L. Gibb, R. Huerta, and M. I. Rabinovich: Biol Cybernetics. **89**, 214–226 (2003).

12. Rubin, J. E., R. C. Gerkin, G.-Q. Bi, C. C. Chow: J. Neurophysiol. **93**, 2600–2613 (2005).

13. Yang, S.N., Y. G. Tang and R. S. Zucker: J Neurophysiol **81**, 781–787, (1999).

14. O'Conner, D., G. Wittenberg and S. S. Wang: Proc. Natl. Acad. Sci. USA. **102** 9679–9684 (2005); O'Conner, D., G. Wittenberg, and S. S. Wang: J. Neurophysiol. **94**, 1565–1573 (2005).

15. Wittenberg, G. Learning and memory in the hippocampus: Events on millisecond to week times scales. PhD Dissertation, Princeton University, (2003).

16. Haas, J., T. Nowotny and H. D. I. Abarbanel: J. Neurophys. **96**, 3305–3313 (2006).

17. Abarbanel, H.D.I., S. S. Talathi, L. Gibb, M.I. Rabinovich: Phys. Rev. E, **72**, 031914 (2005).

18. Drazin, P. G.: *Nonlinear Systems*, (Cambridge University Press, Cambridge, 1992).

19. Carr, C. E. and M. Konishi: Proc. Nat. Acad. Sci. USA. **85**, 8311–8315 (1988).

20. Carr, C. E. and M. Konishi: J. Neurosci. **10**, 3227–3246 (1990).

21. Knudsen E.L. and M. Konishi: J. Neurophysiol. **41**, 870–884 (1996).

22. Köppel, C: J. Neurosci. **17**, 3312–3321 (1997).

23. Forger, D. B. and C. S. Peskin: Proc. Natl. Acad. Sci. USA **100**, 14806–14811 (2003).

24. Forger, D. B. and C. S. Peskin: J. Theor. Biol. **230**, 533–539 (2004).

25. Welker, C: J. Comp. Neurol. **166**, 173–189 (1976).

26. Aarabzadeh, E., S. Panzeri and M. E. Diamond: J. Neurosci. **24**, 6011–6020 (2004).

27. Sugase, Y., S. Yamane, S. Ueno and K. Kawano: Nature, **400**, 869–873 (1999).

28. Buracas, G.T., A. M. Zador, M. R. DeWeese, T. D. Albright: Neuron. **9**, 59–69 (1998).

29. Lewicki, M. and B. Arthur: J. Neurosci. **16**, 6897–6998 (1996).

30. Margoliash, D: J. Neurosci. **3**, 1039–1057 (1983).

31. Margoliash, D: J. Neurosci. **6**, 1643–1661 (1986).

32. Coleman, M. and R. Mooney: J. Neurosci. **24**, 7251–7265 (2004).

33. Janata, P and D. Margoliash: J. Neurosci. **19**, 5108–5118 (1999).

34. Carr, C. E.: Ann. Rev. Neurosci. **16**, 23–243 (1993).

35. Cardin, J. A., J. N. Raskin and M. F. Schmidt: J. Neurophysiol. **93**, 2157–2166 (2005).

36. Rosen, M. J. and R. Mooney: J. Neurophysiol. **95**, 1158–1175 (2006).

37. Kimpo, R. R., F. E. Theunissen and A. J. Doupe: J. Neurosci. **23**, 5750–5761 (2003).

38. Abarbanel, H. D. I., S. S. Talathi, G. B. Mindlin, M. I. Rabinovich and L. Gibb: Phys. Rev. E. **70**, 051911, (2004).

# Part II

Complex Networks

# 3

# Structural Characterization of Networks Using the Cat Cortex as an Example

Gorka Zamora-López, Changsong Zhou and Jürgen Kurths

Nonlinear Dynamics Group, University of Potsdam
gorka@agnld.uni-potsdam.de

**Summary.** In this chapter, *Graph Theory* will be introduced using cat corticocortical connectivity data as an example. Distinct graph measures will be summarized and examples of their usage shown, as well as hints about the kind of information one can obtain from them. Special attention will be paid to *conflicting* points in graph theory that often generate confusion and some algorithmic tips will be provided. It is not our aim to introduce graph theory to the reader in a detailed manner, nor to reproduce what other authors have written in several extensive reviews (see Sect. 3.8).

Some of the examples placed in this chapter referring to the cat cortex are unpublished material and thus, not to be regarded as established scientific results. Otherwise, references will be provided.

## 3.1 Introduction

A network is an abstract manner of representing a broad range of real systems in order to be mathematically tractable. Elements of that system are represented by vertices and their interactions by links, often giving rise to complex topological structures. Links could illustrate some real physical connection: in roadmaps, cities are represented by dots (vertices) and roads by lines (links); neurons (vertices) connect to each other through axons and synapses (links); the Internet is formed by computers or servers (vertices) connected by cables (links). Abstract concepts are also suitable to be translated into networks: in physics, regular solids are represented as crystal lattices; in the social sciences, vertices might represent people and a link between them could be placed when two persons are friends; the World Wide Web is also an abstract network where vertices are web pages linked by hyperlinks pointing from one web page to another; in ecological food webs, species (vertices) are linked depending on their hierarchy in the web.

In the universe of networks, three basic types of graphs are found: *simple graphs*, whose vertices are connected by edges without directional information; *digraphs* (directed graphs), whose directed links are drawn by arrows (arcs);

and *weighted graphs*, when links represent some scalar magnitude. The weights given to a link depend on the specific system under study. In a transportation network, weights could either represent the physical distance between two cities or the number of passengers, cars, trains, etc. that travel from one city to another. In order to simplify the analytical treatment, most theoretical work has been focused in the study of "graphs". In this chapter on the contrary, the general properties of both directed and weighted networks will be introduced because corticocortical connectivity data belong to this class.

### 3.1.1 Brief Historical Review

Historically, the study of networks has been the domain of a branch of discrete mathematics known as *graph theory*. Since its birth in 1736, when the Swiss mathematician Leonhard Euler published the solution to the Königsberg bridge problem (consisting in finding a round trip that traversed each of the bridges of the Prussian city of Königsberg exactly once), the study of networks turned out to be useful in many different contexts. In the social sciences, the practical use of graph theory started as early as the 1920s. During the 1990s thanks to the advances in computation, the handling of very large data sets became affordable for the first time, and thus the study of the interconnectivity of many real systems became possible. The field experienced a rapid growth and nowadays is mainly known under the name of *Complex Networks*; its influence can be seen in different disciplines like sociology, life sciences, technology, physics, economics, politics, etc. We will keep in mind that network theory is just a data analysis toolkit flexible enough to be applied in many different contexts.

Apart from structural characterization, special attention has been paid in recent years to dynamical processes of networks in an attempt to understand the bridge between structure and function, which is important for the study of neuroscience. On one hand, dynamical processes can happen *within* a network, like electrical current flowing through a electrical circuit or when vertices represent some dynamical system as oscillators, neurons, etc (cf. Chaps. 1, 5–14). In this case, we might raise the question of how the complex interconnectivity of elements affects both individual and collective behavior. On the other hand, the structure of the network itself might change in time (i.e. *plasticity processes* happening in neural networks), affecting its dynamical properties (cf. Chaps. 2, 5, 7). Understanding the interrelation between structure and dynamics (function) could allow, for example, the design of flexible technological networks where deliberate change of connections could optimize the "maximum service/minimum cost" problem. In the life sciences, it could provide understanding of how internal dynamical processes drive and regulate structural changes leading to *self-organization* in the system.

## 3.1.2 Complexity, Networks and the Nervous System

Complex systems are typically characterized by a large number of elements that interact nonlinearly. Successful mathematical methods to treat such systems are still a demanding challenge for current and near future research. Classically, nature has been abstracted and broken up into different pieces called systems and further partitioned into semi-independent subsystems susceptible to being separately studied. Once different pieces are understood, the answers are summed up to provide a more global understanding (what is known as the *principle of linear superposition.*) The small contribution of nonlinear interactions and other unknown contributions are simply classified as noise. The Fourier transformation and its applications (a method to decompose a function as a linear superposition of some other functions) is a good example of what can be mathematically achieved thanks to the assumption of linearity. Even for systems composed of very many elements, provided linear interactions dominate, statistical methods allow their description by means of "macroscopic" properties.

But, what if a system cannot be broken up into such independent pieces? Or, what if the system can be partitioned but its overall behavior cannot be described as the linear sum of its components? We are then facing a *complex system.* We are here to show another ingredient that causes complexity in real systems: *the intricate connectivity of interactions among its elements.* Solid crystals have regular structures providing symmetries that simplify their macroscopic description. In many other systems, like star clusters or clouds of charged particles, all elements interact with each other allowing a macroscopic description in terms of "mean fields". The intricate connectivity of many real systems, on the contrary, makes such simplifications impossible. It is rather true that the complexity of the nervous system arises from many different aspects: the coexisting spatio-temporal scales of its dynamics, genetic regulation, molecular organization, the mixture of electrical and molecular signals for communication, intricate connectivity among elements at different scales etc. Each of these features require the use of specific tools and scientific methodologies, making modern neuroscience a highly interdisciplinary field.

What we are trying to emphasize in this book, and especially in this chapter, is that the complex inter-connectivity among the elements comprising the brain (at different scales) is an important aspect to be understood. Therefore, we present Graph Theory as a suitable data analysis toolkit.

## 3.2 The Cat Cortical Connectivity Network

One of the principal enigmas of biology, and the central issue in physiology, is the intrinsic relationship between the physical geometry of biological structures and their function. The study of structures of biological systems at different scales has been enabled by advances in optical devices, imaging and spectroscopy. However, knowledge about structure at different levels of organization in the nervous system and interconnectivity is still far from being satisfactory [1]. Knowledge about mesoscopic structures and microscopic connectivity is largely missing. Even at the macroscopic scale, interconnectivity between different cortical regions is only known for few mammal species: macaque and cat (partially that of rats also), although further work is still necessary to confirm and improve the existing data. In the case of humans, no harmless tracing technique is available yet to obtain a comprehensive map of the anatomic cortical connectivity, although potential use of non-invasive techniques such as DTI (Diffusion Tensor Imaging) from Magnetic Resonance Imaging (MRI) data is already under study, as well as post-mortem tracing studies (see Chap. 4).

For the exercises and scientific tasks of the 5th *Helmholtz Summer School on Supercomputational Physics*, the corticocortical connectivity data of the cat was chosen because it is, for the moment, the most complete of its type.[1] This data summarizes the corticocortical connections, where links represent the bundles of axons projecting between distant cortical areas through the white matter. Connections are classified as weak "1", medium "2" or strong "3" depending on the diameter of the fibres. The current data is a collation of previous reports performed by Jack W. Scannell during his Ph.D. and presented in various versions [3,4] including the thalamo-cortical connections. The network has been extensively analyzed. It has been found to be clustered and hierarchically organized both in its structural [5–7] and in its functional [8] connectivity. This network also has "small-world properties" [9]. More about its network properties will be studied in this chapter and in Chap. 4.

In Fig. 3.1, the parcelation scheme used by Scannell et al. [3] for both lateral and medial views of a single cortical hemisphere are shown as

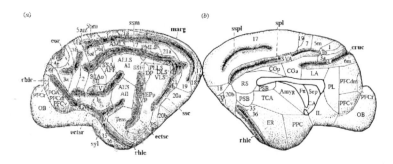

**Fig. 3.1.** Parcelation of a single hemisphere of the cat cortex. Reprinted from [2] with permission. Lateral view (*left*) and medial view (*right*)

---

[1] It is important to mention current efforts to improve the connectivity data of the macaque cortex, see http://www.cocomac.org, that will become the best reference in the near future.

presented in [2]. In Fig. 9.2 of Chap. 9, the corticocortical connectivity data of the cat is shown in *adjacency matrix* form. $A_{ij}$ means that area $i$ (row) projects a bundle of axons into area $j$ (column). The four major clusters on the diagonal represent internal connections between areas within the visual, auditory, somatosensory-motor and frontolimbic cortex, while "off-diagonal" connections represent information exchange paths between different functional clusters. We show here the network with 53 cortical areas since this was the version used for computation during the Summer School.

## 3.3 Network Characterization and the Cat Cerebral Cortex

Vertices of networks are labeled with numbers from 1 to $N$ and links between vertices $i$ and $j$ are represented as entries of a matrix: $A_{ij} = 1$ if $i$ connects to $j$, and $A_{ij} = 0$ otherwise. This is known as the *adjacency matrix* of the network (Fig. 3.2(a)). When computing large networks, matrices become very inefficient in terms of memory allocation. A network of $N$ vertices requires a matrix of $N^2$ elements, where most of the entries will be zero because real networks tend to be sparse (look at Table 1 of [10]). Thus, large amounts of memory are being wasted. Another way to represent networks is through *adjacency lists* (Fig. 3.2(b)), which consist of $N$ lists containing only the vertices to which vertex $i$ projects. In this case, the amount of required memory is proportional to the number of connections $m$ in the network. All we need is a list with the $N$ vertices and the $N$ lists of the neighbors of each vertex. This makes a total memory requirement of $\mathcal{O}(N + \sum_i k_i) = \mathcal{O}(N + m)$, where $k_i$ is the number of neighbors of each vertex.

For example, a network with $10,000$ vertices requires a matrix of $10^8$ elements translating into $\sim 95$ MB if entries are taken as 8 bit (1 Byte) integers. Imagine our network has a density of connections $\rho = 0.1$ ($z \approx 1000$ links per vertex). The number of connections is then $m = \rho \cdot N \cdot (N - 1) = 9,999,000$.

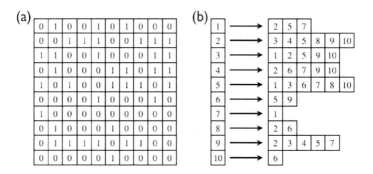

**Fig. 3.2.** Representation of networks: **a)** adjacency matrix and; **b)** adjacency lists

Note that we now need 16 bit (2 Byte) integers since the indices of vertices go up to 10000, thus the memory required in an *adjacency list* representation is only $\sim 20$ MB.

Although adjacency lists reduce memory requirements, performance may be sacrificed. Some operations run faster with adjacency lists, i.e. to calculate the degree of each node we only need to ask the system to return the size of each list while in a matrix, all elements have to be read and nonzero elements counted. However, the calculation of most graph theoretic measures requires one to enquire whether two vertices are connected or not, and this operation on a matrix is very fast because when we code something like "if net($n_1$,$n_2$) == 1" the system only needs to visit the specific memory location for ($n_1$,$n_2$) and read the value. Using adjacency lists, the system has to look through all values in the list for node $n_1$ and compare each value to $n_2$ until it is found or the end of the list is reached. In Chap. 11, more will be discussed about suitable data structures for representing networks and performing simulations of dynamical networks.

### 3.3.1 Degree Distributions and Degree Correlations

The most basic property of nodes is the *vertex degree* $k_i$ which represents the number of connections a vertex $i$ has; this is the building block for other structural measures. In networks without *self-loops* (links going out of a node and returning to itself) and *multiple links* (more than one link connecting two vertices), the degree equals the number of neighbors of $i$. In *directed graphs* (digraphs), the number of connections leaving from $i$ and the number of links that enter $i$ are not necessarily equal, so the degree splits up into the *input degree* (in-$k_i$) and *output degree* (out-$k_i$); most degree-based structural measures also split accordingly. Given a network with $N$ vertices and $M$ connections:

$$\sum_{i=1}^{N} in\text{-}k_i = \sum_{i=1}^{N} out\text{-}k_i = M$$

In the context of weighted graphs, the natural counterpart of vertex degree is the *vertex intensity* $S_i$, defined as the sum of weights of $i$'s connections. The directed versions in-$S_i$ and out-$S_i$ can equally be defined summing the weights of the input connections that $i$ receives or the weights of the output connections leaving from $i$.

In order to obtain statistical information about the degrees in large networks, the *degree distribution* $P(k)$ is measured and defined as the probability that a randomly chosen vertex has degree $k$. Quantitatively, it is measured as the fraction of nodes in the network that have degree $k$, or estimated from a histogram of the degrees. However, for many real networks with scale-free or exponential distribution, a histogram provides poor statistics at high degree vertices and the *cumulative degree distribution*, $P_c(k)$, is recommended. It is defined as the fraction of nodes in the network with degree larger than $k$.

$P_c(k)$ requires no binning of data and is a monotonous decreasing function of $k$, which makes it a better measure to estimate the exponent of scale-free and exponential distributions. The only difference is that if a network has a scale-free degree distribution with exponent, $P(k) \sim k^{-\gamma}$, then the measured exponent in the cumulative distribution is $P_c(k) \sim k^{-\alpha}$ where $\alpha = \gamma - 1$.

Degree distribution alone does not tell one very much about the internal structure of a network, and other measures are required. It is interesting, for example, to look for correlations between degrees of vertices. A network will be called *assortative* when high degree vertices connect preferentially to each other, and low degree vertices to each other (correlation is thus an increasing function of $k$). A network is called *disassortative* when high degree vertices preferentially connect to low degree vertices (correlation is a decreasing function of $k$). Interestingly, as pointed out in [10], social networks are observed to show assortative behavior while many other networks (technological networks, biological networks) tend to be disassortative. Formally, degree-degree correlations are expressed by the *conditional probability* $P(k|k')$ that may be problematic to evaluate due to finite size effects.

A popular measure to evaluate degree correlations is the *average neighbors' degree*, $k_{nn}(k)$ introduced in [11]. Firstly, for each vertex $i$, the average degree of its neighbors is calculated ($k_{nn,i}$). Then, these values are again averaged for all nodes having degree $k$. In the case of directed networks, things become confusing and we may not know what to look for. Below we show only two cases:

1. is the *out-$k_i$* of vertex $i$ correlated to its output neighbors' *in-$k_j$* degree? (Fig. 3.3(a)).
2. is the *in-$k_i$* of vertex $i$ correlated to its input neighbors' *out-$k_j$*? (Fig. 3.3(b)).

The extension of these measurement to weighted networks consists in replacing the degrees by intensities and adjacency matrices by their weighted counterparts as presented in [12].

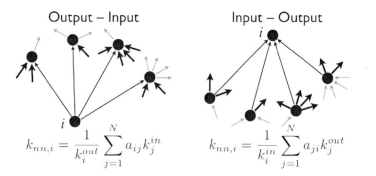

Output – Input

Input – Output

$$k_{nn,i} = \frac{1}{k_i^{out}} \sum_{j=1}^{N} a_{ij} k_j^{in}$$

$$k_{nn,i} = \frac{1}{k_i^{in}} \sum_{j=1}^{N} a_{ji} k_j^{out}$$

**Fig. 3.3.** Schematic representation of two possible combinations to calculate $k_{nn}(k)$ in directed networks

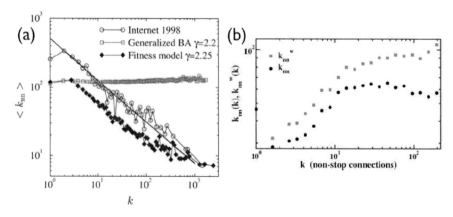

**Fig. 3.4.** Neighbors' degree as a measure for degree correlation: (a) $k_{nn}(k)$ of the Internet and the Barabási-Albert model. Reprinted with permission from [11]; (b) Comparison of $k_{nn}(k)$ to its weighted version for the WAN (World Airport Network) as defined in [12]. Reprinted with permission

Figure 3.4(a) (from [11]) shows the average neighbors' degree of the Internet in the year 1998 to be *disassortative*. The Internet is known to have a scale-free degree distribution with an exponent $\gamma \approx 2.2$. However, a modified Barabási-Albert model (BA) that constructs a network with similar exponent displays uncorrelated degrees, failing to catch the internal structure of the Internet. In Fig. 3.4(b) (from [12]), average neighbors' degree of the World Airport Network (WAN) is shown. In the unweighted case, connection flights between two airports are considered, while in the weighted case, the intensity of the connections represents the number of passengers flying from one airport to another in direct (non-stop) flights. The network has assortative behavior, more pronounced in the weighted case. This example illustrates the loss of information when, for simplicity, the unweighted version of a weighted network is considered. Note that in Fig. 3.4(b) the weighted $k_{nn}^w(k)$ is plotted against $k$ only for comparative reasons, but it is more natural, for obvious reasons, to plot it against the intensity $S_i$ of the vertex.

Figures 3.5(a) and (b) show neighbors' degree of the cat corticocortical network for the two cases previously depicted in Fig. 3.3. It is surprising to observe the asymmetry between the two cases. While *out-$k_i$* happens to be independent of their neighbors' *in-$k_j$* (Fig. 3.5(a)), the opposite case exhibits a nontrivial behavior (Fig. 3.5(b)): it looks assortative for low degree areas and then saturates to become disassortative at high degrees. The meaning of this asymmetry and its functional consequences is as yet unclear.

### 3.3.2 Clustering Coefficient

The concept of clustering coefficient is well illustrated by using social networks as an example: *"If person A is a friend of person B and a friend of*

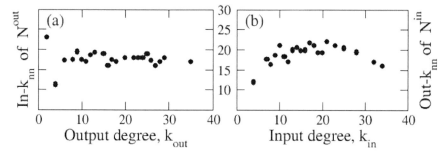

**Fig. 3.5.** Degree correlations of cat cortex. (a) Correlation between $out$-$k_i$ of node $i$ and its neighbors' $in$-$k_n$ (b) opposite case, correlation between $in$-$k_i$ of node $i$ and its neighbors' $out$-$k_n$

*person C, what is then the probability that persons B and C are also friends?"* Clustering coefficient is, thus, a measure to quantify the conditional probability $P(BC\,|\,(AB \cap AC))$ (Fig. 3.6(a)). In general, given a set of vertices, the *average clustering* will quantify the cohesiveness of that set.

The most popular way of measuring $C$ in *graphs* is to count the number of triangles. The number of paths of length 2 $\{a, b, c\}$ gives the total number of possible triangles. Taking into account that a triangle formed by vertices $a$, $b$ and $c$ contains three such paths ($\{a, b, c\}$, $\{b, c, a\}$ and $\{c, a, b\}$), the average clustering of the network is then *measured as*:

$$C = \frac{3 \times \text{number of triangles}}{\text{number of paths of length 2}}$$

Higher order versions of this method are possible by counting the number of squares and so on.

In *directed graphs*, however, it is not clear how to define a triangle. A common approach is to define first the *local clustering coefficient* $C_i$ of a single vertex and then average over all vertices to find the *average clustering*

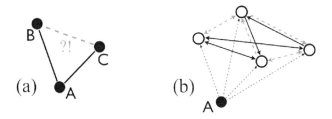

**Fig. 3.6.** Illustration of clustering coefficient: (a) Clustering quantifies the conditional probability of two vertices to be connected provided they share a common neighbor; (b) Local $C_A$ measure as used in directed networks: the fraction of existing connections between neighbors of $A$ to all the possible connections between them

*C*. Interestingly, $C_i$ of a single vertex is not directly a structural measure of $i$ itself but a property of *its neighbors* that measures "how well are the neighbors of $i$ are connected together". Defining $\mathcal{N}_i$ as the set of neighbors of vertex $i$, $C_i$ is then measured (illustrated in Fig. 3.6(b)) as the ratio between the number of *existing* links among vertices in $\mathcal{N}_i$ and the number of *possible* links in the set:

$$C_i = \frac{\sum_{j,k \in \mathcal{N}_i}(A_{jk} + A_{kj})}{k_i(k_i - 1)}$$

where $k_i$ is the degree of vertex $i$. It is very important to stress that, as pointed out in [10], this approach is not exactly a measure of the conditional probability $C_i = \sum_{j,k} P(jk\,|\,(ij \cap ik))$, and it tends to overestimate the contribution of low-degree vertices due to the smaller denominator.

Again, we find different possibilities in the case of directed networks. Clustering of the "output neighbors" (using *out-*$k_i$, $\mathcal{N}_i^{\mathrm{out}}$, Fig. 3.6(b)) can be measured or that of the "input neighbors" (using *in-*$k_i$, $\mathcal{N}_i^{\mathrm{in}}$). Another possibility is to use both the input and output neighbors of $i$. However, it is erroneous to define $k_i = in\text{-}k_i + out\text{-}k_i$ because reciprocal links will be counted twice. $k_i$ has to be now the number of *all* vertices that $i$ is connected to (whether input or output). When reciprocal connections are highly present in the network, output and input versions should give similar results.

Once the $C_i$ are calculated it is interesting to look for their correlations with other local properties. $C_i$ of many real networks has been found to anticorrelate with $k_i$ due to their modular structure [10]. This relationship has also been captured by theoretical models [13]. The cat cortex provides an illustrative example as shown in Fig. 3.7. Those areas with few neighbors have higher $C_i$ than those with larger degree. The reason becomes clear when observing the modular structure of the adjacency matrix in Fig. 9.2 of Chap. 9. Low degree cortical areas do preferentially connect to other areas in the same functional cluster (visual, auditory, etc.) while high degree areas have connec-

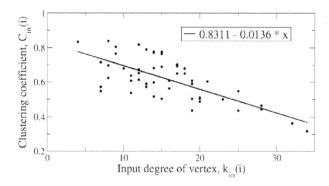

**Fig. 3.7.** Clustering coefficient of each cortical area $C_i$ as function of area degree $k_i$. The negative slope is a signature of modular structure in the network

**Table 3.1.** Average clustering coefficients of the cat cortex and of its anatomical communities

| Whole cortex | Visual | Auditory | Somatosensory-motor | Frontolimbic |
|---|---|---|---|---|
| 0.61 | 0.64 | 0.87 | 0.79 | 0.72 |

tions all over the network. Thus, neighbors of high degree areas are gathered into semi-independent groups that rarely connect with each other.

The *average clustering coefficient* $C$ provides a quantitative estimation of how cohesive a network is (or just a set of neighbors). Table 3.1 summarizes the internal $C$ values of the whole cat cortical network and of its four structural clusters. As expected, the internal $C$ of clusters is higher than that of the whole network, with auditory cortex being the most cohesive of them all. In order to measure $C$ of a subset of vertices, the set must be first isolated from the rest of the network and then $C$ of this sub-network measured using only internal vertices and the connections among them.

### 3.3.3 Distance

In networks, topological distance measures the minimum number of links crossed in order to go from a given vertex $a$ to another vertex $b$. If there is a direct link from $a$ to $b$, then $d(a, b) = 1$. If there is no shorter path than going from $a$ to $c$ and from here to $b$, then $d(a, b) = 2$, and so on. Usually, more than one shortest path from $a$ to $b$ exists. This will give rise to important measures meaningful in the context of flows in networks, whatever the flow represents: water flow in a pipeline, information flow, traffic in transportation networks, etc. In graphs, $d(a, b)$ equals $d(b, a)$, so if the *distance matrix* is defined as the matrix whose elements $d_{ab} = d(a, b)$, then it is symmetric. The *average pathlength* $l$ of a network is defined as the average of all values in the $d_{ij}$ matrix.

When there is no path going from $a$ to $b$, then $d(a, b) = \infty$ and $b$ is said to be *disconnected* from $a$. It is easier to find disconnected vertices in digraphs due to the directed nature of connections. In food-webs for example, a directed link "worms" $\rightarrow$ "birds" exists but no arrow is to be drawn like "birds" $\rightarrow$ "worms" because worms do not eat birds. Thus, $d(\text{worm}, \text{bird}) = 1$ but $d(\text{bird}, \text{worm}) = \infty$. (Note: in food webs arrows point in the direction of the energy flow.) An important matter in networks is the presence of *connected components*: isolated groups of vertices, internally connected, that cannot reach other components.

Taking connection weights into account while measuring distance is only meaningful when weights represent real metric distance between vertices, like in a roadmap. An interesting example is given by the *Via-Michelin* web page and similar services. This web page allows one to, for example, find the road trip between any two addresses within Europe. One option is to perform the search by *shortest distance* for which only metric distance information is used. Another possibility is to perform a search by *fastest trip*. In this case, other kinds of information must be used in order to give a proper "weight" to each road-segment (type of road, average traffic, etc.) that measure how quickly one can drive through them, and combined with the metric distance in order to calculate the "shortest graph distance", that in this case refers to "time".

There is an extensive literature about algorithms to find shortest paths between vertices in a network. The two most popular ones are the *Dijkstra* and the *Floyd-Warshall* algorithms. Both of them are useful for weighted digraphs since they are designed to return the *lowest cost path*, whatever the weight means in the network. The Dijkstra algorithm finds the shortest path between a source vertex $s$ and all other vertices in time $\mathcal{O}(N^2)$, where $N$ is the size of the network. In the case of sparse matrices $(m \ll N^2)$, it can be improved up to $\mathcal{O}(N + m \cdot \log m)$. But if we want to calculate the distance between *all* pairs of vertices, then Dijkstra's algorithm has to be repeated for each vertex, $\mathcal{O}(N^3)$, the length of each path calculated and the distance matrix created. Thus, Dijkstra is the algorithm of choice when we want the shortest path (or distance) between a given pair of vertices, but when *all-to-all* distances are to be calculated, then the Floyd-Warshall algorithm is the choice. This algorithm takes the adjacency matrix of the network as an input and returns the distance matrix in time $\mathcal{O}(N^3)$, but faster than applying Dijkstra $N$ times. It can also be implemented to return the shortest paths between each pair.

### 3.3.4 Centrality Measures

There is a set of measures requiring to initially calculate the distance matrix. Given a graph $G$ and its distance matrix, the *eccentricity* $e_i$ of a vertex $i$ is defined by the maximum distance from $i$ to any other vertex in the network. The *radius* of the network $\rho(G)$ is then the minimal eccentricity of all vertices and the *diameter*, diam$(G)$, the maximal.

| | |
|---|---|
| eccentricity | $e_i = \max(d_{ij}) : j \in V$ |
| radius | $\rho(G) = \min(d_{ij}) = \min(e_i)$ |
| diameter | $\mathrm{diam}(G) = \max(d_{ij}) = \max(e_i)$ |

The *center of the network* is composed by the set of vertices whose $e_i = \rho(G)$. The name "center" is given because these vertices are closest to any other in the network. If a signal is to be sent so that it reaches all vertices as fast as possible, then the signal should be sent from one of the vertices in the center. If an emergency center is to be placed in a city, it is desirable to strategically place it so that ambulances or firemen will arrive at any corner of the city in minimal time.

Such definitions give rise to a discrete spectrum of values, and it would be desirable to have a continuous spectrum in order to correlate them to other vertex properties. We suggest that the reader instead uses the following definition of eccentricity:

$$e_i = \frac{1}{N} \sum_{j=1}^{N} d_{ij}$$

this is nothing but the average pathlength from vertex $i$ to any other vertex in the network.

We could also ask the opposite question: "which vertices can be most quickly reached from anywhere in the network?" In other words, where should a shopping mall be placed such that all inhabitants in a city can reach it as quickly as possible? The *status* $s_i$ of a vertex is defined as the average of the values in the $i$th column of $d_{ij}$,

$$s_i = \frac{1}{N} \sum_{j=1}^{N} d_{ji}$$

and quantifies how quickly, on average, vertex $i$ is reached from other vertices.

Another important measure is the *betweenness centrality* that can be defined both for vertices or links. For brevity, only the case of vertices will be described here. The betweenness centrality $(BC_i)$ of vertex $i$ is the count of how often $i$ is present in all the shortest paths between all pairs of vertices in the network. Note that, usually, there is more than one shortest path between two vertices, thus, *all* shortest paths between a given pair are to be accounted for. It is computationally tricky and expensive to find them all. The Dijkstra and Floyd-Warshall algorithms, for reasons of efficiency, look for only one shortest path.

The approach personally followed by the authors of this chapter was to use a modified version of the *depth-first-search algorithm* (DFS) in order to stop the tree search at a desired depth. Once the distance matrix has been calculated, the DFS algorithm is called for each pair of vertices $i$, $j$. The known distance $d_{ij}$ is introduced into the DFS algorithm so that it will search through the whole tree but no deeper than $d_{ij}$ steps, as depicted in Fig. 3.8(b). If $d_{ij} = \infty$, the search must be skipped. When all shortest paths in the network are found, the number of times every vertex appears as an *intermediate* vertex is counted.

$BC_i$ is an important measure because it quantifies how much of the flow (information transmission, water flow, traffic, etc.) goes through $i$. Besides, $BC_i$ does not necessarily correlate to vertex degree $k_i$. Imagine two independent networks that are connected by a single link. No matter how small the degree of the two vertices is at the end of that link, their $BC$ will be very high because any path connecting the two sub-networks necessarily includes them. Thus, in robustness analysis, selective attack on vertices with high $BC_i$ is much more relevant than attack on those with high $k_i$.

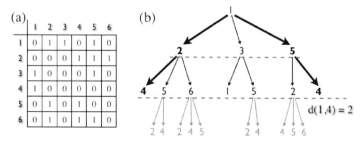

**Fig. 3.8.** Finding all shortest paths between a pair of vertices. (**a**) Adjacency matrix of example network; (**b**) Tree of vertex 1. Once $d(1,4) = 2$ is known, DFS algorithm is modified in order to stop the tree search at desired depth. Two distinct paths are found from 1 to 4, say $\{1 \rightarrow 2 \rightarrow 4\}$ and $\{1 \rightarrow 5 \rightarrow 4\}$, giving rise to BC(2) += 1 and BC(5) += 1

### 3.3.5 Matching Index

Intuitively, two vertices having the same neighbors might be performing similar functions. The matching index $MI(i,j)$ of vertices $i$ and $j$ is the count of common neighbors shared by both vertices, and provides an estimation of their "functional similarity". Directionality of the network brings again different possibilities since we could look for common *input* neighbors, common *output* neighbors or *both*. In order to properly normalize the measure, the number of common neighbors should be divided by the total number of distinct neighbors that $i$ and $j$ have. Normalizing over $(k_i + k_j)$ returns misleading results as illustrated in the following example:

$$1 \rightarrow \{4,5,6\}$$
$$2 \rightarrow \{4,5,7,9\}$$
$$3 \rightarrow \{1,2,5,8,9,11,12\}$$
$$4 \rightarrow \{1,2,5,8,9,11,12\}$$

Vertices 1 and 2 share two $\{4,5\}$ out of five different neighbors $\{4,5,6,7,9\}$, thus $MI(1,2) = 2/5 = 0.4$. Vertices 3 and 4 share all of their seven neighbors giving rise to $MI(3,4) = 1$. In the case where we would normalize over $(k_i + k_j)$, $MI(1,2) = 2/(3+4) = 0.286$ and $MI(3,4) = 7/(7+7) = 0.5$, clearly faulty expressing the probability of common neighbors.

The matrix representation of $MI$ is symmetric. Diagonal elements are simply ignored and receive null values. The matching index of the cat cortical network in Fig. 3.9 reflects its modular structure and includes some new information. A pair of cortical areas do not need to be connected in order to have a large overlap of neighbors, and thus similar functionality (although in the case of cat cortex, $MI$ tends to provide higher values for connected areas). $MI$ values within a cluster lie in general between 0.35 and 0.6 with some higher value exceptions (mainly in the sensory-motor cortex), while inter-cluster connections tend to have values below 0.35 with some exceptions dominated by the

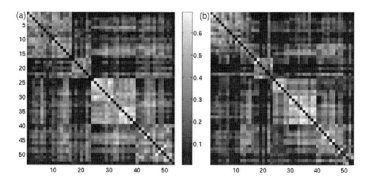

**Fig. 3.9.** Matching index of the cat cortical network: **a)** $MI$ of input connections. **b)** $MI$ of output connections

connections of high degree areas. $MI$ versions for input neighbors in Fig. 3.9(a) and for output neighbors Fig. 3.9(b) show different internal structures of the anatomical clusters. While for the input version the four anatomical clusters (visual, auditory, somatosensory-motor and frontolimbic) are clearly observed, the output version shows subdivisions of the somatosensory-motor cortex into two clusters, and some visual areas are kicked out.

## 3.4 Random Graph Models

In order to understand the natural mechanisms underlying the observed structural properties of real networks, creating network models is essential. Different network classes are known to exist, e.g. technological or biological networks that share similar properties within each class and differences across them. Modeling helps us to understand both the similarities and the differences. There is a very extensive literature about random graph models and readers are strongly recommended to read some of the reviews listed in the last section of this chapter. The goal of this section is just to give a very brief but hopefully clear description of the main network models whose modification have motivated many more detailed and realistic models. We will stick to their original descriptions as *graphs* and leave directed connections and weights aside, since this simplification has permitted an extensive analytical treatment.

### 3.4.1 Erdős-Rényi (ER) Networks

The Erdős-Rényi (ER) model is probably the simplest random network that can be built (cf. Chaps. 5, 7). Starting from an empty set of $n$ vertices without connections, $m$ links are randomly added one by one so that every pair of vertices has the same probability of getting a new link. $G_{n,m}$ is the ensemble of all possible random graphs of size $n$ and $m$ links constructed by this procedure. Typical restrictions are 1) only one link is allowed between a pair of vertices and 2) no self-loops are allowed (self-connections leaving a vertex and re-entering itself; by contrast, see Chap. 7 for an example of an Erdős-Rényi network exhibiting self-loops). The simplicity of this model has permitted extensive analytical work of its properties, started by Erdős and Rényi themselves in the 1960s.

Its degree distribution follows a binomial distribution, becoming Poissonian in the limit of large networks. The probability of a vertex to have degree $k$ is then:

$$P(k) = \binom{N}{k} p^k (1-p)^{N-k} \simeq \frac{z^k e^{-z}}{k!} \quad : N \to \infty$$

where $p$ is the uniform probability of a vertex to get a link. This distribution is peaked around the average degree $z = \langle k \rangle$, which allows the description of the network in terms of $z$, simplifying the analytical approach.

The model drew much attention due to its percolation properties, equivalent to continuous phase transitions studied in statistical physics and thermodynamics. At the beginning of the random process, when few links are present, many independent and small connected groups of vertices appear called *connected components*, while remaining disconnected from each other (find 3 of them in Fig. 3.10). When more links are added, components grow and merge together into larger components making their *size distribution* approach a scale-free distribution — very many small components and few large ones. There is a critical number of links when the largest components merge suddenly giving rise to a *giant component*, composed of most of the nodes of the network (around 80% of them). This transition has been analytically proved and numerically corroborated to happen when $z = \langle k \rangle \approx 1$, so that $m \approx n$. Chap. 7 relates this percolation transition with another one where oscillations emerge in the networks's dynamics when super-cycles are merged from isolated ones.

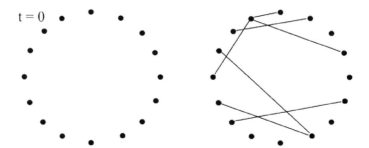

**Fig. 3.10.** Generation of ER random graphs. An initially empty set of $n$ vertices is linked with uniform probability $p$

But, how well does the ER model resemble real networks? A universal property of real complex networks captured by this model (above the percolation threshold) is the *small-world property*. The distance between any pair of nodes, and thus the network average pathlength, is observed to be very short. However, what does "short" mean? Obviously, the larger the network, (keeping $z$ constant) the longer the average pathlength $l$ will become. In order to quantify what *short* means, the *small-world property* has been defined as the following upper scaling limit of the average pathlength in the network:

$$l_{\text{upper}} = \ln N$$

in the case of the ER model, average pathlength is proven to scale as:

$$l = \frac{\ln N}{\ln z} < \ln N$$

under the following conditions, $N \gg z \gg \ln N \gg 1$. This means that the network is connected (well above the percolation threshold) but connection density is not too large.

On the contrary, the ER model resembles real networks very poorly because it does not reproduce any of the other typical properties like clustering coefficient, degree correlations, etc. Real networks do not show Poissonian degree distributions. Many modifications of the ER model have been presented trying to reproduce these other properties. One to be mentioned, since it will be discussed in the next section, is the so-called *configuration model* that produces maximally random networks with a prescribed degree distribution or degree sequence [14–16]. Thus, the ER model is just an special case of the configuration-model.

### 3.4.2 The Watts-Strogatz (W-S) Model of "Small-world" Networks

Motivated by the high clustering observed in many real networks and by the highly unrealistic networks (lattices or random) previously used to model dynamical processes, Watts and Strogatz proposed the following model: starting from a regular lattice (a 1-d ring in this case) whose vertices all have degree $z$, a link is selected with uniform probability $p_{\text{rew}}$ and randomly rewired, conserving one of its original ends (Fig. 3.11). The initial lattice had large clustering coefficient, $C = (3z - 3)/(4z - 2)$, and large average pathlength, $l \approx N/4z$ for large $N$. The introduction of a few shortcuts significantly decreases $l$ while $C$ remains nearly constant. This is true for a range of small $p_{\text{rew}} \leq 0.1$. As $p_{\text{rew}} \to 1$ (more links are rewired), the closer the network is to a random graph of the ER type. Thus, the W-S model is a $p_{\text{rew}}$-mediated transition between regular and ER random networks.

The following restrictions are applied to the rewiring: 1) only one of the end nodes of the link is rewired, 2) no self-loops are accepted after rewiring

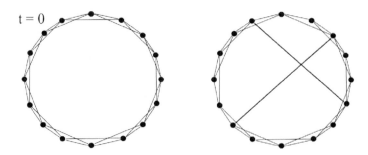

t = 0

**Fig. 3.11.** Generation of W-S random networks. Starting from an initial regular lattice where each vertex has $z$ neighbors, one end of the links is rewired with probability $p_{\text{rew}}$ giving rise to shortcuts

and 3) only one link is allowed between each pair of vertices. Other methods have been later suggested relaxing some of these restrictions in order to simplify analytical work. In fact, analytical results of the scaling properties of the model are difficult since they depend not only on network size and density of connections but also on the given rewiring probability $p_{\text{rew}}$. However, Barrat and Weigt [17] did estimate an analytical expression for the scaling of the clustering coefficient (calculated using their own definition of clustering coefficient) to be:

$$C(p) = \frac{3(z-1)}{2(2z-1)}(1 - p_{\text{rew}})^3$$

In Sect. 3.5, we will display numerical comparison of the scaling properties for different $p_{\text{rew}}$.

Again, while the WS model captures some of the properties of real networks, it fails to resemble their degree distribution and correlations. Initially, as all vertices in the ring have the same degree, $p(k) = \delta(k - z)$. During the rewiring process, some vertices gain a few connections and others lose them making the distribution wider, creating in the end a binomial distribution exactly like the ER model (see Fig. 3.14).

### 3.4.3 Barabási-Albert (BA) Model of Scale-free Networks

During the 1990s, advances in computing power permitted for the first time the analysis of large real networks. Many of them exhibit scale-free (SF) degree distributions, $p(k) \sim k^{-\gamma}$, with exponents $\gamma$ between 2 and 3. This discovery contrasted to the properties of the ER random model, extensively studied for decades as models for real networks. Barabási and Albert proved that two ingredients trigger the emergence of SF distributions in real networks: *growing of networks* adding new vertices in time and *preferential attachment* of new vertices with high degree vertices. A model including these ingredients was first introduced by Price in 1965 [18] trying to account for properties of citation networks, but it was a simpler version published in 1999 by Barábasi and Albert that became popular and provoked an avalanche of new papers and models.

The BA model, as depicted in Fig. 3.12, starts from an initially small and empty network of $n_0$ without connections. At every time step, a new vertex is included that makes $m \leq n_0$ connections to existing vertices, being the probability of connection proportional to their current degree:

$$\Pi(k_i) = \frac{k_i}{\sum_j k_j}$$

Older vertices tend to accumulate more and more connections and thus a higher probability to link to newly introduced vertices (rich-gets-richer phenomenon). Meanwhile, new vertices have few connections and thus a lower probability of gaining links in the future.

Networks generated by the BA model are found to have shorter average pathlength than ER and W-S networks of the same size and density with distance scaling as logarithm of size, $l \sim \ln(N)$. Analytical estimations predict even shorter correlation $l \sim \ln(N)/\ln(\ln(N))$ [19]. Clustering coefficient has been found to scale as $C \sim (\ln(N))^2/N$ [13], and it exhibits no degree-degree correlations. An interesting property of BA networks is that the procedure always generates connected networks in a single component, in large contrast to SF networks generated by the configuration model and similar methods where percolation processes are present. Another very important property of SF networks (not exclusive to the BA model) is that of robustness (or resilience) under attack or failures. As a few vertices accumulate most of the connections (hubs) and most vertices make few links, SF networks are very robust to the random removal of vertices, but selective attack of hubs produces large damage.

The BA is a very simple model intended to capture the main ingredients giving rise to SF degree distributions. This does not mean that it approximates

t = 0

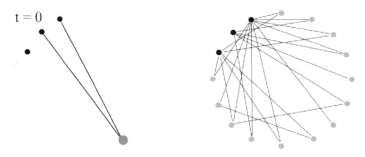

**Fig. 3.12.** Generation of BA scale-free random networks. Starting from an empty and small set of $n_0$ vertices, new vertices are included at each time step that make $m \leq n_0$ links to existing vertices with probability proportional to their current degree (*preferential attachment rule*)

other properties observed in real networks like degree-degree correlations and clustering coefficients. However, it is a flexible model and many modifications have been proposed to reproduce other properties. To only mention some of them, in [11] a mechanism to generate the degree-degree correlations of the WWW is presented (Fig. 3.4) and in [13, 20] mechanisms to introduce large $C$ to evolving SF networks are discussed.

Another limitation of the BA model is that it generates networks with a unique exponent $\gamma = 3$ (when $N \to \infty$) whatever the initial conditions and for all allowed parameters $m \leq n_0$. Besides, in real networks the SF scaling does not generally happen over the whole range of degrees. Some networks display a SF distribution truncated by exponential decays due to "ageing" of vertices (receive no new connections after some time) or saturation (limited capacity for a vertex to host new connections) [21].

## 3.5 Comparison of Random Graph Models

The random models introduced in the previous section share some similarity properties. Besides, nomenclature makes differentiation confusing since the so-called "small-world networks" are not the only ones obeying the *small-world property*. Therefore, in this section we will study and compare the scaling characteristics of the three models in more detail.

### 3.5.1 Average Pathlength

All three random network models described in this chapter generate small-world networks. What is then the difference between them? Which model generates the smallest networks? In Sect. 3.4.2, W-S networks were defined as a transition between lattices and random graphs by a process of rewiring links. For small rewiring probabilities $p_{\text{rew}}$, the networks conserve high clustering while their average pathlength decays very fast approaching that of ER graphs. Thus, lattices have the longest $l$ of all, increasing linearly with network size as $l \sim N/(4z) > \ln N$, faster than the "small-world upper limit". On the contrary, $l$ of ER graphs scale in the large network limit as $l \sim \ln N/\ln z < \ln N$, where $z$ is the average number of links per vertex. There is no analytical estimate for the scaling of W-S graphs, since this depends on $N, z$ and $p_{\text{rew}}$.

In Figs. 3.13(a) and (b), this transition is depicted for generated networks of $N = 500$ vertices and different number of connections (represented here as connection densities instead of $z$). Figure 3.13(a) shows the fast decay of $l$: for a small rewiring probability of $p_{\text{rew}} = 0.01$, average pathlength of W-S lies very close to that of ER networks. Interestingly, it is observed that the differences between the models are only meaningful for rather sparse networks. At connection densities $\rho \approx 0.3$, lattices, random networks and W-S-networks have very similar average pathlengths; at $\rho \approx 0.5$, they are actually equal.

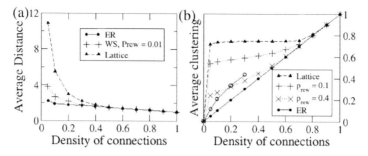

**Fig. 3.13.** W-S networks as a transition between regular lattices ($p_{\text{rew}} = 0$) and ER random graphs ($p_{\text{rew}} = 1$): (a) average pathlength $l$ and; (b) average clustering coefficient $C$ of generated networks with $n = 500$ vertices and increasing number of connections. BA model (o). All models differ at low densities. With increasing number of connections, average measures of all models become indistinguishable. Each data point is an average of 10 realizations

Scaling properties of $l$ in SF networks are not trivial. Bollobás et al. [22] performed a strictly mathematical demonstration that the BA model scales as: $l \sim \ln N / \ln \ln N$. Independently, Cohen et al. [19] studied $l$ of "random" SF networks, using the Molloy-Reed version of the configuration model. It consists of assigning each vertex a given degree probability $P(k_i)$ of getting a link and then introduce connections at random (The ER model is the special case where $P(k)$ is uniform: $P(k_i) = P(k_j)$ for all $i, j \in N$). They found that the scaling of the average pathlength depends on the exponent $\gamma$ and for $\gamma < 3$, SF networks possess "ultra-small" diameter (again, in the large and sparse network limit).

$$2 < \gamma < 3 \quad \rightarrow \quad l \sim \ln \ln N$$
$$\gamma = 3 \quad \rightarrow \quad l \sim \frac{\ln N}{\ln \ln N}$$
$$\gamma > 3 \quad \rightarrow \quad l \sim \ln N$$

Note that the scaling found for $\gamma = 3$ in the Molloy-Reed SF networks co-incides with that of the BA model. Although both methods generate graphs with scale-free degree distribution, they have different internal structure.

Summarizing, SF networks are the smallest of all while ER graphs are smaller than W-S graphs.

### 3.5.2 Clustering Coefficient

Figure 3.13(b) shows the transition of $C$ in W-S networks, between lattices and ER graphs, to be slower with increasing $p_{\text{rew}}$ than the transition of $l$ (Fig. 3.13(a)). At small $p_{\text{rew}} = 0.01$, average pathlength of W-S networks is very close to that of ER graphs, while average clustering coefficient at $p_{\text{rew}} = 0.4$ starts to approach the curve of random graphs. The combination of these features, fast decrease of $l$ while keeping high $C$, was the reason to coin the W-S graphs as "small-world networks" in analogy to social networks where the *"friends of my friends are also my friends"* effect causes high clustering and helps average pathlength between members of the network to be shorter.

Lattices have the largest $C$ of all and W-S graphs have, for any $p_{rew} < 1.0$, larger $C$ than ER graphs. Note that $C$ of ER graphs equals its density of connections since this density is equal to the probability $p$ of two vertices to be linked. Again, there is a critical density of connections ($\rho \approx 0.7 - 0.8$) where $C$ of all models becomes indistinguishable. BA graphs have a relatively low $C$ (solid line with open circles) but still larger than ER graphs. Unfortunately, it is not possible to generate BA graphs of larger densities because they are grown out of an initial empty graphs of finite size $n_0$ and new vertices make a maximum of $m \leq n_0$ connections. An analytical estimation of its scaling with network size was found by Klemm and Eguíluz in [13]. In Fig. 3.14, analytical estimations for the clustering coefficients of the different models are summarized.

### 3.5.3 Other Structural Differences

As mentioned in Sect. 3.3.1, the degree distribution $p(k)$, although important, does not explain very much about the internal structural organization of a network. Random graphs of the same size, connection density and degree distribution may possess very distinct internal structure. While W-S graphs should conserve much of the regularity of their parent lattices, degrees of ER and BA networks are uncorrelated, as depicted in Fig. 3.4(a) (even if BA networks are generated by a "preferential attachment" rule forcing new vertices to preferentially link to those with highest degree). This absence of degree-degree correlations is not a general property of scale-free networks, but intrinsic of the BA model.

Another important structural feature of real complex networks, not captured by any of the models presented here, is the presence of *modules* and *hierarchies*. In many real networks, and specially in biological ones, the combination of modular and hierarchical organization is believed to be a very important consequence of self-organisation: elements specialized in similar function are arranged into modules that hierarchically interact with each other. This requires a complex topology supporting the very rich range of functional capacities exhibited by living organisms. Imagine genetic networks where genes involved in similar regulatory processes form clusters, or in the cat cortex (Fig. 9.2 of Chap. 9) where cortical areas performing similar function form the visual, auditory, somatosensory-motor and frontolimbic clusters observed. After recognizing its importance, Ravasz et al. [23] studied such structures in metabolic networks and presented a simple model that generates networks with both scale-free degree distribution and hierarchical/modular architecture.

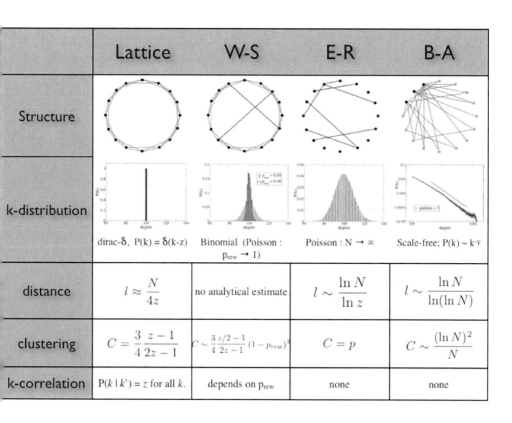

| | Lattice | W-S | E-R | B-A |
|---|---|---|---|---|
| Structure | | | | |
| k-distribution | dirac-$\delta$, P(k) = $\delta$(k-z) | Binomial (Poisson : $p_{rew} \to 1$) | Poisson : N $\to \infty$ | Scale-free; P(k) ~ k$^{-\gamma}$ |
| distance | $l \approx \dfrac{N}{4z}$ | no analytical estimate | $l \sim \dfrac{\ln N}{\ln z}$ | $l \sim \dfrac{\ln N}{\ln(\ln N)}$ |
| clustering | $C = \dfrac{3}{4}\dfrac{z-1}{2z-1}$ | $C \sim \dfrac{3}{4}\dfrac{z/2-1}{2z-1}(1-p_{rew})^3$ | $C = p$ | $C \sim \dfrac{(\ln N)^2}{N}$ |
| k-correlation | P(k \| k') = z for all k. | depends on $p_{rew}$ | none | none |

**Fig. 3.14.** Comparison of random graph models

## 3.6 Randomizing Networks and Comparison with Expected Properties

In Sect. 3.5, we compared random network models in terms of their scaling properties. In practice, however, scaling information is rarely available because usually we have one, and only one real network to be studied. In some cases (for example, the Internet or WWW), its structure might be known at different time stamps and thus its growing properties are available.

Then, how does one classify a real network? How expected, or surprising, are its properties and organization compared to those of random networks? Given any real network, a first approach is to generate a set of random ER networks of the same size $N$ and number of connections $M$ and compare their properties. But if the real network has a scale-free degree distribution, how representative is this comparison? A better approach is to generate a set of SF networks of the same size, number of connections and scaling factor $\gamma$ and again, compare. In such a case, generation of static SF networks is suggested, using the configuration model and similar methods like the one presented in [24]. These methods are computationally more efficient than the BA model and, very importantly, the scaling factor $\gamma$ is tunable while the BA model always returns $\gamma = 3$ exponent.

In reality, however, the degree distribution of real networks will rarely match that of any model, so, what should we expect the measures to be, given the degree distribution? By rewiring the connections, an ensemble of maximally random networks of the same size, density and degree sequence can be generated to be used as null hypotheses. The method is depicted in Fig. 3.15 and summarized as the following:

*Select two connections at random, say, $(a_1 \to b_1)$ and $(a_2 \to b_2)$, and switch them if and only if:*

1. *Neither the $(a_1 \to b_2)$ and $(a_2 \to b_1)$ links previously exist — otherwise double links would be introduced.*
2. *$b_2 \neq a_1$ and $b_1 \neq a_2$ exist — otherwise self-loops would be introduced.*

**Fig. 3.15.** Schematic representation of randomizing networks while conserving degree distribution. After randomly selecting two links, they are exchanged provided some restrictions are fulfilled. This procedure generates a maximally random network of same size, $N$, number of connections $m$ and degree-distributions $P(k_{\text{in}})$ and $P(k_{\text{out}})$ as its parent network

When applied to matrices, the rewiring procedure will also conserve input intensity *in-S* (but not *out-S*). Given a digraph in adjacency list form, the pseudo-code for the algorithm might read like:

```
counter = 0
```

```
while counter < m_rew :
    1) Select one node at random, a₁, and one of its neighbors, b₁.
    2) Select another vertex a₂.
       Check that a₂ ≠ a₁, a₂ ≠ b₁ and NOT a₂ → b₁, otherwise
       start again.
    3) Select a neighbor of a₂, b₂.
       Check that b₂ ≠ a₁ and NOT a₁ → b₂, otherwise start again.
    4) include b₂ in the list net (a₁)    Swap the connections
       include b₁ in the list net (a₂)
       remove b₁ from net (a₁)            delete the old links
       remove b₂ from net (a₂)
       counter += 1
```

where $m_{\text{rew}}$ is the maximum number of times the rewiring process will run (note that at each step two connections are rewired). The two links must be randomly chosen with uniform probability $1/M$ but the algorithm is taking first nodes $a_1$ and $a_2$ (with probability $1/N$) and then one of their neighbors. Individual links are thus selected with probability $p(b|a) = p(a)\,p(b) = \frac{1}{N}\frac{1}{out\text{-}k(a)}$. As a result, links of high degree nodes have a lower probability of being rewired. A very easy way to balance the situation is to initially generate a list of size $M$ containing each node $out\text{-}k(a)$ times (as in the configuration model) and select every time the nodes $a_1$ and $a_2$ occur in this list. The probability of selecting any link is now:

$$p(b\,|\,a) = p(a)\,p(b) = \frac{out\text{-}k(a)}{M}\frac{1}{out\text{-}k(a)} = \frac{1}{M}$$

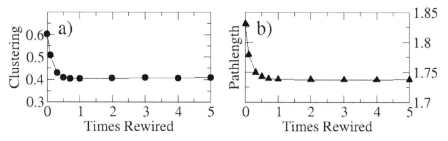

**Fig. 3.16.** Rewiring the cat cortical network towards generation of maximally random digraphs of same $N$, $M$ and degree distributions: **a)** Average clustering coefficient and **b)** average pathlength of the rewired networks. Each data point is the average after 100 realizations

An important question one faces when rewiring networks is how long the process should run so that we are sure of having really randomized networks. In this case, as shown in Figs. 3.16(a) and (b), $C$ and $l$ of the cat cortex become stable after rewiring $2\,M$ connections.

The rewiring process will destroy the internal structure of the network: degree-degree correlations, clustering, communities, hierarchies etc. while conserving a basic property like the degree distribution. However, it is also possible to generate maximally random networks with given degree correlations or clustering coefficients, etc. Different levels of approximation are available and it is up to the readers to decide which level of detail is enough in their specific case.

## 3.7 Classification of the Cat Cortical Network

The cortical networks of cat and macaque are known to have clustered organization as well as small-world properties: high clustering coefficient and short average pathlength [4,5,9]. In an effort to understand how cortical organization has evolved towards such structures, Tononi and Sporns found that similar topology could emerge from the balance between *functional integration* and *segregation* [25].

In Table 3.2 the clustering coefficient $C$ and average pathlength $l$ of the cat cortex and of different generated random networks ($N = 53$, $M = 826$) are presented. $C$ of cat cortex is much higher than that of ER random digraphs and thus, apparently very "surprising" based on what should be expected in a random network of its size and number of connections. $l$ is always very small due to the high density of connections ($\sim 0.3$). As pointed out in the previous section, we need to define a proper ensemble of random networks to be used as a "null hypothesis". Comparison to ER digraphs might be erroneous, as it is in this case. The degree distribution of the cat cortex in Fig. 3.17(b) shows no relation to any of the models introduced in this chapter. The distribution of ER networks is localized around the mean number of links per vertex $z$ ($z_{cat} \approx 16$) but the distribution of cat is very wide, some areas have degree up to $k = 35$ while $N$ is only 53. Using the rewiring procedure described in Sect. 3.6, a set of maximally random networks of the same size, number of connections *and* degree distribution was obtained. $C$ and $l$ of the cat cortex are still higher than those of rewired networks, but the differences are not so large as when compared to ER digraphs, and thus, not so "surprising".

$C$ and $l$ of the generated W-S networks are very close to those of the cat, suggesting that cat cortex is similar to a W-S network with $p_{rew}$ between 0.05 and 0.1. In Fig. 3.17(a), the cumulative degree distribution of the cat cortex is shown together with that of W-S networks ($-+$ line) and "scale-free" networks generated by the configuration model with $\gamma = 3.5$ (solid line). As expected, the distribution of the W-S digraphs is very narrow (decays very quickly) and thus, very unlikely to be a representative model for cortical networks. Even

**Table 3.2.** Comparison of average clustering and average shortest path between the cat cortical network and random networks of same size and number of connections. "Rewired Cat" also possesses the same degree sequence. Each value is the average over 10 realizations

|   | Cat cortex | ER random | Rewired Cat | W-S ($p_{rew} = 0.05$) | W-S ($p_{rew} = 0.1$) |
|---|---|---|---|---|---|
| $C$ | 0.62 | 0.30 | 0.41 | 0.63 | 0.57 |
| $l$ | 1.83 | 1.71 | 1.74 | 1.89 | 1.80 |

if the average properties of the W-S model closely reproduce those of the cat cortex, the internal structural organization is still very different. Besides, it is very difficult to imagine how cortical networks might have evolved as anything similar to a rewired regular lattice.

Wide range degree-distributions are typical of scale-free networks. Obviously, speaking of SF distribution in a network as small as 53 vertices is meaningless. However, the cumulative degree distribution of generated "SF networks" with $N = 53$ and $M = 826$ (solid line in Fig. 3.17(a)) closely follows the real distribution of the cat cortex. It is true that the existence of a SF distribution does not necessarily imply any mechanism of network construction, but SF properties usually emerge out of self-organized systems and the BA model presents an intuitive and likely evolutionary scenario. There is yet another observation suggesting that mammalian cortex has a SF nature: in [26], robustness of cortical networks is shown to behave like SF networks under random or selective attack.

Apart from its modular structure, mammalian cortical networks are also known to be hierarchically organized [6, 27]. Recently, various authors in this book found, by means of correlations in dynamical simulations, that functional connectivity of cat cortical network also follows a modular and

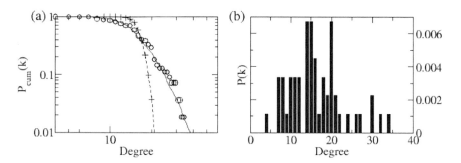

**Fig. 3.17.** Degree distribution of *in-k* of cat cortical network. (a) Cumulative degree distribution of cat cortex (○), generated W-S $p_{rew} = 0.3$ (−+ line) and scale-free $\gamma = 3.5$ (solid line) networks of same size and number of connections as the cat cortical network. Notice that W-S networks produce poor approximations; (b) Histogram of *in-k*

hierarchical pattern very close to that of its structural organization [8, 28]. Internal connections between areas within the same community (visual, auditory, somatosensory-motor and frontolimbic) are the first links being functionally expressed, while later the connections between areas in different communities are expressed. This separation unveils two hierarchical levels representing specialized information processing and integration of multisensorial information.

In the development of realistic models for the evolution of mammalian cortex, the idea of balancing functional integration and segregation proposed by Tononi and Sporns may be useful as a "macroscopic" driving force, yet more natural and detailed mechanisms need to be found to cause the system to self-organism into a modular and hierarchical structure. The broad degree distribution, suggesting a SF-type system, comes to the rescue since power laws are characteristic of self-organized systems. Although the model by Ravasz et al. [23], which generates hierarchical networks with a scale-free degree distribution by self-replication of structures, is an interesting starting point, it seems unlikely as a model for cortical evolution. A recipe for such an evolutionary model should at least reproduce the ingredients we have presented in this last section.

## 3.8 Further Reading

Readers looking for a first introduction to networks (or better said, graph theory) are encouraged to look for online resources. A web page called *An Interactive Introduction to Graph Theory* by Chris K. Caldwell provides an illustrative and educational introduction of basic concepts for the non-initiated (http://www.utm.edu/cgi-bin/caldwell/tutor/departments/math/graph/intro), while in his web page *Algorithmic Graph Theory*, Rashid Bin Muhammad provides a more concise introduction to mathematical graph theory: (http://www.personal.kent.edu/~rmuhamma/GraphTheory/graph Theory.htm)

Books on classical graph theory are usually difficult to read and far from our current interests in complex networks, but are a necessary reference if serious research on networks is to be done. Classical books treat the general case of directed graphs (digraphs) poorly, with the exception of a recent book by Jørgen Bang-Jensen and Gregory Gutin titled *Digraphs: Theory, Algorithms and Applications* [29].

There are some books dedicated to the '*new science*' of complex networks; however, we would recommend reading through some of the excellent reviews available. Newman [10] published the first major review dedicated to complex networks, their topology and random graph models. It lacks many of the current developments and challenges of complex networks research but it is still a "must" to anyone starting in the field. Two recent reviews [30, 31] come to the rescue and after going through structural properties of networks, they jump into dynamical processes occurring in networks. They also provide an overview of current fields where network theory has been applied, such as epidemiology, neuroscience, economics, etc. and summarize the challenges to be faced in each field.

Newman, Barabási and Watts recently published a compilation of selected papers which have been central in the development of complex networks theory [32].

# References

1. O. Sporns, G. Tononi, K. Kötter: *The human connectome: A structural description of the human brain.* PLoS. Comput. Biol. 1(4), 0245–0251, 2005.
2. R. Kötter, F.T. Sommer: *Global relationship between anatomical connectivity and activity propagation in the cerebral cortex.* Phil. Trans. R. Soc. Lond. B, 355 (1393), 127–134, 2000.
3. J. W. Scannell, C. Blakemore, M. P. Young: *Analysis of connectivity in the cat cerebral cortex.* J. Neurosc. 15(2), 1463–1483, 1995.
4. J. W. Scannell, G. A. P. C. Burns, C. C. Hilgetag, M. A. O'Neill, M. P. Young: *The connectional organization of the cortico-thalamic system of the cat.* Cer. Cortex 9(3), (277–299), 1999.
5. C. C. Hilgetag, G. A. P. C. Burns, M. A. O'Neill, J. W. Scannell, M. P. Young: *Anatomical connectivity defines the organization of clusters of cortical areas in the macaque monkey and the cat.* Phil. Trans. R. Soc. Lond. B 355, 91–110, 2000.
6. C.C. Hilgetag, M.A. O'Neill, M.P. Young: *Hierarchical organization of macaque and cat cortical sensory systems explored with a novel network processor.* Phil. Trans. R. Soc. Lond. B 355, 71–89, 2000.
7. C.C. Hilgetag, M. Kaiser: *Clustered organization of cortical connectivity.* Neuroinf. 2, 353–360, 2004.
8. C.S. Zhou, L. Zemanová, G. Zamora, C. C. Hilgetag, J. Kurths: *Hierarchical organization unveiled by functional connectivity in complex brain networks.* Phys. Rev. Lett. 97, 238103, 2006.
9. O. Sporns, J.D. Zwi: *The small world of the cerebral cortex.* Neuroinf. 2, 145–162, 2004.
10. M. E. J. Newman: *The structure and function of complex networks.* SIAM Review 45(2), 167–256, 2003.
11. R. Pastor-Satorras, A. Vázquez, A. Vespignani: *Dynamical and correlation properties of the Internet.* Phys. Rev. Lett. 87, 258701, 2001.
12. A. Barrat, M. Barthélemy, R. Pastor-Satorras, A. Vespignani: *The architecture of complex weighted networks.* PNAS 101(11), 3747–3752, 2004.
13. K. Klemm, V. M. Eguíluz: *Highly clustered scale-free networks.* Phys. Rev. E 65, 036123, 2002.
14. E. A. Bender, E. R. Canfield: *The asymptotic number of labelled graphs with given degree sequences.* J. Combin. Theory A 24 (3),296–307, 1978.
15. M. Molloy, B. Reed: *A critical point for random graphs with given degree sequence.* Random. Structures Algorithms 6, 161–179, 1995.
16. M. E. J. Newman, S. H. Strogatz, D. J. Watts: *Random graphs with arbitrary degree distributions and their applications.* Phys. Rev. E 64, 026118 2001.

17. A. Barrat, M. Weigt: *On the properties of small-world network models.* Eur. Phys. J. B 13, 547–560, 2000.
18. D. J. de S. Price, *Networks of scientific papers.* Science, 149, 510–515, 1965.
19. R. Cohen, S. Havlin, D. ben-Avraham. *Structural properties of scale-free networks.* In Handbook of Graphs and Networks. Editors S. Bornholdt, H. G. Schuster, Wisley-VCH, Berlin, ISBN: 3-527-40336-1, 2002.
20. P. Holme, B.J. Kim: *Growing scale-free networks with tunable clustering.* Phys Rev. E 65, 026107, 2002.
21. L. A. N. Amaral, A. Scala, M. Barthélemy, E. Stanley: *Classes of small-world networks.* PNAS 97(21), 11149–11152, 2000.
22. B. Bollobás, O. Riordan, J. Spencer, G. Tusnády: *The degree sequence of a scale-free random graph process.* Random Structures and Algorithms 18, 279–290, 2001.
23. E. Ravasz, A. L. Somera, D. A. Mongru, Z. N. Ottvai, A.L. Barabási: *Hierarchical organization of modularity in metabolic networks.* Science 297, 1551–1555, 2002.
24. K.I. Goh, B. Kahng, D. Kim: *Universal behaviour of load in scale-free networks.* Phys. Rev. Lett. 87(27), 278701, 2001.
25. O. Sporns, G. Tononi: *Classes of network connectivity and dynamics.* Complexity 7(1), 28–38, 2001.
26. M. Kaiser, R. Martin, P. Andras, M. P. Young: *Simulation of robustness against lesions of cortical networks.* Eur. J. Neurosc. 25, 3185–3192, 2007.
27. L. da F. Costa, O. Sporns: *Hierarchical features of large-scale cortical connectivity.* Eur. Phys. J B 48, 567–573, 2005.
28. L. Zemanová, C.S. Zhou, J. Kurths: *Structural and functional clusters of complex brain networks.* Physica D 224, 202212, 2006.
29. J. Bang-Jensen, G. Gutin: *Digraphs: Theory, Algorithms and Applications.* Springer-Verlag, London. Springer Monographies in Mathematics, ISBN: 1-85233-268-9, 2000.
30. S. Boccaletti, V. Latora, Y. Moreno, M. Chavez, D.-U. Hwang: *Complex networks: structure and dynamics.* Physics Reports 424, 175–308, 2006.
31. L. da F. Costa, F. A. Rodrigues, G. Travieso, P. R. V. Boas: *Characterization of Complex Networks: A Survey of measurements.* arXiv:cond-mat 0505185, 2005.
32. M. E. J. Newman, A-L. Barabási, D. J. Watts: *The Structure and Dynamics of Networks.* Princeton University Press, 2006. ISBN: 0691113572.

# 4

# Organization and Function of Complex Cortical Networks

Claus C. Hilgetag[1] and Marcus Kaiser[2]

[1] International University Bremen, School of Engineering and Science, Bremen,
Germany
C.Hilgetag@iu-bremen.de,
http://www.iu-bremen.de/schools/ses/chilgetag/
[2] School of Computing Science, Newcastle University, Newcastle upon Tyne,
United Kingdom
M.Kaiser@ncl.ac.uk,
http://www.biological-networks.org

**Summary.** This review gives a general overview of the organization of complex brain networks at the systems level, in particular in the cerebral cortex of the cat brain. We identify fundamental parameters of the structural organization of cortical networks, illustrate how these characteristics may arise during brain development and how they give rise to robustness of the cortical networks against damage. Moreover, we review potential implications of the structural organization of cortical networks for brain function.

## 4.1 Introduction

The network organization of the mammalian brain underlies its diverse stable and plastic functions. Experimental approaches from various directions have suggested that the specific organization of nerve fibre networks, particularly in the massively interconnected cerebral cortex of the brain, is closely linked to their function. However, the exact relationship between cortical network structure and brain function is still poorly understood, mainly due to the complexity of the available experimental data on brain networks, both with respect to their great volume and formidable intricacy. This complexity requires theoretical analyses as well as computational modeling to characterize the network organization and deduce functional implications of particular network parameters. In this review, we specifically focus on the structural and functional organization of brain networks, which are interconnecting neural elements at the large-scale systems level of the brain, and we use the cortico-cortical network of the cat as an example to illustrate different aspects of organization and function.

### 4.1.1 A Systems View of Brain Networks

The brain is a networked system of extraordinary complexity. Considering, for instance, the cellular organization of the human brain, one is faced with approximately $10^{10}$ single elements (e.g. [1], each of which on average is connected to more than 1,000 other elements [2]). To complicate matters, these elements are neither completely nor randomly connected, e.g. [3]. Fortunately, there is some regularity in the brain that might help to reduce the number of objects and interconnections that need to be considered. Neurons that possess similar connectional and functional features tend to group into large assemblies of several thousands to millions of cells that are regionally localized (forming cortical 'areas' or subcortical 'nuclei'). A first attempt to understand brain organization and function can, therefore, start with investigating the structure, connectivity and function of these assemblies of cells, rather than the complete cellular substance of the brain, e.g. [4]. It is this so-called systems level approach that we pursue here.

This review focuses on corticocortical connectivity at the systems level, specifically networks formed by long-range projections among cortical areas. There are two main reasons that motivate this approach. First, more extensive and reliable databases are currently available for systems level networks than for cellular neuronal circuits. There have been pioneering studies about the interconnections of different types of neurons at the level of cellular circuits, for instance, [3, 5–9]. However, detailed information about connectivity at the cellular level, based on systematic sampling of different cortical regions, is still largely missing. Second, systems connectivity data play an important role in many models of the brain (e.g. [4, 10–14]). These data have been used to derive conclusions about the global organization, development and evolution of the brain and to suggest modes of information processing. In particular, systems level connectivity may be responsible for important aspects of brain function, such as the neural activation patterns observed in functional imaging studies of perception and cognition (e.g. [15]), functional diversity and complexity [16], as well as other functional aspects reviewed here.

The systems level concept can be readily formalized and treated with the help of graph theoretical approaches, considering cortical areas as nodes and their interconnections by long-ranging nerve fibres as edges of directed or undirected graphs. In the theoretical systems level concept, areas and nuclei of the brain are well-defined, intrinsically uniform entities with sharp borders. It needs to be kept in mind, however, that experimental data present a more complicated picture. For instance, numerous, partly incongruent, mapping schemes exist for describing the parcellation of the cerebral cortex into different areas [17].

### 4.1.2 Types of Brain Connectivity

Different aspects of brain connectivity can be distinguished using the following, widely accepted classification [18]:

- Anatomical or *structural connectivity* denotes the set of physical connections linking neural units (cells and populations) at a given time. Structural connectivity data can range over multiple spatial scales, from area-intrinsic circuits to large-scale networks of inter-regional pathways. Anatomical connection patterns are relatively static over shorter time scales (seconds to minutes), but can be dynamic over longer time scales (hours to weeks); for example, during learning or development.

- *Functional connectivity* [19] captures patterns of statistical dependence between distributed neural units, which may be spatially remote, measuring their correlation/covariance, spectral coherence or phase-locking. Functional connectivity is time-dependent (typically using time series containing hundreds of milliseconds) and 'model-free', that is, it measures statistical interdependence (mutual information) without explicit reference to causal effects. Different methodologies for measuring brain activity may result in different statistical estimates of functional connectivity [20].

- *Effective connectivity* describes the set of causal effects one neural unit exerts over another [19]. Thus, unlike functional connectivity, effective connectivity is not 'model-free', but requires the specification of a causal model including structural connection parameters. Experimentally, effective connectivity can be inferred through network perturbations [21], or through the observation of the temporal ordering of neural events. Other measures estimating causal interactions can also be used (e.g. [22]). Effective connectivity is also time-dependent. Statistical interactions between brain regions change rapidly, reflecting the participation of varying subsets of brain regions and pathways in different cognitive tasks [23–26], behavioral or attentional states [24], and changes within the structural substrate related to learning [27].

Importantly, structural, functional and effective connectivity are mutually interrelated. Clearly, structural connectivity is an essential condition for the kinds of patterns of functional or effective connectivity that can be generated in a network. Structural inputs and outputs of a given cortical region, its connectional fingerprint [28], are major determinants of its functional properties. Conversely, functional interactions can contribute to the shaping of the underlying anatomical substrate [29], either directly through activity (covariance)-dependent synaptic modification, or, over longer time scales, through affecting an organism's perceptual, cognitive or behavioral capabilities, and thus its adaptation and survival.

### 4.1.3 A Case Study of Structural Connectivity: Inter-area Connections in the Cat Cerebral Cortex

The specific analyses described in this and other chapters in this volume are based on a global collation of cat cortical connectivity (892 interconnections of 55 areas) [30]. This collation of cat cortical data was developed from the data set described in [14] and forms part of a larger database of thalamocortical connectivity of the cat [31]. The database was created by the interpretation of a large number of reports of tract-tracing experiments from the anatomical literature. All tract-tracing experiments are based on the same general design. The investigated brain region of an anaesthetized animal is injected with a tracer substance (chemical or microrganismic), which is directly brought into the cells or taken up by neurons close to the injection site. The tracer, which is typically taken up at the axonal or dendritic branch endpoints, is then transported along the neuron's axon, *retrogradely* from the axonal terminals to the soma, or *anterogradely* in the opposite direction, or in both directions. After a method-dependent survival time, the animal is sacrificed and its brain is sectioned, histochemically processed, and analyzed under a microscope to show the distribution of transported tracer. Mathematically speaking, the experimental result of a tract-tracing experiment is a three-dimensional map of tracer concentration, contained in two-dimensional sections, at the moment of the animal's death. From the systematic analyses of the label distribution in the sections, in combination with a given parcellation of the cortical sheet into distinct areas, conclusions can be drawn about the specific interconnections of different areas by neural fiber projections. The invasive nature of these experiments explains why they cannot be applied to the human brain. As alternative non-invasive approaches with similar resolution and reliability are still missing, our knowledge about structural brain connectivity is largely restricted to non-human brains [32].

While the distribution of retrograde labelling may be quantified by systematic counts of labelled cells of projection origin (providing a numerical measure of the number of axons in a particular projection), the strength of anterograde label at a given location can often only be determined as an ordinal measure (e.g. 'sparse', 'moderate', 'dense'). The quantification of the number of projection origins in retrograde experiments is laborious as well; for this reason, the strength of fiber pathways is frequently only reported in ordinal terms (as for the specific cat data shown in Fig. 9.2 of Chap. 9: ones represent sparse, twos moderate and threes dense projections). It should be noted that the absent entries in a connectivity matrix can stand for connections that were investigated and were found to be absent, or potential projections that were not investigated in the first place. The potential impact of future additions to connectivity compilations needs to be carefully considered. However, previous simulations of connectivity matrices in which all entries with unknown information were assumed to exist did not result in principally different findings [33, 34].

In the remainder of this review, we discuss the organization of structural brain connectivity and its implication for the functions of the cerebral cortex, using as a particular example the complex network of interconnections between regions of the cerebral cortex of the cat. First, we explore the spatial layout of cortical networks in Sect. 4.2, and then their topological organization in Sect. 4.3, before presenting aspects of functional implications in Sect. 4.4. We conclude by summarizing the main conclusions and formulating open problems for future research.

## 4.2 Spatial Organization

The organization of neural systems is shaped by multiple constraints, ranging from limits placed by physical and chemical laws to diverse functional requirements. In particular, it is of interest to identify factors influencing the spatial layout of neural connectivity networks. Although spatial coordinates of cortical areas in the cat are not readily available at the moment, it may be expected that the spatial organization of the cat cortex is close to that in other mammals. In particular, we have analyzed detailed information about the spatial organization of the cerebral cortex in primates, such as the macaque monkey [35, 36].

One prominent guiding idea is that the establishment and maintenance of neural connections carries a significant metabolic cost that should be reduced wherever possible [37]. As a consequence, wiring length should be globally minimized in neural systems. A trend toward wiring minimization is apparent in the distributions of projection lengths for various neural systems, which show that most neuronal projections are short [2, 35, 36]. However, wiring length distributions also indicate a significant number of longer-distance projections, which are not formed between immediate neighbors in the network.

Alternatively, it has been suggested that wiring length reductions in neural systems are achieved not by minimal rewiring of projections within the networks, but by suitable spatial arrangement of the components. Under these circumstances, the connectivity patterns of neurons or regions remain unchanged, maintaining their structural and functional connectivity, but the layout of components is perfected such that it leads to the most economical wiring. In the sense of this 'component placement optimization' (CPO) [37], any rearrangement of the position of neural components, while keeping their connections unchanged, would lead to an increase of total wiring length in the network.

Using extensive connectivity datasets for systems and cellular neural networks combined with spatial coordinates for the network nodes, we found that optimized component rearrangements could substantially reduce total wiring length in all tested neural networks [36]. Specifically, total wiring length between 95 primate (macaque) cortical areas could be decreased by 32%, and wiring of neuronal networks in the nematode *Caenorhabditis elegans* could be shortened by 48% on the global level, and by 49% for neurons within frontal ganglia. The wiring length distribution before and after optimization as well as the reduction for the macaque cortical connectivity are shown in Fig. 4.1. Wiring length reductions were possible due to the existence of long-distance

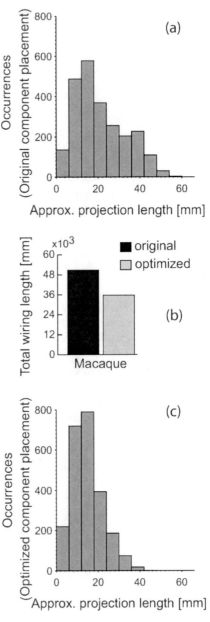

**Fig. 4.1.** Projection length distribution and total wiring length for original and optimally rearranged neural networks: (**a**) Approximated projection length distribution in the macaque monkey cortical connectivity network with 95 areas and 2,402 projections; (**b**) Reduction in total wiring length in rearranged layouts yielded by simulated annealing; (**c**) Approximated projection length distribution in neural networks with optimized component placement. The number of long distance connections is substantially reduced compared to the original length distribution in (**a**)

projections in the neural networks. Thus, biological neural networks feature shorter average pathlengths than networks lacking long-distance connections. Moreover, the average shortest pathlengths of neural networks, corresponding to the average number of processing steps along the shortest signalling paths, were close to pathlengths in networks optimized for minimal paths.

Minimizing average pathlength — that is, reducing the number of intermediate transmission steps in neural integration pathways — has several functional advantages. First, the number of intermediate nodes that may introduce interfering signals and noise is limited. Second, by reducing transmission delays from intermediate connections, the speed of signal processing and, ultimately, behavioral decisions is increased. Third, long-distance connections enable neighboring as well as distant regions to receive activation nearly simultaneously [35, 38] and thus facilitate synchronous information processing in the system (compare [39]). Fourth, the structural and functional robustness of neural systems increases when processing pathways (chains of nodes) are shorter. Each further node introduces an additional probability that the signal is not transmitted, which may be substantial (e.g. failure rates for transmitter release in individual synapses are between 50% and 90% [40]). Even when the signal survives, longer chains of transmission may lead to an increased loss of information. A similar conclusion, on computational grounds, was first drawn by John von Neumann [41] when he compared the organization of computers and brains. He argued that, due to the low precision of individual processing steps in the brain, the number of steps leading to the result of a calculation ('logical depth') should be reduced, and highly parallel computing would be necessary.

## 4.3 Topologic Organization

Various approaches can be used to investigate the topology of cortical networks, at the local node-based level (e.g. Koetter & Stephan, [42]), for small circuits of connected nodes (also called network motifs), or at the global level of the network. Several of these approaches are summarized in [43, 44] and are also reviewed elsewhere in this volume (Chap. 3). Moreover, generic approaches for the investigation of complex networks may also be applied to brain connectivity [45]. Methodologically, analyses have used either techniques from graph theory, or multivariate methods using clustering or scaling techniques to extract statistical structure.

All studies of cortical structural connectivity have confirmed that cerebral cortical areas in mammalian brains are neither completely connected with each other nor randomly linked; instead, their interconnections show a specific and characteristic organization. Various global connectivity features of cortical networks have been described and characterized with the help of multivariate analysis techniques, such as multidimensional scaling or hierarchical cluster analysis [43]. For example, clusters or *streams* of visual cortical areas have

(a)

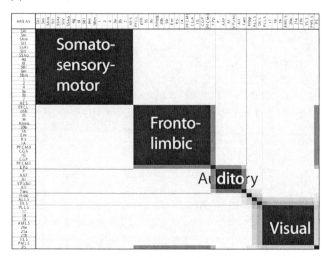

(b)

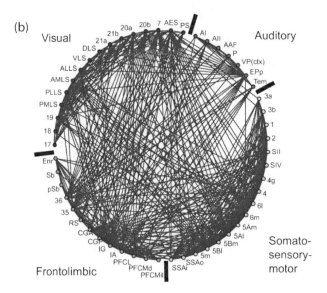

**Fig. 4.2.** Clustered organization of cat cortical connectivity: (**a**) Cluster count plot, indicating the relative frequency with which any two areas appeared in the same cluster, computed by stochastic optimization of a network clustering cost function [30]. Functional labels were assigned to the clusters based on the predominant functional specialization of areas within them, as indicated by the physiologic literature; (**b**) Cat cortical areas are arranged on a circle in such a way that areas with similar incoming and outgoing connections are spatially close. The ordering by structural similarity is related to the functional classification of the nodes, which was assigned as in (**a**)

been identified that are known to be segregated functionally [46] as well as in terms of their inputs, outputs and mutual interconnections [12]. Topological sequences of areas can be distinguished that might provide the layout for signaling pathways across cortical networks [47]. Alternatively, hierarchies of cortical areas can be constructed, based on the laminar origin and termination patterns of interconnections [48, 49].

Significantly, all large-scale cortical connection patterns examined to date, including global connectivity in cat and monkey brains as well as their subdivisions, exhibit small-world attributes with short pathlengths and high clustering coefficients [16, 30, 44]. These properties are also found in intermediate-scale connection patterns generated by probabilistic connection rules, taking into account metric distance between neuronal units [50]. The findings suggest that high clustering and short pathlengths can be found across multiple spatial scales of cortical organization.

To identify the clusters that are indicated by the high clustering coefficients of cortical networks, a computational approach based on evolutionary optimization can be used [30]. This stochastic optimization method delineated a small number of distinctive clusters in global cortical networks of cat and macaque [51] (Fig. 4.2). These clusters contained areas which were more frequently linked with each other than with areas in the remainder of the network, and the clusters followed functional subdivisions (e.g., containing predominantly visual or somatosensory-motor areas), as discussed in Sect. 4.6 below. The algorithm could also be tuned to identify clusters that no longer contained any known absent connections, and thus produced maximally dense clusters of areas, interpretable as network 'building blocks'.

A clustered organization of cortical networks was also indicated by application of the matching index, demonstrating distinct groups of cortical areas with similar input and output [43]. The matching index captures the pairwise similarity of areas in terms of their specific afferents and efferents from other parts of the network [43, 44] (as well as in Chap. 3), following one of the central assumptions of systems neuroscience that the functional roles of brain regions are specified by their inputs and outputs. In agreement with this concept, one finds that pairs of areas with high matching index also share functional properties [43].

The ubiquitous feature of a segregation of cortical networks into multiple, interconnected network clusters (or 'communities' in network analysis parlance) is explored from various perspectives in the following sections.

## 4.4 Network Development

The previously reviewed spatial and topologic analyses demonstrate that cortical networks are predominantly, but not exclusively, connected by short projections, and that cortical connections link areas into densely connected clusters. How does such an organization arise during the development of the brain?

It is known that neural systems on several scales show distance dependence for the establishment of projections. In systems ranging from connectivity between individual neurons in *Caenorhabditis elegans* [36] and rat visual cortex [52] to connectivity between mouse [2,53], macaque [35] (see Fig. 4.1(a)) or human [54] cortical areas, there is a higher tendency to establish short-distance than long-distance connections. This feature can be incorporated into models of cortical connection development.

### 4.4.1 Models for Network Development

Whereas various models exist for network development or evolution, only few of them consider spatial constraints. The standard model for generating scale-free networks, for example, uses growth and preferential attachment [55]. Starting with $m_0$ initial nodes, a new node establishes a connection with an existing node $i$ with the probability

$$P_i = \frac{k_i}{\sum k},$$

that is, the number of edges of node $i$ ($k_i$) divided by the total number of edges already established in the network. The resulting network consists of one cluster, and the degree distribution shows a power law. This model may result in degree distributions comparable with real biological networks but lacks multiple clusters or modules.

In order to generate scale-free networks with a modular organization, a hierarchical model for network development was designed [56, 57]. Starting with one root node, during each step two units are added that are identical to the network generated in the previous iteration. Then the bottom nodes of these two units are linked with the root of the network. While this algorithm resulted in a modular organization, it did not support a large variety of module sizes within the same network, as seen in real-world networks.

Waxman [58] proposed a connection establishment algorithm for the Internet in which the probability of a connection between two nodes decays exponentially with the spatial distance between them. In this way, the high costs for the wiring and maintenance of long-range connections can be represented. Initially, the nodes are distributed at random. Thereafter, edges are attached to the graph. The probability that an edge is established between two nodes decays exponentially with the distance between them. In contrast to the previous models, the location of the nodes is determined from the start, therefore, there was no growth in terms of the size of the network or the number of nodes.

A decay with distance, however, is also likely for growing and expanding biological systems, as the concentration of a chemical substance such as a growth factor decays with the distance from the place of production or emission [59]. We therefore explored different mechanisms for spatial and topological network development through computational modeling, considering (i) a simple dependence of projection formation on spatial distance and enclosing spatial borders, and (ii) dependence on distance as well as on developmental time windows. We also compared the algorithms with previously suggested topological mechanisms for network development.

### 4.4.2 Growth Depending on Spatial Distance

We simulated mechanisms of spatial growth, in such a way that connections between nearby nodes (i.e. areas) in the cortical network were more probable than projections to spatially distant nodes [58]. Such a distribution could, for instance, result from the concentration of unspecific factors for axon guidance decaying exponentially with the distance to the source [59].

At each step of the algorithm, a new area was added to the network until reaching the target number of nodes (55 areas for simulated cat cortical networks). New areas were generated at randomly chosen positions of the embedding space. The probability for establishing a connection between a new area $u$ and existing areas $v$ was set as

$$P(u,v) = \beta\, e^{-\alpha\, d(u,v)} , \qquad (4.1)$$

with $d(u,v)$ being the distance between the nodes and $\alpha$ and $\beta$ being scaling coefficients. Areas that did not establish connections were disregarded. A more detailed presentation of the network growth model is given elsewhere [60,61].

We generated 50 networks of the size of the cat cortical network, through limited spatial growth in a fixed modeling space, and using parameters $\alpha = 5$ and $\beta = 2.5$. The spatial limits imposed during the simulations might represent internal restrictions of growth (e.g. by apoptosis [62]) as well as external factors (e.g. skull borders). The simulated networks yielded clustering coefficients and averaged shortest pathlength (ASP; see Chap. 3), shown in Table 4.1, similar to the cortical network. Moreover, the degree distribution of cortical and simulated limited growth networks showed a significant correlation (Spearman's rank correlation $\rho = 0.77$, $P < 3 \times 10^{-3}$).

Note that a small-world topology with similar ASP and clustering coefficient as in the biological networks could only be generated for limited spatial growth where the growing network quickly reached the borders of the embedding space. For unlimited growth, the ASP was much larger whereas the clustering coefficient was much lower than for the original cortical network.

**Table 4.1.** Comparison of cat cortical and simulated networks. Shown are the clustering coefficient $C_{\text{brain}}$ and $ASP_{\text{brain}}$ of the cat network as well as the average clustering coefficient and ASP of 50 generated limited and unlimited spatial growth networks with respective standard deviations

| | $C_{\text{brain}}$ | $C_{\text{limited}}$ | $C_{\text{unlimited}}$ | $ASP_{\text{brain}}$ | $ASP_{\text{limited}}$ | $ASP_{\text{unlimited}}$ |
|---|---|---|---|---|---|---|
| cat | 0.55 | $0.50 \pm 0.02$ | $0.29 \pm 0.05$ | 1.8 | $1.70 \pm 0.04$ | $3.86 \pm 0.47$ |

This is in accordance with experimental findings in which the lack of growth limits given by prohibiting apoptosis [62] resulted in a different layer architecture and network topology.

The presented spatial growth model proceeds independently of network activity. Such an approach is supported by experimental studies that show that activity is not necessary for the establishment of global connectivity. For example, after blocking neurotransmitter release and thus activity propagation during development, the global connectivity pattern of the brain remained unchanged [63].

We also investigated an alternative growth model, using a developmental mechanism of growth and preferential attachment, in which new nodes were more likely to establish links to existing nodes that already had many connections [55]. This model was also able to yield density and clustering coefficients similar to those in cortical networks. However, it failed to generate multiple clusters seen in the biological systems, as only one main cluster could be generated by this approach.

### 4.4.3 Growth Depending on Distance and Developmental Time Windows

Whereas the spatial growth model described above could replicate the small-world topology of the cat cortical network, there was no guarantee that multiple network clusters, as found in the cortical connectivity of the mammalian brain, would arise. Moreover, in cases where multiple clusters did occur, their size could not be controlled by the model parameters.

In order to explore the essential multiple-cluster feature of cortical connectivity, we modified the previous model and included one additional factor of cortical development, the formation of cortical areas and their interconnections during specific, overlapping *time windows*. Time windows arise during cortical development [64,65], as the formation of many cortical areas overlaps in time but ends at different time points, with highly differentiated sensory areas (for example, Brodmann area 17) finishing last. Based on this experimental finding, we explored a modified wiring rule in which network nodes were more likely to connect if they were (i) spatially close and (ii) developed during the same time window.

The following algorithm was used for network growth depending on distance combined with time windows (cf. Fig. 4.3(a)). First, three seed nodes were placed at spatially distant locations (cf. Fig. 4.3(b)). New nodes were placed randomly in space. The time window of a newly forming node was the same as that of the nearest seed node, as it was assumed to originate from, or co-develop with, that node. Second, the new node $u$ established a connection with an existing node $v$ with probability $P(u,v) = P_{temp}(u) \times P_{temp}(v) \times P_{dist}(u,v)$. The dependence $P_{dist}$ decayed exponentially with the distance between the two nodes (cf. [60]). Third, if the newly formed node failed to establish connections, it was removed from the network.

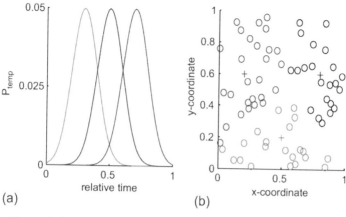

(a)                                    (b)

**Fig. 4.3.** Time windows and initial seed nodes: (a) Temporal dependence $P_{temp}$ of projection establishment depending on node domain. Relative time was normalized such that '0' stands for the beginning of development and '1' for the end of network growth. The three seed nodes had different time windows, which were partially overlapping; (b) Two-dimensional projection of the 73 three-dimensional node positions. The gray level coding represents the time window corresponding to one of the three seed nodes (+)

Although the following results show networks with 73 nodes, comparable to the primate networks, networks similar to the cat cortical network also could be generated (not shown). The timed adjacency matrix shows the development of connections over time (Fig. 4.4(a)). Different gray levels represent the respective time windows of the nodes. The reordered matrix represents the original network

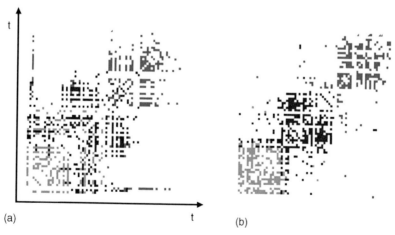

(a)                             t        (b)

**Fig. 4.4.** (a) Timed adjacency matrix (the first nodes are in the left lower corner); (b) Clustered adjacency matrix. The matrix is the same as in (a), but nodes with similar connections are arranged adjacent in the node ordering

with different node order, in such a way that nodes with similar connectivity were placed nearby in the adjacency matrix(Fig. 4.4(b)).

Therefore, the inclusion of developmental time windows into the spatial growth algorithm generated multiple network clusters, with their number being identical to the number of different time windows that governed development. In addition to the number of clusters, the size of clusters could be varied by changing the width of the corresponding time window.

These results demonstrate that simple mechanisms of spatial growth, in combination with constraints by spatial borders or developmental time windows, can account for many of the structural features of corticocortical connectivity, in particular the formation of multiple network clusters.

## 4.5 Network Robustness

The brain can be remarkably robust against physical damage. Significant loss of neural tissue may be compensated in a relatively short time by large-scale adaptation of the remaining brain parts (e.g. [66–68]. For example, there has been a case in which almost an entire hemisphere was removed from an 11 year-old boy who had medically intractable seizures. After three years of rehabilitation training in the hospital, however, the patient was left with few remaining functional deficits [69]. Similarly, on a more local scale for damage within specific brain regions, Parkinson's disease only becomes apparent after half of the pigmented cells in the affected substantia nigra are lost [70]. In other cases, however, the removal of small amounts of tissue (e.g. in regions specialized for language functions) can lead to severe functional deficits. These findings provide a somewhat contradictory picture of the robustness of the brain and highlight several questions. Can we formally evaluate robustness given the variability in the effects of brain lesions? Are severity and nature of the effects of localized damage predictable? And finally, how can robustness against the loss of large amounts of tissue be explained?

In the following sections, we explore network robustness through the effects of lesions of structural network components. Lesions can affect nodes or edges, and can be applied either randomly or in a targeted way, in which case specific components are eliminated by target criteria which assess the perceived importance of the components.

### 4.5.1 Impact of Node Lesions

Following Barabási and Albert [71], we assessed the impact of lesions on network connectivity and integrity by measuring the average shortest path (ASP) or characteristic pathlength. As described in Chap. 3, the ASP between any two nodes in the network is the number of sequential connections that are required, on average, to link one node to another by the shortest possible route [72]. In case a network becomes disconnected in the process of removing edges or nodes, and no path exists between two particular nodes, this pair of nodes is ignored. If no connected nodes remain, the average shortest path is set to zero.

In two separate approaches for sequential node removal, we removed nodes either randomly or by targeted elimination. During random removal, nodes were selected randomly, with a uniform probability distribution, and deleted from the graph. In the case of targeted removal, the nodes with the highest node degree were subsequently eliminated. After each deletion, the ASP of the resulting graph was calculated, and the removal of nodes was continued until all nodes were removed from the network.

To provide benchmarks for comparisons, random, small-world and scale-free networks of the same size as the cat network were created and lesioned in an analogous way. We used rewiring to generate small-world networks [73] and a modified version of growth and preferential attachment for scale-free networks [55]. Fifty benchmark networks were created for each of the conditions. Moreover, the process of random removal of nodes in the cat cortical networks was repeated fifty times.

For the cat brain network (Fig. 4.5), the random and small-world benchmark networks show a different behavior for targeted node removal when compared to the cortical network. The cat network's response to targeted node removal is largely within the 95% confidence interval for the scale-free benchmark networks; however, the peak ASP value and the fraction of deleted nodes where the peak occurs are comparatively lower for the cat cortical network. Thus, in terms of random and targeted node lesion behavior, the cortical networks most closely resemble scale-free networks.

The decline in ASP at a later stage during the elimination process, as observed for the brain and scale-free networks, deserves special attention. It can be for two reasons. First, it could be that the network becomes fragmented into different disconnected components. Each of these is smaller, and likely to have a shorter ASP. Second, the overall decrease in network size with successive eliminations can lead to a decrease in shortest path. This is, however, likely to be a slow process, as it will usually be offset by an increase in ASP due to the targeted nature of the elimination.

In conclusion, this shows that structural properties of cortical networks are quite robust towards the random elimination of nodes from the network. In contrast, the targeted removal of nodes, by removing the most highly-connected nodes first, leads to a rapid fragmentation of the network. Therefore, cortical networks are similar to scale free networks in their response to the random or targeted removal of nodes. Indeed, as for scale-free networks, brain networks contain nodes which are almost connected to all other nodes of the system. Examples of such 'hubs' for the cat would be amygdala and hippocampus in the subcortical domain and anterior ectosylvian sulcus (AES), agranular insula (Ia), and area 7 for cortical regions.

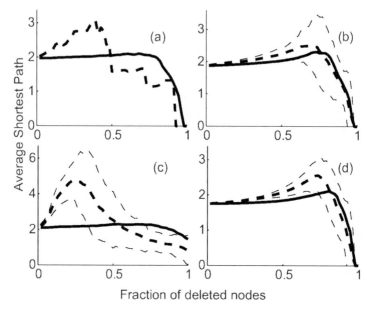

**Fig. 4.5.** Sequential node elimination in cat cortical and benchmark networks. The fraction of deleted nodes (zero for the intact network) is plotted against the average shortest path (ASP) after node removals. Nodes were removed randomly, or starting with the most highly connected nodes (targeted elimination): (**a**) Cat cortical network during targeted (*dashed*) and random (*solid line*) elimination. In the subsequent plots (**b**), (**c**) and (**d**), the dashed line shows the average effect of targeted elimination and the thin dashed lines the 95% confidence interval for the generated same-size benchmark networks. The solid line represents the average effect of random elimination; (**b**) Small-world benchmark network; (**c**) Scale-free benchmark network; (**d**) Random benchmark network

### 4.5.2 Impact of Edge Lesions

In many networks, the failure of single connections may be more likely than the extinction of entire nodes. We tested several measures for identifying vulnerable edges and compared their prediction performance for the cat cortical network. Among the tested measures, edge frequency in all shortest paths of a network yielded a particularly high correlation with vulnerability, and identified inter-cluster connections in biological networks [35].

### Measures for Predicting Edge Vulnerability

We tested four candidate measures for predicting vulnerable edges in networks. First, the product of the degrees (PD) of adjacent nodes was calculated for each edge. A high PD indicates connections between two hubs which may represent potentially important network links. Second, the absolute difference in the adjacent node degrees (DD) of all edges was inspected. A large degree difference signifies connections between hubs and more sparsely connected network regions which may be important for linking central with peripheral regions of a network. Third, the matching index (MI) [43] was calculated as the number of matching incoming and outgoing connections of the two nodes adjacent to an edge, divided by the total number of the nodes' connections (excluding direct connections between the nodes [44]). A low MI identifies connections between very dissimilar network nodes which might represent important 'short cuts' between remote components of the network. Finally, edge frequency (EF), a measure similar to 'edge betweenness' [74, 75], indicates how many times a particular edge appears in all pairs shortest paths of the network. This measure focuses on connections that may have an impact on the characteristic path length by their presence in many individual shortest paths [35].

**Prediction Performance**

In the present calculation, both increase and decrease of ASP indicate an impairment of the network structure. Therefore, we took the deviation from the ASP of the intact network as a measure for structural impairments. We evaluated the correlation between the size of the prediction measures and the damage (shown in Table 4.2). While most of the local measures exhibited good correlation with ASP impact in real-world networks, the highest correlation was consistently reached by the EF measure. Also, the measures of matching index and difference of degrees show a high correlation.

After identifying a measure to predict *which* edges were most vulnerable, we looked at *where* in the network these edges resided. We generated 20 test networks; each consisting of three randomly wired clusters and six fixed inter-cluster connections (Fig. 4.6(a)). The inter-cluster connections (light gray) occurred in many shortest paths (Fig. 4.6(b)) leading to an assignment of the highest EF value, as no alternative paths of the same length were available. Furthermore, their elimination resulted in the greatest network damage as shown by increased ASP.

**Table 4.2.** Density, clustering coefficient CC, average shortest path ASP and correlation coefficients $r$ for different vulnerability predictors of the analyzed networks (the index refers to the number of nodes). Tested prediction measures were the product of degrees (PD), absolute difference of degrees (DD), matching index (MI), and edge frequency (EF)

| | Density | CC | ASP | $\vert$ $r_{PD}$ | $r_{DD}$ | $r_{MI}$ | $r_{EF}$ |
|---|---|---|---|---|---|---|---|
| Cat$_{55}$ | 0.30 | 0.55 | 1.8 | $\vert$ 0.08* | 0.48** | −0.34** | 0.77** |

* Significant Pearson Correlation, 2-tailed 0.05 level.
** Significant Pearson Correlation, 2-tailed 0.01 level.

(a)                                    (b)

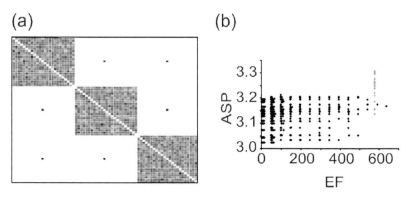

**Fig. 4.6.** Edges with high edge frequency form essential inter-cluster connections:
(a) Connectivity of test networks with three clusters and six pre-defined inter-cluster
connections. The network has comparable density as primate brain connectivity. The
gray-level shading of a connection in the adjacency matrix indicates the relative
frequency of an edge in 20 generated networks. White entries stand for edges absent
in all networks; (b) Edge frequencies in the all-pairs shortest paths against ASP
after elimination of edges. Light gray data points represent the values for the inter-
cluster connections in all 20 test networks. Inter-cluster connections not only have
the largest edge frequency, but also cause most damage after elimination

### 4.5.3 Conclusions of Network Lesion Studies

Cortical systems are known to consist of several distinct, linked clusters with a
higher frequency of connection within than between the clusters (e.g. Fig. 4.2).
Inter-cluster connections have also been considered important in the context of
social contact networks, as 'weak ties' between individuals [76] and separators
of communities [74]. We, therefore, speculated that connections between clusters might be generally important for predicting vulnerability. Whereas many
alternative pathways exist for edges within clusters, alternative pathways for
edges between clusters can be considerably longer. Interestingly, previously
suggested growth mechanisms for scale-free networks, such as preferential
attachment [55], or strategies for generating hierarchical networks [56] did
not produce distributed, interlinked clusters. Consequently, the low predictive value of EF in the scale-free benchmark networks was attributable to
the fact that scale-free networks grown by preferential attachment consisted
of one central cluster, but did not possess a multi-cluster organization. This
suggests that alternative developmental models may be required to reproduce
the specific organization of biological networks (see Sect. 4.4 above).

An analysis of the similarity between the cat network and random, rewired,
small-world, and scale-free benchmark networks shows that cat cortical connectivity is most similar to scale-free networks in several respects. Most importantly, cortical as well as scale-free networks show a huge disparity between
random elimination of nodes or edges and targeted elimination in which the
most vulnerable parts of the network are removed. Similarly, an analysis of

edge vulnerability has shown that the targeted removal of connections between clusters has a large effect compared to the removal of connections within a cluster.

Is the brain optimized for robustness against random removal of nodes or edges? Although this question is highly debatable, there are several arguments why the robust architecture could be explained as a side effect of other functional constraints. For example, the formation of functional clusters is necessary to exclude signals with different modality and highly-connected areas could function as integrators of (multi-modal) information or spreaders of information to multiple clusters or many nodes within one cluster. Therefore, in the future it will be interesting to compare the function of areas with their number of projections and the function of their directly connected neighbors.

The effect of the structural network organization on the functional impact after lesions is also important for neurological lesion analysis [68]. The degree of connectedness of neural structures can affect the functional impact of local and remote network lesions, and this property might also be an important factor for inferring the function of individual regions from lesion-induced performance changes [21].

## 4.6 Functional Implications

Several experimental studies have demonstrated a close link between the organization of structural brain connectivity and functional connectivity. The analysis of neuronographic connection data for the monkey and cat cortex, for example, has revealed dense interconnections among visual, auditory and particularly somatosensory-motor areas, arranged in similar network clusters as those formed by structural connectivity [77]. Neuronographic data are produced by the disinhibition of local populations of cortical neurons through the application of strychnine, and by recording the resulting steady-state activation of remote areas. Thus, this kind of approach reveals eleptiform functional connectivity , which also can be described by a simple propagation model [78].

Similar findings were obtained for another type of functional connectivity independent from particular tasks or stimuli, the slow-frequency coupling among cortical areas in human fMRI resting state data [79]. These resting-state networks are very similar to the structural connectivity from the cat or monkey, in that most interactions proceed only across short distances and run in local clusters, which follow regional and functional subdivisions of the brain [80]. However, some functional interactions also exist across longer distances, particularly those between homotopic regions in the two hemispheres. However, the actual neural or metabolic mechanisms underlying resting-state coupling are still poorly understood. Thus, this type of connectivity currently needs to be interpreted with caution.

In the next subsections, we review particular functional implications of the clustered organization of cortical networks, based on various computational approaches.

### 4.6.1 Functional Motif Diversity

The approach of motif analysis [81] can be adapted to draw conclusions on functional diversity in cortical networks, based on an analysis of structural connectivity. This approach starts from a simple premise: areas need to be connected in order to interact, and the kind and number of different local circuits that can be formed within a given structural network may reflect the diversity of functions performed by this network. This idea was applied to a motif analysis of corticocortical connectivity, by distinguishing between structural and functional motifs. Structural motifs were defined in the conventional sense, as the set of different connection patterns involving $n = 2, 3, 4 \ldots$ nodes found in the given network, while functional motifs represented all possible subsets of identified structural motifs. By comparison with benchmark networks and through network evolution, Sporns and Koetter [82] found that, in cortical networks, the number of structural motifs is small, while the number of functional motifs is large, suggesting that cortical networks are organized as to achieve high functional diversity with a small number of different structural elements. However, this finding may be partly due to the global organization of the cortical networks. Since the studied networks are organized into densely connected clusters, the motif analysis resulted in a small number of structural motifs which are also completely, or almost completely, connected. Naturally, these structural motifs allow a large number of different functional submotives. Moreover, motifs are difficult to interpret in terms of building blocks of development or function. For example, current knowledge about the development or evolution of cortical networks makes it appear unlikely that brain networks develop by adding circuits of three or four areas. Moreover, the functional interactions within such ensembles may be too complex to represent a truly basic unit of cortical functioning.

### 4.6.2 Functional Complexity

Central aspects of cortical functioning are provided by the structural and functional specialization of cortical regions on the one hand, and their integration in distributed networks, on the other. This relationship has been formalized as an expression of functional complexity, with complexity being defined as

$$C(X) = H(X) - \sum_i H(\boldsymbol{x}_i | X - \boldsymbol{x}_i), \tag{4.2}$$

where $H(X)$ is the entropy of the system and the second term on the right-hand side of the equation denotes the conditional entropy of each element, given the entropy of the rest of the system [16]. Thus, this measure describes the balance between the functional independence of a system's elements and their functional integration.

The complexity of cortical networks, as defined in (4.2), was explored through a simple functional model of the network, created by injecting the areas with white noise and investigating their functional coupling. Random rewiring of the actual cortical networks reduced their functional complexity, while graph evolution and selection for maximum complexity tended to produce small-world networks with a similar organization of multiple interconnected clusters as seen in the actual networks. Thus, the clustered organization of the networks appears to be well suited to support functional complexity, by facilitating high integration within clusters, yet high independence between clusters.

### 4.6.3 Critical Range of Functional Activations

In complex neural networks stable activation patterns within a critical functional range are required to allow function to be represented (cf. Chap. 7). In this critical range, the activity of neural populations in the network persists, falling between the extremes of quickly dying out or activating the whole network. We used a basic percolation model to investigate how functional activation spreads through a cortical network which has a clustered organization across several levels of organization. Specifically, cortical networks not only form clusters at the level of connected areas, but neural populations within areas are also more strongly connected with each other than they are with neurons in other cortical areas. Similar clustering can be observed at the even finer levels of hypercolumns and columns [83] (see Chap. 8). Our simulations demonstrated that hierarchical cluster networks were more easily activated than random networks, and that persistent and scalable activation patterns could be produced in hierarchically clustered networks, but not in random networks of the same size. This was due to the higher density of connections within the clusters facilitating local activation, in combination with the sparser connectivity between clusters which hindered the spreading of activity to the whole network. The critical range in the hierarchical networks was also larger than that in simple, same-sized small-world networks (Fig. 4.7). A detailed description of these results will be published elsewhere. The findings indicate that a hierarchical cluster architecture may provide the structural basis for the stable functional patterns observed in cortical networks.

While the reviewed results give ground for optimism that the structural organization of cortical networks provides the basis for their diverse, complex and stable functions, the actual mechanisms linking structural and functional connectivity are still speculative. Further computational modeling may be helpful in exploring some of the potential mechanisms, in particular by taking into account different types of organization at the level of area-intrinsic connectivity.

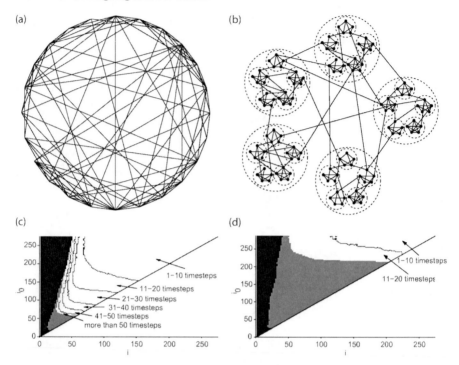

**Fig. 4.7.** Functional criticality in small-world and hierarchical cluster networks: Schematic views of (**a**) a small-world network and; (**b**) a hierarchical multiple-cluster network; (**c**) Critical range (gray) of persistent activity for small-world and; (**d**) hierarchical networks. In the critical parameter range, activity neither dies out nor spreads through the whole network

## 4.7 Conclusions and Open Questions

A number of preliminary conclusions can be drawn from the findings presented here. First, cerebral cortical fibre networks balance short overall wiring with short processing paths. This combination of desirable spatial and topologic network features results in high functional efficiency, providing a reduction of conduction delays along fibre tracts in combination with a reduction of transmission delays at node relays. Second, a central aspect of the topological organization of cortical networks is their segregation into distributed and interconnected multiple clusters of areas. Such a modular organization can also be observed at smaller levels of the cortical architecture, for instance, in densely intraconnected cortical columns. Experimental physiologic studies have demonstrated that the area clusters correspond to functional subdivisions of the cerebral cortex, suggesting a close relationship between global connectivity and function. Third, biologically plausible growth mechanisms for spatial network development can be implemented in simple computational models. Networks resulting from the simulations show the observed distance distribution of cortical connectivity as well its multi-cluster network organization. Fourth, the clustered organization of cortical connectivity may also be the reason for the robustness of cortical networks against damage, in particular for randomly inflicted impairments. In their response to damage and attack on nodes, cortical networks show behavior similar to scale-free networks. Due to this feature, they are robust to random lesions of nodes, but react critically to the removal of highly-connected nodes. Finally, the outlined structural organization of cortical networks has various functional implications: networks of hierarchically organized, inter-linked clusters provide the circuitry for diverse functional interactions and may lead to increased functional complexity as well as a wider critical range of activation behavior.

Nonetheless, many open questions remain. For example, what determines the specific layout of long-range cortical projections, given that their layout is not completely specified by spatial proximity? Is the organization of cortical networks into clusters already determined during ontogenetic development, or are they formed later on by activity-dependent rewiring of the networks? How are long-range connections integrated with the intrinsic micro-circuitry of cortical areas? What determines how widely brain functions are distributed in cortical networks? More broadly, what is the exact relationship between structural and functional connectivity at the systems level? Can general rules be formulated that describe the functional interactions of areas based on their structural connectivity? Progress in answering these questions will depend on a close and fruitful interaction between quantitative anatomical and physiological brain research and new approaches in data analysis and network modeling.

**Acknowledgements**

M.K. acknowledges financial support from the German National Academic Foundation and EPSRC (EP/E002331/1).

# References

1. Braendgaard, H., Evans, S.M., Howard, C.V., Gundersen, H.J.: The total number of neurons in the human neocortex unbiasedly estimated using optical disectors. J. Microsc. **157**(Pt 3) (1990) 285–304.
2. Braitenberg, V., Schüz, A.: Cortex: Statistics and Geometry of Neuronal Connectivity. 2nd edn. Springer (1998).
3. Binzegger, T., Douglas, R.J., Martin, K.A.C.: A quantitative map of the circuit of cat primary visual cortex. J. Neurosci. **24** (2004) 8441–8453.
4. Young, M.P.: The organization of neural systems in the primate cerebral cortex. Phil. Trans. R. Soc. **252** (1993) 13–18.
5. Binzegger, T., Douglas, R.J., Martin, K.A.C.: Axons in cat visual cortex are topologically self-similar. Cereb Cortex **15**(2) (2005) 152–165.

6. Silberberg, G., Grillner, S., LeBeau, F.E.N., Maex, R., Markram, H.: Synaptic pathways in neural microcircuits. Trends. Neurosci. **28**(10) (2005) 541–551.
7. Markram, H., Toledo-Rodriguez, M., Wang, Y., Gupta, A., Silberberg, G., Wu, C.: Interneurons of the neocortical inhibitory system. Nat. Rev. Neurosci. **5**(10) (2004) 793–807.
8. Schubert, D., Kotter, R., Luhmann, H.J., Staiger, J.F.: Morphology, electrophysiology and functional input connectivity of pyramidal neurons characterizes a genuine layer va in the primary somatosensory cortex. Cereb. Cortex **16**(2) (2006) 223–236.
9. Thomson, A.M., Morris, O.T.: Selectivity in the inter-laminar connections made by neocortical neurones. J. Neurocytol. **31**(3–5) (2002) 239–246.
10. Zeki, S., Shipp, S.: The functional logic of cortical connections. Nature **335**(6188) (1988) 311–317.
11. Van Essen, D.C., Anderson, C.H., Felleman, D.J.: Information processing in the primate visual system: an integrated systems perspective. Science **255**(5043) (1992) 419–423.
12. Young, M.P.: Objective analysis of the topological organization of the primate cortical visual system. Nature **358**(6382) (1992) 152–155.
13. Crick, F., Koch, C.: Are we aware of neural activity in primary visual cortex? Nature **375**(6527) (1995) 121–123.
14. Scannell, J., Blakemore, C., Young, M.: Analysis of connectivity in the cat cerebral cortex. J. Neurosci. **15**(2) (1995) 1463–1483.
15. Friston, K.J.: Models of brain function in neuroimaging. Annu. Rev. Psychol. **56** (2005) 57–87.
16. Sporns, O., Tononi, G., Edelman, G.M.: Theoretical neuroanatomy: relating anatomical and functional connectivity in graphs and cortical connection matrices. Cereb. Cortex **10** (2000) 127–141.
17. Stephan, K.E., Zilles, K., Kotter, R.: Coordinate-independent mapping of structural and functional data by objective relational transformation (ORT). Philos. Trans. R. Soc. Lond. B Biol. Sci. **355**(1393) (2000) 37–54.
18. Sporns, O., Chialvo, D.R., Kaiser, M., Hilgetag, C.C.: Organization, development and function of complex brain networks. Trends Cogn. Sci. **8** (2004) 418–425.
19. Friston, K.J.: Functional and effective connectivity in neuroimaging: a synthesis. Hum. Brain Mapp. **2** (1994) 56–78.
20. Horwitz, B.: The elusive concept of brain connectivity. Neuroimage **19**(2 Pt 1) (2003) 466–470.
21. Keinan, A., Sandbank, B., Hilgetag, C.C., Meilijson, I., Ruppin, E.: Fair attribution of functional contribution in artificial and biological networks. Neural Comp. **16** (2004) 1887–1915.
22. Seth, A.K.: Causal connectivity of evolved neural networks during behavior. Network **16**(1) (2005) 35–54.
23. Varela, F., Lachaux, J.P., Rodriguez, E., Martinerie, J.: The brainweb: phase synchonization and large-scale integration. Nature Rev. Neurosci. **2** (2001) 229–239.
24. Büchel, C., Friston, K.J.: Modulation of connectivity in visual pathways by attention: cortical interactions evaluated with structural equation modelling and fMRI. Cereb. Cortex. **7** (1997) 768–778.
25. Bressler, S.L.: Large-scale cortical networks and cognition. Brain Res. Brain Res. Rev. **20**(3) (1995) 288–304.

26. Buchel, C., Friston, K.J.: Dynamic changes in effective connectivity character-ized by variable parameter regression and Kalman filtering. Hum. Brain. Mapp. **6**(5–6) (1998) 403–408 Clinical Trial.
27. Buchel, C., Coull, J.T., Friston, K.J.: The predictive value of changes in effective connectivity for human learning. Science **283**(5407) (1999) 1538–1541.
28. Passingham, R.E., Stephan, K.E., Kötter, R.: The anatomical basis of functional localization in the cortex. Nat. Rev. Neurosci. **3** (2002) 606–616.
29. Izhikevich, E.M., Gally, J.A., Edelman, G.M.: Spike-timing dynamics of neu-ronal groups. Cereb. Cortex **14** (2004) 933–944.
30. Hilgetag, C.C., Burns, G.A.P.C., O'Neill, M.A., Scannell, J.W., Young, M.P.: Anatomical connectivity defines the organization of clusters of cortical areas in the macaque monkey and the cat. Phil. Trans. R. Soc. Lond. B **355** (2000) 91–110.
31. Scannell, J.W., Burns, G.A., Hilgetag, C.C., O'Neil, M.A., Young, M.P.: The connectional organization of the cortico-thalamic system of the cat. Cereb. Cortex **9**(3) (1999) 277–299.
32. Crick, F., Jones, E.: Backwardness of human neuroanatomy. Nature **361**(6408) (1993) 109–110.
33. Young, M.P., Scannell, J.W., O'Neill, M.A., Hilgetag, C.C., Burns, G., Blakemore, C.: Non-metric multidimensional scaling in the analysis of neu-roanatomical connection data and the organization of the primate cortical visual system. Phil. Trans. R. Soc. **348** (1995) 281–308.
34. Hilgetag, C.C., O'Neill, M.A., Young, M.P.: Indeterminancy of the visual cortex. Science **271**(5250) (1996) 776–777.
35. Kaiser, M., Hilgetag, C.C.: Modelling the development of cortical networks. Neurocomputing **58–60** (2004) 297–302.
36. Kaiser, M., Hilgetag, C.C.: Nonoptimal component placement, but short pro-cessing paths, due to long-distance projections in neural systems. PLoS. Com-put. Biol. (2006) e95.
37. Cherniak, C.: Component placement optimization in the brain. J. Neurosci. **14**(4) (1994) 2418–2427.
38. Masuda, N., Aihara, K.: Global and local synchrony of coupled neurons in small-world networks. Biol. Cybern. **90** (2004) 302–309.
39. von der Malsburg, C.: The correlation theory of brain function. Technical report, Max-Planck-Institute for Biophysical Chemistry (1981).
40. Laughlin, S.B., Sejnowski, T.J.: Communication in neuronal networks. Science **301** (2003) 1870–1874.
41. von Neumann, J.: The Computer and the Brain. Yale University Press (1958).
42. Kötter, R., Stephan, K.E.: Network participation indices: characterizing com-ponent roles for information processing in neural networks. Neural Networks **16** (2003) 1261–1275.
43. Hilgetag, C.C., Kötter, R., Stephan, K.E., Sporns, O.: Computational meth-ods for the analysis of brain connectivity. In: Computational Neuroanatomy. Humana Press, Totowa, NJ (2002) 295–335.
44. Sporns, O.: Graph theory methods for the analysis of neural connectivity pat-terns. In: Neuroscience Databases. Kluwer Academic, Dordrecht (2002) 169–183.
45. da Fontura Costa, L., Rodrigues, F.A., Travieso, G., Villas Boas, P.R.: Charac-terization of complex networks: A survey of measurements. cond-mat **0505185** (2005) v3.

46. Ungerleider, L., Mischkin, M.: Two cortical visual systems. In Ingle, M., Goodale, M., Mansfield, R., eds.: The New Cognitive Neurosciences. MIT Press, Cambridge, MA (1982).

47. Petroni, F., Panzeri, S., Hilgetag, C.C., Kötter, R., Young, M.P.: Simultaneity of responses in a hierarchical visual network. Neuroreport 12 (2001) 2753–2759.

48. Felleman, D.J., van Essen, D.C.: Distributed hierarchical processing in the primate cerebral cortex. Cereb. Cortex 1 (1991) 1–47.

49. Hilgetag, C.C., O'Neill, M.A., Young, M.P.: Hierarchical organization of macaque and cat cortical sensory systems explored with a novel network processor. Philos. Trans. R. Soc. Lond. Ser. B 355 (2000) 71–89.

50. Sporns, O., Zwi, J.D.: The small world of the cerebral cortex. Neuroinformatics 2 (2004) 145–162.

51. Hilgetag, C.C., Kaiser, M.: Clustered organisation of cortical connectivity. Neuroinformatics 2 (2004) 353–360.

52. Hellwig, B.: A quantitative analysis of the local connectivity between pyramidal neurons in layers 2/3 of the rat visual cortex. Biol. Cybern. 82 (2000) 111–121.

53. Schuz, A., Chaimow, D., Liewald, D., Dortenman, M.: Quantitative aspects of corticocortical connections: a tracer study in the mouse. Cereb. Cortex 16 (2005) 1474–1486.

54. Schuez, A., Braitenberg, V.: The human cortical white matter: quantitative aspects of corticocortical long-range connectivity. In Schuez, A., Miller, R., eds.: Cortical areas: unity and diversity. CRC Press, London (2002) 377–385.

55. Barabási, A.L., Albert, R.: Emergence of scaling in random networks. Science 286 (1999) 509–512.

56. Barabási, A.L., Ravasz, E., Vicsek, T.: Deterministic scale-free networks. Physica A 3–4 (2001) 559–564.

57. Ravasz, E., Somera, A.L., Mongru, D.A., Oltvai, Z.N., Barabási, A.L.: Hierarchical organization of modularity in metabolic networks. Science 297 (2002) 1551–1555.

58. Waxman, B.M.: Routing of multipoint connections. IEEE J. Sel. Areas Commun. 6(9) (1988) 1617–1622.

59. Murray, J.D.: Mathematical Biology. Springer, Heidelberg (1990).

60. Kaiser, M., Hilgetag, C.C.: Spatial growth of real-world networks. Phys. Rev. E 69 (2004) 036103.

61. Kaiser, M., Hilgetag, C.C.: Edge vulnerability in neural and metabolic networks. Biol. Cybern. 90 (2004) 311–317.

62. Kuida, K., Haydar, T.F., Kuan, C.Y., Gu, Y., Taya, C., Karasuyama, H., Su, M.S., Rakic, P., Flavell, R.A.: Reduced apoptosis and cytochrome c-mediated caspase activation in mice lacking caspase. Cell 94(3) (1998) 325–337.

63. Valverde, S., Cancho, R.F., Solé, R.V.: Scale-free networks from optimal design. Europhys. Lett. 60(4) (2002) 512–517.

64. Sur, M., Leamey, C.A.: Development and plasticity of cortical areas and networks. Nature Rev. Neurosci. 2 (2001) 251–262.

65. Rakic, P.: Neurogenesis in adult primate neocortex: an evaluation of the evidence. Nature Rev. Neurosci. 3 (2002) 65–71.

66. Spear, P., Tong, L., McCall, M.: Functional influence of areas 17, 18 and 19 on lateral suprasylvian cortex in kittens and adult cats: implications for compensation following early visual cortex damage. Brain Res. 447(1) (1988) 79–91.

67. Stromswold, K.:   The cognitive neuroscience of language acquisition.   In Gazzaniga, M., ed.: The New Cognitive Neurosciences. 2nd edn. MIT Press, Cambridge, MA (2000) 909–932.

68. Young, M.P., Hilgetag, C.C., Scannell, J.W.: On imputing function to structure from the behavioural effects of brain lesions. Phil. Trans. R. Soc. **355** (2000) 147–161.

69. Traufetter, G.: Leben ohne links. Spiegel Special (4) (2003)  30.

70. Pakkenberg, B., Moller, A., Gundersen, H.J., Mouritzen, D.A., Pakkenberg, H.: The absolute number of nerve cells in substantia nigra in normal subjects and in patients with parkinson's disease estimated with an unbiased stereological method. J. Neurol. Neurosurg. Psychiatry **54**(1) (1991) 30–33.

71. Albert, R., Jeong, H., Barabási, A.L.: Error and attack tolerance of complex networks. Nature **406** (2000) 378–382.

72. Diestel, R.: Graph Theory. Springer, New York (1997).

73. Watts, D.J., Strogatz, S.H.: Collective dynamics of 'small-world' networks. Nature **393** (1998) 440–442.

74. Girvan, M., Newman, M.E.J.:  Community structure in social and biological networks. Proc. Natl. Acad. Sci. **99**(12) (2002) 7821–7826.

75. Holme, P., Kim, B.J., Yoon, C.N., Han, S.K.: Attack vulnerability of complex networks. Phys. Rev. E **65** (2002) 056109.

76. Granovetter, M.S.:  The strength of weak ties.  Am. J. Sociol. **78**(6) (1973) 1360–1380.

77. Stephan, K.E., Hilgetag, C.C., Burns, G.A.P.C., O'Neill, M.A., Young, M.P., Kötter, R.: Computational analysis of functional connectivity between areas of primate cerebral cortex. Phil. Trans. R. Soc. **355** (2000) 111–126.

78. Kötter, R., Sommer, F.T.: Global relationship between anatomical connectivity and activity propagation in the cerebral cortex. Phil. Trans. Roy. Soc. Lond. Ser. B **355** (2000) 127–134.

79. Achard, S., Salvador, R., Whitcher, B., Suckling, J., Bullmore, E.: A resilient, low-frequency, small-world human brain functional network with highly connected association cortical hubs. J. Neurosci. **26** (2006) 63–72.

80. Salvador, R., Suckling, J., Coleman, M.R., Pickard, J.D., Menon, D., Bullmore, E.: Neurophysiological architecture of functional magnetic resonance images of human brain. Cereb. Cortex **15**(9) (2005) 1332–1342.

81. Milo, R., Shen-Orr, S., Itzkovitz, S., Kashtan, N., Chklovskii, D., Alon, U.: Network motifs: simple building blocks of complex networks. Science **298** (2002) 824–827.

82. Sporns, O., Kötter, R.: Motifs in brain networks. PLoS Biol. **2** (2004) 1910–1918.

83. Buzsaki, G., Geisler, C., Henze, D.A., Wang, X.J.: Interneuron diversity series: circuit complexity and axon wiring economy of cortical interneurons. Trends Neurosci. **27**(4) (2004) 186–193.

# 5

# Synchronization Dynamics in Complex Networks

Changsong Zhou, Lucia Zemanová and Jürgen Kurths

Institute of Physics, University of Potsdam PF 601553, 14415 Potsdam, Germany,
`cszhou@agnld.uni-potsdam.de`

**Summary.** Previous chapters have discussed tools from graph theory and their contribution to our understanding of the structural organization of mammalian brains and its functional implications. The brain functions are mediated by complicated dynamical processes which arise from the underlying complex neural networks, and synchronization has been proposed as an important mechanism for neural information processing. In this chapter, we discuss synchronization dynamics on complex networks. We first present a general theory and tools to characterize the relationship of some structural measures of networks to their synchronizability (the ability of the networks to achieve complete synchronization) and to the organization of effective synchronization patterns on the networks. Then, we study synchronization in a realistic network of cat cortical connectivity by modeling the nodes (which are cortical areas composed of large ensembles of neurons) by a neural mass model or a subnetwork of interacting neurons. We show that if the dynamics is characterized by well-defined oscillations (neural mass model and subnetworks with strong couplings), the synchronization patterns can be understood by the general principles discussed in the first part of the chapter. With weak couplings, the model with subnetworks displays biologically plausible dynamics and the synchronization pattern reveals a hierarchically clustered organization in the network structure. Thus, the study of synchronization of complex networks can provide insights into the relationship between network topology and functional organization of complex brain networks.

## 5.1 Introduction

Real-world complex networks are interacting dynamical entities with an interplay between dynamical states and interaction patterns, such as the neural networks in the brain. Recently, the complex network approach has been playing an increasing role in the study of complex systems [1]. The main research focus has been on the topological structures of complex systems based on simplified graphs, paying special attention to the global properties of complex networks, such as the small-world (small-world networks (SWNs) [2]) scale-free (scale-free networks (SFNs) [3]) features, or to the presence or absence of some very small subgraphs, such as network motifs [4]. Such topological studies have revealed important organizational principles in the structures of many realistic network systems [1]. The structural characterization of networks has been discussed in Chap. 3.

However, a more complete understanding of many realistic systems would require characterizations beyond the interaction topology. A problem of fundamental importance is the impact of network structures on the dynamics of the networks. The elements of many complex systems display oscillatory dynamics. Therefore, synchronization of oscillators is one of the widely studied dynamical behavior on complex networks [5]. It is important to emphasize that synchronization is especially relevant in brain dynamics [6]. Synchronization of neuronal dynamics on networks with complex topology thus has received significant recent attention [7–10].

Most previous studies have focused on the influence of complex network topology on the ability of the network to achieve synchronization. It has been shown that SWNs provide a better synchronization of coupled excitable neurons in the presence of external stimuli [7]. In pulse-coupled oscillators, synchronization becomes optimal in a small-world regime [8], and it is degraded when the degree becomes more heterogeneous with increased randomness [9]. Investigation of phase oscillators [11] or circle maps [12] on SWNs has shown that when more and more shortcuts are created at larger rewiring probability $p$, the transition to the synchronization regime becomes easier [11]. These observations have shown that the ability of a network to synchronize (synchronizability) is generally enhanced in SWNs as compared to regular chains. Physically, this enhanced synchronizability was attributed to the decreasing of the average network distance due to the shortcuts. On the other hand, it has been shown that the synchronizability also depends critically on the heterogeneity of the degree distribution [13]. In particular, random networks with strong heterogeneity in the degree distribution, such as SFNs, are more difficult to synchronize than random homogeneous networks [13], despite the fact that heterogeneity reduces the average distance between nodes [14]. The synchronizability in most previous studies is based on the linear stability of the complete synchronization state using spectral analysis of the network coupling matrix [5].

These studies focusing on the impacts of network topology assumed that the coupling strength is uniform. However, most complex networks in nature where synchronization is relevant are actually weighted, e.g. neural networks [15], networks of cities in the synchronization of epidemic outbreaks [16], and communication and other technological networks whose functioning relies on the synchronization of interacting units [17]. The connection weights of many real networks are often highly heterogeneous [18]. It has been shown that weighted coupling has significant effects on the synchronization of complex networks [19–22].

In this chapter, we discuss the synchronization of nonlinear oscillators coupled in complex networks. Our emphasis is to demonstrate how the network topology and the connection weights influence the synchronization behavior of the oscillators. The theory and method are mainly based on general models of complex networks, and we also study synchronization and the relationship between dynamical clusters and anatomical communities in a realistic complex network of brain cortex.

The chapter is organized as follows. In Sect. 5.2, we present the general dynamical equations and the linear stability analysis for the complete synchronization state when the oscillators are identical. Then, we demonstrate the leading parameters that universally control the synchronizability of a general class of random weighted networks in Sect. 5.3. In Sect. 5.4, we carry out simulations of hierarchical synchronization in SFNs outside the complete synchronization regimes. We demonstrate the influence of small-world connections in Sect. 5.5. Section 5.6 is devoted to synchronization analysis in cat cortical networks. We discuss the possible relevance of the analysis of dynamical complex neural networks and meaningful extensions in Sect. 5.7.

## 5.2 Dynamical Equations and Stability Analysis

The dynamics of a general network of $N$ coupled oscillators is described by:

$$\dot{\boldsymbol{x}}_j = \tau_j \boldsymbol{F}(\boldsymbol{x}_j) + \sigma \sum_{i=1}^{N} A_{ji} W_{ji} [\boldsymbol{H}(\boldsymbol{x}_i) - \boldsymbol{H}(\boldsymbol{x}_j)] \tag{5.1}$$

$$= \tau_j \boldsymbol{F}(\boldsymbol{x}_j) - \sigma \sum_{i=1}^{N} G_{ji} \boldsymbol{H}(\boldsymbol{x}_i), \quad j = 1, \dots, N , \tag{5.2}$$

where $\boldsymbol{x}_j$ is the state of oscillator $j$ and $\boldsymbol{F} = \boldsymbol{F}(\boldsymbol{x})$ governs the dynamics of each individual oscillator. The parameter $\tau_j$ controls the time scales of the oscillators, which are not identical in general. $\boldsymbol{H} = \boldsymbol{H}(\boldsymbol{x})$ is the output function, and $\sigma$ is the overall coupling strength. $A = (A_{ji})$ is the adjacency matrix of the underlying network of couplings, where $A_{ji} = 1$ if there is a link from node $i$ to node $j$, and $A_{ji} = 0$ otherwise. Here, we assume that the coupling is bidirectional so that $A_{ij} = A_{ji}$, i.e., $A$ is symmetric. The number of connections of a node, the degree $k_j$, is just the row sum of the adjacency matrix $A$, i.e., $k_j = \sum_i A_{ji}$. More details about network characterization are found in Chap. 3. $W_{ji}$ is the weight of the incoming strength for the link from node $i$ to node $j$. Note that the incoming and output weights can be in general asymmetric, $W_{ji} \neq W_{ij}$. Here $G = (G_{ji})$ is the coupling matrix combining both topology [adjacency matrix $A = (A_{ji})$] and weights [weight matrix $W =$

$(W_{ji})$]: $G_{ji} = -W_{ji}$ for $i \neq j$ and $G_{jj} = \sum_i W_{ji} A_{ji}$. By definition, the rows of matrix $G$ have zero sum, $\sum_{i=1}^{N} G_{ji} = 0$.

As we mentioned in the introduction, much previous work characterizes the synchronizability of networks using graph spectral analysis. The framework of this analysis is based on the *master stability function*. The readers are referred to the references [5,23] for the details. Here, we outline the main idea, which is to consider the ideal case of identical oscillators, i.e., $\tau_1 = \tau_2 = \cdots = \tau_N = 1$. In this case, it is easy to see that the completely synchronized state, $x_1(t) = x_2(t) = \cdots = x_N(t) = s(t)$, is a solution of (5.2), i.e., all the oscillators follow the same trajectory in the phase space, and the trajectory belongs to the attractor of the isolated oscillator. So this solution is also called the invariant *synchronization manifold*. However, synchronization in the network can only be observed when the synchronization state is robust against desynchronizing perturbations. Now the crucial question is: Is this solution stable? And under what conditions is it stable?

To study the stability of the synchronization state, we consider small perturbations of the synchronization state $s$, $\delta \dot{x}_j = x_j - s$, which are governed by the linear variational equations

$$\delta \dot{x}_j = DF(s)\delta x_j - \sigma DH(s) \sum_{i=1}^{N} G_{ji}\delta x_i, \quad j = 1, \cdots, N, \qquad (5.3)$$

where $DF(s)$ and $DH(s)$ are the Jacobians on $s$.

The main idea of the master stability function is to project $\delta x$ into the eigenspace spanned by the eigenvectors $v$ of the coupling matrix $G$. By doing so, (5.3) can be diagonalized into $N$ decoupled blocks of the form

$$\dot{\xi}_l = [DF(s) - \sigma \lambda_l DH(s)]\xi_l, \quad l = 1, \cdots, N, \qquad (5.4)$$

where $\xi_l$ is the eigenmode associated to the eigenvalue $\lambda_l$ of the coupling matrix $G$. Here, $\lambda_1 = 0$ corresponds to the eigenmode parallel to the synchronization manifold, and the other $N-1$ eigenvalues $\lambda_l$ represent the eigenmodes transverse to the synchronization manifold.

Note that all the variational equations in (5.4) have the same form:

$$\dot{\xi} = [DF(s) - \epsilon DH(s)]\xi. \qquad (5.5)$$

They differ only by the parameter $\epsilon = \sigma \lambda_l$. From this, we understand that the stability of each mode is determined by the property of the master stability of the normal form in (5.5) and the eigenvalue $\lambda_l$. In this chapter, we focus on the cases where $G$ has real eigenvalues, ordered as $0 = \lambda_1 \leq \lambda_2 \cdots \leq \lambda_N$. The largest Lyapunov exponent $\Lambda(\epsilon)$ of (5.5) as a function of the parameter $\epsilon$ is called the *master stability function*. If $\Lambda(\epsilon) < 0$, it follows that $\xi(t) \sim \exp(\Lambda t) \to 0$ when $t \to \infty$ and the mode is stable, otherwise, small perturbations will grow with time $t$ and the mode is unstable. For many oscillatory dynamical systems [23], (5.5) is stable (e.g., $\Lambda(\epsilon) < 0$) in a single, finite interval $\epsilon_1 < \epsilon < \epsilon_2$, where the thresholds $\epsilon_1$ and $\epsilon_2$ are determined only by $\boldsymbol{F}$, $\boldsymbol{H}$, and $\boldsymbol{s}$. A transverse mode is damped if the corresponding eigenvalue satisfies $\epsilon_1 < \sigma\lambda_i < \epsilon_2$, and the complete synchronization state is stable when all the transverse modes are damped, namely,

$$\epsilon_1 < \sigma\lambda_2 \le \sigma\lambda_3 \le \cdots \le \sigma\lambda_N < \epsilon_2 . \tag{5.6}$$

This condition can only be fulfilled for some values of $\sigma$ when the eigenratio $R$ meets

$$R \equiv \lambda_N/\lambda_2 < \epsilon_2/\epsilon_1 . \tag{5.7}$$

It is impossible to synchronize the network completely if $R > \epsilon_2/\epsilon_1$, since there is no $\sigma$ value for whom the solution is linearly stable. The eigenratio $R$ depends only on the network structure, as defined by the coupling matrix $G$. If $R$ is small, in general the condition in (5.7) will be more easily satisfied. It follows that the smaller the eigenratio $R$ the more synchronizable the network and vice versa [5], and we can characterize the synchronizability of the networks with $R$, without referring to specific oscillators. In some special cases, $\epsilon_2 = \infty$, and the synchronization state is stable when the overall coupling strength $\sigma$ is larger than a threshold $\sigma_c = \epsilon_1/\lambda_2$. More detailed characterization of the synchronizability in this case can be found in [21].

In the following section, we characterize the synchronizability of weighted networks using only the eigenratio $R$.

## 5.3 Universality of Synchronizability

### 5.3.1 Universal Formula

How the synchronizability, the eigenratio $R$, depends on the structure of networks is one of the major questions in previous studies. Based on spectral graph theory, previous work has obtained bounds for the eigenvalues of *unweighted* networks ($W_{ji} = 1$) [13, 24]. For arbitrary networks [24], the eigenvalues are bounded as $\frac{4}{ND_{\max}} \le \lambda_2 \le \frac{N}{N-1}k_{\min}$ and $\frac{N}{N-1}k_{\max} \le \lambda_N \le \max(k_i + k_j) \le 2k_{\max}$, where $k_{\min}$ and $k_{\max}$ are the minimum and maximum degrees, respectively, $k_i$ and $k_j$ are the degrees of two connected nodes, and $D_{\max}$, the diameter of the graph, is the maximum of the distances between nodes. From this, one gets

$$k_{\max}/k_{\min} \le R \le ND_{\max}\max(k_i + k_j)/4 . \tag{5.8}$$

The reader can find more details concerning how these bounds are obtained in [24]. Such bounds can provide some insights into the synchronizability of the networks. For example, networks with heterogeneous degrees have low synchronizability, since $R$ is bounded away from 1 by $k_{\max}/k_{\min}$. However, such bounds are not tight. The upper bound, as a function of the network size $N$, can be orders of magnitude larger than the lower one even for small networks (especially for random networks), thus providing limited information about the actual synchronizability of the networks.

In the following, we present tighter bounds for more general weighted networks, including unweighted networks as special cases. We restrict ourselves to sufficiently random networks. Our analysis is based on the combination of a mean field approximation and new graph spectral results [21].

First, in random networks with a large enough minimal degree $k_{\min} \gg 1$, (5.1) $(\tau_j = 1, \forall j)$ can be approximated as

$$\dot{\boldsymbol{x}}_j = \boldsymbol{F}(\boldsymbol{x}_i) + \sigma \frac{S_j}{k_j} \sum_{i=1}^{N} A_{ji}[\boldsymbol{H}(\boldsymbol{x}_i) - \boldsymbol{H}(\boldsymbol{x}_j)] \,. \tag{5.9}$$

The reason is that each oscillator $i$ receives signals from a large and random sample of other oscillators in the network and $\boldsymbol{x}_i$ is not affected directly by the individual output weights $W_{ji}$. Thus, we may assume that $W_{ji}$ and $\boldsymbol{H}(\boldsymbol{x}_i)$ are statistically uncorrelated and the following approximation holds

$$\sum_{i=1}^{N} W_{ji} A_{ji} \boldsymbol{H}(\boldsymbol{x}_i) \approx \frac{1}{k_j} \sum_{i=1}^{N} A_{ji} \boldsymbol{H}(\boldsymbol{x}_i) \sum_{i=1}^{N} W_{ji} A_{ji} = \bar{\boldsymbol{H}}_j S_j \tag{5.10}$$

if $k_i \gg 1$. Here, $\bar{\boldsymbol{H}}_j = (1/k_j) \sum_{i=1}^{N} A_{ji} \boldsymbol{H}(\boldsymbol{x}_i)$ is the local mean field, and $S_j$ is the intensity of node $j$. Defined as the total input weight of node,

$$S_j = \sum_{i=1}^{N} A_{ji} W_{ji} \,, \tag{5.11}$$

the intensity $S_j$ is a significant measure integrating the information of connectivity and weights [18].

Now, if the network is sufficiently random, the local mean field $\bar{\boldsymbol{H}}_j$ can be approximated by the global mean field of the network, $\bar{\boldsymbol{H}}_j \approx \bar{\boldsymbol{H}} = (1/N) \sum_{i=1}^{N} \boldsymbol{H}(\boldsymbol{x}_i)$, since the information that a node obtains from its $k_j \gg 1$ connected neighbors is well distributed in the network and the averaged signal is close enough to the average behavior of the whole network. Moreover, close to the synchronized state $\boldsymbol{s}$, we may assume $\bar{\boldsymbol{H}}_j \approx \boldsymbol{H}(\boldsymbol{s})$, and the system is approximated as

$$\dot{\boldsymbol{x}}_j = \boldsymbol{F}(\boldsymbol{x}_j) + \sigma S_j [\boldsymbol{H}(\boldsymbol{s}) - \boldsymbol{H}(\boldsymbol{x}_j)], \quad j = 1, \ldots, N \,. \tag{5.12}$$

We call this the *mean field approximation*, which indicates that the oscillators are decoupled and forced by a common oscillator $\boldsymbol{s}$ with a forcing strength proportional to the intensity $S_j$ (cf. Chap. 1). The variational equations (5.12) have the same form as (5.4), except that $\lambda_l$ is replaced by $S_l$. If there is some $\sigma$ satisfying

$$\epsilon_1 < \sigma S_{\min} \leq \cdots \leq \sigma S_l \leq \cdots \leq \sigma S_{\max} < \epsilon_2 \,, \tag{5.13}$$

then all the oscillators are synchronizable by the common driving $H(s)$, corresponding to a complete synchronization of the whole network. These observations suggest that the eigenratio $R$ can be approximated as

$$R \approx S_{\max}/S_{\min} , \tag{5.14}$$

where $S_{\min}$, $S_{\max}$ are the minimum and maximum intensities $S_j$ in (5.11), respectively.

Next, we present tight bounds for the above approximation. Equation (5.9) means that the coupling matrix $G$ is replaced by the new matrix $G^a = (G^a_{ji})$, with $G^a_{ji} = \frac{S_j}{k_j}(\delta_{ji}k_j - A_{ji})$. $G^a$ can be written as $G^a = S\hat{G} = SD^{-1}(D - A)$, where $S = (\delta_{ji}S_j)$ and $D = (\delta_{ji}k_j)$ are the diagonal matrices of intensities and degrees, respectively, and $\hat{G}$ is the normalized Laplacian matrix [25]. Importantly, now the contributions from the topology and weight structure are separated and accounted for by $\hat{G}$ and $S$, respectively. We can show that in large enough complex networks, such as SFNs in many realistic complex systems, the largest and smallest nonzero eigenvalues of the matrix $G^a$ are bounded by the eigenvalues $\mu_l$ of $\hat{G}$ as

$$S_{\min}\mu_2 c \leq \lambda_2 \leq S_{\min}c', \quad S_{\max} \leq \lambda_N \leq S_{\max}\mu_N , \tag{5.15}$$

where $c$ and $c'$ can be approximated by 1 for most large complex networks of interest, such as realistic SFNs. The upper bound of $\lambda_N$ in the inequality (5.15) follows from

$$\lambda_N = \max_{||v||=1} ||S\hat{G}v|| \leq \max_{||v||=1} ||Sv|| \max_{||v||=1} ||\hat{G}v|| = S_{\max}\mu_N \tag{5.16}$$

where $|| \cdot ||$ is the Euclidean norm and $v$ denotes the normalized eigenvector of $G^a$. The other bounds in (5.15) are obtained in a similar spirit. If the network is sufficiently random, the spectrum of $\hat{G}$ tends to the *semicircle law* for large networks with arbitrary expected degrees [25]. The semicircle law says that if $k_{\min} \gg \sqrt{K}$, where $K$ is the mean degree of the network, the distribution function of the eigenvalues of $\hat{G}$ follows a semicircle with the center at 1.0 and the radius $r = 2/\sqrt{K}$ when $N \to \infty$. In particular, we have $\max\{1 - \mu_2, \mu_N - 1\} = [1 + o(1)]\frac{2}{\sqrt{K}}$ for $k_{\min} \gg \sqrt{K}\ln^3 N$. From these, it follows that

$$\mu_2 \approx 1 - 2/\sqrt{K}, \quad \mu_N \approx 1 + 2/\sqrt{K} , \tag{5.17}$$

which we find to provide also a good approximation under the weaker condition $k_{\min} \gg 1$, regardless of the degree distribution of the random network. From (5.15) and (5.17), we have the following approximation for the bounds of $R$:

$$\frac{S_{\max}}{S_{\min}} \leq R \leq \frac{S_{\max}}{S_{\min}}\frac{1 + 2/\sqrt{K}}{1 - 2/\sqrt{K}} . \tag{5.18}$$

We stress that this result applies to unweighted networks ($S_i = k_i$) and the upper bound in the inequality (5.18) is much tighter than that in (5.8).

The bounds in (5.18) show that the contribution of the network topology is mainly captured by the mean degree $K$. Therefore, for a given $K$, the synchronizability of random networks with a large $k_{\min}$ is expected to be well approximated by the following universal formula:

$$R = A_R \frac{S_{\max}}{S_{\min}} , \qquad (5.19)$$

where the pre-factors $A_R$ are expected to be close to 1. In the case of uniform intensity ($S_i = 1 \ \forall i$), they are given by the upper bounds, $A_R = \frac{1+2/\sqrt{K}}{1-2/\sqrt{K}}$, and $A_R \to 1$ in the limit $K \to \infty$. Equation (5.19) is consistent with the approximation in (5.14) and indicates that the synchronizability of these networks is primarily determined by the heterogeneity of the intensities, regardless of the degree distribution.

In simpler words, the universal formula says that if you give me a random enough but arbitrary network, we then can measure the intensities and tell you whether or not the network is synchronizable for a given oscillator (which specifies $\epsilon_1$ and $\epsilon_2$). If $S_{\max}/S_{\min}$ is clearly smaller than $\epsilon_2/\epsilon_1$, then we can predict that the suitable coupling strength for complete synchronization is $\epsilon_1/S_{\min} < \sigma < \epsilon_2/S_{\max}$.

### 5.3.2 Numerical Confirmation of the Universality

In the following, we present some results of numerical simulations of the synchronizability of various weighted and unweighted networks. These results confirm the above universal formula obtained based on physical arguments and graph spectral analysis. For example, let us consider the following weighted coupling scheme:

$$W_{ji} = S_j/k_j , \qquad (5.20)$$

in which the intensities $S_j$ follow an arbitrary distribution not necessarily correlated with the degrees $k_j$. This means that the intensities $S_j$ of a node $j$ are equally distributed into the $k_j$ input connections of this node, which serves as an approximation of realistic networks. However, nonuniform weights of the input links does not change our conclusion according to the approximation in (5.9). In [21], we presented results for more realistic weighted networks and showed that the universal formula also applies.

With the weighted coupling in (5.20), (5.9) and (5.1) are identical and $G^a = G$. This weighted coupling scheme includes many previously studied systems as special cases. If $S_j = k_j \ \forall j$, it corresponds to the widely studied case of unweighed networks [5, 13, 26]. In the case of fully uniform intensity

$(S_j = 1 \ \forall j)$, it accommodates a number of previous studies about synchronization of coupled maps [27,28]. The weighted scheme studied in [19], $W_{ji} = k_j^\theta$, is another special case of (5.20) where $S_j = k_j^{1+\theta}$. We have applied the weighted scheme to various network models:

(i) *Growing SFNs with aging* [29]. This model of complex networks extends the Barabási-Albert (BA) model [3]. Starting with $2m + 1$ fully connected nodes, at each time step we connect a new node to $m$ existing nodes according to the probability $\Pi_i \sim k_i \tau_i^{-\alpha}$, where $\tau_i$ is the age of the node. The minimum degree is then $k_{\min} = m$ and the mean degree is $K = 2m$. For the aging exponent $-\infty < \alpha \le 0$, this growing rule generates SFNs with a power-law tail $P(k) \sim k^{-\gamma}$ and the scaling exponent in the interval $2 < \gamma \le 3$ [29], as in most real SFNs. For $\alpha = 0$, we recover the usual Barabási-Albert (BA) model [3], which has $\gamma = 3$.

(ii) *Random SFNs* [30]. Each node is assigned to have a number $k_i \ge k_{\min}$ of "half-links" according to the distribution $P(k) \sim k^{-\gamma}$. The network is generated by randomly connecting these half-links to form links, prohibiting self- and repeated links.

(iii) *K-regular random networks.* Each node is randomly connected to $K$ other nodes.

We now present results for two different distributions of intensity $S_i$ that are uncorrelated with the distribution of the degree $k_i$: (1) a uniform

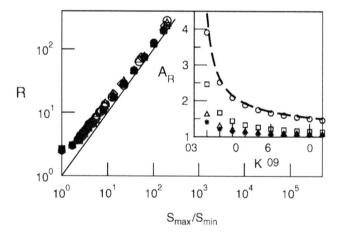

**Fig. 5.1.** $R$ as a function of $S_{\max}/S_{\min}$. Filled symbols: uniform distribution of $S_i \in [S_{\min}, S_{\max}]$. Open symbols: power-law distribution of $S_i$, $P(S) \sim S^{-\Gamma}$ for $2.5 \le \Gamma \le 10$. Different symbols are for networks with different topologies: BA growing SFNs (*circles*), growing SFNs with aging exponent $\alpha = -3$ (*squares*), random SFNs with $\gamma = 3$ (*diamonds*), and $K$-regular random networks (*triangles*). The number of nodes is $N = 2^{10}$ and the mean degree is $K = 20$. Inset: $A_R$ as a function of $K$ for $S_{\max}/S_{\min} = 1$ (*circles*), 2 (*squares*), 10 (*triangles*), and 100 (*stars*), obtained with a uniform distribution of $S_i$ in $K$-regular networks. The dashed lines are the bounds. Solid line: (5.19) with $A_R = 1$

distribution in $[S_{min}, S_{max}]$; and (2) a power-law distribution, $P(S) \sim S^{-\Gamma}$, $S \geq S_{min}$, where $S_{min}$ is a positive number. Consistently with the prediction of the universal formula, if $k_{min} \gg 1$, the eigenratio $R$ collapses into a single curve for a given $K$ when plotted as a function of $S_{max}/S_{min}$, irrespective of the distributions of $k_j$ and $S_j$, as shown in Fig. 5.1. The behavior of the fitting parameter $A_R$ is shown in the inset of Fig. 5.1. For a uniform intensity, it is very close to the upper bounds. It approaches 1 very quickly when the intensities become more heterogeneous ($S_{max}/S_{min} > 3$). Therefore, (5.19) with $A_R = 1$ (Fig. 5.1, solid line) provides a good approximation of the synchronizability for any large $K$ if the intensities are not very homogeneous.

## 5.4 Effective Synchronization in Scale-Free Networks

So far, we have presented an analysis of the stability of the complete synchronization state and the synchronizability of the networks based on the spectrum of the weighted graphs. The main conclusion is that, for random networks, the ability of the network to achieve complete synchronization is determined by the maximal and minimal values of the intensities $S_j$. The intensity of a node in (5.11), defined as the sum of the strengths of all input connections of that node, incorporates both topological and weighted properties of the network.

Now we carry out simulations on concrete dynamical systems. We would like to demonstrate that the intensities $S_j$ are still the important parameter for the organization of *effective synchronization* on the network outside the complete synchronization regime, e.g., when the network is perturbed by noise, or when the oscillators are non-identical, which are typical cases in more realistic systems.

In the following we give a summary of the main results (for more details, see [31]). For this purpose, we analyze the paradigmatic Rössler chaotic oscillator $x = (x, y, z)$:

$$\dot{x} = -0.97x - z, \tag{5.21}$$

$$\dot{y} = 0.97x + 0.15y, \tag{5.22}$$

$$\dot{z} = x(z - 8.5) + 0.4. \tag{5.23}$$

The oscillations are chaotic, and its time-dependent phase can be defined as $\phi = \arctan(y/x)$ [32,33], as illustrated in Fig. 5.2.

Without loss of generality, we consider SFNs generated with the BA model [3] and use the coupling scheme in (5.20). For simplicity, we also sorted the label of the nodes according to the degrees, $k_1 \geq k_2 \geq \cdots \geq k_N$. We compare unweighted networks (UN) ($W_{ij} = 1$) by taking $S_j = k_j$ and weighted networks (WN) ($W_{ji} = 1/k_j$) by taking uniform intensities $S_j = 1$ for all the

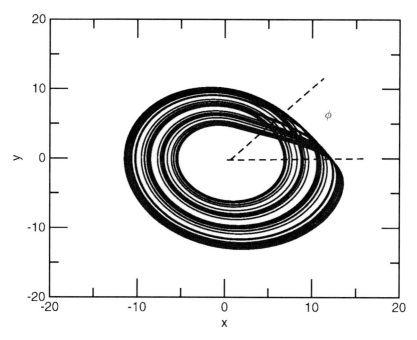

**Fig. 5.2.** Chaotic attractor of the Rössler oscillator. $\phi$ indicates the phase of the oscillation

nodes. Here, we use the output function $\boldsymbol{H}(\boldsymbol{x}) = \boldsymbol{x}$ in our simulations. In this case, $\epsilon_1 = \Lambda_F$, the largest Lyapunov exponent of the isolated Rössler chaotic oscillator, and $\epsilon_2 = \infty$.

Numerical simulations confirm that synchronization is achieved when $\sigma > \sigma_c = \Lambda_F/\lambda_2$. The transition to synchronization is shown in Fig. 5.3 as a function of the normalized coupling strength $g = \sigma\langle S\rangle$, where $\langle S\rangle$ is the average intensity of the networks. Here, we have plotted the average synchronization error $E = (1/N)\sum_{j=1}^{N}\Delta X_j$, where $\Delta X_j = \langle|x_j - X|\rangle_t$ is the time-averaged distance between the oscillator $x_j$ and the mean activity of the whole network, $X = (1/N)\sum_{j=1}^{N}x_j$. When complete synchronization is achieved at $g > g_c$, one has $E = 0$ after a sufficiently long transient. Additionally, we show the oscillation amplitude $A_X$ of the mean field $X$, calculated as the standard deviation of $X$ over time. As expected, the WN with uniform intensity achieves complete synchronization at a critical coupling strength $g_c$ smaller than that of the UN having the same mean degree $K$. It is important to emphasize that the network already maintains collective oscillations when the coupling strength $g$ is much smaller than the threshold value $g_c$ for complete synchronization. This is manifested by an amplitude $A_X$ of the collective oscillations, which has the same level as that of the completely synchronized state at $g > g_c$.

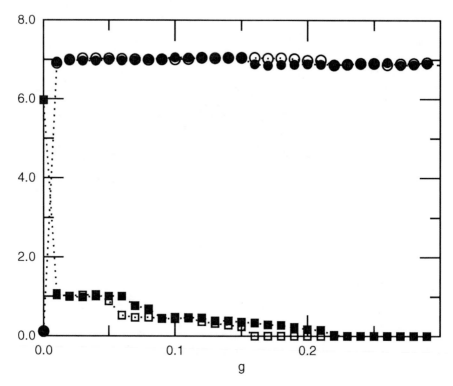

**Fig. 5.3.** Transition to synchronization in the UN and WN, indicated by the synchronization error $E$ (*squares*) and the amplitude $A_X$ of the mean field $X$ (*circles*). The filled symbols are for the UN and the open symbols for the WN. In both networks, $N = 1000$ and $K = 10$ ($k_{\min} = 5$)

### 5.4.1 Hierarchical Synchronization

Now we look into the different behavior of the two networks outside the complete synchronization regime, when the coupling is too weak ($g < g_c$) or when the synchronization state at $g > g_c$ is perturbed by noise. Noise is simulated by adding independent Gaussian random perturbations $D\eta_j(t)$ with a standard deviation $D$ to the variables of the oscillators, i.e., $\langle \eta_j(t)\eta_i(t - \tau) \rangle = \delta_{ji}\delta(\tau)$. We examine the synchronization difference $\Delta X_j$ of an individual oscillator with respect to the collective oscillations $X$ of the whole network. The typical results are as shown in Figs. 5.4(a) and (b) for weak coupling and noise perturbation, respectively.

It is seen that $\Delta X_j$ is almost the same for the nodes in the WN (only slightly smaller for nodes with larger degrees), since in this case, the intensity is fully uniform ($S_j = 1$) and independent of the degrees. In sharp contrast, the synchronization difference is strongly heterogeneous in the UN and is negatively correlated with the intensity ($S_j = k_j$) of the nodes. To get a clear

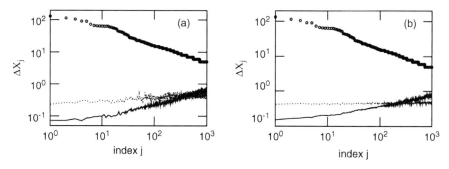

**Fig. 5.4.** Synchronization difference $\Delta X_j$ of the oscillators with respect to the global mean field $X$ in the UN (*solid line*) and WN (*dotted line*). The symbol ($\circ$) denotes the degree $k_j$ of the nodes. Note the log-log scales of the plots: (a) The coupling strength is weak ($g = 0.1$); (b) The synchronized state ($g = 0.3$) is perturbed by noise ($D = 0.5$)

dependence of $\Delta X$ on the degree $k$, we calculate the average value $\Delta X(k)$ among all nodes with degree $k$, i.e.,

$$\Delta X(k) = \frac{1}{N_k} \sum_{k_j = k} \Delta X_j , \qquad (5.24)$$

where $N_k$ is the number of nodes with degree $k_j = k$ in the SFN network. Now a pronounced dependence can be observed for the UN, as shown in Fig. 5.5. For both weak couplings and noise perturbations, the dependence is characterized by a power-law scaling

$$\Delta X(k) \sim k^{-\alpha} , \qquad (5.25)$$

with the exponent $\alpha \approx 1$. These results demonstrate that in the UN, where the intensities ($S_j = k_j$) are heterogeneous due to the power-law distribution of the degrees, a small portion of nodes with large intensities synchronize more

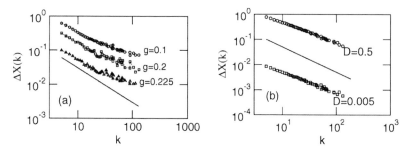

**Fig. 5.5.** The average values $\Delta X(k)$ as a function of $k$ at various coupling strength $g$ in the UN: (a) The coupling strength is weak; (b) The synchronized state ($g = 0.3$) is perturbed by noise. The solid lines with slope $-1$ are plotted for reference

closely to the mean field $X$, while most of the nodes with small intensities are still rather independent of $X$ outside the synchronization regime. The effective synchronization patterns of the networks are controlled by the distribution of the intensities $S_j$, while complete synchronization is determined by the maximal and minimal values.

### 5.4.2 Effective Synchronization Clusters

We have shown that the synchronization behavior of the individual oscillators in the UN is highly nonuniform; in particular, the nodes with large degrees, i.e., the hubs, are close to the mean field. As a result, the synchronization difference between them should also be relatively small. We define an effective synchronization cluster for those oscillators that synchronize to each other within some threshold. For this purpose, we have calculated the pairwise synchronization difference $\Delta X_{ij} = \langle |x_i - x_j| \rangle_t$. A pair of oscillators $(i \neq j)$ is considered to be synchronized effectively when their synchronization difference is smaller than a threshold: $\Delta X_{ij} \leq \Delta_{th}$. Since the synchronization difference is heterogeneous, there is no unique choice of the threshold value $\Delta_{th}$. What we can expect is that with smaller values of $\Delta_{th}$, the size of the effective cluster is smaller. The effective synchronization clusters for different values of the threshold $\Delta_{th}$ are shown in Figs. 5.6(a) and (b). The same clusters are also represented in the space of degrees $(k_i, k_j)$ in Figs. 5.6(c) and (d), respectively. Note that almost all the oscillators forming the clusters have a degree $k_j > k_{th}$, where $k_{th}$ is the threshold degree satisfying $\Delta X(k_{th}) = \Delta_{th}$; or correspondingly, the effective cluster is formed by nodes with $j < J_{th}$, where $J_{th}$ is the mean index of nodes with degree $k_j = k_{th}$. The triangular shape of the effective clusters in Fig. 5.6 is well described by the relation $i + j \leq J_{th}$. Above the solid line $i + j = J_{th}$, those oscillators ($i \leq J_{th}$ and $j \leq J_{th}$) having large enough degrees, i.e., $k_i \geq k_{th}$ and $k_j \geq k_{th}$, are close to the mean field with $\Delta X_i \leq \Delta_{th}$ and $\Delta X_j \leq \Delta_{th}$, but the pairwise distance is large, $\Delta X_{ij} > \Delta_{th}$. These results demonstrate clearly that the nodes with the largest intensities are the dynamical core of the networks.

### 5.4.3 Non-identical Oscillators

Now we consider *non-identical* oscillators by assuming that the time scale parameters $\tau_j$ are heterogeneous in (5.1), so that the oscillators have different mean oscillation frequencies $\Omega_j$. In our simulations, we use a uniform distribution of $\tau_j$ in an interval $[1 - \Delta\tau, 1 + \Delta\tau]$, with $\Delta\tau = 0.1$. We define the phases of the oscillations as indicated in Fig. 5.2. The average frequency can be computed as $\Omega_j = \langle \dot{\phi}_j \rangle$.

Let us first examine the collective oscillations in the network. Fig. 5.7 shows the amplitude $A_X$ of the mean field $X$ as a function of the coupling

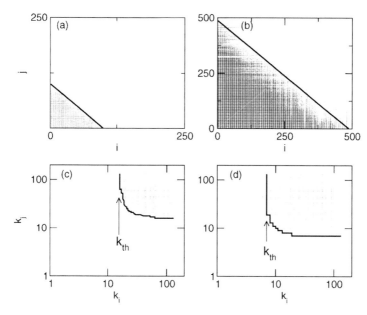

**Fig. 5.6.** The effective synchronization clusters in the synchronized UN in the presence of noise ($g = 0.3$, and $D = 0.5$), represented simultaneously in the index space $(i, j)$ (**a, b**) and in the degree space $(k_i, k_j)$ (**c, d**). A dot is plotted when $\Delta X_{ij} \leq \Delta_{th}$. (**a**) and (**c**) for the threshold value $\Delta_{th} = 0.25$, and (**b**) and (**d**) for $\Delta_{th} = 0.50$. The solid lines in (**a**) and (**b**) denote $i + j = J_{th}$ and are also plotted in (**c**) and (**d**) correspondingly. Note the different scales in (**a**) and (**b**) and the log-log scales in (**c**) and (**d**)

strength $g$ for the UN and WN. It is seen that both networks generate a coherent collective oscillation when the coupling strength is larger than a critical value $g_{cr} \approx 0.08$. However, the UN generates a weaker degree of collective synchronization as indicated by a smaller amplitude $A_X$ of the mean field.

Now we study in more detail synchronization behavior in the weak, intermediate and strong coupling regimes, indicated by the three vertical dashed lines in Fig. 5.7.

## I. Weak Coupling: Non-synchronization Regime

We start with the weak coupling regime with $g = 0.05$. Here, neither the UN nor the WN display significant collective oscillations. The frequencies of the oscillators are still distributed and the phases of the oscillators are not locked. However, interesting dynamical changes can be already expected in the UN. Based on the mean field approximation described in Sect. 3.1, we have

$$\dot{x}_j = \tau_j F(x_j) + g \frac{S_j}{\langle S \rangle}(X - x_j), \quad k_j \gg 1 . \tag{5.26}$$

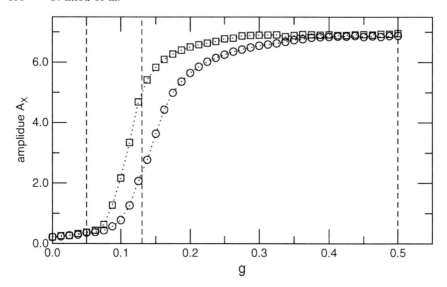

**Fig. 5.7.** The amplitude of the mean field as a function of the coupling strength $g$ in the UN ($\circ$) and WN ($\square$). The networks have mean degree $K = 10$ and size $N = 1000$

For the UN, $S_j = k_j$, and this approximation means that the oscillators are forced by a common signal, the global mean field $\boldsymbol{X}$, with the forcing strength being proportional to their degree $k_j$. Now even though the overall coupling strength $g$ is still small, the oscillators with large degrees are already strongly forced by the common signal $\boldsymbol{X}$. In this weak coupling regime, the global mean field displays some small fluctuations around the unstable fixed point of the isolated oscillator, $\boldsymbol{X} \approx \boldsymbol{x}_F$ ($\boldsymbol{F}(\boldsymbol{x}_F) = \boldsymbol{0}$), thus it has only a very small amplitude. These oscillators should somewhat synchronize to $\boldsymbol{X}$. As shown in Figs. 5.8(a) and (b), oscillators with $k > 10$ already display some degree of synchronization, indicated by a decreasing $\Delta X$ for larger $k$, while all the oscillators in the WN are distant from $X$, since $S_j = 1$ for all of them. A small distance of an oscillator $j$ from $X$, which has an almost vanishing amplitude, shows that the oscillation amplitude $A_j$ of the oscillator is small. We have calculated $A_j$ as the standard deviation of the time series $x_j$. We can see from Fig. 5.8(c) that $A_j$ indeed displays almost the same behavior as $\Delta X_j$. This becomes even more evident when we compare the average value $A(k)$, similar to (5.24), (Fig. 5.8(d)), with $\Delta X(k)$, (Fig. 5.8(b)). The changes in the amplitudes can be understood as follows: taking $\boldsymbol{X} \approx \boldsymbol{x}_F$, from (5.29) one gets for the UN

$$\dot{\boldsymbol{x}}_j = \tau_j \boldsymbol{F}(\boldsymbol{x}_j) - g\frac{k_j}{K}(\boldsymbol{x}_j - \boldsymbol{x}_F), \quad k_j \gg 1 , \tag{5.27}$$

which yields that hubs (nodes with the largest degrees) are experiencing a strong negative self-feedback, so that the trajectory is stabilized at the

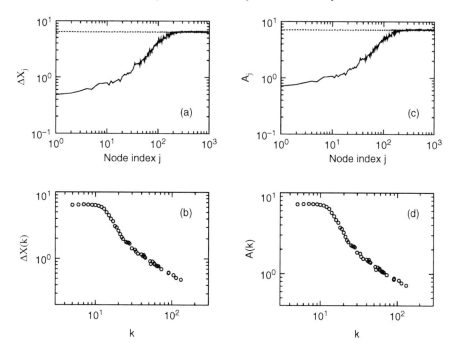

**Fig. 5.8.** (a) Synchronization difference $\Delta X_j$ of the oscillators with respect to the global mean field $X$ in the UN (*solid line*) and WN (*dotted line*); (b) The average values $\Delta X(k)$ as a function of $k$ in the UN; (c) and (d): as in (a) and (b), but for the oscillation amplitude $A_j$ and the average value $A(k)$ of the oscillators. The results are averaged over 50 realizations of the random time scale parameters $\tau_j$. The coupling strength is $g = 0.05$

originally unstable fixed point $x_F$, but with some fluctuations due to small non-vanishing perturbations from the mean activity of the neighbors.

To summarize: in the weak coupling regime, where even frequency and phase synchronization are not yet established, the heterogeneous UN already displays a form of hierarchical synchronization expressed by a change in the oscillation amplitudes.

## II. Intermediate Coupling: Phase Synchronization

Next, we take an intermediate coupling strength $g = 0.13$, where both networks are in the regime of transition to strong collective oscillations (Fig. 5.7). In this regime, frequency and phase synchronization become evident, while the absolute distance $\Delta X$ is still large. In Fig. 5.9, we show the mean oscillation frequencies $\Omega_j$ of all oscillators. In the UN, we find that about 70% of the nodes are locked to a common frequency $\Omega = 0.99$, forming a frequency

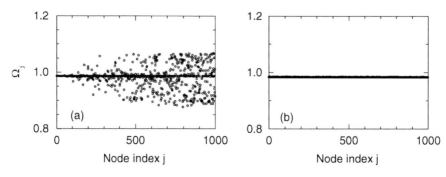

**Fig. 5.9.** The mean oscillation frequencies $\Omega_j$ of the oscillators in the UN: (a) and the WN; (b) at the coupling strength $g = 0.13$

synchronization cluster. Note that almost all nodes with the largest degrees $k_j$ are synchronized in frequency, while many nodes with small degrees are not yet locked. In the WN, on the contrary, the frequencies of all nodes are locked so that the network is globally synchronized in frequency. The nodes that are not frequency locked in the UN are largely uncorrelated with each other and they do not generate significant contributions to the collective oscillations, while all the nodes in the WN have a significant contribution; as a result, the amplitude of the collective oscillation is much smaller in the UN.

Now we examine phase synchronization of the nodes with respect to $X$. We measure phase synchronization by the time-averaged order parameter (Kuramoto parameter)

$$r_j = \langle \sin(\Delta\phi_j) \rangle^2 + \langle \cos(\Delta\phi_j) \rangle^2 \,, \tag{5.28}$$

where $\Delta\phi_j = \phi_j - \phi_X$ is the difference of the phases of an individual oscillator $j$ and the mean field $X$. Here, the phases are defined as $\phi_j = \arctan(y_j/x_j)$ and $\phi_X = \arctan(Y/X)$ for an individual oscillator $j$ and the mean field, respectively. Note that $r_j \approx 0$ when there is no phase locking and $r_j \approx 1$ when the phases are locked with an almost constant phase difference. Consistent with Fig. 5.9, we find that $r_j = 1$ for all oscillators in the WN; while $r_j < 1$ for many nodes with small degrees in the UN (Fig. 5.10(a)). To get a clear dependence of $r$ on the degree $k$, we again calculate the average value $r(k)$ between all nodes with degree $k$. Now there is a more pronounced dependence between $r(k)$ and $k$ (Fig. 5.10(b)).

We also calculate the absolute distances $\Delta X_j$. They are not small on average in both networks in spite of phase synchronization (Fig. 5.10(c)), because phase locked oscillators may have significant (but bounded) phase differences. However, $\Delta X_j$ again displays the hierarchical structure in the UN (Fig 5.10(d)). The nodes with large degrees are not only locked in frequency, but also have small phase differences. So, in this regime, the hierarchical synchronization is manifested by different degrees of frequency and phase locking.

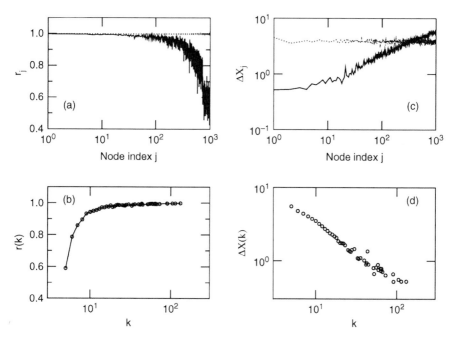

**Fig. 5.10.** (a) Phase synchronization order parameter $r_j$ of node $j$ with respect to the mean field $X$ in the UN (*solid line*) and WN (*dotted line*). (b) Average value $R(k)$ of nodes with degree $k$ as a function of $k$ in the UN; (c) and (d) as in (a) and (b), but for the distance $\Delta X_j$ and its average value $\Delta X(k)$, respectively. The results are averaged over 50 realizations of random distribution of the time scale parameter $\tau_j$. The coupling strength is $g = 0.13$

## III. Strong Coupling: Almost Complete Synchronization

Now we consider the strong coupling regime where both networks have a large and saturated amplitude in their collective oscillations (Fig. 5.7).

We take $g = 0.5$, at which the amplitude of $X$ is almost the same for both networks. The frequencies of all the oscillators are locked mutually as well as locked to the mean field; as a result, the phase synchronization order parameter is $r_j = 1$ for all oscillators in both UN and WN, i.e., the networks are globally phase synchronized. In the WN network, the phase difference $\Delta\phi_j$ between an oscillator and the mean field, averaged over time and over different realizations of random distribution of the time scale parameters $\tau_j$, is small and on average rather homogeneous for all the oscillators (Fig. 5.11(a)). This implies that the oscillators are almost completely synchronized in the sense that $\Delta X \approx A_X \sin(\Delta\phi) \approx A_X \Delta\phi$ is also small and uniform on average (Fig. 5.11(c)). In the UN, however, many nodes with a degree smaller than the mean value $K$ is not as strongly connected to the

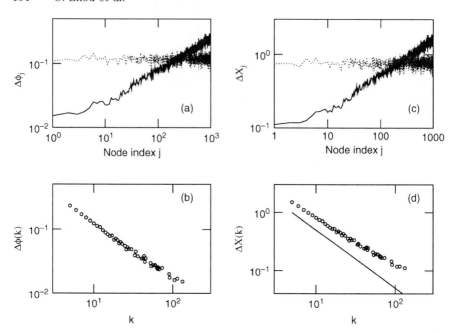

**Fig. 5.11.** (a) Averaged phase difference $\Delta\phi_j$ between a node $j$ and the mean field $X$ in the UN (*solid line*) and WN (*dotted line*); (b) Average value $\Delta\phi(k)$ of nodes with degree $k$ as a function of $k$ in the UN; (c) and (d) as in (a) and (b), but for the absolute difference $\Delta X_j$ and its average value $\Delta X(k)$, respectively. The solid line in (d) with slope $\alpha = 1$ is plotted for reference. The results are averaged over 50 realizations of the random time scale parameters $\tau_j$. The coupling strength is $g = 0.5$

mean field, and on average they have phase differences larger than that of the WN (Fig. 5.11(a)), as is shown evidently by the average value $\Delta\phi(k)$ over nodes with degree $k$ (Fig. 5.11(b)). Consequently, the synchronization difference $\Delta X_j$ is still heterogeneous (Fig. 5.11(c)) and $\Delta X(k) \sim k^{-\alpha}$ with $\alpha \approx 1$ (Fig. 5.11(d)).

### 5.4.4 Analysis of Hierarchical Synchronization

We have shown with unweighted SFNs that the effective synchronization displays a hierarchical organization according to the intensities, when the coupling is not strong enough, when there is noise or when the oscillators are non-identical. Here, we present an analysis of this hierarchical synchronization. The analysis is based on a mean field approximation of (5.1) with (5.26). For the case of UN, it reads

$$\dot{x}_j = \tau_j F(x_j) + \frac{gk_j}{K}(X - x_j), \quad k_j \gg 1 . \tag{5.29}$$

This approximation means that the oscillators are forced by a common signal $X$, with the forcing strength being proportional to their degree $k_j$.

For identical oscillators ($\tau_j = 1, \forall j$), the linear variational equations of (5.29) are

$$\dot{\xi}_j = \left[ \mathbf{D}\boldsymbol{F}(\boldsymbol{X}) - \frac{g}{K} k_j \mathbf{I} \right] \xi_j, \quad k_j \gg 1 , \tag{5.30}$$

which have the same form as (5.4), except that $\lambda_j$ is replaced by $k_j$ and $\mathbf{D}\boldsymbol{H}(\boldsymbol{s})$ is replaced by the identity matrix $\mathbf{I}$ when $\boldsymbol{H}(\boldsymbol{x}) = \boldsymbol{x}$. The largest Lyapunov exponent $\Lambda(k_j)$ (master stability function) of this linear equation is a function of $k_j$, i.e.,

$$\Lambda(k_j) = \Lambda_F - g k_j / K . \tag{5.31}$$

Remember that $\Lambda_F$ is the largest Lyapunov exponent of the isolated oscillator $\boldsymbol{F}(\boldsymbol{x})$. $\Lambda(k_j)$ becomes negative for $\frac{g}{K} k_j > \Lambda_F$. For large values of $k$ satisfying $\frac{g}{K} k \gg \Lambda_F$, we have $\Lambda(k) \approx -\frac{g}{K} k$.

Now suppose that the network is close to being completely synchronized, when the coupling strength $g$ is below the threshold $g_c$, or when there is noise present in the system. For nodes with a large degree $k$ so that $\Lambda(k) \approx -\frac{g}{K} k$ is sufficiently negative, the dynamics of the averaged synchronization difference $\Delta X(k)$ over large time scales can be expressed as

$$\frac{\mathrm{d}}{\mathrm{d}t} \Delta X(k) = \Lambda(k) \Delta X(k) + c , \tag{5.32}$$

where $c > 0$ is a constant denoting the level of perturbation with respect to the complete synchronization state, which depends on the noise level $D$ or the coupling strength $g$. For the case of non-identical oscillators, the perturbation level (constant $c$) is due to the disorder in the time scale $\tau_j$ of the oscillators. From this, we get the asymptotic result $\Delta X(k) = c / |\Lambda(k)|$, giving

$$\Delta X(k) \sim k^{-1} , \tag{5.33}$$

which explains qualitatively the numerically observed scaling in Figs. 5.5, 5.10 and 5.11. The slight deviation of the scaling exponents from the linear result $\alpha = 1$ may result from the mean field approximation and significant nonlinearity, since the linear analysis in (5.32) is only a first order approximation.

For a general weighted random network, the degree $k$ in (5.33) should be replaced by the intensity $S$, and we have

$$\Delta X(S) \sim S^{-1} . \tag{5.34}$$

## 5.5 Phase Synchronization in Small-World Networks of Oscillators

So far, we have analyzed networks that are random, where the nodes do not have spatial properties. However, in many realistic networks, the oscillators are arranged in space, and this spatial arrangement has a significant impact on the connection patterns. A good example of this type is the network of interacting neurons in local areas of the brain cortex. Here, the neurons have sparse connections to neighboring neurons, neither in a fully regular, nor in a completely random manner, but somewhere in between. A simple model describing this type of network is the SWN model proposed in [2, 34]. It is based on a regular array of oscillators, each coupled to its $k$ nearest neighbors. With a probability $p$, a link is added (rewired) to a randomly selected pair of the oscillators, i.e., some long-range connections are introduced. For small $p$ values, the resulting networks display both the properties of regular networks (high clustering) and of random networks (short pathlength), i.e., they are SWNs.

In this section, we demonstrate the important impact of such random long-range interactions on the synchronization of non-identical oscillators. We start with a regular ring of $N$ nodes, each connected to its two nearest neighbors, i.e., $k = 2$. Shortcuts are then added between randomly selected pairs of nodes, with probability $p$ per link of the basic regular ring, so that typically there are $pN$ shortcuts in the resulting networks. In this way, the total number of connections also increases with $p$. Again, we use the Rössler chaotic oscillators $F(x)$ and output function $H(x)$ as in Sect. 4. The dynamical equation is

$$\dot{x}_j = \tau_j F(x_j) + \frac{g}{k_j} \sum_{i=1}^{N} A_{ji}(x_i - x_j) , \quad j = 1, \ldots, N , \qquad (5.35)$$

Note that the coupling strength $g$ is normalized by the degree $k_j$ of each node, so that we can scale out effects of the increasing average degree $K = \langle k_j \rangle$ when more and more shortcuts are added in the network at larger probability $p$. As in Sect. 4.3, we consider a uniform distribution of $\tau_j$ in the interval $[1 - \Delta\tau, 1 + \Delta\tau]$, and we fix $\Delta\tau = 0.4$ in the simulations.

We now discuss the synchronization behavior of (5.35) for networks with different shortcut probabilities $p$. The degree of synchronization is quantified by the amplitude $A_X$ of the mean field oscillation $X = (1/N)\sum_{j=1}^{N} x_j$ as a function of the coupling strength $g$ (Fig. 5.12(a)). We also examine the variation of the oscillation amplitudes in individual oscillators with respect to $g$ by measuring the average value of the standard deviation of $x_j(t)$, $\langle A_j \rangle$ (Fig. 5.12(b)). We observe the following types of synchronous behavior:

i) When the shortcut probability $p$ is very small ($p = 0.01$), the network is still dominated by local coupling and it does not display obvious collective synchronization effects over a broad range of $g$, as indicated by an almost vanishing mean field $X$. However, the oscillation amplitude of individual oscillators changes with $g$.

ii) With a larger number of shortcuts at $p = 0.1$, the network starts to synchronize and generates a coherent collective oscillation at a strong enough coupling strength.

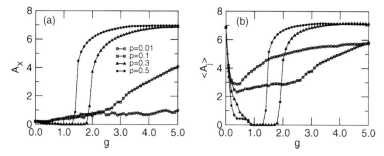

**Fig. 5.12.** Transition to oscillation death and synchronization in various SWNs of chaotic Rössler oscillators: (**a**) The amplitude $A_X$ of the mean field $X$ as a function of the coupling strength $g$; (**b**) The average value of the amplitudes $A_j$ of all the individual oscillators. The network size is $N = 1024$

iii) At even larger values of $p$, e.g., $p = 0.3$ and $p = 0.5$, the networks display three dynamical regimes: (1) When the coupling strength is increased from very small values, the trajectory of each oscillator draws closer and closer to the unstable steady state $\boldsymbol{x}_F$ ($\boldsymbol{F}(\boldsymbol{x}_F) = \boldsymbol{0}$), as seen by a rapid decrease of the amplitude $\langle A_j \rangle$ of individual oscillators (Fig 5.12(b)). The oscillation frequencies $\Omega_j$ are still distributed in this regime (Fig 5.13(a,b)). (2) When a critical value $g_1$ is reached, all oscillators become stable at the same steady

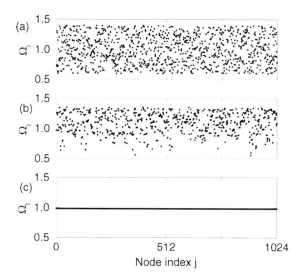

**Fig. 5.13.** Oscillation frequency $\Omega_j$ of the oscillators in SWNs with the shortcut probability $p = 0.5$ for different coupling strength: (**a**) $g = 0$; (**b**) $g = 0.1$; and (**c**) $g = 1.5$

state $x_F$, so that $A_X = 0$ and $\langle A_j \rangle = 0$, and we observe oscillation death in SWNs, i.e., all the oscillators stop oscillating (Fig 5.12(b)). (3) When $g$ is further increased to exceed another critical value $g_2$, the steady state $x_F$ becomes unstable again, and the oscillation is restored. Importantly, the whole network is now in a global synchronization regime: the frequencies and phases of all oscillators are locked (Fig 5.13(c)). Comparing the critical value $g_2$ of the coherent synchronization regime for $p = 0.3$ and $p = 0.5$, one can see that networks with more shortcuts achieve this coherent synchronization with a smaller coupling strength.

For a fixed value of the coupling strength $g$, the two regimes of oscillation death and global synchronization can also be obtained by adding a suffi-cient number of shortcuts (Fig. 5.14). The system behavior is not sensitive to increasing $p$ when $p < 0.02$. With a further increase of $p$, the oscillation amplitudes of the oscillators are reduced and finally the regime of oscillation death is reached, which is stable for networks in a certain range of $p$, and afterwards a coherent collective oscillation is observed due to global synchro-nization, which becomes more pronounced as more shortcuts are added to the network.

We have shown that the coupling topology in the SWNs has significant ef-fects on the synchronization of strongly non-identical nonlinear oscillators. Compared to regular networks with local coupling ($p \approx 0$), SWNs with many shortcuts display enhanced synchronization as expressed by the regimes

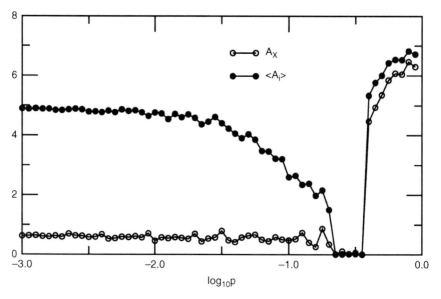

**Fig. 5.14.** Synchronization behavior vs. the shortcut probability $p$ for a fixed coupling strength ($g = 1.5$)

of oscillation death and global synchronization similar to globally coupled networks [35].

## 5.6 Hierarchical Synchronization and Clustering in Complex Brain Networks

Synchronization of distributed brain activity has been proposed as an important mechanism for neural information processing [6]. Experimentally observed brain activity, characterized by synchronization phenomena over a wide range of spatial and temporal scales [36], reflects a hierarchical organization of the dynamics. Such an organization arises through a hierarchy of complex cortical networks: the microscopic level of interacting neurons, the mesoscopic level of mini-columns and local neural circuits, and the macroscopic level of nerve fiber projections between brain areas [15]. While details at the first two levels are still largely missing, extensive information has been collected about the latter level in the brain of animals, such as the cat and the macaque monkey [37]. The complex topology of cortical networks has been the subject of many recent analyses [37]. See Chaps. 3, 4, or 9 for a complete review. Analyses of the anatomical connectivity of the mammalian cortex [37] and the functional connectivity of the human brain [38] have shown that the two share typical features of many complex networks. However, the relationship between anatomical and functional connectivities remains one of the major challenges in neuroscience [6].

Conceptually modeling the dynamics of the neural system based on a realistic network of corticocortical connections and investigating the synchronization behavior should provide meaningful insights into this problem. Here we consider the cortical network of the cat. The cortex of the cat can be parcellated into 53 areas, linked by about 830 fibers of different densities [15] into a weighted complex network as shown in Fig. 5.15(a). This network displays typical small-world properties, i.e., short average pathlength and high clustering coefficient, indicating an optimal organization for an effective inter-area communication and for achieving high functional complexity [39,40]. The degrees of the nodes are heterogeneous, for example, some nodes have only two or three links, while some others have up to 35 connections. Due to the small number of areas, it is difficult to claim a scale-free distribution [40], nevertheless, analyses comparing this network to scale-free network models with the same size and connectivity density does suggest a scale-free distribution (see Sect. 3.7 of Chap. 3).

Different from random networks models, the cortical network of the cat exhibits hierarchically clustered organization [40,41]. There are a small number of clusters that broadly agree with the four functional cortical sub-divisions, i.e., visual cortex (V, 16 areas), auditory (A, 7 areas), somatosensory-motor (SM, 16 areas) and frontolimbic (FL, 14 areas). To distinguish these from the dynamical clusters in the following discussion, we refer to the topological

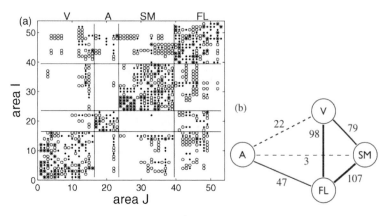

**Fig. 5.15.** (a) Connection matrix $M^A$ of the cortical network of the cat brain. The different symbols represent different connection weights: 1 (• *sparse*), 2 (○ *intermediate*) and 3 (∗ *dense*). The organization of the system into four topological communities (functional sub-systems, V, A, SM, FL) is indicated by the dashed lines; (b) The number of connections between the four communities

clusters as *communities* [42]. The inter-community connections in Fig. 5.15(b) show that A is much less connected while V, SM and FL are densely connected with each other.

Next, we analyze synchronization dynamics of this network by simulating each cortical area with (i) periodic neural mass oscillators for modeling neural rhythms [43, 44] and (ii) a subnetwork of interacting excitable neurons [45, 46]. While the model with neural mass oscillators typically displays the synchronization behavior explained in Sect. 5.4, the model with subnetworks shows that the dynamics is also hierarchically organized and reveals different scales in the hierarchy of the network topology. In particular, in the biologically plausible regime, the most prominent dynamical clusters coincide closely with anatomical communities that agree broadly with the functional sub-divisions V, A, SM and FL.

### 5.6.1 Neural Mass Model

The mean activities of a population of neurons in the brain often exhibit rhythmic oscillations with well defined frequency bands, as seen in EEG measurements (cf. Chaps. 7 and 8). Such oscillations can be captured by realistic macroscopic models of EEG generation proposed in the early 1970s [43] (see Chap. 1 for a discussion). In this section, we use the neural mass model and parameters presented in [44]. A population of neurons contains two subpopulations: subset 1 consists of pyramidal cells receiving excitatory or inhibitory feedback from subset 2. Subset 2 is composed of local interneurons receiving excitatory input. The neural mass model describes the evolution of the macroscopic variables, i.e., mass potentials $v^e$, $v^i$ and $v^d$ for the excitatory, inhibitory and interneurons, respectively. A static nonlinear sigmoid function $f(v) = 2e_0/(1 + e^{r(v_0-v)})$ converts the average membrane potential into an average pulse density of action potentials. Here, $e_0$ is the firing rate at the mass potential $v_0$ and $r$ is the steepness of the activation. The input from another group of neurons and from external signals is fed into the population of interneurons. The dynamical equations for $I = 1, \ldots, N$ multiple coupled populations read

$$\ddot{v}_I^e = Aaf(v_I^d - v_I^i) - 2a\dot{v}_I^e - a^2 v_I^e \, , \tag{5.36}$$

$$\ddot{v}_I^i = BbC_4 f(C_3 v_I^e) - 2b\dot{v}_I^i - b^2 v_I^i \, , \tag{5.37}$$

$$\ddot{v}_I^d = Aa\left[C_2 f(C_1 v_I^e) + p_I(t) + \frac{g}{\langle S \rangle} \sum_J^N W_{IJ} f(v_J^d - v_J^i)\right] \tag{5.38}$$

$$- 2a\dot{v}_I^d - a^2 v_I^d \, ,$$

where $v_I^e$, $v_I^i$ and $v_I^d$ are the mass potentials of the area $I$. Here, $A$ and $B$ are the average synaptic gain, $a$ and $b$ are the characteristic time constants of the EPSP and IPSP, respectively; $C_1$ and $C_2$, $C_3$ and $C_4$ are the average number of synaptic contacts, for the excitatory and inhibitory synapses, respectively. More detailed interpretation and standard values of these model parameters can be found in [44]. The coupling strength $g$ is normalized by the mean intensity $\langle S \rangle$ as in Sect. 5.4.

Here we model the cat cortical network by simulating each cortical area (a large ensemble of neurons) by such a macroscopic neural mass oscillator, i.e., by taking the cortical network in Fig. 5.15(a) as the coupling matrix $W_{IJ}$ in (5.39). As in [44], in our simulations we take $p_I(t) = p_0 + \xi_I(t)$ where $\xi_I(t)$ is a Gaussian white noise with standard deviation $D = 2$. We fix $p_0 = 180$ so that the system is in the periodic regime corresponding to alpha waves. A typical time series of the output, the average potential $V_I = v_I^d - v_I^i$, is shown in Fig. 5.16(a). Synchronization between the areas is measured by the linear correlation coefficient $R(I, J)$ between the outputs $V_I$ and $V_J$. The average correlation $\langle R \rangle$ among all pairs of areas is shown in Fig. 5.16(b) as a function of the coupling strength $g$.

According to our analysis in Sects. 5.3 and 5.4, in a sufficiently random network, each oscillator is influenced by the mean activity of the whole network with a coupling strength proportional to the intensity $S_I = \sum_{J=1}^N W_{IJ}$. Not much direct relationship between the pair-wise coupling strength $W_{IJ}$ and the strength of synchronization $R(I, J)$ is expected. We find that this still roughly holds for the cat cortical network although it is not very random due to the clustered organization. To demonstrate this, we distinguish three cases for any pair of nodes in the network: reciprocal projections (P2), uni-directional couplings (P1) and non-connection (P0), and compute the distribution of the correlation $R(I, J)$ for these cases separately. As seen in Fig. 5.17, when the coupling is weak (e.g., $g = 2$), the distributions for P0, P1 and P2 pairs

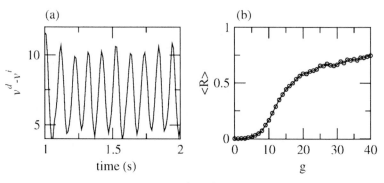

**Fig. 5.16.** (a) Typical activity $V = v^d - v^i$ of the uncoupled neural mass model; (b) The average correlation coefficient $\langle R \rangle = \frac{1}{N(N-1)} \sum_{I \neq J} R(I, J)$ ($N = 53$) vs. the coupling strength $g$ in (5.39)

coincide and display a Gaussian shape around zero and no significant correlation is established. At a stronger coupling (e.g., $g = 5$), the P2 pairs have slightly stronger correlation than P1, however, the distributions still overlap significantly, as for strong coupling.

The dynamical pattern is not structured with very weak coupling, but with stronger coupling ($g \geq 5$), the system forms a major cluster including most of the areas from V, SM and FL, while the auditory system A remains relatively independent (Fig. 5.18). This is consistent with the inter-community connectivity shown in Fig. 5.15(b). Also, some other areas with the smallest degrees and intensities are also relatively independent. The correlation coefficient $R_X$ between the activity $V_I$ of an area and the global mean field $\bar{X} = (1/N) \sum^N V_I$ is shown in Fig. 5.19. It is roughly an increasing function of $S$, which basically reproduces the behavior presented in Sect. 5.4 (e.g., Figs. 5.5, 5.10 and 5.11).

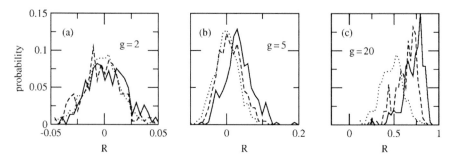

**Fig. 5.17.** (a) Distribution of the correlation $R$ for P2 (*solid line*), P1 (*dashed line*) and P0 (*dotted line*) pairs at various values of the coupling strength $g$, (a) $g = 2$, (b) $g = 5$ and (c) $g = 20$

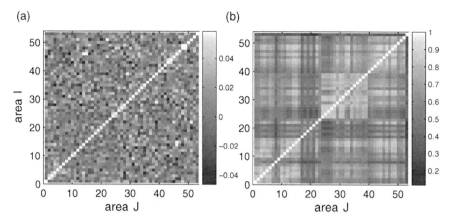

**Fig. 5.18.** Correlation matrices $R(I, J)$ for (a) weak coupling $g = 2$; (b) For strong coupling $g = 20$. Note the different gray-scales in the colorbars

## 5.6.2 Subnetworks of Interacting Neurons

Each brain area is composed of a large ensemble of neurons coupled in a complex network topology having several levels of organization; the detailed connectivity, however, is still largely unclear. In the following section, we model each cortical area with a sub-network of $N_a$ interacting neurons. We use the SWN model [2] to couple the neurons. Specifically, a regular array of $N_a$ neurons with a mean degree $k_a$ is rewired with a probability $p$. Such a topology incorporates the basic biological feature that neurons are mainly connected to their spatial neighbors, but also have a few long-range synapses [47]. The small-world topology has been shown to improve the synchronization of interacting neurons [7–9]. Our model also includes other realistic, experimentally observed features, i.e., 25% of the $N_a$ neurons are inhibitory and only a small number of neurons (about 5%) of one area receive excitatory synapses from another connected area [48]. To our knowledge, no information is available about the output synapses of each area, and for simplicity, we assume that the output signal from one area to another one is the mean activity of the output area. Individual neurons are described by the FitzHugh-Nagumo (FHN) excitable model [49] with non-identical excitability (cf. Chap. 1). A weak Gaussian white noise (with strength $D = 0.03$) is added to each neuron to generate sparse, Poisson-like irregular spiking patterns in isolated FHN neurons, as in realistic neurons.

Thus, our model of the neural network of a cat cortex is composed of a large ensemble of noisy neurons connected in a *network of networks*, and the dynamics of the neuron $i$ in the area $I$ is specified as:

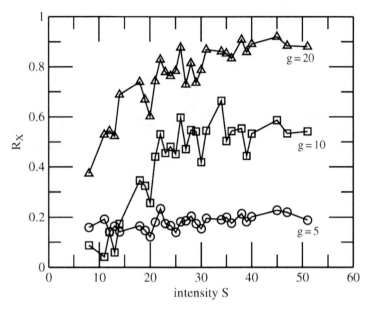

**Fig. 5.19.** Correlation between an area and the global mean field, as a function of the intensity $S$ (averaged over nodes with the same intensity $S$) at various coupling strengths $g$

$$\epsilon \dot{x}_{I,i} = f(x_{I,i}) + \frac{g}{k_a} \sum_{j}^{N_a} M_I^L(i,j)(x_{I,j} - x_{I,i})$$

$$+ \frac{g}{\langle w \rangle} \sum_{J}^{N} M^A(I,J) L_{I,J}(i)(V_J - x_{I,i}) \, , \qquad (5.39)$$

$$\dot{y}_{I,i} = x_{I,i} + a_{I,i} + D\xi_{I,i}(t) \, , \qquad (5.40)$$

where

$$f(x_{I,i}) = x_{I,i} - \frac{x_{I,i}^3}{3} - y_{I,i} \, . \qquad (5.41)$$

Here, the matrix $M^A$ represents the corticocortical connections in the cat network as in Fig. 5.15(a). $M_I^L$ denotes the local SWN of the $I$-th area $(M_I^L(i,j): (i,j) = 1, \ldots, N_a))$. A neuron $j$ is inhibitory if $M_I^L(i,j) = -1$ for all of its connected neighbors. The label $L_{I,J}(i) = 1$ if the neuron $i$ is among the 5% within the area $I$ receiving the mean field signal $V_J = (1/N) \sum_l^{N_a} x_{J,l}$ from the area $J$, otherwise, $L_{I,J}(i) = 0$. The diffusive coupling, describing electrical synapses (gap junctions; see Chap. 2) and not being the most typical case in mammalian cortex, is mainly used for the simplicity of simulation at this stage. Normalized by the mean degree $k_a$ of the SWNs within the areas, and normalized by the average weight $\langle w \rangle$ of inter-area connections, $g$ represents the average coupling strength between any pair of neurons and is the control parameter in our simulations. (Note that we assume $g$ to be equal for couplings within and between subnetworks).

The system is simulated with $N_a = 200$, $k_a = 12$ and $p = 0.3$ for the subnetworks. Our focus is to study the synchronization behavior at the systems level, i.e., the synchronization behavior between the mean activity $V_I$ of the subnetworks and its relationship with the underlying cortical network in Fig. 5.15(a). The behavior demonstrated below does not depend critically on the parameters of the subnetworks, while the detailed synchronization behavior *within* the subnetwork does depend on them [7–9].

The coupling strength $g$ controls the mutual excitation between neurons. At small $g$ (e.g. $g = 0.06$), a neuron is not often excited by the noise-induced spiking of its connected neighbors, so the synchronization within and between the subnetworks is weak. This is shown by small fluctuations of the mean activity $V_I$ of each area (Fig. 5.20(a)) and a small average correlation coefficient $\langle R \rangle$ among $V_I$ (Fig. 5.20(d)). Weak synchronization in the subnetwork of an area is manifested by some clear peaks in $V_I$ (Fig. 5.20(a)). When we increase $g$, the synchronization becomes stronger with more frequent and larger peaks in $V_I$ (Fig. 5.20(b)) and at large enough $g$, the neurons are mutually excited achieving both strongly synchronized and regular spiking behavior (Fig. 5.20(c)). $\langle R \rangle$ approaches 1 (Fig. 5.20(d)), indicating an almost global synchronization of the network.

The patterns of the correlation matrix $R(I, J)$ are shown in Fig. 5.21. The behavior at strong couplings is very similar to that of the neural mass model in Fig. 5.18, since both models display well defined oscillations. However, Fig. 5.21(a) suggests that the dynamics of the present model with weak coupling has a nontrivial organization and an intriguing relationship with the underlying network topology. The distribution of $R$ over all pairs of areas

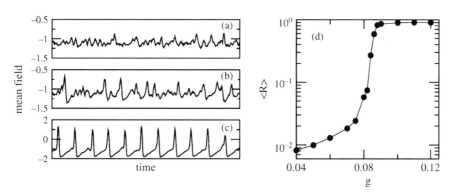

**Fig. 5.20.** Typical mean activity $V_I X$ of one area at various coupling strengths: (a) $g = 0.06$; (b) $g = 0.082$; (c) $g = 0.09$; (d) The average correlation coefficient $\langle R \rangle = \frac{1}{N(N-1)} \sum_{I \neq J} R(I, J)$ ($N = 53$) plotted vs. $g$

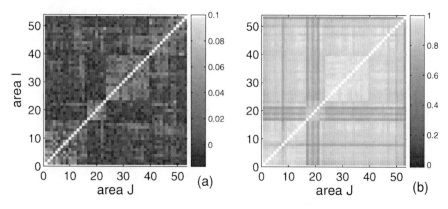

**Fig. 5.21.** Correlation matrices $R(I, J)$ at weak coupling $g = 0.06$: (**a**) and strong coupling $g = 0.12$; (**b**) Note the different gray-scales in the colorbars

displays a Gaussian peak around zero, but with a long tail for large values (Fig. 5.22(a), solid line). Although the correlations are relatively small, we find that the large values are significant when compared to the distribution of $R$ of surrogate data by random shuffling of the time series $V_I$ (Fig. 5.22(a), dash-dotted line). The weak coupling regime is biologically more realistic since here, the neurons only have a low frequency of irregular spiking and irregular mean activities (Fig. 5.20(a)), similar to those observed experimentally (e.g., EEG data [50]). The propagation of a signal between connected areas is mediated by synchronized activities (peaks in $V$) and a temporal correlation is most likely established when receivers produce similar synchronized activities from one input, or when two areas are excited by strongly correlated signals from common neighbors. Due to the weak coupling and the existence of subnetworks, such a synchronized response does not always occur and a local signal (excitation) does not propagate through the whole network. As a result, the correlation patterns are closely related to the network topology, although the values are relatively small due to infrequent signal propagation. With strong coupling, the signal can propagate through the whole network, corresponding to pathological situations, such as epileptic seizure [51].

Let us now characterize the dynamical organization and its relationship to the network topology. Based on an argument of signal propagation, we expect that the correlations for the P2, P1 and P0 areas should be different. Indeed, the distributions of $R$ for these three cases display well-separated peaks in the weak coupling regime (Fig. 5.22(a)). Note especially that, all the P2 pairs have significant correlations compared to the surrogate data. With strong coupling (e.g. $g > 0.09$), where the excitation propagates through the whole network, the distribution is very similar to the neural mass model in Fig. 5.17(c), and the separation is no longer pronounced (not shown).

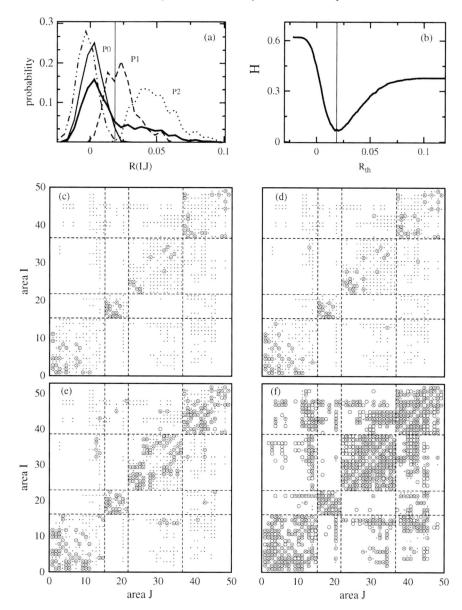

**Fig. 5.22.** (a) Distribution of the correlation $R$ $(g = 0.07)$ for all nodes (*solid line*), P2 (*light solid line*), P1 (*dashed line*) and P0 (*dotted line*). The dash-dotted line denotes the results for the surrogate data; (b) Hamming distance $H$ vs. $R_{th}$. The vertical solid lines in (a) and (b) indicate the natural threshold $R_{th} = 0.019$. The functional networks (∘) with thresholds $R_{th} = 0.070, 0.065, 0.055$, and $0.019$ are shown in (c), (d), (e), and (f), respectively. The small dots indicate the anatomical connections

We extract a *functional network* $M^F$ [38] by applying a threshold $R_{th}$ to the correlation $R$, i.e., a pair of areas is considered to be functionally connected if $R(I, J) \geq R_{th}$ ($M^F(I, J) = 1$). We can then compare the topological structures of the anatomical network $M^A$ and the functional networks $M^F$ with varying $R_{th}$ and examine how the various levels of synchronization reveal different scales in the network topology. focusing on the biologically meaningful weak coupling regime.

We focus on the biologically meaningful weak coupling regime and take $g = 0.07$ as the typical case. When $R_{th}$ is very close to the maximal value of $R$, only a few P2 areas in the auditory system A are functionally connected, because of their strong anatomical links and sharing of many common neighbors. With lower values, e.g. $R_{th} = 0.07$, about 2/3 of the areas but only 10% of the P2 links are active in the functional network (Fig. 5.22(c)). Interestingly, within each anatomical community V, A, SM, and FL, a core subnetwork is functionally manifested in the form of connected components without inter-community connections. At lower values, e.g., $R_{th} = 0.065$, more areas from the respective communities are included into these components and a few inter-community connections appear to join the components from V, SM, and FL (Fig. 5.22(d)). This observation suggests that a core subnetwork coupled more strongly and communicating more frequently among the areas within the respective community is most likely to perform specialized functions of this community. Going to an even lower threshold, e.g., $R_{th} = 0.055$, all areas become involved and form a single connected functional network, but this network contains only about 1/3 of the anatomical P2 links and very few P1 links. However, the communication of the whole network is still mediated only by a small number of inter-community connections while most of the connections are within V, A, SM and FL, i.e., the functional network is highly clustered and agrees well with the anatomical communities. With further reduction of $R_{th}$, still more anatomical links are expressed as functional connections. For example, at $R_{th} = 0.019$, all P2 links are just fully expressed and about 70% of P1 links too. Meanwhile, about 4% of non-connected pairs (P0) establish significant functional connections (the significance level $\approx 0.004$ at $R = 0.019$), since they have many common neighbors. Thus, the functional network reveals the anatomical network rather faithfully (Fig. 5.22(f)). To compare the matrices $M^F$ (symmetrical) and $M^A$ (asymmetrical) in a more quantitative way, we take the binary matrix of $M^A$, symmetrize all P1 links and compute the Hamming distance $H$, i.e., the percentage of elements between $M^F$ and the binary and symmetrized $M^A$ that are different. The closeness between them is confirmed by a very small Hamming distance $H = 0.074$, which is almost minimal for varying $R_{th}$ (Fig. 5.22(b)). It is interesting to note that this value of the threshold $R_{th}$ is exactly where the full distribution of $R$ starts to deviate from the Gaussian and the distribution of P2 areas separates from that of the surrogate data (Fig. 5.22(a), solid line). We find that such a natural choice of $R_{th}$ always reproduces the network topology well with $H \approx 0.06$ for different coupling strength $0.04 \leq g \leq 0.08$.

We have analyzed the dynamical clusters using the algorithm for hierarchical clustering in Matlab with the dissimilarity matrix $d = [d(I, J) = 1 - R(I, J)]$. Typical hierarchical trees for the weak and strong synchronization regimes are shown in Fig. 5.23. Figure 5.24 displays the most prominent clusters for the weak coupling regime. The functional clusters closely resemble the four communities obtained by using graphical tools based on anatomical structures [40, 41]. The four dynamical clusters sufficiently correspond to the functional sub-division of the cortex– $C_1$ (V), $C_2$ (A), $C_3$ (SM), $C_4$ (FL). However, it is also important to notice that there are a few nodes that belong to one anatomical community but join another dynamical cluster. For example, the area $I = 49$ (anatomically named as 36 in the cat cortex) of the fronto-limbic system is in the dynamical cluster $C_2$ mainly composed of areas from the auditory system (Fig. 5.24 ($C_2$)). A close inspection shows that these nodes bridging different anatomical communities and dynamical clusters are exactly the areas sitting in one anatomical community but in close connectional association with the areas in other communities [15]. In the strong synchronization regime, V, SM and FL join to form a major cluster (Fig. 5.25 ($C_3$)), while the auditory system A remains as a distinct cluster (Fig. 5.25 ($C_2$)). The formation of a cluster from community A both in the weak and the strong synchronization regimes is due to almost global connections within A. The cluster formation behavior in the strong coupling regime is also in good accordance with the inter-community connectivity shown in Fig. 5.15(b). There are also two single areas showing themselves as independent clusters. It turns out that these are the nodes with the minimal intensities in the network. In [46], we have shown that the clustering patterns remain almost the same in randomized networks that preserve the sequence of the

(a)                                    (b)

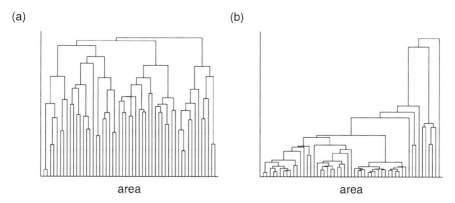

area                                    area

**Fig. 5.23.** Typical hierarchical tree of the dynamical clusters in the weak coupling regime; (**a**) $g = 0.07$ and strong coupling regime: (**b**) $g = 0.12$

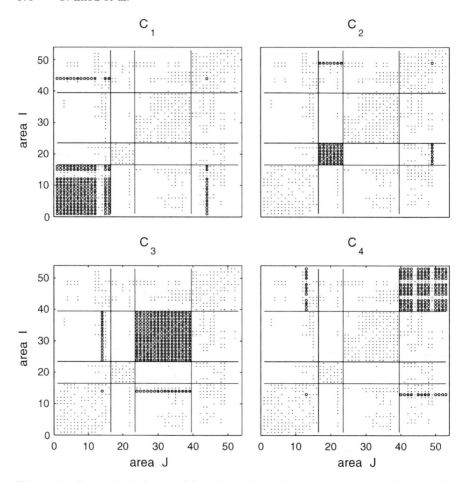

**Fig. 5.24.** Dynamical clusters (∘) with weak coupling strength $g = 0.07$, overlaid on the underlying anatomical connections (·)

intensities $S_I$ as in the cat cortical network; the auditory system A no longer forms a distinct cluster when the pronounced intra-community connections are destroyed in the randomized networks. This demonstrates that our understanding of synchronization based on weighted random networks in Sects. 5.3 and 5.4 can be applied when the node dynamics (mean activity of the subnetwork in this case) display a well-defined oscillatory behavior.

The comparison between models with subnetworks and those with neural mass oscillators indicates that self-sustained oscillator models may not be as appropriate for the understanding of the interplay between dynamics and structure in the brain as a hierarchical network of *excitable* elements.

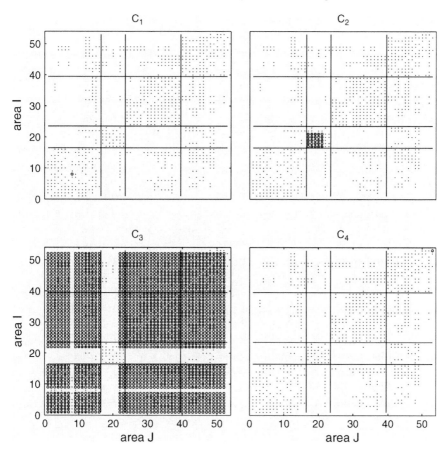

**Fig. 5.25.** Dynamical clusters (∘) with strong coupling strength $g = 0.12$, overlaid on the underlying anatomical connections (·).

## 5.7 Conclusion and Outlook

In this chapter, we discussed synchronization dynamics on complex networks. Firstly, we analyzed the relationship of some structural measures (such as degree and intensity of the nodes, fraction of random shortcuts etc.) in general network models to the synchronization behavior of the networks. Sections 5.2 and 5.3 considered the ideal case of complete synchronization allowing us to characterize the synchronizability of the network based solely on the spectral properties of the network. The main result is that the synchronizability, as measured by the ratio between the maximal and the minimal eigenvalues, is mainly determined by the maximal and minimal intensities in sufficiently random networks. In more general situations, the dynamics is perturbed away from the complete synchronization state, and we showed that the effective synchronization is hierarchically organized according to the distribution of the intensities. We also demonstrated that random long-range interactions in spatially extended networks (small-world networks) can induce different synchronization regimes, such as oscillation death and global synchronization with highly non-identical oscillators.

Secondly, we studied synchronization in a realistic network of cat cortical connectivity. We demonstrated that if well-defined oscillatory dynamics is assumed for the nodes (which represent cortical areas composed of large ensembles of neurons), for example, by employing a neural mass model, or by a subnetwork of rather strongly coupled neurons, the synchronization patterns can be understood by using general principles discussed in the first part of this chapter, i.e., the synchronization is mainly controlled by the global structural statistics (intensities) of the network. However, with weak coupling, the model with subnetworks displays biologically plausible dynamics and the synchronization pattern exhibits a close relationship with the hierarchically clustered organization in the network structure, i.e., the dynamics is mainly controlled by the local structures in the network.

Here, we have only focused on the highest structural level and modeled each large cortical area with one level of subnetwork and simple neuron dynamics. The maximal correlation (0.1–0.2) is low in a biologically plausible regime. The model displays a large region of frequent and regular spiking in the neurons and strong synchronization even without strong external influence. The model can be extended and improved in several ways, in order to address more realistic information processing in the brain:

(i) Biologically, a system of $10^5$ neurons corresponding to a cubic millimeter of cortex is the minimal system size at which the complexity of the cortex can be represented, e.g., the number of synapses a neuron receives is $10^4$ [52]. Thus, large sub-networks with other biologically realistic features, e.g., additional hierarchically clustered organization and more detailed spatial structure of neural circuits, should be considered. Such an extension is important for the modeling and simulation of experimentally observed hierarchical activity characterized by synchronization phenomena over a wide range of spatial and temporal scales.

(ii) Cortical neurons display rich dynamics which would require more subtle neuron models.

(iii) Biologically more realistic coupling by chemical synapses should be used and synaptic plasticity considered.

An extension of our model by additionally including the hierarchy of clustered structures reflecting the connectivity at the level of local neuronal circuits would allow localized and strong synchronization in some low-level clusters and naturally organize dynamics at higher scales. This will significantly broaden the biologically plausible regimes with stronger correlations, as observed experimentally [38].

Synchronization of distributed brain activity has been believed to be an important mechanism for neural information processing [6]. A carefully extended model could be used to investigate the relative contributions of network topology and task-related network activations to functional brain connectivity and information processing. The dynamics of the model could be then compared to the observed spread of activity in the cortex [53] and to the functional connectivity [38] at suitable spatio-temporal scales.

Simulations of such large, complex neural network models of the cortex and investigations of the relationship between network structure, dynamics organization and function of the system crossing various levels in the hierarchy would require significant developments both in neurophysics, in theory of dynamical complex networks and in algorithms of parallel computing [52].

## Acknowledgements

The authors thank A.E. Motter and C.C. Hilgetag for helpful discussions.

## References

1. See, e.g., reviews: S. H. Strogatz, Nature (London) **410**, 268 (2001); R. Albert and A.-L. Barabási, Rev. Mod. Phys. **74**, 47 (2002); S. Boccaletti et al., Phys. Rep. **424**, 175 (2006).
2. D. J. Watts and S. H. Strogatz, Nature (London) **393**, 440 (1998).
3. A.-L. Barabási and R. Albert, Science **286**, 509 (1999).
4. R. Milo, S. Shen-Orr, S. Itzkovitz, N. Kashtan, D. Chlkovskii and U. Alon, Science **298**, 824 (2002).
5. A partial list, e.g., P. M. Gade and C. K. Hu, Phys. Rev. E **62**, 6409 (2000); J. Jost and M. P. Joy, Phys. Rev. E **65**, 016201 (2001); M. Barahona and L. M. Pecora, Phys. Rev. Lett. **89**, 054101 (2002); A. E. Motter, C. S. Zhou and J. Kurths, Europhys. Lett. **69**, 334 (2005); Phys. Rev. E **71**, 016116 (2005); L. Donetti, P. I. Hurtado and M. A. Munoz, Phys. Rev. Lett. **95**, 188701 (2005); A. Arenas, A. Diáz-Guilera and C. J. Peréz-Vicentz, Phys. Rev. Lett. **96**, 114102 (2006).
6. E. Salinas and T. J. Sejnowski, Nature Neurosci. **2**, 539 (2001); P. Friés, Trends Cogn. Sci. **9**, 474 (2005); A. Schnitzler and J. Gross, Nature Neurosci. **6**, 285 (2005).
7. L. F. Lago-Fernández, R. Huerta, F. Corbacho and J. A. Sigüenza, Phys. Rev. Lett. **84**, 2758 (2000).
8. N. Masuda and K. Aihara, Biol. Cybern. **90**, 302 (2004).
9. X. Guardiola, A. Diaz-Guilera, M. Llas and C. J. Peréz, Phys. Rev. E **62**, 5565 (2000).
10. M. Timme, F. Wolf and T. Geisel, Phys. Rev. Lett. **92**, 074101 (2004); M. Denker, M. Timme, M. Diesmann, F. Wolf and T. Geisel, Phys. Rev. Lett. **92**, 074103 (2004); V. N Belykh, E. de Lange and M. Hasler, Phys. Rev. Lett. **94**, 188101 (2005).
11. H. Hong, M. Y. Choi and B. J. Kim, Phys. Rev. E **65**, 026139 (2002).
12. A. M. Batista, S. E. D. Pinto, R. L. Viana and S. R. Lopes, Physica A **322**, 118 (2003).

13. T. Nishikawa, A. E. Motter, Y.-C. Lai and F. C. Hoppensteadt, Phys. Rev. Lett. **91**, 014101 (2003).

14. F. Chung and L. Lu, Proc. Natl. Acad. Sci. U.S.A. **99**, 15879 (2002); R. Cohen and S. Havlin, Phys. Rev. Lett. **90**, 058701 (2003).

15. J. W. Scannell, G. A. P. C. Burns, C. C. Hilgetag, M. A. O'eil and M. P. Yong, Cereb. Cortex **9**, 277 (1999).

16. B. T. Grenfell, O. N. Bjornstad and J. Kappey, Nature (London) **414**, 716 (2001).

17. G. Korniss, M. A. Novotny, H. Guclu, Z. Toroczkai and P. A. Rikvold, Science **299**, 677 (2003).

18. A. Barrat, M. Barthélemy, R. Pastor-Satorras and A. Vespignani, Proc. Natl. Acad. Sci. U.S.A. **101**, 3747 (2004).

19. A. E. Motter, C. S. Zhou and J. Kurths, Europhys. Lett. **69**, 334 (2005); Phys. Rev. E **71**, 016116 (2005).

20. M. Chavez, D.-U. Hwang, A. Amann, H. G. E. Hentschel and S. Boccaletti, Phys. Rev. Lett. **94**, 218701 (2005).

21. C. S. Zhou, A.E. Motter and J. Kurths, Phys. Rev. Lett. **96**, 034101 (2006).

22. C. S. Zhou and J. Kurths, Phys. Rev. Lett. **96**, 164102 (2006).

23. L. M. Pecora and T. L. Carroll, Phys. Rev. Lett. **80**, 2109 (1998).

24. L. M. Pecora and M. Barahona, Chaos and Complexity Lett. **1**, 61 (2005).

25. F. Chung, L. Lu and V. Vu, Proc. Natl. Acad. Sci. U.S.A. **100**, 6313 (2003).

26. X. F. Wang, Int. J. Bifurcation Chaos Appl. Sci. Eng. **12**, 885 (2002).

27. J. Jost and M. P. Joy, Phys. Rev. E **65**, 016201 (2001).

28. S. Jalan and R. E. Amritkar, Phys. Rev. Lett. **90**, 014101 (2003).

29. S. N. Dorogovtsev and J. F. F. Mendes, Phys. Rev. E **62**, 1842 (2000).

30. M. E. J. Newman, S. H. Strogatz and D. J. Watts, Phys. Rev. E **64**, 026118 (2001).

31. C. S. Zhou and J. Kurths, Chaos **16**, 015104 (2006).

32. M. Rosenblum, A. Pikovsky and J. Kurths, Phys. Rev. Lett. **76**, 1804 (1996).

33. A. S. Pikovsky, M. Rosenblum and J. Kurths, *Synchronization – A universal concept in nonlinear sciences*, Cambridge University Press, 2001; S. Boccaletti, J. Kurths, G. Osipov, D. L. Valladares and C.S. Zhou, The Synchronization of Chaotic Systems, Phys. Rep. **366**, 1–101 (2002).

34. M. E. J. Newman, C. Moore and D. J. Watts, Phys. Rev. Lett. **84**, 3201 (2000).

35. G. V. Osipov, J. Kurths and C. S. Zhou, *Synchronization in Oscillatory Networks,* Spring, Berlin, 2007.

36. C. J. Stam and E. A. de Bruin, Hum. Brain Mapp. **22**, 97 (2004).

37. See a recent review: O. Sporns, D. R. Chialvo, M. Kaiser and C. C. Hilgetag, Trends Cogn. Sci. **8**, 418 (2004).

38. C. J. Stam, Neurosci. Lett. **355**, 25 (2004); V. M. Eguíluz, D. R. Chialvo, G. Cecchi, M. Baliki, and A. V. Apkarian, Phys. Rev. Lett. **94**, 018102 (2005); R. Salvador et al., Cereb. Cortex **15**, 1332 (2005).

39. O. Sporns and J. D. Zwi, Neuroinformatics **2**, 145 (2004).

40. C. C. Hilgetag and M. Kaiser, Neuroinformatics **2**, 353 (2004).

41. C. C. Hilgetag, G. A. Burns, M. A. O'Neill, J. W. Scannell and M. P. Young, Phil. Trans. R. Soc. Lond. B. **355**, 91 (2000).

42. M. E. J. Newman and M. Girvan, Phys. Rev. E. **69**, 026113 (2004).

43. F. H. Lopes da Silva, A. Hoeks, H. Smits and L. H. Zetterberg, Kybernetik **15**, 27 (1974).

44. F. Wendling, J. J. Bellanger, F. Bartolomei and P. Chauvel, Biol. Cybern. **83**, 367 (2000).
45. C. S. Zhou, L. Zemanová, G. Zamora, C. C. Hilgetag and J. Kurths, Phys. Rev. Lett. **97**, 238103 (2006).
46. L. Zemanová, C. S. Zhou, J. Kurths, Physica D **224**, 202 (2006).
47. G. Buzsaki, C. Geisler, D. A. Henze and X. J. Wang, Trends Neurosci. **27**, 186 (2004).
48. M. P. Young, Spat. Vis. **13**, 137 (2000).
49. R. FitzHugh, Biophys. J. **1**, 445 (1961).
50. E. Niedermeyer and F. Lopes da Silva, *Electroencephalography: Basic principles, clinical applications, and related fields*, Williams & Wilkins, 1993; R. Kandel, J. H., Schwartz, and T. M. Jessell, *Principles of Neural Science*, McGraw-Hill, 2000.
51. P. Kudela, P. J. Franaszczuk and G. K. Bergey, Biol. Cybern. **88**, 276 (2003).
52. A. Morrison, C. Mehring, T. Geisel, A. Aertsen and M. Diesmann, Neural Comput. **17**, 1776 (2005).
53. R. Kötter and F. T. Sommer, Phil. Trans. R. Soc. Lond. B **355**, 127 (2000).

# 6

# Synchronization Analysis
# of Neuronal Networks by Means
# of Recurrence Plots

André Bergner and Maria Carmen Romano, Jürgen Kurths and Marco Thiel

Nonlinear Dynamics Group, University of Potsdam
bergner@agnld.uni-potsdam.de

**Summary.** We present a method for synchronization analysis, that is able to handle large networks of interacting dynamical units. We focus on large networks with different topologies (random, small-world and scale-free) and neuronal dynamics at each node. We consider neurons that exhibit dynamics on two time scales, namely spiking and bursting behavior. The proposed method is able to distinguish between synchronization of spikes and synchronization of bursts, so that we analyze the synchronization of each time scale separately. We find for all network topologies that the synchronization of the bursts sets in for smaller coupling strengths than the synchronization of the spikes. Furthermore, we obtain an interesting behavior for the synchronization of the spikes dependent on the coupling strength: for small values of the coupling, the synchronization of the spikes increases, but for intermediate values of the coupling, the synchronization index of the spikes decreases. For larger values of the coupling strength, the synchronization index increases again until all the spikes synchronize.

## 6.1 Introduction

Networks are ubiquitous in nature, biology, technology and in the social sciences (see [1] and references therein). Much effort has been made to describe and characterize them in different fields of research. One key finding of these studies is that there are unifying principles underlying their behavior. In the past, two major approaches have been pursued to deal with networks. The first approach considers networks of regular topology, such as arrays or rings of coupled systems with nonlinear and complex dynamics on each node. The second approach concentrates on the topology of the network and sets aside the dynamics or at most considers a rather simple one at each node. Some of the prototypical types of network architectures that have been considered are random, small-world, scale-free and generalized random networks [2].

Recently, the study of complex dynamics on the nodes has been extended from regular to more complex architectures [3]. However, in most previous work, each node is still considered to be a phase oscillator (system with one predominant time scale), often pulse-coupled to each other. Much is left, however, to understand about network behavior with more realistic complex dynamics on the nodes of networks of complex architecture, such as chaotic and stochastic dynamics, which is found in many real application systems, such as in neural networks. The influence of the topology of the network on the dynamical properties of the complex systems is currently being investigated in the context of synchronisation [4–6].

Synchronization of complex systems has been intensively studied during the last years [7] and it has been found to be present in numerous natural and engineering systems [8]. Chaotic systems defy synchronization due to their sensitivity to slight differences in initial conditions. However, it has been demonstrated that these kind of systems are able to synchronize. In the case of two interacting non-identical chaotic systems (which is more likely to occur in nature than if they were identical), several types of synchronization might occur, dependent on the coupling strength between the systems. For rather weak coupling strength, phase synchronisation (PS) might set in. In this case, the phases and frequencies of the complex systems are locked, i.e. $|\phi_1(t) - \phi_2(t)| <$ const. and $\omega_1 \approx \omega_2$, whereas their amplitudes remain uncorrelated. If the coupling strength is further increased, a stronger relationship between the interacting systems might occur, namely generalized synchronization (GS). In this case, there is a functional relationship between both systems. Finally, for very strong coupling, both systems can become almost completely synchronized. Then, their trajectories evolve almost identically in time [7].

In the case of phase synchronization, the first step in the analysis is to determine the phases $\phi_1(t)$ and $\phi_2(t)$ of the two interacting systems with respect to the time $t$. If the chaotic systems have mainly one characteristic time scale, i.e. a predominant peak in the power spectrum, the phase can be estimated as the angle of rotation around one center of the projection of the trajectory on an appropriate plane. Alternatively, the analytical signal approach can be used [9]. However, for most of the complex systems found in nature, there is more than one characteristic time scale [10]. Hence, the approaches mentioned above to estimate the phase are not appropriate. Recently, a new method, based on the recurrence properties of the interacting systems [11], has been introduced to overcome this problem. By means of this technique, it is possible to analyze systems with a rather broad spectrum, as well as systems strongly contaminated by noise or subjected to non-stationarity [12].

In this chapter, we extend the recurrence based technique for phase synchronization analysis to systems with two predominant time scales, so that it is possible to obtain one synchronization index for each time scale. Moreover, we apply this method to large networks of different architectures with neuronal dynamics on their nodes.

The outline of this chapter is as follows: in Sect. 6.2, we introduce the concept of recurrence, as well as the synchronization index based on the recurrence properties of the system. In Sect. 6.2.2, we present the method to analyze the synchronization for two different time scales separately. In Sect. 6.3, we apply the method to complex networks of neurons and present the obtained results.

## 6.2 Phase Synchronization by Means of Recurrences

First, we show the problem of defining the phase in systems with rather broad power spectrum by using the paradigmatic system of two coupled non-identical Rössler oscillators:

$$\dot{x}_{1,2} = -\omega_{1,2}y_{1,2} - z_{1,2}$$
$$\dot{y}_{1,2} = \omega_{1,2}x_{1,2} + ay_{1,2} + \mu(y_{2,1} - y_{1,2}) \qquad (6.1)$$
$$\dot{z}_{1,2} = 0.1 + z_{1,2}(x_{1,2} - 8.5) \,,$$

where $\mu$ is the coupling strength and $\omega_{1,2}$ determine the mean intrinsic frequency of the (uncoupled) oscillators in the case of phase coherent attractors. In our simulations, we take $\omega_1 = 0.98$ and $\omega_2 = 1.02$. The parameter $a \in [0.15, 0.3]$ governs the topology of the chaotic attractor. When $a$ is below a critical value $a_c$ ($a_c \approx 0.186$ for $\omega_1 = 0.98$ and $a_c \approx 0.195$ for $\omega_2 = 1.02$), the chaotic trajectories always cycle around the unstable fixed point $(x_0, y_0) \approx (0, 0)$ in the $(x, y)$ subspace (Fig. 6.1(a)). In this case, the rotation angle

$$\phi = \arctan \frac{y}{x} \qquad (6.2)$$

can be defined as the phase which increases almost uniformly. The oscillator has coherent phase dynamics, i.e. the diffusion of the phase dynamics is very low $(10^{-5}–10^{-4})$. In this case, other phase definitions, e.g. based on the Hilbert transform or on the Poincaré section, yield equivalent results [9]. However, beyond the critical value $a_c$, the trajectories no longer completely cycle around $(x_0, y_0)$ – the attractor becomes the so-called funnel attractor. Such earlier returns in the funnel attractor happen more frequently with increasing $a$ (Fig. 6.1(b)). It is clear that for the funnel attractors, usual (and rather simple) definitions of phase, such as (6.2), are no longer applicable [9].

Another problematic case arises if the systems under consideration have two predominant time scales, which is common in many real systems, e.g. neurons with spiking and bursting dynamics. In such cases, the definition of the phase given by (6.2) is also not appropriate.

Figure 6.2 shows these problems with the time series of a Hindmarsh-Rose neuron.[1]

Rosenblum et al. [13] have proposed the use of an ensemble of phase coherent oscillators that is driven by a non-phase-coherent oscillator in order to estimate the frequency of the latter and hence detect PS in such kind

---

[1] For the definition of the Hindmarsh-Rose neuron see Sect. 6.3 and also Chap. 1.

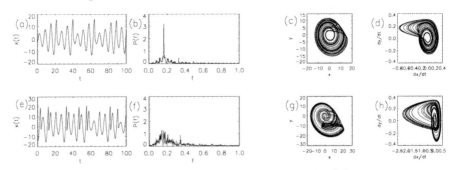

**Fig. 6.1.** (**a,e**): Segment of the $x_1$-component of the trajectory of the Rössler systems (6.1); (**b,f**): periodogram of the x-component of the trajectory; (**c,g**): projection of the attractor onto the $(x, y)$ plane; (**d,h**): projection onto the $(\dot{x}, \dot{y})$ plane. Upper panel (**a,b,c,d**) computed for $a = 0.16$ and lower panel (**e,f,g,h**) computed for $a = 0.2925$

of systems. However, depending on the component one uses to couple the non-phase-coherent oscillator to the coherent ones, the result of the obtained frequency can be different.

Furthermore, Osipov et al. [10] have proposed another approach which is based on the general idea of the curvature of an arbitrary curve. For any two-dimensional curve $\boldsymbol{r} = (u, v)$ they propose that the phase $\phi$ be defined as $\phi = \arctan \frac{\dot{v}}{\dot{u}}$. By means of this definition, the projection $\dot{\boldsymbol{r}} = (\dot{u}, \dot{v})$ is a curve cycling monotonically around a certain point.

This definition of $\phi$ holds in general for any dynamical system if the projection of the phase trajectory onto some plane is a curve with a positive curvature. This approach is applicable to a large variety of chaotic oscillators, such as the Lorenz system [14], the Chua circuit [15] or the model of an ideal four-level laser with periodic pump modulation [16].

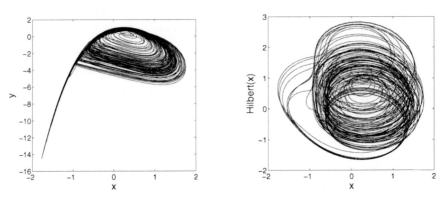

**Fig. 6.2.** Projection of the $x$-$y$-plane; (**a**) and the plot of the Hilbert Transform of $x$ versus $x$; (**b**) for the Hindmarsh-Rose-neuron

This is clear for phase-coherent as well as funnel attractors in the Rössler oscillator. Here, projections of chaotic trajectories on the plane $(\dot{x}, \dot{y})$ always rotate around the origin (Figs. 6.1(c) and (d)) and the phase can be defined as

$$\phi = \arctan \frac{\dot{y}}{\dot{x}} . \tag{6.3}$$

Although this approach works well in non-phase-coherent model systems, we have to consider that one is often confronted with the computation of the phase in experimental time series, which are usually corrupted by noise. In this case, some difficulties may appear when computing the phase given in (6.3), because derivatives are involved in its definition.

## 6.2.1 Cross-correlation of the Probability of Recurrence

We use a different approach, based on recurrences in phase space, to detect PS indirectly. We define a recurrence of the trajectory of a dynamical system $\{x_i\}_{i=1}^N$ in the following way: we say that the trajectory has returned at time $t=j$ to the former point in phase space visited at $t=i$ if

$$R_{i,j}^{(\varepsilon)} = \Theta(\varepsilon - \|x_i - x_j\|) = 1 , \tag{6.4}$$

where $\varepsilon$ is a pre-defined threshold and $\Theta(\cdot)$ is the Heaviside function. A "1" in the matrix at $i, j$ means that $x_i$ and $x_j$ are neighboring, a "0" that they are not. The black and white representation of this binary matrix is called a recurrence plot (RP). This method has been intensively studied in the last years [11]: different measures of complexity have been proposed based on the structures obtained in the RP and have found numerous applications for example, in physiology and earth science [17]. Furthermore, it has been even shown that some dynamical invariants can be estimated by means of the recurrence structures [18].

Based on this definition of recurrence, one is able to tackle the problem of performing a synchronization analysis in the case of non-phase-coherent systems. We avoid the direct definition of the phase and use instead the recurrence properties of the systems in the following way: the probability $P^{(\varepsilon)}(\tau)$ that the system returns to the neighborhood of a former point $x_i$ of the trajectory[2] after $\tau$ time steps can be estimated as follows:

$$P^{(\varepsilon)}(\tau) = \frac{1}{N-\tau} \sum_{i=1}^{N-\tau} \Theta(\varepsilon - \|x_i - x_{i+\tau}\|) = \frac{1}{N-\tau} \sum_{i=1}^{N-\tau} R_{i,i+\tau}^{(\varepsilon)} . \tag{6.5}$$

This function can be regarded as a generalized autocorrelation function, as it also describes higher order correlations between the points of the trajectory

---

[2] The neighborhood is defined as a box of size $\varepsilon$ centered at $x_i$, as we use the maximum norm.

dependent on the time delay $\tau$. A further advantage with respect to the linear autocorrelation function is that $P^{(\varepsilon)}(\tau)$ is defined for a trajectory in phase space and not only for a single observable of the system's trajectory.

For a periodic system with period $T$, it can be easily shown that $P^{(\varepsilon)}(\tau)=1$ if $\tau = T$ and $P^{(\varepsilon)}(\tau) = 0$ otherwise. For coherent chaotic oscillators, such as (6.1) for $a = 0.16$, $P^{(\varepsilon)}(\tau)$ has well-expressed local maxima at multiples of the mean period, but the probability of recurrence after one or more rotations around the fixed point is less than one (Fig. 6.3(b,d)).

Analyzing the probability of recurrence, it is possible to detect PS for non-phase-coherent oscillators as well. This approach is based on the following idea: Originally, a phase $\phi$ is assigned to a periodic trajectory $x$ in phase space, by projecting the trajectory onto a plane and choosing an origin, around which the trajectory oscillates all the time. Then, an increment of $2\pi$ is assigned to $\phi$ when the point of the trajectory has returned to its starting position, i.e. when $\|x(t + T) - x(t)\| = 0$. Analogously to the case of a periodic system,

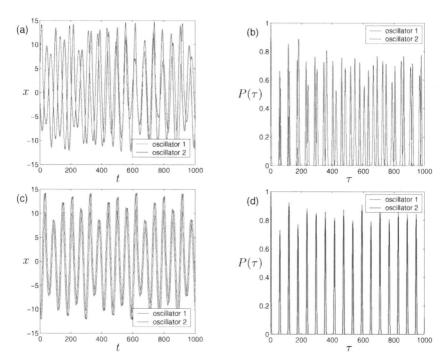

**Fig. 6.3.** Time series (**a** and **c**) and the probability of recurrence (**b** and **d**) of the Rössler system with parameters $a = 0.15$, $b = 0.2$, $c = 8.5$, $\omega_1 = 1$ and $\omega_2 = 1.05$. The coupling strength for the non-PS case (**a** and **b**) is $\mu_{\mathrm{nonPS}} = 0.01$ and $\mu_{\mathrm{PS}} = 0.07$ for the PS case (**c** and **d**), respectively. The values for CPR that have been calculated are $\mathrm{CPR}_{\mathrm{nonPS}} = 0.0102$ and $\mathrm{CPR}_{\mathrm{PS}} = 0.9995$. The figures show clearly how the peaks drift apart from each other in the absence of PS and coincide in the case of PS

we can assign an increment of $2\pi$ to $\phi$ for a complex non-periodic trajectory $\boldsymbol{x}(t)$ when $\|\boldsymbol{x}(t+T) - \boldsymbol{x}(t)\| \sim 0$, or equivalently when $\|\boldsymbol{x}(t+T) - \boldsymbol{x}(t)\| < \varepsilon$, where $\varepsilon$ is a predefined threshold. That means that a recurrence $R^{(\varepsilon)}_{t,t+\tau} = 1$ can be interpreted as an increment of $2\pi$ of the phase in the time interval $\tau$.[3]

$P^{(\varepsilon)}(\tau)$ can be viewed as a statistical measure of how often $\phi$ in the original phase space has increased by $2\pi$ or multiples of $2\pi$ within the time interval $\tau$. If two systems are in PS, on the average, the phases of both systems increase by $2\pi k$, with $k$ a natural number, within the same time interval $\tau$. Hence, looking at the coincidence of the positions of the maxima of $P^{(\varepsilon)}(\tau)$ for both systems, we can quantitatively identify PS (from now on, we omit $(\varepsilon)$ in $P^{(\varepsilon)}(\tau)$ to simplify the notation). The proposed algorithm then consists of two steps:

- Compute $P_{1,2}(\tau)$ of both systems based on (6.5).
- Compute the cross-correlation coefficient between $P_1(\tau)$ and $P_2(\tau)$ (Correlation between probabilities of recurrence)

$$\mathrm{CPR}_{1,2} = \frac{\langle \bar{P}_1(\tau)\bar{P}_2(\tau)\rangle_\tau}{\sigma_1\sigma_2} , \qquad (6.6)$$

where the bar above $\bar{P}_{1,2}$ denotes that the mean value has been subtracted and $\sigma_1$ and $\sigma_2$ are the standard deviations of $P_1(\tau)$ and $P_2(\tau)$, respectively.

If both systems are in PS, the probability of recurrence is maximal simultaneously and $\mathrm{CPR}_{1,2} \approx 1$. In contrast, if the systems are not in PS, the maxima of the probability of recurrence do not occur jointly and we would expect low values of $\mathrm{CPR}_{1,2}$.

In Figs. 6.3 and 6.4, we illustrate the performance of the method with two examples of the Rössler system.

## 6.2.2 The Problem of Separating the Time Scales

As already mentioned, neurons can exhibit dynamics on several distinct time scales (spiking and bursting) and are also able to synchronize on both scales separately. To perform a synchronisation analysis of such a system, one has to segregate the two scales of each other. Figure 6.5 shows the RP of a Hindmarsh-Rose neuron.[4] In Fig. 6.5(a), the structures that emerged from the recurrence of the bursts can be identified quite clearly, namely the "swelling diagonal lines". In Fig. 6.5(b), one of those "swellings" is presented magnified. Here, the recurrences of the spike dynamics can be noticed as diagonal lines on a smaller scale in the RP.

Separating the scales is a non-trivial task. Filtering the time series could be one approach, but this is not recommended as the attractor of the filtered time series will be distorted, which will change the recurrence behavior.

---

[3] This can be considered as an alternative definition of the phase to (6.2) and (6.3).
[4] For the definition of the Hindmarsh-Rose neuron see Sect. 6.3 and Chap. 1, again.

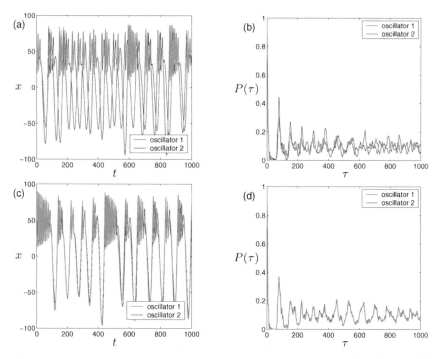

**Fig. 6.4.** Time series (**a** and **c**) and the probability of recurrence (**b** and **d**) of the Rössler system in a bursting regime with the parameters $a = 0.38$, $b = 0.4$, $c = 50$, $\omega_1 = 1$ and $\omega_2 = 1.05$. The coupling strength for the non PS case (**a** and **b**) is $\mu_{\mathrm{nonPS}} = 0.005$ and $\mu_{\mathrm{PS}} = 0.23$ for the PS case (**c** and **d**), respectively. The values for CPR that have been calculated are $\mathrm{CPR}_{\mathrm{nonPS}} = 0.0258$ and $\mathrm{CPR}_{\mathrm{PS}} = 0.9684$. Clearly, the peaks do not coincide in the non-PS case and do so in the presence of PS. This example shows quite well that the algorithm is able to detect PS for systems with a very complicated flow of the phase

Therefore, separating the time scales after calculating the RP or $P(\tau)$ is a better approach. We separate the time scales in two ways: The first one requires the choice of an appropriate recurrence rate and the second one is the application of some filter to $P(\tau)$.

In Fig. 6.6(a) the recurrence probability $P(\tau)$ of a Hindmarsh-Rose neuron is presented. The large peaks correspond to the recurrence of the bursts. The arrows indicate the smaller peaks generated by the recurrence of the spikes. There are many methods for separating both scales, e.g. wavelets, etc. In this analysis, an infinite impulse response (IIR) filter has been used, which can be implemented easily by simple difference equations.

Figure 6.6(b) shows the highpass filtered $P(\tau)$. The cutoff has been chosen to be $0.2 \times$ sampling rate. The broad peaks, originated by the burst, are filtered out and the smaller peaks corresponding to the spike recurrence become clearer. Note that the filtered $P(\tau)$ cannot be interpreted as a probability of

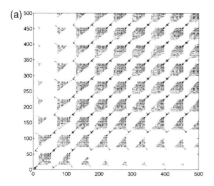
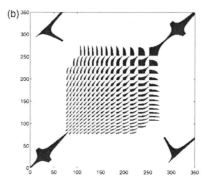

**Fig. 6.5.** Recurrence plot of the time series of a Hindmarsh-Rose neuron on a large scale (**a**) and zoomed in to show a small scale features (**b**)

recurrence any more, since it also assumes negative values. However, it still captures all the relevant information about the recurrence of the spiking dynamics. Thus, a separate synchronization analysis of the spike scale can now be accomplished by computing the index CPR of the filtered functions.

The recurrence rate is the parameter that specifies the number of black points in the RP and determines the threshold $\varepsilon$ in (6.4). This parameter also influences the patterns obtained in the recurrence plot. Hence, by varying the recurrence rate, we can enhance or suppress certain information.

Figure 6.7(a) shows the RP of a Hindmarsh-Rose neuron time series, computed for a high recurrence rate of 0.5. Comparing this plot with the one in Fig. 6.5(a), it can be observed that the shorter lines originating from the recurrence of the spikes are "smeared out" The corresponding probability of recurrence $P(\tau)$ in Fig. 6.7(b) shows only the oscillations that are caused by

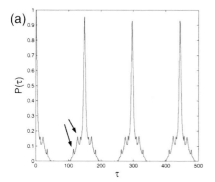
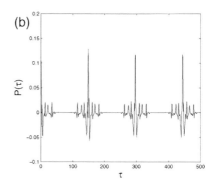

**Fig. 6.6.** Probability of recurrence $P(\tau)$ for the Hindmarsh-Rose neuron: (**a**) The original and; (**b**) highpass filtered. The arrows indicate the features created by the recurrence of the spikes

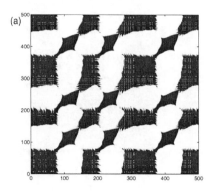
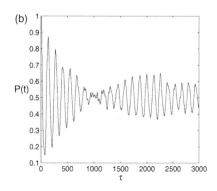

**Fig. 6.7.** RP with $RR = 0.5$; (**a**) and corresponding $P(\tau)$; (**b**) for an exemplary Hindmarsh-Rose neuron

the recurrence of the bursts. Consequently, the recurrence rate can be used to analyze the synchronization of the slow time scale (bursts), since the influence of the fast scale is automatically removed.

Analogously, choosing a rather low value for the recurrence rate causes the fine structures of the spike recurrences to appear more clearly. Therefore, it is advisable to use a rather low recurrence rate to analyze the synchronization of the spikes. In Fig. 6.8, the RP and the corresponding high-pass filtered $P(\tau)$ are presented. This example demonstrates quite well, how the large peaks, which are usually created by the recurrence of the bursts, are suppressed, so that the recurrence of the spikes is clearer than for higher values of the recurrence rate.

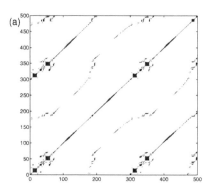
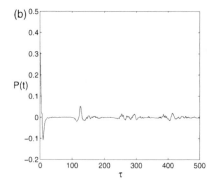

**Fig. 6.8.** RP with $RR = 0.05$: (**a**) and corresponding highpass filtered $P(\tau)$; (**b**) for an exemplary Hindmarsh-Rose neuron

**A Few Notes on the Parameters**

The RP based method has several parameters that need to be chosen in an appropriate way. These parameters are the already discussed recurrence rate and cutoff frequency of the filter, the averaging length $N$ in (6.5), and the maximum recurrence time $\tau_{\max}$ when calculating CPR.

On the one hand, small values of $N$ and $\tau_{\max}$ are desirable, such that the analysis can operate as locally as possible and with as small as possible computational cost. On the other hand, the values cannot be too small, since the analyses requires averaging and thus needs a large number of points for a correct calculation. Therefore, one has to determine the minimum values of $N$ and $\tau_{\max}$ to serve both requirements. This can be done by calculating $CPR$ for different values of these parameters. For large values, one can expect some kind of asymptotic behavior.

## 6.3 Application of the Algorithm

In this section, we present a few results that have been obtained by applying the proposed algorithm to networks of coupled neurons with different topologies. The neuron model that has been used is a (modified) four-dimensional Hindmarsh-Rose system (for details, see [19] and [20]),

$$
\begin{aligned}
\dot{x}_n &= \omega_{\text{fast},n}(y_n + 3x_n^2 - x_n^3 - 0.99z_n + I_n) + \mu \sum_{m=1}^{N} A_{nm}(x_m - x_n) \\
\dot{y}_n &= \omega_{\text{fast},n}(1.01 - y_n - 5.0128x_n^2 - 0.0278w_n) \\
\dot{z}_n &= \omega_{\text{slow1},n}(-z_n + 3.966(x_n + 1.605)) \\
\dot{w}_n &= \omega_{\text{slow2},n}(-0.9573w_n + 3(y_n + 1.619)),
\end{aligned}
\tag{6.7}
$$

where $x_n$ is the membrane potential, and $y_n$, $z_n$, and $w_n$ represent inner degrees of freedom of neuron $n$, with $n = 1, \ldots, N$. Whereas $y_n$ is responsible for the fast dynamics of the spikes, $z_n$ and $w_n$ represent the slow dynamics of the bursts. $I_n$ is the external input current of neuron $n$, $\omega_{\text{fast},n}$ determines the firing rate, and $\omega_{\text{slow1},n}$ and $\omega_{\text{slow2},n}$ determine the duration of the bursts. The neurons are electrically coupled, while the coupling topology of the neurons is given by the adjacency matrix $A_{nm}$ (see Chap. 3). The parameter $\mu$ is the coupling strength of the whole network.

### 6.3.1 Analysis of Two Coupled Neurons

First, we apply the algorithm to a pair of coupled Hindmarsh-Rose neurons. We consider different parameter sets for the two neurons (see Table 6.1), so that we have three possibilities for the dynamical regime of the neurons: (i) both neurons in regular bursting regime with different frequencies, (ii) both neurons in chaotic bursting regime with different frequencies, and (iii) one neuron in spiking regime and one in regular bursting regime, both neurons with the same frequencies.

**Table 6.1.** A list of parameters in the examined pair of Hindmarsh-Rose neurons

|  | $w_{slow1,1}$ | $w_{slow2,1}$ | $w_{fast1}$ | $I_1$ | $w_{slow1,2}$ | $w_{slow2,2}$ | $w_{fast2}$ | $I_2$ |
|---|---|---|---|---|---|---|---|---|
| regular bursting | 0.0015 | 0.019 | 1.1 | 3.0 | 0.0018 | 0.0012 | 0.9 | 2.9 |
| chaotic bursting | 0.0050 | 0.0010 | 1.1 | 3.1 | 0.0022 | 0.0007 | 0.9 | 3.1 |
| one bursting, one spiking | 0.0015 | 0.0009 | 1.0 | 5.0 | 0.0015 | 0.0009 | 1.0 | 2.5 |

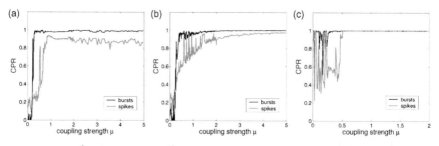

**Fig. 6.9.** $CPR^{bursts}$ and $CPR^{spikes}$ vs. coupling strength $\mu$ for a pair of Hindmarsh-Rose neurons with parameters according to Table 6.1: **(a)** regular bursting **(b)** chaotic bursting; **(c)** one spiking, one regular bursting

Then, we compute the synchronisation indices $CPR^{bursts}$ and $CPR^{spikes}$ for each case dependent on the coupling strength (see Fig. 6.9). For all three cases the spikes need higher coupling strengths to become phase synchronized than the bursts. This result is in good accordance with [21].

### 6.3.2 Analysis of Networks of Neurons

Different network topologies (random, small-world and scale-free) with Hindmarsh-Rose neurons at each node have been analyzed. Each network had $N = 200$ nodes and an average degree $\langle d \rangle$ of 10. The parameters of the neurons have been chosen as follows: $I_n \in \mathcal{N}(3.1, 0.05)$ (chaotic bursting regime), $\omega_{fast,n} \in \mathcal{N}(1, 0.05)$, $\omega_{slow1,n} \in \mathcal{N}(0.002, 0.0005)$, and $\omega_{slow,n} = 0.001$, where $\mathcal{N}(\tilde{\mu}, \sigma)$ denotes a Gaussian normal distribution with mean $\tilde{\mu}$ and variance $\sigma$. The coupling strength has been chosen as $\mu = g/\langle d \rangle$. The synchronization indices $CPR^{bursts}$ and $CPR^{spikes}$ have been calculated for each pair of nodes from the networks for increasing values of the coupling parameter $g$. Thus, we obtain two matrices $(CPR_{nm}^{bursts})$ and $(CPR_{nm}^{spikes})$, where $n, m = 1, \ldots, 200$ indicate the nodes.

In Fig. 6.10, we present a few snapshots of those CPR-matrices for different values of the coupling strength $g$ for the scale free network. We have found that with an increasing coupling strength, the hubs (nodes with largest degree, see Chap. 3 for details) will synchronize first, while the rest of the nodes need a higher coupling strength to become synchronized. This is in good accordance with [22], where this has been shown for a scale free network of

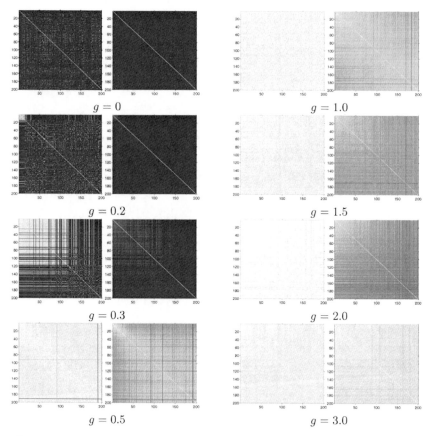

**Fig. 6.10.** Several snapshots of the CPR matrix of a network of 200 Hindmarsh-Rose neurons are presented for different coupling strengths. The left hand plot of each pair corresponds to the bursts, the right hand one to the spikes, respectively. Several phenomena stand out: 1. the hubs synchronize first, "attracting" the remaining nodes when the coupling increases further; 2. the spikes synchronize for a higher coupling strength than the bursts and; 3. there is a collapse of the spike synchronization in a certain domain of the coupling strength

Rössler oscillators. Furthermore, we have found for all three networks, as in the case of two coupled neurons, that the synchronization of the spikes sets in for higher values of the coupling strength than for the bursts.

To quantify the degree of phase synchronization of the whole network, we count the number of values in $(CPR_{nm})$ that are above a certain threshold and we call this number "area of synchronization". The threshold has been chosen as 0.8. In Fig. 6.11, those areas of synchronization are plotted versus the coupling strength.

An interesting result can be observed in the plot of the area of synchronization, as well in the snapshots of the CPR-matrices: there is a collapse

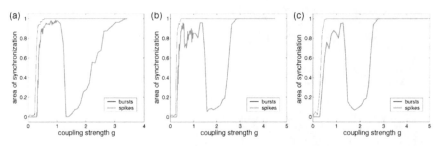

**Fig. 6.11.** Area of synchronization for; (**a**) random network; (**b**) small-world network; and (**c**) scale-free network for bursts and spikes, respectively

of the synchronization of the spikes for intermediate values of the coupling strength. In contrast, the synchronization of the bursts remains unchanged. This could be due to a change of the dynamics, namely the coherence of the oscillators with increasing values of the coupling strength $g$.

## 6.4 Conclusions

In this chapter, we have analyzed phase synchronization in networks with complex topology and complex dynamics. In particular, we have concentrated on dynamics on two time scales, as is typically observed in neurons with spiking and bursting dynamics. In order to analyze the synchronization behavior of such systems we extended an existing method, which is based on the concept of recurrence [11], to treat the two time scales separately. We have applied the proposed method to complex networks of Hindmarsh-Rose neurons. Our results are in accordance with [21], where it has been shown that the spikes need higher values of the coupling strength than the bursts in order to phase synchronize. Moreover, we have found that in a scale-free network of Hindmarsh-Rose neurons, the hubs synchronize first with increasing coupling strength, while the rest of the nodes need a higher coupling to synchronize, as has been reported in [22] for a scale-free network of Rössler oscillators. In addition, the most interesting result of our analysis is that we have found a collapse in the synchronization of the spikes in those complex networks for an intermediate coupling strength. This effect will be discussed in detail in a forthcoming paper.

## References

1. S. H. Strogatz, Nature **410**, 268, 2001; M. E. J. Newman, SIAM Rev. **45**, 167, 2003; S. Boccaletti et al., Phys. Rep. **424**, 175, 2006; R. Albert, A.-L. Barabási, *Statistical mechanics of complex networks*, Rev. Mod. Phys., **74**, 47–97, 2002.
2. P. Erdös, and A. Rényi, Publ. Math. Inst. Hung. Acad. Sci. **5**, 17, 1960; D. J. Watts, and S. H. Strogatz, Nature **393**, 440,1998;

L. Barabási, and R. Albert, Science **286**, 509, 1999;
M. Molloy, and B. Reed, Random Struct. Algorithms **6**, 161, 1995.

3. L. Donetti et al., Phys. Rev. Lett. **95**, 188701, 2005.

4. Y. Moreno, and A. F. Pacheco, Europhys. Lett. **68 (4)**, 603, 2004;
J. G. Restrepo et al., Phys. Rev. E **71**, 036151, 2005.

5. F. M. Atay et al., Phys. Rev. Lett. **92 (14)**, 144101, 2004; W. Lu, and T. Chen, Physica D **198**, 148, 2004;
Y. Jiang et al., Phys. Rev. E **68**, 065201(R), 2003.

6. C. Zhou, and J. Kurths, Chaos **16**, 015104 2006.

7. N. F. Rulkov et al., Phys. Rev. E, **51** (2), 980, 1995;
L. Kocarev, and U. Parlitz, Phys. Rev. Lett. **76** (11), 1816, 1996;
S. Boccaletti et al., Phys. Rep. **366**, 1, 2002.

8. B. Blasius et al., Nature **399**, 354, 1999; P. Tass et al., Phys. Rev. Lett. **81 (15)**, 3291, 1998;
M. Rosenblum et al., Phys. Rev. E. **65**, 041909, 2002;
D. J. DeShazer et al., Phys. Rev. Lett. **87 (4)**, 044101, 2001.

9. A. Pikovsky, M. Rosenblum, J. Kurths, *Synchronization - A universal concept in nonlinear science*, Cambridge University Press, 2001.

10. G. V. Osipov, B. Hu, C. Zhou, M. V. Ivanchenko, and J. Kurths, Phys. Rev. Lett. **91**, 024101, 2003.

11. N. Marwan, M. C. Romano, M. Thiel, and J. Kurths, Phys. Rep. **438**, 237, 2007.

12. M. C. Romano, M. Thiel, J. Kurths, I. Z. Kiss, J. L. Hudson, *Detection of synchronization for non-phase-coherent and non-stationary data*, Europhys. Lett., 71 (3), 466, 2005.

13. M. G. Rosenblum, A. S. Pikovsky, J. Kurths, G. V. Osipov, I. Z. Kiss, and J. L. Hudson, Phys. Rev. Lett. **89**, 264102, 2002.

14. C. Sparrow, *The Lorenz equations: Bifurcations, chaos, and strange attractors*, Springer-Verlag, Berlin, 1982.

15. R. N. Madan, *Chua circuit: A paradigm for chaos*, World Scientific, Singapore, 1993.

16. W. Lauterborn, T. Kurz, and M. Wiesenfeldt, *Coherent Optics. Fundamentals and Applications*, Springer-Verlag, Berlin, Heidelberg, New York, 1993.

17. C. L. Weber Jr., and J. P. Zbilut, J. Appl. Physiology **76 (2)** 965, 1994;
N. Marwan, N. Wessel, U. Meyerfeldt, A. Schirdewan, and J. Kurths, Phys. Rev. E **66 (2)**, 026702, 2002;
N. Marwan, and J. Kurths, Phys. Lett. A **302 (5–6)**, 299, 2002; M. Thiel et al., Physica D **171**, 138, 2002.

18. M. Thiel, M. C. Romano, P. Read, J. Kurths, *Estimation of dynamical invariants without embedding by recurrence plots*, Chaos, **14** (2), 234–243, 2004.

19. J. L. Hindmarsh, R. M. Rose, *A model of neuronal bursting using three coupled first order differential equations*, Proc. Roy. Soc. Lond. B **221**, 87–102, 1984.

20. R. D. Pinto, P. Varona, A. R. Volkovskii, A. Szücs, H. D. I. Abarbanel, M. I. Rabinovich *Synchronous behavior of two coupled electronic neurons*, Phys. Rev. Lett. E **62**, nr. 2, 2000.

21. M. Dhamala, V. K. Jirsa, M. Ding *Transitions to synchrony in coupled bursting neurons*, Phys. Rev. Lett. **92**, nr. 2, p. 028101, 2004.

22. C. Zhou, J. Kurths, *Hierarchical synchronization in complex networks with heterogenous degrees* Chaos **16**, 015104, 2006.

# Part III

Cognition and Higher Perception

# 7

# Neural and Cognitive Modeling with Networks of Leaky Integrator Units

Peter beim Graben[1,3], Thomas Liebscher[2] and Jürgen Kurths[3]

[1] School of Psychology and Clinical Language Sciences,
   University of Reading, United Kingdom
   p.r.beimgraben@reading.ac.uk
[2] Bundeskriminalamt, Wiesbaden, Germany
[3] Institute of Physics, Nonlinear Dynamics Group, Universität Potsdam, Germany

**Summary.** After reviewing several physiological findings on oscillations in the electroencephalogram (EEG) and their possible explanations by dynamical modeling, we present neural networks consisting of leaky integrator units as a universal paradigm for neural and cognitive modeling. In contrast to standard recurrent neural networks, leaky integrator units are described by ordinary differential equations living in continuous time. We present an algorithm to train the temporal behavior of leaky integrator networks by generalized back-propagation and discuss their physiological relevance. Eventually, we show how leaky integrator units can be used to build oscillators that may serve as models of brain oscillations and cognitive processes.

## 7.1 Introduction

The electroencephalogram (EEG) measures the electric fields of the brain generated by large formations of certain neurons, the *pyramidal cells*. These nerve cells roughly possess an axial symmetry and they are aligned in parallel perpendicular to the surface of the cortex [1–4]. They receive excitatory input at the superficial apical dendrites from thalamic relay neurons and inhibitory input at the basal dendrites and at their somata from local interneurons [1,3–5]. Excitatory and inhibitory synapses cause different ion currents through the cell membranes thus leading to either depolarization or hyperpolarization, respectively. When these synapses are activated, a single pyramidal cell behaves as a microscopic electric dipole surrounded by its characteristic electric field [1,6].

According to the inhomogeneity of the cortical gray matter, a mass of approximately 10,000 synchronized pyramidal cells form a dipole layer whose fields sum up to the *local field potentials* that polarize the outer tissues of the scalp, which acts thereby as a low pass filter [1,3,5,6]. These filtered sum potentials are macroscopically measurable as the EEG at the surface of a subject's head (cf. Chap. 1).

Some of the most obvious features of the EEG are oscillations in certain frequency bands. The *alpha waves* are sinusoidal-like oscillations between 8–14 Hz, strongly pronounced over parietal and occipital recording sites that reflect a state of relaxation during wakefulness, with no or only low visual attention. Figure 7.1 shows a characteristic power spectrum for the alpha rhythm: There is one distinguished peak superimposed to the $1/f$ background EEG. When a subject starts paying attention, the powerful slow alpha waves disappear, while smaller oscillations with higher frequencies around 14–30 Hz (the *beta waves*) arise [2, 7, 8]. We will refer to this finding, sometimes called *desynchronization* of the EEG, as to the *alpha blocking* [7].[4] Alpha waves are assumed to be related to awareness and cognitive processes [11–14]. Experimental findings suggest that thalamocortical feed-back loops are involved in the origin of the alpha EEG [1, 2, 4, 8, 15, 16].

The $1/f$-behavior and the existence of distinguished oscillations in the EEG such as the alpha waves are cornerstones in the evaluation of computational models of the EEG. Indeed, modeling these brain rhythms has a long tradition. Wilson and Cowan [17] were the first to use populations of excitatory and inhibitory neurons innervating each other (see Sect. 7.3.2). They introduced a two-dimensional state vector whose components describe the proportion of firing McCulloch-Pitts neurons [18] within a unit volume of neural tissue at an instance of time. This kind of ensemble statistics leads to the

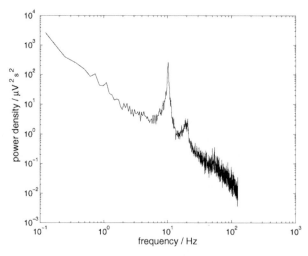

**Fig. 7.1.** Power spectrum of the alpha EEG at one parietal electrode site (PZ)

---

[4] The term "desynchronization" is misleading since it has no direct relation to synchronization in the sense of, for example, [9, 10]. From the viewpoint of data analysis it simply means: decreasing power in the alpha band of the spectrum. However, biophysical theories of the EEG explain the loss of spectral power by a loss of coherence of neuron activity, i.e. a reduction of synchronization [7, 8, 11].

well-known sigmoidal activation functions for neural networks [19] through the probability distributions of either synapses or activation thresholds (see also [5]). The model further includes the refractory time in which a neuron that has been activated just before cannot be activated again, and the time-course of postsynaptic potentials as *impulse response functions*. This model has been strongly simplified by Wilson [20], leading to a network of only two recurrently coupled *leaky integrator units* (see Sect. 7.2). Wilson reported limit cycle dynamics of this system for a certain range of the excitatory input, playing the role of a control parameter. However, this network does not exhibit an equivalent to the alpha blocking, because the frequency of the oscillations becomes slower for increasing input.

Lopez da Silva et al. [21] pursued two different approaches: a distributed model of the thalamus where relay cells and interneurons are considered individually, and a "lumped" model analogous to the one of Wilson and Cowan [17] but without refractory time and with even more complicated postsynaptic potentials. In order to determine the sum membrane potential of each population as a model EEG, one has to compute the convolution integral of the postsynaptic impulse response functions with the spike rate per unit of volume. Linearizing the activation functions allows a system-theoretic treatment of the model by means of the Laplace transform, thus allowing the analytical computation of the power spectrum. Lopez da Silva et al. [21,22] have shown that their model of thalamical or cortical feedback loops actually exhibits a peak around 10 Hz, i.e. alpha waves, in the spectrum, although they were not able to demonstrate alpha blocking. This population model [21] has been further developed by Freeman [23], Jansen et al. [24,25], Wendling et al. [26,27], and researchers from the Friston group [28–30] in order to model the EEG of the olfactory system, epileptic EEGs, and event-related potentials (ERP), respectively.

A further generalization of the Lopez da Silva et al. model [21] led Rotterdam et al. [31] to a description of spatio-temporal dynamics by considering a chain of coupled cortical oscillators. A similar approach has been pursued by Wright and Liley [32,33] who discussed a spatial lattice of coupled unit volumes of excitatory and inhibitory elements obeying cortical connectivity statistics. The convolution integrals of the postsynaptic potentials with the spike rates were substituted by convolution sums over discrete time. The most important result for us is that the power spectrum shows the alpha peak, and, that there is a "shift to the right" (towards the beta band) of this peak with increasing input describing arousal, i.e. actually alpha blocking.

Additionally, Liley et al. [34] also suggested a distributed model of cortical alpha activity using a compartmental description of membrane potentials [35]. In such an approach, nerve cells are thought to be built up of cylindrical compartments that are governed by generalized Hodgkin-Huxley equations [36] (see also Chap. 1). Liley et al. [34] reported two oscillatory regimes of this dynamics: one having a broad-band spectrum with a peak in the beta range and the other narrowly banded with a peak around the alpha frequency.

There are also field theoretic models of neural activity [37–41] (see Chap. 8). In these theories, the unit volumes of cortical tissue are considered to be infinitesimally small. Thus, the systems of coupled ordinary differential equations are substituted by nonlinear partial differential equations. Robinson et al. [41] have proposed such a theory in order to describe thalamocortical interactions and hence the alpha EEG.

Another approach that could lead to the explanation of the EEG is Hebb's concept of a *cell assembly* [42], where *reverberatory circles* form neural oscillators. We shall see in Sect. 7.4.3 how such circles may emerge in an evolving neural network.

On the other hand, Kaplan et al. [43], van der Velde and de Kamps [44], Wennekers et al. [45], and Smolensky and Legendre [46] argue how neural networks could bridge the gap between the sub-symbolic representation of single neurons and "a symbol-like unit of thought" in models of cognitive processes. Kaplan et al. proposed that the cell assembly be an assembly of neural units that are recurrently connected to exhibit reverberatory circles, in which information needs to cycle around until the symbolic meaning is fully established. They presented a series of experiments in which they made use of physiological principles that should be present in the functioning of cell assemblies: temporally structured input, dependency on prior experience, competition between assemblies and control of its activation. A main result is that after a cell assembly is provided with input, its activation gradually increases until an asymptotic activation is reached or the input is removed. After removal of the input, the activation gradually decreases until it comes back to its resting level.

## 7.2 Leaky Integrator Networks

### 7.2.1 Description of Leaky Integrator Units

When neural signals are exchanged between different cell assemblies that are typically involved in brain functions, oscillations caused by recurrent connections between the neurons should become visible. A possible way to model this behavior is by describing each cell assembly by a leaky integrator unit [47], which integrates input over time while the internal activation is continuously decreased by a dampening leakage term. We shall present the relationship between cell assemblies and leaky integrator units in Sect. 7.3.2. However, also single neurons can be described by a leaky integrator unit, though with quite different leakage constants, as we shall see in Sect. 7.3.1. In terms of standard units (as e.g. used by Rumelhard et al. [48]), a leaky integrator unit looks like the one depicted in Fig. 7.2.

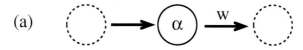

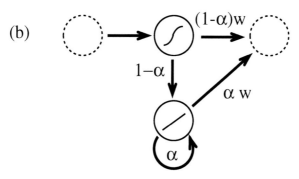

**Fig. 7.2.** Simulation of a leaky integrator unit (**a**) and a recurrent combination of two standard units (**b**). The function of the leakage rate $\alpha$ is mimicked by two parallel standard units with a logistic and a linear activation function, respectively. The synaptic weights to subsequent units are denoted by $w$ (cf. (7.3))

The activation of this leaky integrator unit is described by

$$\frac{\mathrm{d}x_i(t)}{\mathrm{d}t} = -x_i(t) + (1 - \alpha_i)\, x_i(t) + \alpha_i f(y_i(t))$$
$$= -\alpha_i x_i(t) + \alpha_i f(y_i(t)) \,, \qquad (7.1)$$

or, in another form:

$$\tau_i \frac{\mathrm{d}x_i(t)}{\mathrm{d}t} + x_i(t) = f(y_i(t)) \,. \qquad (7.2)$$

The symbols have the following meanings:

| | |
|---|---|
| $\frac{\mathrm{d}x_i(t)}{\mathrm{d}t}$ | change of activation of unit $i$ at time $t$ |
| $x_i(t)$ | activation of unit $i$ at time $t$ |
| $y_i(t)$ | net input of unit $i$ at time $t$ |
| $\alpha_i$ | leakage rate of unit $i$ |
| $\tau_i = \alpha_i^{-1}$ | time constant of unit $i$ |
| $f$ | activation function of each unit; usually sigmoidal (e.g. logistic as in (7.5)) or linear. |

The leakage rate $\alpha$ tells how much a unit depends on the actual net input. Its value is between 0 and 1. The lower the value of $\alpha$, the stronger the influence of the previous level of activation and the less the influence of the actual net input. If $\alpha = 1$, the previous activation does not have any influence and the new activation is only determined by the net input (this is the case e.g. for the standard units used by the PDP group [48]). By contrast, $\alpha = 0$ means that the actual net input does not have any influence and the activation remains constant. $(1 - \alpha$ could also be regarded as the strength of its *memory* with respect to earlier activations.)

The net input of unit $i$ is given as the sum of all incoming signals:

$$y_i(t) = \sum_j w_{ij} x_j(t) + b_i + I_i^{\text{ext}}(t) . \tag{7.3}$$

With

| | |
|---|---|
| $y_i(t)$ | net input of unit $i$ at time $t$ |
| $w_{ij}$ | weight of connection from unit $j$ to unit $i$ |
| $b_i$ | bias of unit $i$ |
| $I_i^{\text{ext}}$ | external input to unit $i$ |

Equation (7.1) is very similar to the general form of neural networks equations for continuous-valued units (described, for example, in [19]). The difference lies in the presence of the leakage term $\alpha$ that makes the current activation dependent on its previous activation. We motivate (7.1) by the equivalent recurrent network of Fig. 7.2 and we shall use it in Sect. 7.2.2 subsequently to derive a generalized back-propagation algorithm as a learning rule for temporal patterns. On the other hand, (7.2) is well-known from the theory of ordinary differential equations. Its associated homogeneous form

$$\tau_i \frac{\mathrm{d}x_i}{\mathrm{d}t} + x_i = 0$$

simply describes an exponential decay process. Therefore, the inhomogeneous (7.2) can be seen as a forced decay process integrating its input on the right hand side.

Hertz et al. [19, p. 54] discuss a Hopfield network of leaky integrator units which is characterized by (7.2) with symmetric synaptic weights $w_{ij}$. Such a network is a dynamical system whose attractors are the patterns which are to be learned. Moreover, Hertz et al. [19, p. 55] consider another dynamical system

$$\tau_i \frac{\mathrm{d}x_i(t)}{\mathrm{d}t} + x_i(t) = \sum_j w_{ij} f(x_j(t)) + b_i + I_i^{\text{ext}}(t) \tag{7.4}$$

having the same equilibrium solutions as (7.2). As we shall see in Sect. 7.3.1, (7.4) appropriately models small networks of single neurons. The time-course of activation for a leaky integrator unit using a logistic activation function

$$f(x) = \frac{1}{1 + e^{-\beta x}} \tag{7.5}$$

with respect to input and leakage rate is shown in Figs. 7.3(a) and (b).

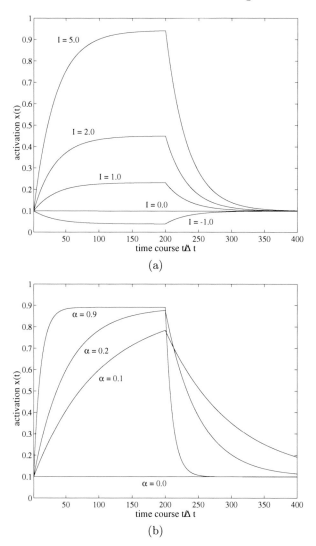

**Fig. 7.3.** Time-course of activation (7.1); (**a**) for different input with $\Delta t = 0.1$, $\alpha = 0.3$ and $b = -2.2$; (**b**) Time-course of activation (7.1) for different leakage rates with $\Delta t = 0.1$, $I^{\text{ext}} = 4.3$ and $b \approx -2.2$

## 7.2.2 Training Leaky Integrator Networks

In order to use leaky integrator units to create network models for simulation experiments, a learning rule that works in continuous time is needed. The following formulation is motivated by [49, 50] and describes how a backpropagation algorithm for leaky integrator units can be derived.

In a first step, Euler's algorithm is used to change the differential equations into difference equations:[5]

$$x_i(t + \Delta t) \approx x_i(t) + \frac{\mathrm{d}x_i(t)}{\mathrm{d}t} \Delta t$$

$$\Rightarrow \qquad \frac{\mathrm{d}x_i(t)}{\mathrm{d}t} \approx \frac{x_i(t + \Delta t) - x_i(t)}{\Delta t} \, . \qquad (7.6)$$

Combining (7.1) and (7.6) yields

$$\tilde{x}_i(t + \Delta t) = (1 - \Delta t)\tilde{x}_i(t) + \Delta t \left\{ (1 - \alpha_i)\tilde{x}_i(t) + \alpha_i f(\tilde{y}_i(t)) \right\}$$
$$= (1 - \Delta t \alpha_i) \tilde{x}_i(t) + \Delta t \alpha_i f(\tilde{y}_i(t)) \, , \qquad (7.7)$$

where tildes above variables (e.g. $\tilde{x}$) denote continuous functions that have been discretized.

Figures 7.3(a) and (b) show the time-course of activation for a leaky integrator unit with different values of external input $I$ and leakage parameters $\alpha$ with $I \neq 0$ for $t \in [0, 20]$. In order to train a network, one needs to define an error function

$$E = \int_{t_0}^{t_1} f_{\mathrm{err}} \left[ \boldsymbol{x}(t), t \right] \mathrm{d}t \, . \qquad (7.8)$$

Here, we choose the least mean square function

$$E = \frac{1}{2} \sum_i \int_{t_0}^{t_1} s_i \left[ x_i(t) - d_i(t) \right]^2 \mathrm{d}t \, , \qquad (7.9)$$

where $d_i(t)$ is the desired activation of unit $i$ at time $t$ and $s_i$ is the relative importance of this activation: $s = 0$ means unimportant and $s = 1$ means most important.

If one changes the activation of unit $i$ at time $t$ for a small amount, one gets a measure of how much this change influences the error function:

$$e_i(t) = \frac{\partial f_{\mathrm{err}} \left[ \boldsymbol{x}(t), t \right]}{\partial x_i(t)} \qquad (7.10)$$

with

$$f_{\mathrm{err}} = \frac{1}{2} \sum_i s_i \left[ x_i(t) - d_i(t) \right]^2 \, .$$

With (7.9) as error function, we get

$$e_i(t) = s_i \left[ x_i(t) - d_i(t) \right] \, . \qquad (7.11)$$

---

[5] Note that the following derivation could also be achieved using the variational calculus well-known from Hamilton's principle in analytical mechanics [49]. We leave this as an exercise for the reader.

Equations (7.10) and (7.11) describe the influence of a change of activation only for $t$. In a neural net that is described by (7.1) and (7.3), each change of activation at $t$ also influences the activation at later times $t'$ ($t < t'$). The amount of this influence can be described by using time-ordered derivatives [51,52]:

$$\tilde{z}_i(t) = \frac{\partial^+ E}{\partial \tilde{x}_i(t)}$$

$$:= \frac{\partial E}{\partial \tilde{x}_i(t)} + \sum_{t' > t} \sum_j \frac{\partial^+ E}{\partial \tilde{x}_j(t')} \frac{\partial \tilde{x}_j(t')}{\partial \tilde{x}_i(t)} \tag{7.12}$$

with        $j = 1, 2, \ldots, n$        $n$ number of units
            $t' = t + \Delta t, t + 2\Delta t, \ldots, t_1$        $t_1$ last defined time

$\tilde{z}_i(t)$ measures how much a change of activation of unit $i$ at time $t$ influences the error function at all times.

Performing the derivations in (7.12) with (7.9), (7.11), (7.7) and (7.3) and setting $t' = t + \Delta t$ gives:

$$\frac{\partial E}{\partial \tilde{x}_j(t)} = \Delta t e_i \tag{7.13}$$

$$\frac{\partial \tilde{x}_i(t + \Delta t)}{\partial \tilde{x}_i(t)} = (1 - \Delta t \alpha_i) + \Delta t \alpha_i w_{ii} f'(\tilde{y}_i(t)) \tag{7.14}$$

$$\frac{\partial \tilde{x}_j(t + \Delta t)}{\partial \tilde{x}_i(t)} = \Delta t \alpha_j w_{ji} f'(\tilde{y}_j(t)) \tag{7.15}$$

for all units $j$ that are connected with unit $i$.

All other derivatives are zero. With this, one gets

$$\tilde{z}_i(t) = \Delta t e_i + (1 - \Delta t \alpha_i) \tilde{z}_i(t + \Delta t)$$
$$+ \sum_j \Delta t \alpha_j w_{ji} f'(\tilde{y}_j(t)) \tilde{z}_j(t + \Delta t). \tag{7.16}$$

The back-propagated error signal $z(t)$ is equivalent to the $\delta$ in standard back-propagation. After the last defined activation $d_i(t_1)$, there is no further change of $E$, so $z_i(t_1 + \Delta t) = 0$.

Making use of Euler's method in the opposite direction, we find that the back-propagated error signal can be described by the following differential equation:

$$\frac{dz_i(t)}{dt} = \alpha_i z_i(t) - e_i - \sum_j \alpha_j w_{ji} f'(y_j(t)) z_j(t). \tag{7.17}$$

With (7.16), it is possible to calculate how the error function changes if one changes the parameters $\alpha_i$ and $w_{ij}$. Each variation also changes the activation $x_i$. The influence of this activation on $E$ can be calculated using the chain rule of derivatives.

If $w_{ij}$ changes over $\Delta t$ by $\partial w_{ij}$, then the influence of this change on the error function can be described by

$$
\frac{\partial E}{\partial w_{ij}}\bigg|_t^{t+\Delta t} := \frac{\partial^+ E}{\partial x_i(t+\Delta t)} \frac{\partial x_i(t+\Delta t)}{\partial w_{ij}}
$$
$$
= z_i(t+\Delta t)\alpha_i x_j(t) f'(y_i(t))\Delta t .
\tag{7.18}
$$

A change of $\partial w_{ij}$ during the *whole* time $t_0 \le t \le t_1$ produces:

$$
\frac{\partial E}{\partial w_{ij}} = \alpha_i \int_{t_0}^{t_1} z_i(t) x_j(t) f'(y_i(t)) dt .
\tag{7.19}
$$

For the influence of a change in $\alpha_i$ on $E$ one finds

$$
\frac{\partial^+ E}{\partial \alpha_i}\bigg|_t^{t+\Delta t} = \frac{\partial E}{\partial x_i(t+\Delta t)} \frac{\partial x_i(t+\Delta t)}{\partial \alpha_i}
$$
$$
= z_i(t+\Delta t) \{ f(y_i(t)) - x_i(t) \} \Delta t .
\tag{7.20}
$$

For the whole time:

$$
\frac{\partial E}{\partial \alpha_i} = \int_{t_0}^{t_1} z_i(t) \{ f(y_i(t)) - x_i(t) \} dt .
\tag{7.21}
$$

Now, we have nearly all the equations that are needed to train a neural network of leaky integrator units. Finally, we must keep in mind the fact that the leakage term $\alpha$ must be between 0 and 1. This can be done by using

$$
\alpha = \frac{1}{1 + e^{-\bar{\alpha}}}
\tag{7.22}
$$

and learning $\bar{\alpha}$ instead of $\alpha$. With this replacement we set

$$
\frac{\partial E}{\partial \bar{\alpha}_i} = \frac{1}{1 + e^{-\bar{\alpha}_i}} \left( 1 - \frac{1}{1 + e^{-\bar{\alpha}_i}} \right) \times
$$
$$
\times \int_{t_0}^{t_1} z_i(t) \{ f(y_i(t)) - x_i(t) \} dt .
\tag{7.23}
$$

### 7.2.3 Overview of the Learning Procedure

To start the training, one needs to have the following information:

(i) topology of the net with number of units ($n$) and connections
(ii) values of the parameters $\boldsymbol{W}(0) = (w_{ij}(0))$ and $\bar{\boldsymbol{\alpha}}(0)$ at $t = 0$
(iii) activations $\boldsymbol{x}(t_0)$ at $t = 0$

(iv) time-course of the input $I^{\text{ext}}(t), t_0 \leq t \leq t_1$
(v) time-course of the desired output $d(t)$
(vi) activation function $f$ for each unit
(vii) error function $E$
(viii) time-step size $\Delta t$ that resembles the required resolution of the time-course ($\Delta t = 0.1$ turned out to be a good default value).

After having fixed these parameters according to the desired learning schedule, the goal is then to find a combination of $W$ and $\bar{\alpha}(0)$ that gives a minimum for $E$. This can be achieved by the following algorithm:

(i) At first one has to calculate the net input (7.3) for each unit successively and for each time-step *forward* in time. Simultaneously, the activations are calculated with (7.7).
(ii) With (7.9), one calculates the main error $E$ and the error vector $e(t)$ using (7.11).
(iii) Then, the error signals are propagated *backwards* through time with (7.16), making use of the condition $\tilde{z}_i(t_1 + \Delta t) = 0$.
(iv) Now, one calculates the gradient of each free parameter with respect to the error function $E$ with the discrete versions of (7.19) and (7.23):

$$\frac{\partial E}{\partial w_{ij}} = \frac{1}{1 + e^{-\bar{\alpha}_i}} \sum_{t=t_0}^{t_1} \tilde{z}_i(t + \Delta t)\tilde{x}_j(t)f'(\tilde{y}_i(t))\Delta t \tag{7.24}$$

$$\frac{\partial E}{\partial \bar{\alpha}_i} = \frac{1}{1 + e^{-\bar{\alpha}_i}} \left(1 - \frac{1}{1 + e^{-\bar{\alpha}_i}}\right)$$

$$\sum_{t=t_0}^{t_1} \tilde{z}_i(t) \left\{f(\tilde{y}_i(t)) - \tilde{x}_i(t)\right\} \Delta t \,. \tag{7.25}$$

(v) After this, the parameters are changed along the negative gradient (*gradient descent*):

$$w_{ij} = w_{ij} - \eta_w \frac{\partial E}{\partial w_{ij}} \tag{7.26}$$

$$\bar{\alpha}_i = \bar{\alpha}_i - \eta_{\bar{\alpha}} \frac{\partial E}{\partial \bar{\alpha}_i} \,, \tag{7.27}$$

with $\eta_w$ and $\eta_{\bar{\alpha}}$ as learning rates. ($\eta = 0.1$ is commonly a suitable starting value.) The gradient can be used for *steepest descent, conjugate gradient* or other numeric approximations (see e.g. [53]).

(vi) Having obtained the new values $W$ and $\bar{\alpha}$, the procedure goes back to step (i) and is followed until the main error falls below a certain value in step (ii) or this criterion is not reached after a maximal number of iterations.

(For a model that uses this type of learning algorithm with leaky integrator units, see [54]). In the context of modeling oscillating brain activity, recurrent networks of leaky integrator units become interesting. Section 7.4 will describe three typical examples.

## 7.3 From Physiology to Leaky Integrator Models

### 7.3.1 Leaky Integrator Model of Single Neurons

Let us consider the somatic membrane of a neuron $i$ in the vicinity of its trigger zone. For the sake of simplicity, we shall assume that the membrane behaves only passively at this site. For further simplification, we do not describe the trigger zone by the complete Hodgkin-Huxley equations [36], but instead as a McCulloch-Pitts neuron [18], i.e. as a threshold device: the neuron fires if its membrane potential $U_i(t)$ exceeds the activation threshold $\theta \approx -50\,\mathrm{mV}$ from below due to the law of "all-or-nothing" [55, 56]. Because of this, the membrane potential $U_i(t)$ becomes translated into a spike train which can be modeled by a sum of delta functions

$$R_i(t) = \sum_{\substack{k:U_i(t_k)=\theta \\ \dot{U}_i(t_k)>0}} \delta(t - t_k)\,. \tag{7.28}$$

Now, we can determine the number of spikes in a time interval $[0, t]$ [35], which is given by

$$N_i(t) = \int_0^t R_i(t')\mathrm{d}t'\,.$$

Thus, from the *spike rate* per unit time, we regain the original signal

$$\frac{\mathrm{d}}{\mathrm{d}t}N_i(t) = R_i(t)\,. \tag{7.29}$$

In the next step, we consider the membrane potential $U_i$ in the vicinity of the trigger zone which obeys Kirchhoff's First Law (see Fig. 7.4), i.e.

$$\sum_j I_{ij} = \frac{U_i - E_m}{r_m} + c_i\,\frac{\mathrm{d}U_i}{\mathrm{d}t}\,, \tag{7.30}$$

here, $E_m$ is the Nernst equilibrium potential of the leakage channels with resistance $r_m$. $c_i$ is the capacitance of the membrane of neuron $i$ and $I_{ij}$ is the current through the membrane at the chosen site coming from the synapse formed by the $j$th neuron with neuron $i$.

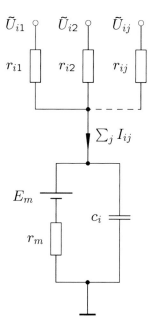

**Fig. 7.4.** Equivalent circuit for the leaky integrator neuron

These synaptic currents depend upon both the potential difference $\tilde{U}_{ij} - U_i$ between the postsynaptic potential $\tilde{U}_{ij}$ at the synapse connecting neuron $j$ to $i$ and the potential $U_i$ at the trigger zone of $i$, and the intracellular resistance along the current's path $r_{ij}$. Therefore

$$I_{ij} = \frac{\tilde{U}_{ij} - U_i}{r_{ij}} \tag{7.31}$$

applies. Inserting (7.31) into (7.30) yields

$$\sum_j \frac{\tilde{U}_{ij} - U_i}{r_{ij}} = \frac{U_i - E_m}{r_m} + c_i \frac{dU_i}{dt} ,$$

and after some rearrangements

$$r_m c_i \frac{dU_i}{dt} + U_i \left(1 + \sum_j \frac{r_m}{r_{ij}}\right) = E_m + \sum_j \frac{r_m}{r_{ij}} \tilde{U}_{ij} . \tag{7.32}$$

After letting $E_m = 0$, without loss of generality and introducing the *time constants*

$$\tau_i = \frac{r_m c_i}{1 + \sum_j \frac{r_m}{r_{ij}}} \tag{7.33}$$

and provisory *synaptic weights*

$$\tilde{w}_{ij} = \frac{\frac{r_m}{r_{ij}}}{1 + \sum_j \frac{r_m}{r_{ij}}}, \tag{7.34}$$

we eventually obtain

$$\tau_i \frac{dU_i}{dt} + U_i = \sum_j \tilde{w}_{ij} \tilde{U}_{ij}. \tag{7.35}$$

Next, the postsynaptic potentials $\tilde{U}_{ij}$ require our attention. We assume that an action potential arriving at the presynaptic terminal of the neuron $j$ releases, on average, one transmitter vesicle.[6] The content of the vesicle diffuses through the synaptic cleft and reacts with receptor molecules embedded in the postsynaptic membrane. After chemical reactions described by kinetic differential equations (cf. Chap. 1, [35]), opened ion channels give rise to a postsynaptic impulse response potential $G_{ij}(t)$. Because we characterize the dendro-somatic membranes as linear systems here, the postsynaptic potential elicited by a spike train $R_j(t)$ is given by the convolution

$$\tilde{U}_{ij}(t) = G_{ij}(t) * R_j(t). \tag{7.36}$$

Let us make a rather crude approximation here by setting the postsynaptic impulse response function proportional to a delta function:

$$G_{ij}(t) = g_{ij} \delta(t), \tag{7.37}$$

where $g_{ij}$ is the *gain* of the synapse $j \rightarrow i$. Then, the postsynaptic potential is given by the product of the gain with the spike rate of the presynaptic neuron $j$.

Finally, we must take the stochasticity of the neuron into account as thoroughly described in Chap. 1. This is achieved by replacing the membrane potential $U_j$ at the trigger zone by its average obtained from the distribution function, which leads to the characteristic sigmoidal activation functions [57], e.g. the logistic function (see (7.5))

$$R_j(t) = f(U_j(t)) = \frac{1}{1 + e^{-\beta[U_j(t) - \theta]}}. \tag{7.38}$$

Collecting (7.35, 7.36) and (7.38) together and introducing the proper *synaptic weights*

$$w_{ij} = g_{ij} \tilde{w}_{ij} \tag{7.39}$$

yields the leaky integrator model of a network of distributed single neurons

$$\tau_i \frac{dU_i}{dt} + U_i = \sum_j w_{ij} f(U_j(t)) \tag{7.40}$$

which is analogous to (7.4).

---

[6] The release of transmitter is a stochastic process that can be approximately described by a binomial distribution [55], and hence, due to the limit theorem of de Moivre and Laplace, is normally distributed (see Chap. 1).

## 7.3.2 Leaky Integrator Model of Neural Populations

According to Freeman [58] (see also [59]), a neuronal *population* ("KI" set) consists of many reciprocally connected neurons of one kind, either excitatory or inhibitory. Let us consider such a set of McCulloch-Pitts neurons [18] distributed over a unit volume $i$ of neural tissue. We introduce the proportions of firing cells (either excitatory or inhibitory, in contrast to [17]) in volume $i$ at the instance of time $t$, $Q_i(t)$, as the state variables [17, 32].

A neuron belonging to volume $i$ will fire if its net input $U_i$ (analogous to the membrane potential at the trigger zone, see Sect. 7.3.1) crosses the threshold $\theta$. But now, we have to deal with an ensemble of neurons possessing randomly distributed thresholds within the unit volume $i$. We therefore obtain an ensemble activation function [5] (cf. Chap. 14) by integrating the corresponding probability distribution density $D(\theta)$ of thresholds [17],

$$f(U_i) = \int_0^{U_i} D(\theta)\mathrm{d}\theta \,. \tag{7.41}$$

Depending upon the modality of the distribution $D(\theta)$, the activation function could be sigmoidal or even more complicated. For unimodal distributions such as Gaussian or Poissonian distributions, $f(U_i)$ might be approximated by the logistic function (7.38). As for the single neuron model, the net input is obtained by a convolution

$$U_i(t) = \int_{-\infty}^{t} G(t - t') \sum_j w_{ij}\, Q_j(t')\mathrm{d}t' \,, \tag{7.42}$$

with "synaptic weights" $w_{ij}$ characterizing the neural connectivity and whether the population is excitatory or inhibitory.

In the following, we shall simplify the model of Wilson and Cowan [17] by neglecting the refractory time. The model equations are then

$$Q_i(t + \tau_i) = f\left( \int_{-\infty}^{t+\tau_i} G(t - t') \sum_j w_{ij}\, Q_j(t')\mathrm{d}t' \right), \tag{7.43}$$

such that $Q_i(t + \tau_i)$ is the proportion of cells being above threshold in the time interval $[t, t + \tau_i]$. Expanding the left hand side into a Taylor series at $t$ and assuming again that $G(t - t') = \delta(t - t')$, we obtain

$$\tau_i \frac{\mathrm{d}Q_i(t)}{\mathrm{d}t} + Q_i(t) = f\left( \sum_j w_{ij}\, Q_j(t) \right), \tag{7.44}$$

a leaky integrator model again, yet characterized by (7.2).

## 7.4 Oscillators from Leaky Integrator Units

### 7.4.1 Linear Model

In this section, we demonstrate that a damped harmonic oscillator can be obtained from a simple model of two recurrently coupled leaky integrator units with linear activation functions [49]. Figure 7.5 shows the architecture of this model.

The network of Fig. 7.5 is governed by (7.4):

$$\tau_1 \frac{dx_1}{dt} + x_1 = w_{11}x_1 + w_{12}x_2 + p \tag{7.45}$$

$$\tau_2 \frac{dx_2}{dt} + x_2 = w_{21}x_1 + w_{22}x_2 , \tag{7.46}$$

where $x_1$ denotes the activity of unit 1 and $x_2$ that of unit 2. Correspondingly, $\tau_1$ and $\tau_2$ are the time constants of the units 1 and 2, respectively. The synaptic weights $w_{ij}$ are indicated in Fig. 7.5. Note that the weights $w_{11}$ and $w_{22}$ describe autapses [60]. The quantity $p$ refers to excitatory synaptic input that might be a periodic forcing or any other function of time.

Equations (7.45) and (7.46) can be converted into two second-order ordinary differential equations

$$\frac{d^2x_1}{dt^2} + \gamma\frac{dx_1}{dt} + \omega_0^2 = p_1 \tag{7.47}$$

$$\frac{d^2x_2}{dt^2} + \gamma\frac{dx_2}{dt} + \omega_0^2 = p_2 , \tag{7.48}$$

where we have introduced the following simplifying parameters:

$$\gamma = \frac{\tau_1(1 - w_{22}) + \tau_2(1 - w_{11})}{\tau_1\tau_2}$$

$$\omega_0^2 = \frac{w_{11}w_{22} - w_{12}w_{21} - w_{11} - w_{22} + 1}{\tau_1\tau_2}$$

$$p_1 = \frac{1}{\tau_1}\frac{dp}{dt} + \frac{1 - w_{22}}{\tau_1\tau_2}$$

$$p_2 = \frac{w_{21}}{\tau_1\tau_2}p$$

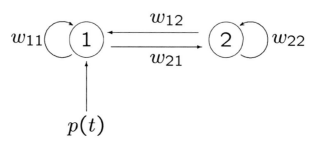

**Fig. 7.5.** Architecture of an oscillator formed by leaky integrator units

Now, (7.47) and (7.48) describe two damped, decoupled, harmonic oscillators with external forcing when $\gamma \geq 0$ and $\omega_0^2 > 0$, i.e. one unit must be excitatory and the other inhibitory.

## 7.4.2 Simple Nonlinear Model

Next, we discuss a simple nonlinear system, consisting of three coupled leaky integrator units, which provides a model of the thalamocortical loop. Figure 7.6 displays its architecture.

According to Fig. 7.6, the model (7.4) are

$$\tau_1 \frac{dx_1}{dt} + x_1 = -\alpha f(x_3(t)) \tag{7.49}$$

$$\tau_2 \frac{dx_2}{dt} + x_2 = \beta f(x_1(t)) \tag{7.50}$$

$$\tau_3 \frac{dx_3}{dt} + x_3 = \gamma f(x_2(t)). \tag{7.51}$$

Setting all $\tau_i = 1$ and rearranging, we get

$$\frac{dx_1}{dt} = -x_1 - \alpha f(x_3(t))$$

$$\frac{dx_2}{dt} = -x_2 + \beta f(x_1(t))$$

$$\frac{dx_3}{dt} = -x_3 + \gamma f(x_2(t)).$$

These equations define a vector field $\boldsymbol{F}$ with the Jacobian matrix

$$\mathbf{D}\boldsymbol{F} = \begin{pmatrix} -1 & 0 & -\alpha f'(x_3(t)) \\ \beta f'(x_1(t)) & -1 & 0 \\ 0 & \gamma f'(x_2(t)) & -1 \end{pmatrix}.$$

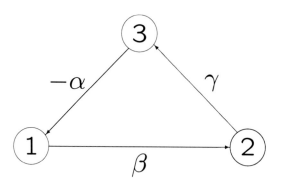

**Fig. 7.6.** Thalamocortical oscillator of three leaky integrator units: (1) pyramidal cell; (2) thalamus cell; (3) cortical interneuron (star cell)

For the activation function, we chose $f(x) = \tanh x$, which can be obtained by a coordinate transformation from the logistic function in (7.38). Therefore, $\boldsymbol{F}(x_1, x_2, x_3) = 0$ and we can look at whether the center manifold theorem [61] can be applied. The Jacobian at $(0, 0, 0)$ is

$$\mathrm{D}\boldsymbol{F}(0) = \begin{pmatrix} -1 & 0 & -\alpha \\ \beta & -1 & 0 \\ 0 & \gamma & -1 \end{pmatrix},$$

having eigenvalues

$$\lambda_1 = -1 - \sqrt[3]{\alpha\beta\gamma}$$
$$\lambda_2 = -1 + \frac{1}{2}(1 - i\sqrt{3})\sqrt[3]{\alpha\beta\gamma}$$
$$\lambda_3 = -1 + \frac{1}{2}(1 + i\sqrt{3})\sqrt[3]{\alpha\beta\gamma}.$$

Since $\lambda_1 < 0$ for $\alpha, \beta, \gamma \geq 0$, we seek for the weight parameters making $\mathrm{Re}(\lambda_{2|3}) = 0$. This leads to the condition

$$\alpha\beta\gamma = 8, \tag{7.52}$$

which can be easily fulfilled, for example, by setting

$$\alpha = 4, \qquad \beta = 1, \qquad \gamma = 2.$$

In this case, the center manifold theorem applies: the dynamics stabilizes along the eigenvector corresponding to $\lambda_1$, exhibiting a limit cycle in the center manifold spanned by the eigenvectors of $\lambda_2$ and $\lambda_3$. Figure 7.7 shows a numerical simulation of this oscillator. It is also possible to train a leaky integrator network using the algorithm described in Sect. 7.2.2 in order to replicate a limit cycle dynamics [49].

### 7.4.3 Random Neural Networks

In this last subsection, we describe a network model that is closely related to those presented in Chaps. 3, 5, 12, 13, and 14, namely a random graph carrying leaky integrator units described by (7.4) or, equivalently, (7.40), at its nodes:

$$\tau_i \frac{\mathrm{d}x_i(t)}{\mathrm{d}t} + x_i(t) = \sum_j w_{ij} f(x_j(t)).$$

We shall see that the onset of oscillatory behavior is correlated with the emergence of super-cycles in the topology of the network provided by an evolving directed and weighted Erdős-Rényi graph of $N$ nodes where all connections between two nodes are equally likely with increasing probability [62–64].

As explained in Chap. 3, a directed Erdős-Rényi graph consists of a set of vertices $V$ that are randomly connected by arrows taken from an edge set

$E \subset V \times V$ with equal probability $q$. The topology of the graph is completely described by its *adjacency matrix* $\boldsymbol{A} = (a_{ij})$ where $a_{ij} = 1$, if there is an arrow connecting the vertex $j$ with the vertex $i$ (i.e. $(j, i) \in E$ for $i, j \in V$) while $a_{ij} = 0$ otherwise. A directed and weighted Erdős-Rényi graph is then described by the *weight matrix* $\boldsymbol{W} = (w_{ij})$ which is obtained by element-wise multiplication of the adjacency matrix with constants $g_{ij}$: $w_{ij} = g_{ij} a_{ij}$. Biologically plausible models must satisfy Dale's law, which says that excitatory neurons only have excitatory synapses while inhibitory neurons only possess inhibitory synapses [56]. Therefore, the column vectors of the weight matrix are constrained to have a unique sign. We achieve this requirement by randomly choosing a proportion $p$ of the vertices to be excitatory and the remainder to be inhibitory.

In our model, the weights become time-dependent due to the following evolution algorithm:

(i) Initialization: $\boldsymbol{W}(0) = 0$.
(ii) At evolution time $t$, select a random pair of nodes $i, j$.
(iii) If they are not connected, create a synapse with weight $w_{ij}(t + 1) = +\delta$ if $j$ is excitatory, and $w_{ij}(t+1) = -\delta$ if $j$ is inhibitory. If they are already connected, enhance the weight $w_{ij}(t + 1) = w_{ij}(t) + \delta$ if $w_{ij}(t) > 0$ and $w_{ij}(t+1) = w_{ij}(t) - \delta$ if $w_{ij}(t) < 0$. All other weights remain unchanged.
(iv) Repeat from (ii) for a fixed number of iterations $L$.

As the "learning rate", we choose $\delta = 1$, while the connectivity increases for $L$ time steps. In order to simplify the simulations, we further set $\tau_i = 1$ for all $1 \leq i \leq N$.

Since (7.40) describes the membrane potential of the $i$th neuron, we can estimate its dendritic field potential by the inhomogeneity of (7.40),

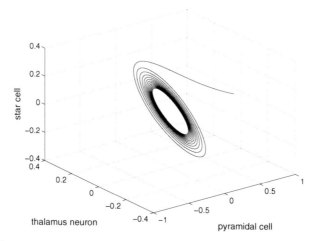

**Fig. 7.7.** Limit cycle of the thalamocortical oscillator in its center manifold plane

$$F_i(t) = \sum_j w_{ij}\, f(x_j)\,. \tag{7.53}$$

Then, the model EEG[7] is given by the sum of the dendritic field potentials of all excitatory nodes

$$E(t) = \sum_{i^+} F_i(t)\,. \tag{7.54}$$

The indices $i^+$ indicate that the neuron $i$ belongs to the population of excitatory neurons, namely the EEG generating pyramidal cells.

We create such random neural networks with size $N = 100, 200, 500,$ and $1000$ nodes. Since about 80% of cortical neurons are excitatory pyramidal cells, $p = 80\%$ of the network's nodes are chosen to be excitatory [66]. For each iteration of the network's evolution, the dynamics of its nodes is calculated. After preparing them with normally distributed initial conditions ($\mu = 0, \sigma = 1$), (7.40) is solved numerically with the activation functions $f_i(x) = \tanh x$ for an ensemble of $K = 10$ time series of length $T = 100$ with a step-width of $\Delta t = 0.0244$. The dendritic field potential and EEG are computed according to (7.53) and (7.54).

From the simulated EEGs, the power spectra are computed and averaged over all $K$ realizations of the network's dynamics. In order to monitor sudden changes in the topologies of the networks, three characteristic statistics are calculated:

(1) The *mean degree* (the average number of vertices attached to the nodes) $\langle k \rangle$ of the associated undirected graphs, described by the symmetrized adjacency matrix $\boldsymbol{A}^s = \Theta(\boldsymbol{A} + \boldsymbol{A}^T)$ ($\Theta$ denotes Heaviside's step function),

(2) the *total distribution*

$$d(l) = \frac{\mathrm{tr}(\boldsymbol{A}^l)}{l\mathcal{N}} \tag{7.55}$$

of cycles of the exact length $l$ [62–64, 67–70]. In (7.55), $\mathrm{tr}(\boldsymbol{A}^l)$ provides the total number of (not necessarily self-avoiding) closed paths of length $l$ through the network. Since any node at such a path may serve as the starting point and there are $l$ nodes, the correct number of cycles is obtained by dividing by $l$. Finally, $\mathcal{N} = \sum_l \mathrm{tr}(\boldsymbol{A}^l)/l$ is a normalization constant. From the cycle distribution (7.55), we derive

(3) an *order parameter* $s$ for topological transitions from the averaged slopes of the envelope of $d(l)$, where the envelopes are estimated by connecting the local maxima of $d(l)$. The above mentioned procedure is repeated for each network size $M = 10$ times where we have chosen $L_{100} = 150, L_{200} = 400, L_{500} = 800,$ and $L_{1000} = 1700$ iterations of network evolution in order to ensure sufficiently dense connectivities.

---

[7] In fact, (7.54) describes better the local field potential (LFP) rather than the EEG. Considering (7.4) as a model of coupled neural populations, instead, seems to be more appropriate for describing the EEG [65].

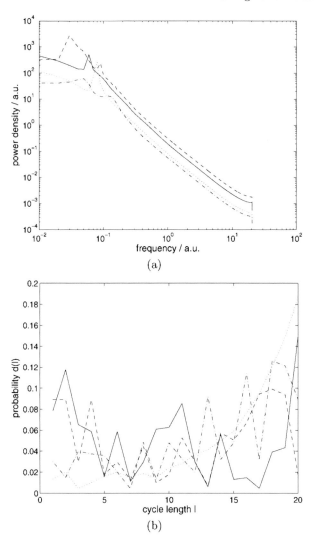

**Fig. 7.8.** (a) Power spectra of representative simulated time series during the oscillatory transition (critical phase) for four different network sizes: $N = 100$ (*dotted*), $N = 200$ (*dashed-dotted*), $N = 500$ (*solid*), and $N = 1000$ (*dashed*); (b) Total distributions of cycles (7.55) for the same networks

Figure 7.8 shows four representative networks in the critical phase characterized by the smallest positive value of the cycle order parameter $s$, averaged over the $M = 10$ network simulations, when sudden oscillations occur in the dynamics of the units, as is visible by the peaks in the power spectra [Fig. 7.8(a)]. The cycle distributions $d(l)$ [Fig. 7.8(b)] for network sizes $N = 200, 500,$ and $1000$ display a transition from geometrically decaying to exponentially growing functions while this transition has already taken place for $N = 100$. As Fig. 7.8(a) reveals, the power spectra display a broad $1/f$ continuum. Superimposed to this continuum are distinguished peaks that can be regarded as the "alpha waves" of the model.

According to random graph theory, Erdős-Rényi networks exhibit a percolation transition when a giant cluster suddenly occurs for $\langle k \rangle = 1$ [62–64]. A second transition takes place for $\langle k \rangle = 2$, indicating the appearance of mainly isolated cycles in the graph. Isolated cycles are characterized by a geometrically decaying envelope of the total cycle distribution. Our simulations suggest the existence of a third transition when super-cycles are composed from merging smaller ones. This is reflected by a transition of the total cycle distribution $d(l)$ from a geometrically decaying to an exponentially growing behavior due to a "combinatorial explosion" of possible self-intersecting paths through the network (super-cycles are common in regular lattices with $\langle k \rangle \geq 3$). We detect this transition by means of a suitably chosen order parameter $s$ derived from $d(l)$ as the averaged slope of its envelope. For decaying $d(l)$, $s < 0$ and for growing $d(l)$, $s > 0$. The appearance of super-cycles is associated with $s \approx 0$ if $d(l)$ is approximately symmetric in the range of $l$. In this case, sustained oscillations emerge in the network's dynamics due to the presence of reverberatory circles. For further details, see [65].

## 7.5 Cognitive Modeling

In this chapter we have reviewed neurophysiological findings on oscillations in the electroencephalogram as well as certain approaches to model these through coupled differential equations. We have introduced the theory of networks of *leaky integrator units* and presented a general learning rule to train these networks in such a way that they are able to reproduce temporal patterns in continuous time. This learning rule is a generalized back-propagation algorithm that has been applied for the first time to model reaction times from a psychological experiment [54]. Therefore, leaky integrator networks provide a unique and physiologically plausible paradigm for neural and cognitive modeling.

Mathematically, leaky integrator models are described by systems of coupled ordinary differential equations that become nonlinear dynamical systems by using sigmoidal activation functions. Networks of leaky integrator units may display a variety of complex behaviors: limit cycles, multistability, bifurcations and hysteresis [17]. They could therefore act as models of perceptional instability [71] or cognitive conflicts [72], as has already been demonstrated by Haken [73, 74] using *synergetic computers*. As Haken [74, p. 246] pointed out, the order parameter equations of synergetic computers are analogous to neural networks whose activation function is expanded into a power series. However, these computers are actually leaky integrator networks as we will see subsequently.

Basically, synergetic computers are time-continuous Hopfield nets [19] governed by a differential equation

$$\frac{\mathrm{d}\boldsymbol{x}}{\mathrm{d}t} - \sum_{k=1}^{K} \eta_k \boldsymbol{v}_k (\boldsymbol{v}_k^+ \boldsymbol{x}) = -B \sum_{k' \neq k}^{K} (\boldsymbol{v}_{k'}^+ \boldsymbol{x})^2 (\boldsymbol{v}_k^+ \boldsymbol{x}) \boldsymbol{x} - C(\boldsymbol{x}^+ \boldsymbol{x}) \boldsymbol{x} \qquad (7.56)$$

where $\boldsymbol{x}(t)$ denotes the activation vector of the network; the $K$ vectors $\boldsymbol{v}_k$ are training patterns with adjoints $\boldsymbol{v}_k^+$ such that the orthonormality relations $\boldsymbol{v}_k^+ \boldsymbol{v}_l = \delta_{kl}$ hold. In this notation, $\boldsymbol{x}^+ \boldsymbol{y} = \sum_i x_i y_i$ means the *inner product* of the row vector $\boldsymbol{x}^+$ with a column vector $\boldsymbol{y}$ yielding a scalar. On the other hand, the outer product $\boldsymbol{y}\boldsymbol{x}^+$ of a column vector $\boldsymbol{y}$ with a row vector $\boldsymbol{x}^+$ is a matrix with elements $y_i x_j$.

Therefore, the second term of the left hand side of (7.56) can be rewritten as

$$\sum_{k=1}^{K} \eta_k \boldsymbol{v}_k (\boldsymbol{v}_k^+ \boldsymbol{x}) = \sum_{k=1}^{K} \eta_k (\boldsymbol{v}_k \boldsymbol{v}_k^+) \boldsymbol{x} = \left( \sum_{k=1}^{K} \eta_k \boldsymbol{v}_k \boldsymbol{v}_k^+ \right) \boldsymbol{x} = \boldsymbol{W} \boldsymbol{x}$$

where

$$\boldsymbol{W} = \sum_{k=1}^{K} \eta_k \boldsymbol{v}_k \boldsymbol{v}_k^+$$

is the synaptic weight matrix obtained by Hebbian learning of the patterns $\boldsymbol{v}_k$ with learning rates $\eta_k$.

The notion "synergetic computer" refers to the possibility of describing the network (7.56) by the evolution of *order parameters*, which are appropriately chosen as the "loads" of the training patterns $\boldsymbol{v}_k$ in a kind of principal component analysis. We therefore separate activation space and time by the ansatz

$$\boldsymbol{x}(t) = \sum_k \xi_k(t) \boldsymbol{v}_k + \boldsymbol{w}(t),$$

where $\xi_k(t) = \boldsymbol{v}_k^+ \boldsymbol{x}(t)$ and $\boldsymbol{w}(t)$ is a fast decaying remainder. Multiplying (7.56) from the left with $\boldsymbol{v}_l^+$ and exploiting the orthonormality relations, we eventually obtain

$$\frac{\mathrm{d}\xi_l}{\mathrm{d}t} - \eta_l \xi_l = -B \sum_{k \neq l}^{K} \xi_k^2 \xi_l - C \left( \sum_k \xi_k^2 \right) \xi_l . \qquad (7.57)$$

Division by $-\eta_l = 1/\tau_l$ then yields the leaky integrator equations for the order parameters with rescaled constants $B', C'$ and a cubic activation function

$$\tau_l \frac{\mathrm{d}\xi_l}{\mathrm{d}t} + \xi_l = B' \sum_{k \neq l}^{K} \xi_k^2 \xi_l + C' \left( \sum_k \xi_k^2 \right) \xi_l . \qquad (7.58)$$

The "time constants" play then the role of *attention parameters* describing the amount of attention devoted to a particular pattern. These parameters might also depend on time, e.g. for modeling habituation.

From a formal point of view, the attention model of Lourenço [75], the *cellular neural networks* (CNN) of Chua [76] (see also [77–79]) and the disease model of Huber et al. [80] can also be regarded as leaky integrator networks.

Also, higher cognitive functions such as language processing and their neural correlates such as event-related brain potentials (ERPs) [72,81,82] can be modeled with leaky integrator networks. Kawamoto [83] used a Hopfield net with exponentially decaying activation and habituating synaptic weights to modeling lexical ambiguity resolution. The activations of the units in his model are governed by the equations

$$x_i(t+1) = f\left(\delta x_i(t) + \sum_j w_{ij} x_j(t)\right). \tag{7.59}$$

Setting $\delta = 1 - \alpha = 1 - \tau^{-1}$ and approximating $f'(x) \approx 1$ for typical activation values yields, after a Taylor expansion of the activation function,

$$f\left(\delta x_i(t) + \sum_j w_{ij} x_j(t)\right) \approx f\left(\sum_j w_{ij} x_j(t)\right) + f'\left(\sum_j w_{ij} x_j(t)\right)\delta x_i(t),$$

the leaky integrator equation (7.2).

Smolensky and Legendre [46] consider Hopfield nets of leaky integrator units that can be described by a Lyapunov function $E$. They call the function $H = -E$ the *harmony* of the network and argue that cognitive computations maximize this harmony function at the sub-symbolic level. Additionally, the harmony value can also be computed at the symbolic level of linguistic representations in the framework of harmonic grammars or optimality theory [84]. By regarding the harmony as an order parameter of the network, one could also model neural correlates of cognitive processes, e.g., ERPs.

This has recently been attempted by Wennekers et al. [83], who built a six-layer model of the perisylvian language cortex by randomly connecting leaky integrator units within each layer (similar to our exposition in Sect. 7.4.3). The network was trained with a Hebbian correlation learning rule to memorize "words" (co-activated auditory and motor areas) and "pseudowords" (activation of the auditory layer only). After training, cell assemblies of synchronously oscillating units across all six layers emerged. Averaging their event-related oscillations in the recall phase then yielded a larger amplitude for the "words" than for the "pseudowords", thus emulating the mismatch negativity (MMN) ERP known from word recognition experiments [45].

## Acknowledgements

We gratefully acknowledge helpful discussions with W. Ehm, P. Franaszcuk, M. Garagnani, A. Hutt, V. K. Jirsa, and J. J. Wright. This work has been supported by the Deutsche Forschungsgemeinschaft (research group "Conflicting

Rules in Cognitive Systems"), and by the Helmholtz Institute for Supercomputational Physics at the University of Potsdam.

# References

1. O. Creutzfeld and J. Houchin. Neuronal basis of EEG-waves. In *Handbook of Electroencephalography and Clinical Neurophysiology*, Vol. 2, Part C, pp. 2C-5–2C-55. Elsevier, Amsterdam, 1974.
2. F. H. Lopes da Silva. Neural mechanisms underlying brain waves: from neural membranes to networks. *Electroencephalography and Clinical Neurophysiology*, 79: 81–93, 1991.
3. E.-J. Speckmann and C. E. Elger. Introduction to the neurophysiological basis of the EEG and DC potentials. In E. Niedermeyer and F. Lopez da Silva, editors, *Electroencephalography. Basic Principles, Clinical Applications, and Related Fields*, Chap. 2, pp. 15–27. Lippincott Williams and Wilkins, Baltimore, 1999.
4. S. Zschocke. *Klinische Elektroenzephalographie*. Springer, Berlin, 1995.
5. W. J. Freeman. Tutorial on neurobiology: from single neurons to brain chaos. *International Journal of Bifurcation and Chaos*, 2(3): 451–482, 1992.
6. P. L. Nunez and R. Srinivasan. *Electric Fields of the Brain: The Neurophysics of EEG*. Oxford University Press, New York, 2006.
7. E. Başar. *EEG-Brain Dynamics. Relations between EEG and Brain Evoked Potentials*. Elsevier/North Holland Biomedical Press, Amsterdam, 1980.
8. M. Steriade, P. Gloor, R. R. Llinás, F. H. Lopes da Silva, and M.-M. Mesulam. Basic mechanisms of cerebral rhythmic activities. *Electroencephalography and Clinical Neurophysiology*, 76: 481–508, 1990.
9. C. Allefeld and J. Kurths. Testing for phase synchronization. *International Journal of Bifurcation and Chaos*, 14(2): 405–416, 2004.
10. C. Allefeld and J. Kurths. An approach to multivariate phase synchronization analysis and its application to event-related potentials. *International Journal of Bifurcation and Chaos*, 14(2): 417–426, 2004.
11. R. Srinivasan. Internal and external neural synchronization during conscious perception. *International Journal of Bifurcation and Chaos*, 14(2): 825–842, 2004.
12. G. Pfurtscheller. EEG rhythms — event related desynchronization and synchronization. In H. Haken and H. P. Koepchen, editors, *Rhythms in Physiological Systems*, pp. 289–296, Springer, Berlin, 1991.
13. E. Başar, M. Özgören, S. Karakaş, and C. Başar-Eroğlu. Super-synergy in brain oscillations and the grandmother percept is manifested by multiple oscillations. *International Journal of Bifurcation and Chaos*, 14(2): 453–491, 2004.
14. W. Klimesch, M. Schabus, M. Doppelmayr, W. Gruber, and P. Sauseng. Evoked oscillations and early components of event-related potentials: an analysis. *International Journal of Bifurcation and Chaos*, 14(2): 705–718, 2004.
15. N. Birbaumer and R. F. Schmidt. *Biologische Psychologie*. Springer, Berlin, 1996.
16. M. Steriade. Cellular substrates of brain rhythms. In E. Niedermeyer and F. Lopez da Silva, editors, *Electroencephalography. Basic Principles, Clinical Applications, and Related Fields*, Chap. 3, pp. 28–75. Lippincott Williams and Wilkins, Baltimore, 1999.

17. H. R. Wilson and J. D. Cowan. Excitatory and inhibitory interactions in localized populations of model neurons. *Biophysical Journal*, 12: 1–24, 1972.
18. W. S. McCulloch and W. Pitts. A logical calculus of ideas immanent in nervous activity. *Bulletin of Mathematical Biophysics*, 5: 115–33, 1943. Reprinted in J. A. Anderson and E. Rosenfeld (1988), pp. 83ff.
19. J. Hertz, A. Krogh, and R. G. Palmer. *Introduction to the Theory of Neural Computation*. Perseus Books, Cambridge (MA), 1991.
20. H. R. Wilson. *Spikes, Decisions and Actions. Dynamical Foundations of Neuroscience*. Oxford University Press, New York (NY), 1999.
21. F. H. Lopes da Silva, A. Hoecks, H. Smits, and L. H. Zetterberg. Model of brain rhythmic activity: the alpha-rhythm of the thalamus. *Kybernetik*, 15: 27–37, 1974.
22. F. H. Lopes da Silva, A. van Rotterdam, P. Bartels, E. van Heusden, and W. Burr. Models of neuronal populations: the basic mechanisms of rhythmicity. In M. A. Corner and D. F. Swaab, editors, *Perspectives of Brain Research*, Vol. 45 of *Progressive Brain Research*, pp. 281–308. 1976.
23. W. J. Freeman. Simulation of chaotic EEG patterns with a dynamic model of the olfactory system. *Biological Cybernetics*, 56: 139–150, 1987.
24. B. H. Jansen and V. G. Rit. Electroencephalogram and visual evoked potential generation in a mathematical model of coupled cortical columns. *Biological Cybernetics*, 73: 357–366, 1995.
25. B. H. Jansen, G. Zouridakis, and M. E. Brandt. A neurophysiologically-based mathematical model of flash visual evoked potentials. *Biological Cybernetics*, 68: 275–283, 1993.
26. F. Wendling, F. Bartolomei, J. J. Bellanger, and P. Chauvel. Epileptic fast activity can be explained by a model of impaired GABAergic dendritic inhibition. *European Journal of Neuroscience*, 15: 1499–1508, 2002.
27. F. Wendling, J. J. Bellanger, F. Bartolomei, and P. Chauvel. Relevance of nonlinear lumped-parameter models in the analysis of depth-EEG epileptic signals. *Biological Cybernetics*, 83: 367–378, 2000.
28. O. David, D. Cosmelli, and K. J. Friston. Evaluation of different measures of functional connectivity using a neural mass model. *NeuroImage*, 21: 659–673, 2004.
29. O. David and K. J. Friston. A neural mass model for MEG/EEG: coupling and neuronal dynamics. *NeuroImage*, 20: 1743–1755, 2003.
30. O. David, L. Harrison, and K. J. Friston. Modelling event-related respones in the brain. *NeuroImage*, 25: 756–770, 2005.
31. A. van Rotterdam, F. H. Lopes da Silva, J. van den Ende, M. A. Viergever, and A. J. Hermans. A model of the spatial-temporal characteristics of the alpha rhythm. *Bulletin of Mathematical Biology*, 44(2): 283–305, 1982.
32. J. J. Wright and D. T. L. Liley. Simulation of electrocortical waves. *Biological Cybernetics*, 72: 347–356, 1995.
33. J. J. Wright and D. T. J. Liley. Dynamics of the brain at global and microscopic scales: neural networks and the EEG. *Behavioral and Brain Sciences*, 19: 285–320, 1996.
34. D. T. J. Liley, D. M. Alexander, J. J. Wright, and M. D. Aldous. Alpha rhythm emerges from large-scale networks of realistically coupled multicompartmental model cortical neurons. *Network: Computational. Neural Systems*, 10: 79–92, 1999.

35. C. Koch and I. Segev, editors. *Methods in Neuronal Modelling. From Ions to Networks.* Computational Neuroscience. MIT Press, Cambridge (MA), 1998.
36. A. L. Hodgkin and A. F. Huxley. A quantitative description of membrane current and its application to conduction and excitation in nerve. *Journal Physiology*, 117: 500–544, 1952.
37. V. K. Jirsa and H. Haken. Field theory of electromagnetic brain activity. *Physical Review Letters*, 77(5): 960–963, 1996.
38. V. K. Jirsa. Information processing in brain and behavior displayed in large-scale scalp topographies such as EEG and MEG. *International Journal of Bifurcation and Chaos*, 14(2): 679–692, 2004.
39. J. J. Wright, C. J. Rennie, G. J. Lees, P. A. Robinson, P. D. Bourke, C. L. Chapman, E. Gordon, and D. L. Rowe. Simulated electrocortical activity at microscopic, mesoscopic, and global scales. *Neuropsychopharmacology*, 28: S80–S93, 2003.
40. J. J. Wright, C. J. Rennie, G. J. Lees, P. A. Robinson, P. D. Bourke, C. L. Chapman, E. Gordon, and D. L. Rowe. Simulated electrocortical activity at microscopic, mesoscopic and global scales. *International Journal of Bifurcation and Chaos*, 14(2): 853–872, 2004.
41. P. A. Robinson, C. J. Rennie, J. J. Wright, H. Bahramali, E. Gordon, and D. L. Rowe. Prediction of electroencephalic spectra from neurophysiology. *Physical Reviews E*, 63, 021903, 2001.
42. D. O. Hebb. *The Organization of Behavior.* Wiley, New York (NY), 1949.
43. S. Kaplan, M. Sonntag, and E. Chown. Tracing recurrent activity in cognitive elements (TRACE): a model of temporal dynamics in a cell assembly. *Connection Science*, 3: 179–206, 1991.
44. F. van der Velde and M. de Kamps. Neural blackboard architectures of combinatorial structures in cognition. *Behavioral and Brain Sciences*, 29:37–108, 2006.
45. T. Wennekers, M. Garagnani, and F. Pulvermüller. Language models based on hebbian cell assemblies. *Journal of Physiology - Paris*, 100: 16–30, 2006.
46. P. Smolensky and G. Legendre. *The Harmonic Mind. From Neural Computation to Optimality-Theoretic Grammar*, Vol. 1: Cognitive Architecture. MIT Press, Cambridge (MA), 2006.
47. R. B. Stein, K. V. Leung, M. N. Oğuztöreli, and D. W. Williams. Properties of small neural networks. *Kybernetik*, 14: 223–230, 1974.
48. D. E. Rumelhart, J. L. McClelland, and the PDP Research Group, editors. *Parallel Distributed Processing: Explorations in the Microstructure of Cognition*, Vol. I. MIT Press, Cambridge (MA), 1986.
49. B. A. Pearlmutter. Learning state space trajectories in recurrent neural networks. *Neural Computation*, 1(2): 263–269, 1989.
50. B. A. Pearlmutter. Gradient calculations for dynamic recurrant neural networks: A survey. *IEEE Transaction on Neural Networks*, 6(5): 1212–1228, 1995.
51. P. Werbos. Back-propagation through time: What it does and how to do it. Vol. 78 of *Proc. IEEE*, pp. 1550–1560, 1990.
52. P. Werbos. Maximizing long-term gas industry profits in two minutes in Lotus using neural network models. *IEEE Transaction on Systems, Man, and Cybernetics*, 19: 315–333, 1989.
53. W. H. Press, S. A. Teukolsky, W. T. Vetterling, and B. P. Flannery. *Numerical Recipies in C.* Cambridge University Press, New York, 1996.

54. T. Liebscher. Modeling reaction times with neural networks using leaky integrator units. In K. Jokinen, D. Heylen, and A. Nijholt, editors, *Proc. 18th Twente Workshop on Language Technology*, Vol. 18 of *TWLT*, pp. 81–94, Twente (NL), 2000. Univ. Twente.

55. E. R. Kandel, J. H. Schwartz, and T. M. Jessel, editors. *Principles of Neural Science*. Appleton & Lange, East Norwalk, Connecticut, 1991.

56. P. Dayan and L. F. Abbott. *Theoretical Neuroscience*. Computational Neuroscience. MIT Press, Cambridge (MA), 2001.

57. D. J. Amit. *Modeling Brain Function. The World of Attractor Neural Networks*. Cambridge University Press, Cambridge (MA), 1989.

58. W. J. Freeman. *Mass Action in the Nervous System*. Academic Press, New York (NY), 1975.

59. W. J. Freeman. How and why brains create meaning from sensory information. *International Journal of Bifurcation and Chaos*, 14(2): 515–530, 2004.

60. C. S. Herrmann and A. Klaus. Autapse turns neuron into oscillator. *International Journal of Bifurcation and Chaos*, 14(2): 623–633, 2004.

61. J. Guckenheimer and P. Holmes. *Nonlinear Oscillations, Dynamical Systems, and Bifurcations of Vector Fields*, Vol. 42 of *Springer Series of Appl. Math. Sciences*. Springer, New York, 1983.

62. R. Albert and A.-L. Barabási. Statistical mechanics of complex networks. *Reviews of Modern Physics*, 74(1): 47–97, 2002.

63. S. Bornholdt and H. G. Schuster, editors. *Handbook of Graphs and Networks. From the Genome to the Internet*. Wiley-VCH, Weinheim, 2003.

64. B. Bollobás. *Random Graphs*. Cambridge University Press, Cambridge (UK), 2001.

65. P. beim Graben and J. Kurths. Simulating global properties of electroencephalograms with minimal random neural networks. *Neurocomputing*, doi:10.1016/j.neucom.2007.02.007, 2007.

66. A. J. Trevelyan and O. Watkinson. Does inhibition balance excitation in neocortex? *Progress in Biophysics and Molecular Biology*, 87: 109–143, 2005.

67. S. Itzkovitz, R. Milo, N. Kashtan, G. Ziv, and U. Alon. Subgraphs in random networks. *Physical Reviews E*, 68, 026127, 2003.

68. G. Bianconi and A. Capocci. Number of loops of size $h$ in growing scale-free networks. *Physical Review Letters*, 90(7), 2003.

69. H. D. Rozenfeld, J. E. Kirk, E. M. Bollt, and D. ben Avraham. Statistics of cycles: how loopy is your network? *Journal of physics A: Mathematical General*, 38: 4589–4595, 2005.

70. O. Sporns, G. Tononi, and G. M. Edelman. Theoretical neuroanatomy: Relating anatomical and functional connectivity in graphs and cortical connection matrices. *Cerebral Cortex*, 10(2): 127–141, 2000.

71. J. Kornmeier, M. Bach, and H. Atmanspacher. Correlates of perceptive instabilities in visually evoked potentials. *International Journal of Bifurcation and Chaos*, 14(2): 727–736, 2004.

72. S. Frisch, P. beim Graben, and M. Schlesewsky. Parallelizing grammatical functions: P600 and P345 reflect different cost of reanalysis. *International Journal of Bifurcation and Chaos*, 14(2): 531–549, 2004.

73. H. Haken. *Synergetic Computers and Cognition. A top-down Approach to Neural Nets*. Springer, Berlin, 1991.

74. H. Haken. *Principles of Brain Functioning*. Springer, Berlin, 1996.

75. C. Lourenço. Attention-locked computation with chaotic neural nets. *International Journal of Bifurcation and Chaos*, 14(2): 737–760, 2004.
76. L. O. Chua. *CNN: A paradigm for complexity*. World Scientific, Singapore, 1998.
77. D. Bálya, I. Petrás, T. Roska, R. Carmona, and A. R. Vázquez. Implementing the multi-layer retinal model on the complex-cell cnn-um chip prototype. *International Journal of Bifurcation and Chaos*, 14(2): 427–451, 2004.
78. V. Gál, J. Hámori, T. Roska, D. Bálya, Zs. Borostyánköi, M. Brendel, K. Lotz, L. Négyessy, L. Orzó, I. Petrás, Cs. Rekeczky, J. Takács, P. Venetiáner, Z. Vidnyánszky, and Á Zarándy. Receptive field atlas and related CNN models. *International Journal of Bifurcation and Chaos*, 14(2): 551–584, 2004.
79. F. S. Werblin and B. M. Roska. Parallel visual processing: A tutorial of retinal function. *International Journal of Bifurcation and Chaos*, 14(2): 843–852, 2004.
80. M. T. Huber, H. A. Braun, and J.-C. Krieg. Recurrent affective disorders: nonlinear and stochastic models of disease dynamics. *International Journal of Bifurcation and Chaos*, 14(2): 635–652, 2004.
81. P. beim Graben, S. Frisch, A. Fink, D. Saddy, and J. Kurths. Topographic voltage and coherence mapping of brain potentials by means of the symbolic resonance analysis. *Physical Reviews E*, 72: 051916, 2005.
82. H. Drenhaus, P. beim Graben, D. Saddy, and S. Frisch. Diagnosis and repair of negative polarity constructions in the light of symbolic resonance analysis. *Brain and Language*, 96(3): 255–268, 2006.
83. A. H. Kawamoto. Nonlinear dynamics in the resolution of lexical ambiguity: A parallel distributed processing account. *Journal of Memory and Language*, 32: 474–516, 1993.
84. A. Prince and P. Smolensky. Optimality: from neural networks to universal grammar. *Science*, 275: 1604–1610, 1997.

# 8

# A Dynamic Model of the Macrocolumn

James J. Wright[1,2]

[1] Liggins Institute, and Department of Psychological Medicine,
University of Auckland, School of Medicine, Auckland, New Zealand
[2] Brain Dynamics Center, Westmead Hospital,
University of Sydney, Sydney, Australia
jj.w@xtra.co.nz

**Summary.** Neurons within a cortical macrocolumn can be represented in continuum state equations that include axonal and dendritic delays, synaptic densities, adaptation and distribution of AMPA, NMDA and GABA postsynaptic receptors, and back-propagation of action potentials in the dendritic tree. Parameter values are independently specified from physiological data. In numerical simulations, synchronous oscillation and gamma activity are reproduced and a mechanism for self-regulation of cortical gamma is demonstrated. Properties of synchronous fields observed in the simulations are then applied in a model of the self-organization of synapses, using a simple Hebbian learning rule with decay. The patterns of connection of maximally stable configuration are compared to real cortical synaptic connections that emerge in neurodevelopment.

## 8.1 Introduction

This chapter gives an account of two complementary approaches to modeling the axo-dendritic dynamics, and of the functionally related synaptic dynamics, within a small volume of cerebral cortex. Choice of the appropriate volume is somewhat arbitrary, but a useful scale is that which has been described using a variety of related criteria as the macrocolumn, or corticocortical column a volume approximately 300 microns in surface diameter [1–3]. The extension of the dendritic and intracortical proximal axonal trees of pyramidal cells within the cortex conform to this 300 micron approximation, as is indicated in Fig. 8.1, and because of the branching structure of both dendrites and local axons, the density of synaptic connections between neurons declines with distance from the cell body [4,5]. Consequently, the strength of interaction of neurons up to 300 microns apart is comparatively high. On the other hand, the largest fraction of synapses within any volume of cortex arises from cell bodies outside the column, and the sparse connectivity of neurons inside the column makes the intermingling of adjacent "columns" inevitable [3,5]. As will be shown, the theoretical modeling reported here may help to provide a new definition

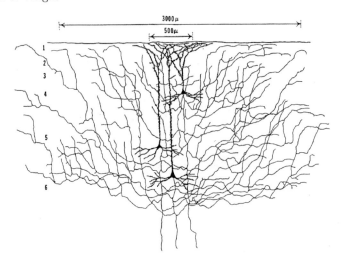

**Fig. 8.1.** Dendritic trees of pyramidal cells. (Braitenberg and Schüz, 1991)

of the macrocolumn, as the scale of a synaptic map of activity relayed from the wider cortex.

To simplify description, the cortex will be considered to be two-dimensional, thus largely ignoring organization in depth.

Analysis of cortical functional anatomy at the macrocolumnar scale has been found of particular utility within the visual cortex [6, 7], because this scale is also the scale of ramification in cortex of terminal axons from the direct visual pathway [2,3,6,7]. Studies of responses of individual cells and of groups of cells to moving objects in the visual field, have shown that neural responses are organized so that responses to moving lines are not only selective for the orientation of the line (orientation preference, or OP), but also to the velocity, angle relative to motion, and extension of the lines [8]. Where binocular vision is present, neurons are organized into bands, each of which are about as broad as a macrocolumn, with alternating bands — the ocular dominance (OD) columns — selectively responding to one or other eye. This yields a unit system — the hypercolumn [1,6] — with the capacity to process information from a specific small part of the visual field [9]. It can also be shown that the function of each such unit is modulated by the contextual activity of the surrounding visual field [10].

It is presently unclear how this anatomical detail is involved in the processing of visual information in neural-network terms. The modeling described here attempts to solve parts of this problem.

Synchronous oscillation is a physiological phenomenon relevant to all considerations of the processing of cortical information. When separate neurons are concurrently activated by discrete stimuli, they begin to fire synchronously, emitting action potentials with maximum cross-correlation at zero lag [11–14]. The emission of action potentials and fluctuation of the local dendritic potential is typically at around 50 Hz — although not uniquely so — and because this frequency falls within the gamma-band of electrocortical activity, it is referred to as "gamma synchrony". Gamma synchrony is believed to underlie the psychological process of perceptual "binding", allowing states of the cortex to code for many different perceptual states, by use of combinations of a smaller set of unit states per neuron [15, 16].

To provide an account of the relation of synchronous oscillation to functional anatomical organization at the scale of the macrocolumn, a continuum model of electrocortical activity is proposed. The model uses discretized integral state equations and includes effects of axonal and dendritic delays, back-propagation in the dendritic tree, reversal potentials, synaptic densities, and kinetics of AMPA, GABA and NMDA receptors, and has been reported in an earlier form in relation to global electrocortical activity [17]. To the extent practicable, all parameters are obtained from independent physiological and anatomical estimates, and all lie in the physiologically plausible range. In this and related models, gamma activity is reproduced, associated with synchronous oscillation [18–22]. As a further step toward realism, the most recent development introduces a mechanism of control of transition into autonomous gamma, which is initiated and suppressed by the level of subcortical reticular activation, and the transcortical synaptic flux originating from outside the macrocolumn.

Having described a means of generation of synchronous fields of activity in cortex, a model [23] for the self-organization of synapses can then be applied, thus extending consideration to learning-related modifications of synapses. According to this model, "local maps" are formed by self-organization of synapses during development, and each local map is analogous to a projection of the primary visual cortex (V1) onto a Möbius strip. The scale of each map is that of a macrocolumn, and they represent the most stable synaptic state in fields of neural synchrony, under conditions of uniform metabolic load and of Hebbian learning with decay. In this maximally stable state, all synapses are either saturated, or have minimum pre/postsynaptic coincidence. Each local map is arrayed as approximately a mirror-image reflection of each of its neighbors, accounting for a number of major features of local anatomical organization. Preliminary consideration is given to the impact of dynamic perturbation upon the stable synaptic configuration, and the implications for perception and cortical information processing.

## 8.2 A Continuum Model of Electrocortical Activity

Electrotonic and pulse activity in the cortex can be treated as activity in a wave medium, as shown by numerous workers [24–28] (cf. Chap. 1). All members of this family of models have strong resemblances, but there are differences in both the details of the state equations and the parameters applied.

The form given here has been developed to be applicable to cortex at a variety of scales, and to permit application alongside related models [17, 29].

### 8.2.1 State Equations

**Synaptic Flux Density**

The distribution of neuron cell bodies sending afferent connections to a cortical point, $r$, is $f(r, r')$, where $\{r'\}$ are all other points in the field. The synaptic flux density, $\varphi_p(r, t)$, the average input pulse rate per synapse at $r$, is given by

$$\varphi_p(r, t) = \int f(r, r') Q_p \left( r', t - \frac{|r - r'|}{v_p} \right) \mathrm{d}^2 r', \qquad (8.1)$$

where the normalized axonal spread $f(r, r')$ satisfies

$$\int f(r, r') \mathrm{d}^2 r = \int f(r, r') \mathrm{d}^2 r' = 1,$$

$Q_p(r', t)$    are the pulse densities of neurons in the afferent field,
$v_p$    is the velocity of axonal conduction,
$p = e, i$    indicates whether the afferent neurons are excitatory or inhibitory.

An alternative to (8.1) is a damped wave equation, for which the implicit axonal spread is approximately a two-dimensional Gaussian [28], is:

$$\left( \frac{\partial^2}{\partial t^2} + 2\gamma_p \frac{\partial}{\partial t} + \gamma_p^2 - v_p^2 \nabla^2 \right) \varphi_p(r, t) = \gamma_p^2 Q_p(r, t), \qquad (8.2)$$

where $\gamma_p$ is the ratio of action potential conduction velocity and the axonal range, for excitatory and inhibitory axons respectively. The wave equation is much more numerically efficient, but for the present work, the integral form has been retained.

**Synapto-dendritic Transformations of Synaptic Flux**

Afferent synaptic activity ultimately gives rise to a change in membrane polarization at the trigger point for the generation of action potentials. The change in membrane polarization depends upon the types of postsynaptic receptor, adaptive changes in receptor configurations, membrane reversal potentials, position of the synapses on the dendritic tree, and the state of the postsynaptic neuron — notably, whether or not it has recently discharged an action potential.

All these processes can be reduced in first approximation to steady-state equations and linear impulse responses. To compress the equations and emphasize analogies, the following conventions apply: $p = e, i$ indicates presynaptic neurons and $q = e, i$ indicates postsynaptic neurons. The superscripts

$[R] = [\text{AMPA}], [\text{NMDA}]$ when $p = e$, and $[R] = [\text{GABA}]$ when $p = i$, indicate receptor type, described further below.

Within the synapse the afferent synaptic flux is modified by changes in the conformation of ion channels [30–33]. A normalized impulse response function, $J^{[R]}(\tau)$, describes the rise and fall of receptor adaptation to a brief afferent stimulus:

$$J^{[R]}(\tau) = \left[ \sum_n \frac{B_n^{[R]}}{\beta_n^{[R]}} - \sum_m \frac{A_m^{[R]}}{\alpha_m^{[R]}} \right]^{-1} \times \left[ \sum_n B_n^{[R]} \exp(-\beta_n^{[R]}\tau) \right. \tag{8.3}$$
$$\left. - \sum_m A_m^{[R]} \exp(-\alpha_m^{[R]}\tau) \right],$$

where $\int_0^\infty J^{[R]}(\tau)\mathrm{d}\tau = 1$, $\{A_n^{[R]}, B_n^{[R]}, \alpha_n^{[R]}, \beta_n^{[R]}\}$ are constants, and $J^{[R]} = 0$ if $\tau = 0$.

The postsynaptic depolarization, $\psi_{qp}^{[R]}(\boldsymbol{r}, t)$, is the time-varying change of membrane voltage produced via synaptic receptors of a specific type, consequent on the synaptic flux density, and is defined without initial regard to the position of specific synapses on the dendritic tree. $\Psi_{qp}^{[R]}(\boldsymbol{r})$ is the steady-state value of $\psi_{qp}^{[R]}(\boldsymbol{r}, t)$:

$$\Psi_{qp}^{[R]} = g_p^{[0]} \exp[-\lambda^{[R]}\varphi_p] \left( \frac{V_p^{rev} - V_q}{V_p^{rev} - V_q^{[0]}} \varphi_p \right), \tag{8.4}$$

where

$g_p^{[0]}$    is the synaptic gain at resting membrane potential,
$\lambda^{[R]}$    is a measure of steady-state synaptic adaptation to $\varphi_p$,
$V_p^{rev}$    is the excitatory or inhibitory reversal potential,
$V_q^{[0]}$    is the resting membrane potential, and
$V_q$    is the average membrane potential.

Another normalized impulse response function, $H(\tau)$, describes the rise and fall of postsynaptic membrane potential

$$H(\tau) = \frac{ab}{b - a}(\mathrm{e}^{-a\tau} - \mathrm{e}^{-b\tau}), \tag{8.5}$$

where $\int_0^\infty H(\tau)\mathrm{d}\tau = 1$, $a, b$, are constants, and $H = 0$ if $\tau = 0$.

Consequently, from (8.1–8.5), and where "$*$" indicates convolution in time,

$$\psi_{qp}^{[R]} = H_p * (J^{[R]} * \Psi_{qp}^{[R]}). \tag{8.6}$$

## Transmission of Postsynaptic Depolarization
## to Initiate Action Potentials

At the release of an action potential at the soma, a retrograde propagation takes place, depolarizing the dendritic membrane throughout the proximal dendritic tree [34]. This must have major implications for the weight of individual synapses in determining any subsequent action potential generation, depending on the recent history of activity in the neuron. Those synapses within the zone of back-propagation can be called "near" synapses, and those more distal in the dendritic trees "far" synapses. It is assumed that when the neuron is fully re-polarized, the greatest weight in the generation of a subsequent action potential can be ascribed to activity at the near synapses, because of their weighting by proximity to the axon hillock. On the release of an action potential, the near synapses have their efficacy reduced to zero during the absolute refractory period, and the distal synaptic trees become partially depolarized, so that whether or not a subsequent action potential is generated at the end of the refractory period is relatively weighted toward activity at the far synapses, conducted via cable properties with delay to the trigger point. Thus, in the continuum formulation, the impact of transmission of total postsynaptic flux to the trigger point depends upon cable delays in near and far dendritic trees, the fraction of neurons which have recently fired, and the relative distribution of synapses and receptor types in the near and far dendritic trees.

Following the normalized format of (8.4) and (8.5), the cable delay, $L^j$, is given by

$$L^j = a^j \exp(-a^j \tau),  \tag{8.7}$$

where $a^j$ are constants, and $j = n, f$ indicate synapses positioned in the near and far dendritic trees, respectively.

Consequently, the fractions $A^f(t), A^n(t)$, of neurons responding primarily to near or far synapses, are

$$A^f(t) = \frac{Q_q}{Q_q^{max}}  \tag{8.8}$$

$$A^n(t) = 1 - \frac{Q_q}{Q_q^{max}},  \tag{8.9}$$

where $Q_q^{max}$ is the maximum firing rate of neurons and reflects the refractory period, while $Q_q$ is the pulse density at $r$.

Since the distribution of postsynaptic receptors differs in near and far trees, fractional distributions $r^{j[R]}$ can be defined with

$$r^{n[R]} + r^{f[R]} = 1.  \tag{8.10}$$

$r^{j[R]}$ can be used to fractionally weight the synaptic numbers, $N_{qp}$, the number of excitatory or inhibitory synapses per neuron.

Equation (8.6) can then be aggregated over types of afferent neuron and number of synapses and types of receptor in the near and far dendritic trees, as

$$V_q(t) = V_q^{[0]} + \sum_p \sum_j \sum_{[R]} r^{j[R]} N_{qp} A^j (L^j * \psi_{qp}^{[R]}),$$  (8.11)

where $V_q(t)$ is the potential of the dendritic membrane at the trigger points, and $V_q^{[0]}$ is the resting membrane potential. The value of $V_e(t)$ — the potential in the excitatory (pyramidal) neurons — scales as the local field potential (LFP; cf. Chap. 1), and $V_q(t)$ can be applied as a surrogate for $V_q$ in (8.4).

### Generated Pulse Density

Generation of action potentials then follows the sigmoidal relation

$$Q_q(t) = \frac{Q_q^{\max}}{1 + \exp[-\pi(V_q - \theta_q)/(\sqrt{3}\sigma_q)]}.$$  (8.12)

$\theta_q$ is the mean value of $V_q$ at which 50% of neurons are above threshold for the emission of action potentials. $\sigma_q$ approximates one standard deviation of probability of emission of an action potential in a single cell, as a function of $V_q$.

### 8.2.2 Parameter Values

Parameter values for the state equations have been obtained from anatomical and physiological measurements, or inferred from direct measurements. They are presented in Tabs. 8.1–8.5. In most instances values are only known approximately, and confidence intervals are unknown or uncertain. Parameters expressed as constants or as linear processes must, in reality, be time-varying and nonlinear to some degree. However, sensitivity studies to be reported elsewhere indicate that despite reservations on the accuracy of individual parameters, the system properties to be reported are relatively robust to parameter

**Table 8.1.** Synaptic numbers and gain factors [3, 5, 21, 35, 36]

| | | | |
|---|---|---|---|
| $N_{ee,cc}$ | Excitatory to excitatory corticocortical synapses/cell | 3710 | dimensionless |
| $N_{ie,cc}$ | Excitatory to inhibitory corticocortical synapses/cell | 3710 | dimensionless |
| $N_{ee,ic}$ | Excitatory to excitatory intracortical synapses/cell | 410 | dimensionless |
| $N_{ei,ic}$ | Inhibitory to excitatory intracortical synapses/cell | 800 | dimensionless |
| $N_{ie,ic}$ | Excitatory to inhibitory intracortical synapses/cell | 410 | dimensionless |
| $N_{ii,ic}$ | Inhibitory to inhibitory intracortical synapses/cell | 800 | dimensionless |
| $N_{ee,ns}$ | Synapses per excitatory cell from subcortical sources | 100 | dimensionless |
| $N_{ie,ns}$ | Synapses per inhibitory cell from subcortical sources | 0 | dimensionless |
| $g_e[0]$ | Excitatory gain per synapse at rest potential | $2.4 \times 10^{-6}$ | Vs |
| $g_i[0]$ | Inhibitory gain per synapse at rest potential | $-5.9 \times 10^{-6}$ | Vs |

**Table 8.2.** Threshold values [21, 37]

| | | |
|---|---|---|
| $Q_e^{max}$ | Maximum firing rate of excitatory cells | $100\,s^{-1}$ |
| $Q_i^{max}$ | Maximum firing rate of inhibitory cells | $200\,s^{-1}$ |
| $V_e^{rev}$ | Excitatory reversal potential | $0\,V$ |
| $V_i^{rev}$ | Inhibitory reversal potential | $-0.070\,V$ |
| $V_{q,p}^{[0]}$ | Resting membrane potential | $-0.064\,V$ |
| $\theta_q$ | Mean dendritic potential when 50% of neurones firing | $-0.035\,V$ |
| $\sigma_q$ | Standard deviation of neuron firing probability, versus mean dendritic potential | $0.0145\,V$ |

variation. Since the state equations are given in terms of scalar gains and normalized impulse responses embedded within convolutions, errors in individual parameters can partially cancel. With the following tables, sources for the values are given as numbered references with each table title, and further qualifications given in the associated text.

The parameters $\mathfrak{a}^j$ have not been specifically sourced, but are approximate physiologically realistic delays.

These parameters were obtained by deriving steady-state and impulse response functions from mass-action models of receptor/transmitter interactions.

Distribution of three receptor types were considered as representative of a much wider group of receptors. These were the principal fast excitatory glutamate receptor (AMPA), the principal fast inhibitory GABA receptor (GABA$_A$) and the principal slow and voltage-dependent glutamate receptor (NMDA). NMDA is distributed predominantly in the distal dendritic tree [38] and the others more uniformly — and their distribution may be subject to dynamic functional variation. Thus, the values applied are rather arbitrary. Sensitivity analyses indicate that robust results may be obtained despite considerable variation of the values applied, and adjustment of these parameters depended upon obtaining match to the average firing rates of cells observed in cortex [39].

**Table 8.3.** Membrane time constants [21]

| | |
|---|---|
| $a_{ee}$ EPSP decay time-constant in excitatory cells | $68\,s^{-1}$ |
| $b_{qp}$ EPSP and IPSP rise time-constants | $500\,s^{-1}$ |
| $a_{ei}$ IPSP decay time-constant in excitatory cells | $47\,s^{-1}$ |
| $a_{ie}$ EPSP decay time-constant in inhibitory cells | $176\,s^{-1}$ |
| $a_{ii}$ IPSP decay time-constant in inhibitory cells | $82\,s^{-1}$ |
| $\mathfrak{a}^j$ Delay attributable to position of near and far synapses on dendritic tree | $\mathfrak{a}^n = 1000\,s^{-1}$ |
| | $\mathfrak{a}^f = 200\,s^{-1}$ |

**Table 8.4.** Receptor adaptation gains and time constants [30–33]

| | | |
|---|---|---|
| $\lambda^{[R]}$ | Receptor adaptation pulse-efficacy decay constants | $[\text{AMPA}] = 0.012\ \text{s}$ |
| | | $[\text{NMDA}] = 0.037\ \text{s}$ |
| | | $[\text{GABA}_A] = 0.005\ \text{s}$ |
| $B_n^{[R]}$ | Receptor onset coefficients | $[\text{AMPA}]_1 = 1.0$ |
| | | $[\text{NMDA}]_1 = 1.0$ |
| | | $[\text{GABA}_A]_1 = 1.0$ |
| | | dimensionless |
| $A_n^{[R]}$ | Receptor offset coefficients | $[\text{AMPA}]_1 = 0.0004$ |
| | | $[\text{AMPA}]_2 = 0.6339$ |
| | | $[\text{AMPA}]_3 = 0.3657$ |
| | | $[\text{NMDA}]_1 = 0.298$ |
| | | $[\text{NMDA}]_2 = 0.702$ |
| | | $[\text{GABA}]_1 = 0.0060$ |
| | | $[\text{GABA}]_2 = 0.9936$ |
| | | dimensionless |
| $\beta_n^{[R]}$ | Receptor onset time-constants | $[\text{AMPA}]_1 = 760.0\,\text{s}^{-1}$ |
| | | $[\text{NMDA}]_1 = 50.5\,\text{s}^{-1}$ |
| | | $[\text{GABA}]_1 = 178.0\,\text{s}^{-1}$ |
| $\alpha_n^{[R]}$ | Receptor offset time-constants | $[\text{AMPA}]_1 = 21.8\,\text{s}^{-1}$ |
| | | $[\text{AMPA}]_2 = 60.3\,\text{s}^{-1}$ |
| | | $[\text{AMPA}]_3 = 684.0\,\text{s}^{-1}$ |
| | | $[\text{NMDA}]_1 = 0.608\,\text{s}^{-1}$ |
| | | $[\text{NMDA}]_2 = 3.3\,\text{s}^{-1}$ |
| | | $[\text{GABA}]_1 = 11.2\,\text{s}^{-1}$ |
| | | $[\text{GABA}]_2 = 127\,\text{s}^{-1}$ |

## 8.2.3 Application to the Macrocolumn

The continuum model was applied numerically in discrete form, using a $20 \times 20$ matrix of "elements", each of which can be considered as situated at the position $r$, surrounded by other elements at positions $\{r'\}$, and each coupled to the others so as to create an approximation of the "Mexican Hat"

**Table 8.5.** Receptor distribution

| | | |
|---|---|---|
| $r^{n[R]}$ | Relative weighting of receptors on near dendritic field | $[\text{AMPA}] = 1 - r^{f[R]}$ |
| | | $[\text{NMDA}] = 1 - r^{f[R]}$ |
| | | $[\text{GABA}] = 1 - r^{f[R]}$ |
| $r^{f[R]}$ | Relative weighting of receptors on far dendritic field | $[\text{AMPA}] = 0.5$ |
| | | $[\text{NMDA}] = 1.0$ |
| | | $[\text{GABA}] = 0.375$ |

configuration [40] of excitatory and inhibitory intracortical connections within a macrocolumn, in accord with a two-dimensional Gaussian version of (8.1). Where $\gamma_p$ represents the standard deviation of axonal range. The connection densities as a function of distance are thus

$$\varphi_p(\boldsymbol{r}, t) = \int \frac{1}{2\pi\gamma_p^2} \exp\left[-\frac{|\boldsymbol{r} - \boldsymbol{r}'|^2}{2\gamma_p^2}\right] Q_p(t - \delta_p)\mathrm{d}^3\boldsymbol{r}', \qquad (8.13)$$

The value of $\gamma_p$ was 4.9 simulation elements for the excitatory couplings, and 4.5 simulation elements for the inhibitory couplings. A wide range of plausible axonal conduction velocities were applied, and results found insensitive to variation for all small conduction lags, consistent with the size of

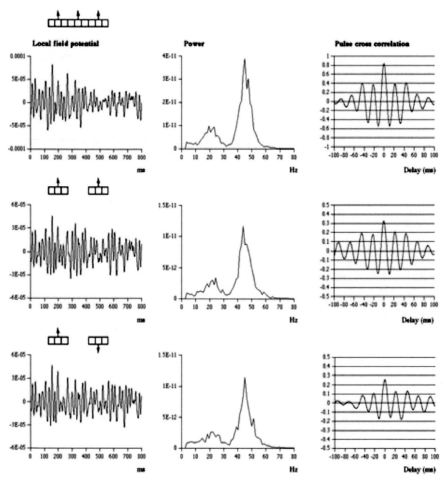

**Fig. 8.2.** Simulation of synchronous oscillation induced by moving bars (size and movement shown as arrowed icons) in the visual field. Plotted are local field potential time-series, power spectral content, and cross-correlations of two sites in the cortical field, when driven by simulated moving bars — with each "bar" a field of zero-mean white noise, uncorrelated in separate bars. (Wright et al., 2000)

the macrocolumn. In the results shown below, axonal delay was 0.4 ms per element. Simulation time-step was 0.1 ms.[3]

The tabulated parameters were applied distinguishing the synapses for nonspecific cortical activation from the reticular formation ($N_{ee,ns}$) and those reaching the macrocolumn from the surrounding cortex ($N_{ee,cc}$ and $N_{ie,cc}$) as the principal sites of external input to the macrocolumn.

The non-specific afferent flux was considered excitatory and terminating on excitatory cortical neurons only (consistent with its predominant input to the upper layers of the cortex, where the pyramidal cell dendritic trees predominate) [3]. The afferent flux from trans-cortical sources terminated on both excitatory and inhibitory neurons.

### 8.2.4 Comparison of Simulation to Experimental Data

Figure 8.2 shows that when the simulation is configured to imitate results representative of synchronous oscillation (differential response to short and long moving bars in the visual field), LFP time-signatures, LFP spectra, and pulse cross-correlations are like those seen in real data [11–15]. (These results were obtained with an earlier simulation having properties identical in the respects shown, to the present simulation.)

Figure 8.3 shows that the balance of the non-specific afferent synaptic flux and the trans-cortical synaptic flux entering the macrocolumn can act

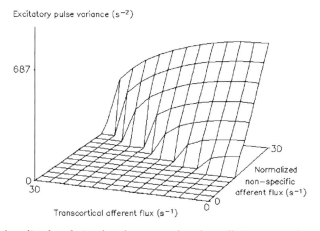

**Fig. 8.3.** Amplitude of simulated gamma band oscillation as a function of the excitatory synaptic flux delivered to pyramidal neurons only (the nonspecific afferent flux) versus the excitatory synaptic flux delivered to both pyramidal and inhibitory neurons (the transcortical afferent flux). (Units of nonspecific afferent flux have been "normalized" to avoid specification of synaptic efficacies of connections from subcortical sources)

---

[3] In computation, due to the serial nature of the algorithm, delay by a time step must be assumed between (8.4) and (8.11), and also between (8.12) and (8.11).

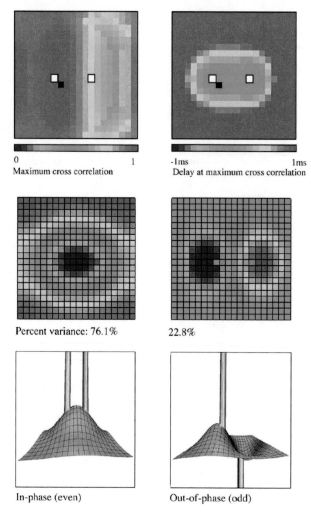

**Fig. 8.4.** Essential properties of synchronous oscillation. **Top figures** A representation of the simulated cortical surface. Open squares represent the sites of input of uncorrelated white noise. The *filled square* is the reference point from which cross-correlations are calculated with respect to the rest of the field. **Top left** Maximum positive cross-correlation (over all lags). **Top right** Delay associated with maximum positive cross-correlation. **Middle figures** The first and second principle eigenmodes of spatial activity on the same simulated surface. **Bottom figures** Schematic "freeze frame" images of local field potentials (or pulse densities) on the simulated cortical surface when the twin inputs are in-phase or anti-phase signals. (Wright et al. 2003)

as a control parameter, respectively initiating and suppressing the onset of oscillation in the macrocolumn according to the balance of excitatory and inhibitory tone and providing an explanation of gamma bursts, with a pulse variance consistent with gamma oscillation [39].

Figure 8.4 shows the basis of the synchronous fields generated within simulations of this type [18–20, 22]. In brief, synchronous oscillation arises from a distinctive property of the cortical wave medium. "Odd" components in any pair of the Fourier components in signals input to dendrites are selectively dissipated — since dendrites are summing junctions. This selective elimination leaves the synchronous "even" components of activity at any two sites predominant.

The ubiquity of synchronous oscillation in the cortical field, and the mechanism of synchrony — which is not confined to the gamma band, but applies to all frequencies [41] — has implications for the self-organization of synapses in the developing brain, as we argue in the next section.

## 8.3 Synaptic Dynamics

The second part of the model summarized here is concerned with self-organization of synapses during antenatal and/or early postnatal visual development [23]. The emphasis here is upon the most stable configuration of synapses that can emerge under a Hebbian learning rule that incorporates "decay" (forgetting) under metabolic constraints, given an initial set of connections of random strength, density and with distribution consistent with anatomical findings, as well as the dynamical properties described above.

### 8.3.1 Initial Connections and the Transmission of Information in Early V1

Decline of synaptic density with distance occurs in the local intracortical connections at the scale of macrocolumns and in the longer intracortical connections spanning a fraction of the extent of the visual cortex (V1) [3–5]. Via polysynaptic transmission, information can potentially reach each macrocolumnar-sized area from the whole, or a substantial part, of V1. Thus, the distribution of terminal axonal ramifications in intracortical axons defines the scale of a *local map* of approximately macrocolumnar size, and associate the scale of V1 with a *global map* — the map of the visual field.

### 8.3.2 Visual Spatial Covariance, and Synchronous Fields

Because of the decline of synaptic density with distance, fields of synchronous oscillation decline in magnitude with distance [18–20, 22] as can be seen in

Fig. 8.4. Visual stimuli themselves exhibit a decline in cross-covariance with distance. Thus, cross-covariance of activity in V1 declines with distance at both the global, V1, scale, and the local, macro-columnar, scale.

### 8.3.3 Learning Rule

As a generic simplification of synaptic plasticity at multiple time-scales, a simple Hebbian rule with decay can be applied. These learning-related synaptic modifications fall outside the mechanisms included in the preceding account, and discussion of their physiological analogs is deferred to the conclusion.

At each synapse, the coincidence of pre- and postsynaptic activity, $r_{Q\varphi}$, is given by a relation of the form

$$r_{Q\varphi} \propto \sum_t Q_e(t) \times \varphi_e(t), \qquad (8.14)$$

where $Q_e(t) \in \{0, 1\}$ is the postsynaptic firing state, and $\varphi_e(t) \in \{0, 1\}$ is the presynaptic firing state. A multiplication factor, $H_s$, operating on the gain of synapses at steady states of pre- and postsynaptic firing is approximately

$$H_s = H_{\max} \exp(-\lambda/r_{Q\varphi}), \qquad (8.15)$$

where $\lambda$ is a suitable constant. With changes in either the pre- or postsynaptic firing state, $H_s$ can increase or fade over time, at rates differing for fast and slow forms of memory storage.

### 8.3.4 Individual Synaptic States of Stability

It can be shown [23] that under these learning rules, synapses can approach a stable, unchanging state only by approaching either one of two extremes — either *saturated* or *sensitive*. In the saturated state, $H_s$ and $r_{Q\varphi}$ are at maxima, while in the sensitive state, $H_s$ and $r_{Q\varphi}$ are at minima. Conversely, $\frac{dH_s}{dr_{Q\varphi}}$, the sensitivity to change in synaptic gain, is at a minimum for saturated synapses and a maximum for sensitive synapses — hence the choice of the names.

### 8.3.5 Metabolic Uniformity

Competition for metabolic resources within axons adds a constraint to Hebbian rules [42]. The metabolic energy supply of all small axonal segments can be presumed to remain approximately uniform, while the metabolic demand of saturated synapses, which have high activity, will be much greater than for sensitive synapses. Therefore, the proportion of saturated and sensitive synapses must be uniform along axons, and consequently, the densities of both saturated and sensitive synapses must decline with the distance of the presynapses from the cell bodies of origin.

### 8.3.6 The Impact of Distance/Density and Saturation/Sensitivity on Overall Synaptic Stability

All positions in V1, $\{P_{j,k}\}$, can be given an ordered numbering in the complex plane, $1\ldots,j,\ldots,k,\ldots,2n$, and all positions within a macrocolumn located at $P_0$, $\{p_{j,k}\}$, can be similarly numbered. The total perturbation of synaptic gains for the synapses from V1 entering the macrocolumn, $\Psi(pP)$, and the total perturbation of synaptic gains within the macrocolumn, $\Psi(pp)$, can thus be written as

$$\Psi(pP) = \sum_{j=1}^{j=n}\sum_{k=1}^{k=n}\sigma_{SAT}(p_jP_k)S_{SAT}(p_jP_k) + \tag{8.16}$$

$$\sum_{j=1}^{j=n}\sum_{k=1}^{k=n}\sigma_{SENS}(p_jP_k)S_{SENS}(p_jP_k)$$

$$\Psi(pp) = \sum_{j=1}^{j=n}\sum_{k=1}^{k=n}\sigma_{SAT}(p_jp_k)S_{SAT}(p_jp_k) + \tag{8.17}$$

$$\sum_{j=1}^{j=n}\sum_{k=1}^{k=n}\sigma_{SENS}(p_jp_k)S_{SENS}(p_jp_k),$$

where $\sigma_{SAT}(p_jP_k, p_jp_k)$ and $\sigma_{SENS}(p_jP_k, p_jp_k)$ are the densities of saturated and sensitive synapses respectively, and $S_{SAT}(p_jP_k,\ p_jp_k)$ and $S_{SENS}(p_jP_k, p_jp_k)$ are the corresponding variations of synaptic gains over a convenient short epoch.

Since the densities of synapses decline with increasing cell separation, then as a simple arithmetic property of sums of products, minimization of $\Psi(pp)$ requires neurons separated by short distances to most closely approach maximum saturation, or maximum sensitivity. Yet, metabolic uniformity requires that both sensitive and saturated synaptic densities must decline with distance from the cell bodies of origin, and remain in equal ratio. An apparent paradox arises, since sensitive synapses must link pre- and postsynaptic neurons with minimal pre- and postsynaptic pulse coincidence, yet the reverse is true for saturated synapses. Also apparently paradoxically, minimization of $\Psi(pP)$ requires that saturated connections afferent to any $p_j$ arise from highly covariant, and therefore closely situated, sites in V1, while sensitive connections afferent to $p_j$ must arise from well-separated sites. Yet, metabolic uniformity requires that both sensitive and saturated presynapses arise from cells at the same site. The paradoxes exist only in the Euclidean plane, and can be resolved as in the next subsection.

### 8.3.7 Möbius Projection, and the Local Map

By re-numbering $\{P_{j,k}\}$ as $\{P_{j1,j2,k1,k2}\}$, and $\{p_{j,k}\}$ as $\{p_{j1,j2,k1,k2}\}$, the subscript numbers $1,\ldots,j1,\ldots,j2,\ldots,n,(n+1),\ldots,k1,\ldots,k2,\ldots,2n$ can be

assigned in the global map so that $j1$ and $j2$ are located diametrically opposite and equidistant from $P_0$, while in the local map $j1$ and $j2$ have positions
analogous to superimposed points located on opposite surfaces of a Möbius
strip. This generates a *Möbius projection* (the *input map*) from global to local,
and a *Möbius ordering* within the local map. That is,

$$\frac{P_{jm}^2}{|P_{jm}|} \to p_{km}, \qquad m \in \{1,2\} \tag{8.18}$$

and

$$p_{jm} \to p_{km} \qquad m \in \{1,2\}. \tag{8.19}$$

In (8.18), the mapping of widely separated points in the global map con-
verge to coincident points on opposite surfaces of the local map's Möbius
representation. In (8.19), the density of saturated synaptic connections now
decreases as $|j1 - k1|$ and $|j2 - k2|$, while the density of sensitive couplings
decreases as $|j2 - k1|$ and $|j1 - k2|$.

The anatomical parallel requires $j1$ and $j2$ in the local map to represent
two distinct groups of neurons. To attain maximum synaptic stability within
the local map, an intertwined mesh of saturated couplings forms, closed after
passing twice around the local map's center, with sensitive synapses locally
linking the two turns of the mesh together. In this fashion, both saturated and
sensitive synapses decline in density with distance as required. The input map
is of corresponding form, conveying an image of the activity in V1 analogous
to projection onto a Möbius strip.

Evolution of these patterns of synaptic connections is shown in Figs. 8.5
and 8.6.

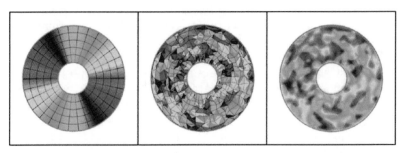

**Fig. 8.5.** Initial conditions for local evolution of synaptic strength. **Left**. The global
field (V1) in polar co-ordinates. Central defect indicates the position of a local area
of macro-columnar size. Polar angle is shown by the color spectrum, twice repeated.
**Middle**. Zones of random termination (shown by color) of lateral axonal projections
from global V1 in the local area. Central defect is an arbitrary zero reference. **Right**.
Transient patterns of synchronous oscillation generated in the local area, mediated
by local axonal connections

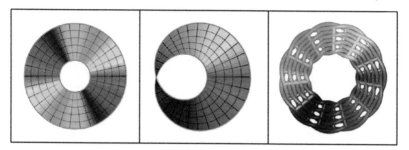

**Fig. 8.6.** Evolution of synaptic strengths to their maximally stable configuration. **Left.** The global field (V1), as represented in Fig. 8.1. **Middle.** Saturated synaptic connections input from the global field now form a Möbius projection of the global field, afferent to the local neuronal field, and forming a local map. **Right.** Saturated local synapses, within the local map, form a mesh of connections closed over $0 - 4\pi$ radians. The central defect now corresponds to the position within the local map, of the local map within the global map. Sensitive synapses (not shown) link adjacent neurons as bridges between the $0 - 2\pi$ and $2\pi - 4\pi$ limbs of the mesh of saturated connections. (Wright et al. 2006)

### 8.3.8 Monosynaptic Interactions between Adjacent Local Maps

The input and local maps can, in principle, emerge with any orientation, and with either left or right-handed chirality. However, chirality and orientation of adjacent local maps is also constrained by a requirement for overall stability. Adjacent local maps should form an approximately mirror image relation, as shown in Fig. 8.7, because in that configuration, homologous points within the local maps have the densest saturated and sensitive synaptic connections, thus meeting minimization requirements analogous to those of (8.16) and (8.17).

### 8.3.9 Projection of Object Motion to the Local Map and Dendritic Integration

Since the emergent input and local maps form a 1:1 representation of points in the global map, they enable the relay of information delivered to V1 by the visual pathway to every local map. This pattern is relayed to the local map according to

$$O\left(\frac{P_{jm}^2}{|P_{jm}|}, t - \frac{P_{jm}}{\nu}\right) \Rightarrow O(p_{jm}, t), \tag{8.20}$$

where $O(P, t)$ is the pattern of neuronal firing generated in V1 by a visual object and $\nu$ is the axonal conduction velocity.

When signals from global V1 are received in the local map, they are subject to integration over time in local dendrites. If we represent local dendritic potentials as $V(p, t)$, average synaptic gain (incorporating the Hebbian gain factor) as $g$, dendritic rise and fall time-constants as $a, b$, then

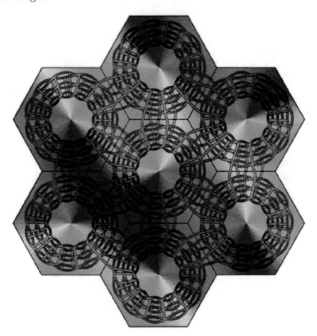

**Fig. 8.7.** Mutual organization of saturated coupling within and between local maps. Orientation and chirality of OP in macrocolumns. (Wright et al. 2006)

$$V(p_{jm}, t) = [g(e^{-a\tau} - e^{-b\tau})] * \left[ O(P_{jm}, t - \frac{P_{jm}}{\nu} - \tau) \right], \qquad t, \tau \geq 0 \quad (8.21)$$

expresses the way moving objects in the visual field exert threshold or sub-threshold effects on action potential generation within the local map. The impact of delayed conduction from the global to the local maps can account for the recently discovered [8] dependence of OP on stimulus velocity, angle of stimulus orientation to direction of motion, and extension of the stimulus [23].

### 8.3.10 Effects of Perturbation

Because all activity in global V1 is projected to each local map, visual stimuli must act to perturb synaptic gains away from the stable configuration. Further, cells at any two corresponding positions on the mesh of saturated connections positioned on the opposite $0 - 2\pi$ and $2\pi - 4\pi$ limbs of the mesh and connected with sensitive synapses of high density are maximally sensitive to perturbation when concurrently stimulated by some visual object. Figure 8.8 shows the effect of a stimulus such as a moving line in the visual field, which will give rise to a strong perturbation, followed by a relaxation back toward the stable configuration as the stimulus is withdrawn. Perturbation interactions within and between local maps on short time scales may account for phenomena of perceptual closure.

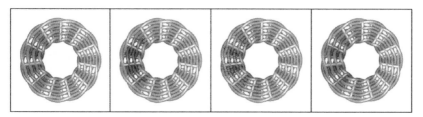

**Fig. 8.8.** Perturbation of synaptic saturation and sensitivity by extended stimuli. **Left.** A representation of the connections formed by a small group of neurons with cell bodies located at 3 o'clock in the local map. Saturated connections (*red*) and sensitive connections (*green*) are arrayed at their maximally stable configuration. **Second from left.** An afferent volley is delivered to neurons at 3 o'clock in the local map, arising from sites on both sides of the position of the local map, so that neurons in the $0 - 2\pi$ and $2\pi - 4\pi$ limbs of the mesh are forced into highly correlated firing. **Second from right.** On withdrawal of the perturbing afferent volley, the synaptic configuration generated by the perturbation begins to decay. **Right.** The maximally stable configuration is again attained

Decay to the maximally stable configuration may occur on multiple time-scales, and with continuing perturbation may be retarded indefinitely if growth mechanisms overcome the prior requirements of metabolic uniformity.

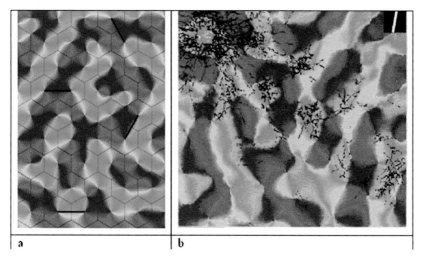

**Fig. 8.9.** Simulated and real maps of orientation preference: (**a**) Final configuration of OP consequent to seeding the development of fields of OP with the local map mirror-image pairs shown joined by solid lines. (Wright et al. 2006); (**b**) Real OP as visualized in the tree shrew by Bosking et al. (1997). Intracortical connections superimposed in black connect zones of like OP

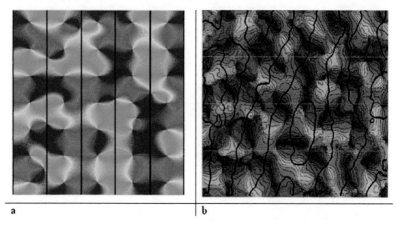

a                                              b

**Fig. 8.10.** (a) Simulated OD columns (Wright et al. 2006); (b) Real OD columns, as visualized by Obermayer & Blasdel (1993)

### 8.3.11 Comparison to Standard Anatomical Findings

Figures 8.9 and 8.10 show the results of simulations [23] based on the topological principles described, matched to experimental data [43, 44]. Other experimental data accounted for include direction preference fractures and the occurrence of OD columns. The OD columns arise as an "exception to the rule". Representation of visual input from each eye separately is required, since images seen with binocular disparity by the two eyes are spatially lag-correlated. This violates the requirement that information be mapped from global to local map with preservation of cross-covariance with distance. Separate representation of the images of each eye is a required compromise to achieve overall maximum stability.

## 8.4 Conclusion

The two pieces of research work described in this chapter offer a contrast and a convergence. The model of axo-dendritic dynamics depends upon the quantitative choice of parameter values obtained from physiological data to the extent possible. The properties of this model are relatively robust to perturbation of the parameter values, but wide variation of parameters leads to dynamics wholly unmatched to real electrocortical activity, the parameters that result in realism needing to be in particular proportions to each other [45]. Moreover, limited aspects of real dynamics can be reproduced by partial models, which utilize apparently different mechanisms [17, 29] (cf. Chaps. 7 and 5). More complete models will require confrontation of simulations with many separate, but concurrent, classes of data.

Conversely, the model of synaptic organization is essentially independent of parameter values, and depends upon topological effects, which emerge in between-scale dynamic interactions in cortex. It appears that dynamic interactions may contribute to the organization of realistic anatomical connections, thus supplementing the actions of the many growth factors, and chemical gradients contributing at biochemical level to the formation and dissolution of connections (e.g. [46]).

The basis for learning-related modifications of synaptic gains was earlier left undiscussed. The mechanisms are likely to be multiple, and occurring on many time scales. A likely major candidate for inclusion is long term potentiation (LTP) and depression (LTD). Recent work [47,48] on learning-rule modeling and experiments in hippocampus have led to the proposal of a learning rule and a link to LTP/D consistent with the requirements for synaptic stability and sensitivity in the stable state and during perturbation (cf. Chap. 2). At a more abstract level, comparison with the information-theoretic coherent-infomax principle is apparent [38]. In a unified version, the two models may provide a framework for more detailed comparisons with experimental data, while also enabling analysis of their information storage and processing properties.

The author presumes, but has not proved, that the two models are mutually compatible and could be combined in a single numerical simulation.

# References

1. P. L. Nunez. *Neocortical Dynamics and Human EEG Rhythms.* Oxford University Press, New York, Oxford. pp. 99–114, 1995.
2. J. Szentagothai. Local neuron circuits of the neocortex. In: *The Neurosciences 4th Study Program.* (F. O. Schmitt and F. G. Worden, Eds.). MIT Press, Cambridge, Mass. pp. 399–415, 1979.
3. V. Braitenberg and A. Schüz. *Anatomy of the Cortex: Statistics and Geometry.* Springer-Verlag, New York. 1991.
4. D. A. Scholl. *The Organization of the Cerebral Cortex.* Wiley, New York. 1956.
5. D. T. J. Liley and J. J. Wright. Intracortical connectivity of pyramidal and stellate cells: estimates of synaptic densities and coupling symmetry. *Network,* 5: 175–189, 1994.
6. V. B. Mountcastle. An organizing principle for cerebral function: the unit module and the distributed system. In: *The Neurosciences 4th Study Program.* (F. O. Schmitt and F. G. Worden, Eds.). MIT Press, Cambridge, Mass., 1979.
7. E. R. Kandel, J. H. Schwartz and T. M. Jessell. *Principles of Neural Science.* 3rd Edition. Prentice-Hall International (UK), London. pp. 421–439, 1991.
8. A. Basole, L. E. White and D. Fitzpatrick. Mapping multiple features of the population response of visual cortex. *Nature,* 423: 986–990, 2003.
9. D. H. Hubel and T. N. Wiesel. Receptive fields, binocular interaction, and functional architecture of the in the cat's visual cortex. *J. Physiol.,* 160: 106–154, 1962.

10. M. Fiorani, M. G. P. Rosa, R. Gattass and C. E. Rocha-Miranda. Dynamic surrounds of receptive fields in primate striate cortex: a physiological basis for perceptual completion? *Proc. Natl. Acad. Sci. USA*, 89: 8547–8551, 1992.

11. R. Eckhorn, R. Bauer, W. Jordon, M. Brosch, W. Kruse, M. Monk, H. J. Reitböck. Coherent oscillations: a mechanism of feature linking in the in the visual cortex? *Biological Cybernetics*, 60: 121–130, 1988.

12. C. M. Gray, P. König, A. K. Engel and W. Singer. Oscillatory responses in cat visual cortex exhibit intercolumnar synchronization which reflects global stimulus properties. *Nature*, 388: 334–337, 1989.

13. C. M. Gray, A. K. Engel, P. König and W. Singer. Synchronization of oscillatory neuronal responses in cat striate cortex: temporal properties. *Visual Neuroscience*, 8: 337–347, 1992.

14. C. M. Gray and W. Singer. Stimulus-specific neuronal oscillations in orientation columns of cat visual cortex. *Proc. Natl. Acad. Sci. USA*, 86: 1698–1702, 1989.

15. W. Singer and C. M. Gray. Visual feature integration and the temporal correlation hypothesis. *Annual Rev. Neuroscience*, 18: 555–586, 1995.

16. M. P. Stryker. Is grandmother an oscillation? *Nature*, 388: 297–298, 1989.

17. J. J. Wright, C. J. Rennie, G. J. Lees, P. A. Robinson, P. D. Bourke, C. L. Chapman, E. Gordon, and D. L. Rowe. Simulated electrocortical activity at microscopic, mesoscopic, and global scales. *Neuropsychopharmacology*, 28: S80–S93, 2003.

18. J. J. Wright. EEG simulation: variation of spectral envelope, pulse synchrony and 40 Hz oscillation. *Biological Cybernetics*, 76: 181–184, 1997.

19. P. A. Robinson, C. J. Rennie and J. J. Wright. Synchronous oscillations in the cerebral cortex. *Physical Review*, E 57: 4578–4588, 1998.

20. J. J. Wright, P. D. Bourke and C. L. Chapman. Synchronous oscillation in the cerebral cortex and object coherence: simulation of basic electrophysiological findings. *Biological Cybernetics*, 83: 341–353, 2000.

21. C. J. Rennie, J. J. Wright and P. A. Robinson. Mechanisms of cortical electrical activity and emergence of gamma rhythmn. *J. Theoretical Biol.*, 205(1): 17–35, 2000.

22. C. L. Chapman, P. D. Bourke and J. J. Wright. Spatial eigenmodes and synchronous oscillation: coincidence detection in simulated cerebral cortex. *J. Math. Biol.*, 45: 57–78, 2002.

23. J. J. Wright, D. M. Alexander and P. D. Bourke. Contribution of lateral interactions in V1 to organization of response properties. *Vision Research*, 46: 2703–2720, 2006.

24. W. J. Freeman. *Mass Action in the Nervous System*. Academic Press, New York. 1975.

25. H. Haken. *Principles of Brain Functioning*. Springer, Berlin. 1996.

26. P. L. Nunez. *Neocortical Dynamics and Human EEG Rhythms*. Oxford University Press, New York, Oxford. 1995.

27. V. K. Jirsa and H. Haken. Field theory of electromagnetic brain activity. *Phys. Rev. Lett.*, 77: 960–963, 1996.

28. P. A. Robinson, C. J. Rennie and J. J. Wright. Propagation and stability of waves of electrical activity in the cortex. *Phys. Rev. E*, 55: 826–840, 1997.

29. P. A. Robinson, C. J. Rennie, D. L. Rowe, S. C. O'Connor, J. J. Wright, E. Gordon et al. Neurophysical modeling of brain dynamics. *Neuropsychopharmacology*, 28: S74 – S79, 2003.

30. R. A. Lester and C. E. Jahr. NMDA channel behavior depends on agonist affinity. *J. Neuroscience*, 12: 635–643, 1992.
31. C. Dominguez-Perrot, P. Feltz and M. O. Poulter. Recombinant GABAA receptor desensitization: the role of the gamma2 subunit and its physiological significance. *J. Physiol.*, 497: 145–159, 1996.
32. W. Hausser and A. Roth. Dendritic and somatic glutamate receptor channels in rat cerebellar Purkinje cells. *J. Physiol.*, 501.1: 77–95, 1997.
33. K. M. Partin, M. W. Fleck and M. L. Mayer. AMPA receptor flip/flop mutants affecting deactivation, desensitization and modulation of cyclothiazide, aniracetam and thiocyanate. *J. Neuroscience*, 16: 6634–6647, 1996.
34. G. J. Stuart and B. Sakmann. Active propagation of somatic action potentials into neocortical cell pyramidal dendrites. *Nature*, 367: 69–72, 1994.
35. A. M. Thompson, D. C. West, J. Hahn and J. Deuchars. Single axon IPSPs elicited in pyramidal cells by three classes of interneurones in slices of rat cortex. *Journal of Physiology* (London) 496.1: 81–102, 1997.
36. A. M. Thompson. Activity-dependent properties of synaptic transmission at two classes of connections made by rat neocortical pyramidal axons *in vitro*. *Journal of Physiology* (London) 502.1: 131–147, 1997.
37. E. R. Kandel, J. H. Schwartz and T. M. Jessell. *Principles of Neural Science*. 3rd Edition. Prentice-Hall International (UK), London. pp. 81–118, 1991.
38. W. A. Phillips and W. Singer. In search of common foundations for cortical computations. *Behavioral and Brain Sciences*, 20: 657–722, 1997.
39. M. Steriade, I. Timofeev and F. Grenier. Natural waking and sleep states: a view from inside neocortical neurons. *J. Neurophysiol.*, 85: 1969–1985, 2001.
40. K. Kang, M. Shelley and H. Sompolinsky. Mexican Hats and pinwheels in visual cortex. *Proc. Natl. Acad. Sci. USA.*, 100: 2848–2853, 2003.
41. S. L. Bressler, R. Coppola and R. Nakamura. Episodic multiregional cortical coherence at multiple frequencies during visual task performance. *Nature*, 366: 153–156, 1993.
42. S. Grossberg and J. R. Williamson. A neural model of how horizontal and interlaminar connections of visual cortex develop into adult circuits that carry out perceptual grouping and learning. *Cerebral Cortex*, 11: 37–58, 2001.
43. W. H. Bosking, Y. Zhang, B. Schofield and D. Fitzpatrick. Orientation selectivity and the arrangement of horizontal connections in tree shrew striate cortex. *J. Neuroscience*, 17(6): 2112–2127, 1997.
44. K. Obermayer and G. G. Blasdel. Geometry of orientation and ocular dominance columns in monkey striate cortex. *J. Neuroscience*, 13(10): 4114–4129, 1993.
45. J. J. Wright. Simulation of EEG: dynamic changes in synaptic efficacy, cerebral rhythms and dissipative and generative activity in cortex. *Biological Cybernetics*, 81: 131–147, 1999.
46. Y. Yin, M. T. Henzl, B. Lorber, T. Nakazawa, T. T. Thomas, F. Jiang, R. Langer and L. Benowitz. Oncomodulin is a macrophage-derived signal for axon regeneration in retinal ganglion cells. *Nature Neuroscience*, 9(6): 843–852, 2006.
47. M. Tsukada and X. Pan. The spatio-temporal learning rule and its efficiency in separating spatio-temporal patterns. *Biological Cybernetics*, 92: 139–146, 2005.
48. T. Aihara, Y. Kobayashi and M. Tsukada. Spatiotemporal visualization of long-term potentiation and depression in the hippocampal CA1 area. *Hippocampus*, 15: 68–78, 2005.

# Part IV

Implementations

# 9

# Building a Large-Scale Computational Model of a Cortical Neuronal Network

Lucia Zemanová, Changsong Zhou and Jürgen Kurths

Institute of Physics, University of Potsdam, Germany
zemanova@agnld.uni-potsdam.de

**Summary.** We introduce the general framework of the large-scale neuronal model used in the 5th Helmholtz Summer School — Complex Brain Networks. The main aim is to build a universal large-scale model of a cortical neuronal network, structured as a network of networks, which is flexible enough to implement different kinds of topology and neuronal models and which exhibits behavior in various dynamical regimes. First, we describe important biological aspects of brain topology and use them in the construction of a large-scale cortical network. Second, the general dynamical model is presented together with explanations of the major dynamical properties of neurons. Finally, we discuss the implementation of the model into parallel code and its possible modifications and improvements.

## 9.1 Introduction

In the last few decades, an innumerable amount of information about the mammalian brain has been collected [1, 2]. The anatomical properties of the cortices of different animal species have been explored in detail with modern imaging techniques revealing the functions of various brain regions and giving insight into the processes of perception and cognition.

Neural modeling represents a powerful and effective tool for the investigation and understanding of the development and organization of the brain, and of the dynamical processes. The wide spectrum of neuronal models captures and describes processes ranging from the behavior of a single cell at the microscopic level to large-scale neuronal population activity. 'Bottom-up' modeling is a common strategy used to design large cortical networks [3–6]. In this approach, the basic dynamical and topological unit of the system is a single neuron. The specific pattern of interconnections between the simple units can be represented as a complicated network. Depending on the network structure, the model can stand for a local neuronal ensemble of a cortical area or for the hierarchically organized architecture of the brain. The selection of the concrete neuronal model should take the main dynamical behaviors, such as spiking or bursting, into account.

**Table 9.1.** Parameters of the network — structure and connections

| Parameter | Description |
|---|---|
| $m$ | Number of areas |
| $n$ | Number of neurons per area |
| $z$ | Number of connections per neuron within an area |
| $p_{\text{ring}}$ | Density of connections inside one area |
| $p_{\text{rew}}$ | Probability of rewiring |
| $p_{\text{inh}}$ | Ratio of inhibitory neurons |
| $p_3$ | Ratio of neurons receiving synapses from a connected area |
| $p_4$ | Ratio of neurons with synapses towards a connected area |
| $g_{1,\text{exc}}$ | Non-normalized strength of intra-areal excitatory synapses |
| $g_{1,\text{inh}}$ | Non-normalized strength of intra-areal inhibitory synapses |
| $g_{2,\text{exc}}$ | Non-normalized strength of inter-areal excitatory synapses |

The main idea of this chapter is to introduce a general framework for building a complex large-scale brain network that can be used to study the relationship between network topology and spreading of activity (see Chaps. 14 and 13) and present a large-scale cortical model using the 'bottom-up' approach. We discuss the neuronal properties of a single unit and the structure of the network connecting these neurons. Our aim is to build a general neuronal model able to capture and mimic various dynamical processes, as well as the wide spectrum of possible neuronal topologies. Furthermore, we would like to use this complex model to investigate the relationship between the structure and the function of the system.

In Sect. 9.2, we introduce the concept of the connectome. Subsequently, the model of the network topology and structural details are presented. All network parameters are summarized in Table 9.1. In Sect. 9.3, we deal with the

**Table 9.2.** Parameters of the neuronal dynamics

| Parameter | Description |
|---|---|
| $I_{\text{base}}$ | Constant base current |
| $V_{\text{exc}}$ | Reversal potential for excitatory synapses |
| $V_{\text{inh}}$ | Reversal potential for inhibitory synapses |
| $D$ | Intensity of the Gaussian white noise |
| $G_{\text{ex}}$ | Strength of Poissonian current (Pc) |
| $N_p$ | Number of Pc |
| $\lambda$ | Frequency of Pc |
| $\tau_{1,\text{exc}}, \tau_{2,\text{exc}}$ | Rise and decay times of excitatory synaptic current |
| $\tau_{1,\text{inh}}, \tau_{2,inh}$ | Rise and decay times of inhibitory synaptic current |
| $A_+, A_-$ | Magnitude of the LTP, LTD |
| $\tau_+, \tau_-$ | Rise and decay rate of the LTP, LTD |
| $t_{\text{del},1,\text{exc}}$ | Delay of intra-areal excitatory synapses |
| $t_{\text{del},1,\text{inh}}$ | Delay of intra-areal inhibitory synapses |
| $t_{\text{del},2,\text{exc}}$ | Delay of inter-areal excitatory synapses |

dynamical characteristics of neurons. The basic neuronal properties are listed and their specific role in the neuronal dynamics is explained. We again present an overview of all dynamical variables used in the model (cf. Table 9.2). Furthermore, the general framework of the large-scale neuronal model is summarized and its implementation into parallel code is described. At the end, we discuss possible improvements and extensions of the model.

## 9.2 Topology

### 9.2.1 Connectome

Mammalian brains consist of a vast number of neurons that are interconnected in complex ways [2]. In recent years, the network of anatomical links connecting neuronal elements, the connectome, has been the subject of intensive investigation. From numerous neurohistological studies, information about the morphology, location and connections of different types of neuronal cells, microcircuits and anatomical areas has been collected and sorted. These data play an important role in creating a global image of the brain. The implementation of such topological information in a large-scale neuronal model might help us to understand the mechanisms of temporal and spatial spreading of the cortical activity.

Although the details of the neuronal network architecture are not fully known, several levels of cortical connectivity can be defined [2].

**Microscopic Connectivity**

In the human brain, approximately $10^{11}$ neurons are linked together by $10^{14}$ to $10^{15}$ connections, which correspond to $10^4$ synapses per neuron. The network is rather sparsely connected, with mainly local connectivity. Neurohistological studies of animal cortical tissue have pointed out that each neuron makes contact to its closest neighbors only by one synapse or not at all [7, 8]. Generally, individual neuronal interconnections are partially predetermined by genetic constraints and later modified by adaptation rules and processes like 'spike-timing-dependent plasticity' (STDP) [9], nutrition, and learning, often happening on the daily base. For many reasons — the high number of neurons, the complex topology, frequent changes in the connectivity, the rapid decrease of living neurons in the dead tissue and invasive histological techniques (staining, neurotracers, etc.) — it is not possible to extract the complete realistic connectivity of neuronal ensembles either for animals or for humans [7]. Thus, the connections, especially at this microscopic scale, have to be modelled as a graph, whose structure ranges from simple networks such as random [5, 10–12], small-world [13, 14], or globally coupled networks [15–17] to more realistic networks reflecting spatial growth of the cortex [18] (cf. Chap. 4).

## Mesoscopic Connectivity

A cortical minicolumn, an ensemble of neurons organized in the vertical direction, is considered to be a basic functional unit processing information in the brain of mammals. Such local circuit consists of only approx. 80–100 neurons, but the exact anatomical details of its structure are still not fully described [2,8]. It is assumed that the minicolumnar architecture is more complex than just random or distance dependent connections patterns. A set of these functionally specialized and precisely rewired small neuronal populations gives rise to the cortical column. Therefore, the minicolumn is deemed to be a basic building block of the complete connectome [2,19].

## Macroscopic Connectivity

In the cerebral cortex, neurons are organized into numerous regions (areas) that differ in cytoarchitecture and function. These areas, originally defined and listed by Brodmann, may be assumed to be basic elements at the macroscale. Several studies have examined the topology of the neuronal fiber connections linking different areas in the animal brain [20–22]. For various species, like rat, cat and monkey, cortical maps were extracted that capture the presence and the strength of cortical connections between the areas. Unfortunately, the current histological techniques using mainly tracer injections have toxic effects on the neuronal tissue and thus it is not possible to perform similar studies on humans. Other imaging methods like Diffusion Tensor Imaging [23] are still under development and do not bring sufficiently satisfactory results.

The detailed knowledge of the anatomical connectivity at the systems level offers a good starting point to explore the undergoing dynamical processes.

## Databases

Even though the human connectome still remains unrevealed, a large amount of information concerning animal anatomy has been already summarized and presented in various databases on several web sites. At the mesoscopic scale the database `Microcircuit` [24] or `Wormatlas` [25] offers insights into local circuit connectivity. The database `Cocomac` [26] contains connectivity maps of macroscopic brain networks of macaque monkey and `BrainMaps` [27] maps the anatomical details of different animal species like domestic mouse, rat, cat, and several types of monkeys.

### 9.2.2 Topology of Network Model

Due to the modular and hierarchical organization of the human connectome (brain), simple models of individual levels do not offer an appropriate insight into the complex dynamics occurring in such a complex topology. Thus, our model combines the microscopic and macroscopic levels into one framework.

The higher level copies the known connectivity of real neuroanatomical data, especially the interconnectivity between 53 cat cortical areas [20, 21]. At the lower level, single cortical areas of the cat brain are modeled by large neuronal ensembles. Implementation of these two layers gives rise to a specific topology — a *network of networks*. Recent analysis has confirmed a crucial role of this type of hierarchical network structure in the uncovering of dynamical properties of the system (see Chaps. 4 and 5). In the following section, we will describe details of the topology of the model and discuss possible modifications.

### Global Cortical Network

As a representation of the large scale connectivity in our model, we chose the cat cortical map, see Fig. 9.1. The cat cortex, together with the cerebral cortex of the macaque monkey, are the most completely described brain systems among the mammals. The first collation of the cat corticocortical connections, including 65 areas and 1139 reported links, was presented by Scannell et al. [20]. The results of the study were later completed and reorganized which led to the origin of a corticocortical network of 53 cortical areas and additional thalamocortical network of 42 thalamic areas [21]. We will consider only corticocortical connections in our modeling.

The corticocortical network of the cat is composed of 53 highly reciprocally interconnected brain areas, see Fig. 9.2. The density of afferent and efferent axonal fibers is expressed in three levels — 3 for the strongest bundles of fibers, 2 for intermediate or unknown density and 1 for the weakest connections. The value 0 characterizes absent or unknown connections. These values convey more the ranks of the links than the absolute density of the fibers, in the sense that a '2' is stronger than a '1' but weaker than a '3'. All together, there are around 830 connections in the corticocortical network with an average of 15 links per area [20, 21]. (For more details, see Chaps. 3 and 4).

Generally, in the network of cat cortical connections, four distinct subsystems can be identified. The three sensory or sensorimotor subsystems — visual

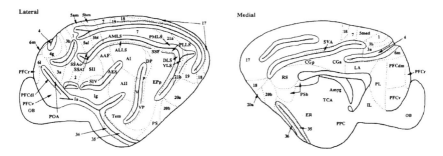

**Fig. 9.1.** Topographical map of cat cerebral cortex (from [20])

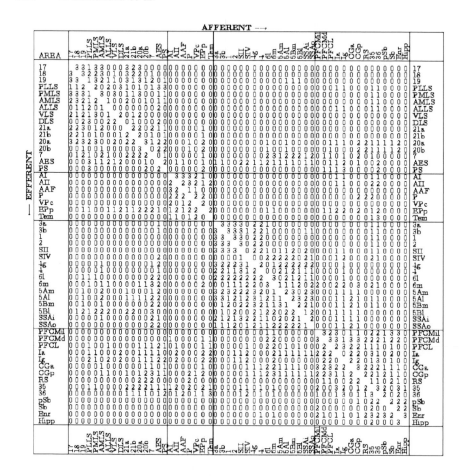

**Fig. 9.2.** Connectivity matrix representing connections between 53 cortical areas of the cat brain

(V, 16 areas), auditory (A, 7 areas), somatosensory-motor (SM, 16 areas) — involve regions participating in the processing of sensory information and execution of motoric function. The fourth subsystem — frontolimbic (FL, 14 areas) — consists of various cortical areas related to higher brain functions, like cognition and consciousness.

The subsystems are defined as sets of cortical areas with specialized function. To obtain the optimal arrangement of the areas into clusters, several methods based on network connectivity were applied [21,28,29]. For example, in the evolutionary optimization algorithm, the number of connections between units of the cluster should be maximized while inter-cluster connections are minimized. The four resulting clusters largely agree with the functional subsystems.

The corticocortical network has been a subject of much detailed analysis based on graph theory (clustering coefficient, average pathlength, matching index and many other statistical properties) [30] and theoretical neuroanatomy (e.g. segregation and integration) [31]. (See more in Chaps. 3 and 4). The knowledge of the topological properties provides a good starting point for our investigation of the relationship of the structure and dynamics. Our model, however, is flexible to allow for the inclusion of any known cortical connectivity or artificially created network of long-range cortical connections. The cat cortical map is, in evolutionary terms, not so closely related to the structure of the human cortex. To minimize this difference, one can replace the cat matrix by the cortical map extracted from macaque monkey or possibly by a map of the human connectome in the future.

**Local Neuronal Network**

As we have already mentioned, the individual areas differ in cytoarchitecture and function. Due to these natural distinctions, we model each area as a local network, i.e. a population of neurons having its own topology. Considering the fact that local connections are more frequent than long range ones (although the exact neuronal topology is unknown), we have chosen a small-world architecture as a minimal model [13]. This type of network, originally proposed by Watts and Strogatz [32], represents a transition between random and regular connectivity. At the beginning, each unit of the network connects to a number $z$ of the nearest neighbors, specified by a connection density parameter $p_{ring}$ as $z = p_{ring} \times n$. Later, links are rewired with a probability $p_{rew}$ to a randomly selected node, which introduces the long-range connections. So, the parameters $p_{ring}$ and $p_{rew}$ are crucial for the selection of specific network character (regular, small-world or random), see [32] or Chap. 3.

The small-world topology disposes of improved structural properties like short average pathlength and large clustering coefficient. From the dynamical point of view, it is known that synchronization is enhanced on such networks because of these two characteristics. Such an improvement in the ability to achieve synchronization plays an important role in neural signaling. Many studies also confirmed the presence of the small-world properties in various biological networks, including cortical networks [30, 33].

Previous Chaps. (3, 5) presented a general overview of different kind of networks, their network properties and the influence of these properties on the network dynamics.

We distinguished two types of neurons — excitatory and inhibitory. It is known that approximately 75–80% of the neurons are excitatory (pyramidal type) and the remaining 20–25% are inhibitory neurons (interneurons) [1,5]. In our simulations, we randomly select the inhibitory neurons with a probability $p_{inh} = 0.25$. Since only pyramidal neurons are involved in the long-range inter-areal connections, we consider all inter-areal links to be excitatory.

In the following part, we are mainly interested in the specification of the strength of the different types of synapses. Generally, due to the smaller amount of inhibitory neurons (and thus inhibitory synapses), these connections are usually stronger than the excitatory ones. We assume different coupling strengths for the excitatory ($g_{1,\mathrm{ex}}$) and inhibitory ($g_{1,\mathrm{in}}$) synapses within a cortical area. The modification of $g_{1,\mathrm{ex}}$ and $g_{1,\mathrm{in}}$ allows us to balance the excitatory and inhibitory inputs to the neurons within a single cortical area and achieve the 'natural' firing rate of neurons in the range of 1–3 Hz. To exclude the dependence of the neuron firing rate on the network size, we additionally normalize the coupling strength by the square root of the number of connections per area ($z$). In Chap. 14, the students present an efficient description of the search for the optimal coupling parameters.

Additionally, we also have to consider signals coming from other cortical areas (inter-areal links). If two areas are connected, only 5% of neurons within each area will receive or send signals to the other area. On average, up to 30–40% of neurons of one area can be involved in communication with other areas [34]. The coupling strength $g_{2,\mathrm{ex}}$ of the inter-areal connections is scaled

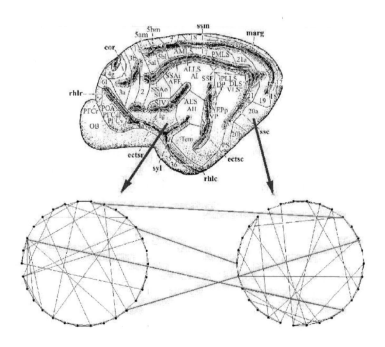

**Fig. 9.3.** The modeled system — a network of networks. Note that local subnetworks have small-world structure

here by the square of the total number of neurons from the distant area sending signals to the specific neuron.

Table 9.1 offers an overview of all network parameters presented in the model of the network topology.

Let us briefly summarize the network structure of the model: The system represents a *network of networks*, see Fig. 9.3. The macroscopic level corresponds to the known anatomical connectivity map of 53 cat cortical areas. At the microscopic level, a single cortical area is modeled by a large neuronal population of excitatory and inhibitory neurons. The topology and the size of the local network can be adjusted by changing the network parameters. Four possible patterns of the local connectivity structure are available — global (all-to-all), regular, small-world and random. Omitting the layered structure of the cortical area and corresponding topological details, we randomly choose 5% of the neurons to receive an input from, and 5% of neurons to send an output to the various connected cortical areas. The coupling strengths are tuned to reproduce the 'natural' firing rate of neurons in their resting state.

## 9.3 Dynamics

Whether the model system will plausibly reflect the biological behavior or not does not only depend on the network structure but also on the presence of other necessary properties of neurons. We are aware that this book and chapter resulting from the Summer School are not able to capture all these properties, so here, we will discuss only the most relevant ones.

Neurons are highly specialized cells of the nervous system responsible for the processing and transmission of information encoded in the form of electrical activity. To handle such a peculiar task, neurons possess complex morphology, including a wide dendritic tree with branches contacting many neighboring cells, and an axon with its special myelin sheath conveying action potential effectively and quickly. Additionally, several types of ion channels incorporated in the cell membrane moderate the ionic currents and flexibly respond to incoming signals (see Chaps. 1 and 2).

An implementation of all these dynamical and spatial properties would lead to a complex structural model of the neuron, computationally expensive and thus unsuitable for large-scale neuronal simulations. (However, this approach is used in the Blue Brain Project, which uses supercomputers to simulate the neuronal dynamics of the brain on different levels [6].) Rather than representing neurons as a spatial unit with a complex geometry, they should be modeled as dynamical systems with emphasis on the various ionic currents. These ionic currents determine the neural response to the stimuli and its excitability, which are of the main importance for the neural dynamics.

The general answer of a neuron to the stimuli is an 'all-or-nothing' activity. Neurons only fire when the total synaptic or external input reaches some threshold. If the inputs are weak, only temporal and spatial summation of such inputs will cause the neuron to fire. After the emission of a spike, the upcoming short refractory period prevents the neuron from firing for a certain time interval, even under the application of strong stimuli. Such a neuronal response can be captured by a simple threshold or excitable model.

Thus, excitable neurons can be modeled by a variety of point spike models, e.g. Integrate-and-fire model, Hindmarsh-Rose model, Izhikevich model (see Chap. 1 and several overviews [35, 36]). In several simulations during the Summer School, we chose the Morris-Lecar model, which is able to mimic different types of the behavior categorized according to the neuron excitability and is computationally efficient [37, 38].

Here we introduce the general global dynamics of the neurons described by two variables $V$ and $W$, see (9.1, 9.2).

$$\dot{V}_i = f(V_i, I^{\text{base}}) + I_i^{\text{syn}}(t) + I_i^{\text{ext}}(t) \tag{9.1}$$
$$\dot{W}_i = h(W_i) + D\xi_i(t) \tag{9.2}$$

The dynamics of the fast variable $V$ imitating the membrane potential are predetermined by a function $f$ of two arguments: the membrane potential $V$ and the basic current $I^{\text{base}}$, which flows into the neuron and sets up the neuronal excitability. Moreover, the membrane potential $V$ is modified by the total synaptic current $I^{\text{syn}}$ coming from other connected neurons and external current $I^{\text{ext}}$, representing perturbations from lower brain parts. The slow recovery variable $W$, modeled by a function $h$, accounts for the activity of various ion channels. Neurons are additionally stimulated by Gaussian white noise $\xi$ of intensity $D$, which simulates inherent neuronal stochastic disturbance.

Now we will describe each individual input and its properties.

- Noise

  In the living brain, neurons in the normal state usually do not exhibit strong activity. According to some estimates, they are silent 99% of the time, just sitting below the critical threshold and being ready to fire [39]. In our model, the neurons are also initially set in the excitable state.

  To mimic the intrinsic stochastic character of neuronal dynamics, caused by stochastic processes like synaptic transmission, Gaussian white noise is included:

$$\langle \xi_i(t)\xi_j(t-\tau) \rangle = \delta_{ij}\delta(\tau).$$

The tunable parameter $D$ in (9.2) scales the intensity of this random input. We would like to emphasize that due to the stochastic term in the system, the Euler method is more appropriate for numerical integration. Furthermore, the excitatory neurons are stimulated by multiple inputs of Poissonian noise $I^{\text{Poiss}}$ (9.3), where $N_p$ is a number of Poissonian inputs of frequency $\lambda = 3$ Hz. Poissonian input simulates external influences, e.g.

from subcortical areas [10, 11, 39, 40]. In order to obtain the natural firing rate of individual neurons (1–3 Hz), we can also vary the strength of the Poissonian current $G_{ex}$ (normalized to the square root of number of Poissonian processes) until the expected firing rate is reached.

$$I_i^{ext}(t) = \frac{G_{ex}}{\sqrt{N_p}} \sum_l^{N_p} I_i^{Poiss,l}(t) \tag{9.3}$$

- Synapses
  Neurons in brain tissue are connected together in a sparse network through two types of synapses: electrical and chemical [1].

  *Electrical coupling* (linear) appears only locally through the close contact of the membrane of the neurons. The information about the change of membrane potential of one neuron is transmitted directly as a current flowing through ion channels, called gap junctions.

  *Chemical synaptic connections* (nonlinear) represent the majority of connections between neurons in the neocortex. The principle of signal transmission is based on the release of chemical messengers from the depolarized presynaptic neuron, which consequently bind to the receptors of the postsynaptic neuron and cause the flow of ions in the cytoplasm. Depending on the type of the receptor, we can distinguish excitatory or inhibitory neurons, which occur in the ratio about 3:1 ($p_{inh} = 0.25$). Several models of chemical coupling have been proposed, varying from simple [41] to more complex ones [1, 14, 15]. In our model, we consider only the chemical type of neuronal coupling (cf. Chap. 1).

The term $I_i^{syn}$ in (9.1) represents a total synaptic current to the $i$th cell, i.e. the sum of signals (spikes) $k = 1, \ldots, m$ from all pre-synaptic neurons, $j = 1, \ldots, n$, as shown in (9.4).

The response from all synapses is modeled by (i) the nonlinear function $\alpha(t)$ describing the neuronal response and (ii) the difference between the membrane potential of the postsynaptic neuron $V_i$ and the reversal potential $V_s$ (see a similar approach in [11]). $V_s$ stands for $V_{exc}$ or $V_{inh}$ depending on whether the neuron is excitatory or inhibitory.

$$I_i^{syn}(t) = \sum_k^m g_{ij}\alpha_j(t - t_{j,spike}^k - t_{del})[V_i(t) - V_s] \tag{9.4}$$

The parameter $g_{ij}$ determines the connectivity and coupling strength between the postsynaptic $i$ and presynaptic $j$ neurons. In the case of disconnected neurons, we have $g_{ij} = 0$; $g_{ij} > 0$ indicates the presence of excitatory links and $g_{ij} < 0$ the presence of inhibitory ones.

The gain function $\alpha(t)$ (9.5) expresses the dynamics of the neural response with $\tau_{1,s}$ and $\tau_{2,s}$ as parameters of rise and decay times, where $s$ symbolizes whether the neurons are excitatory or inhibitory.

Time $t_{j,\text{spike}}^k$ is the spiking time of the $k$-th input spike of neuron $i$ from presynaptic neuron $j$. The variable $t_{\text{del}}$ represents the time delay in the signal transmission typical for the neuronal connection of $i$ and $j$. For more details, see Chaps. 1, 2 and 8.

$$\alpha_j(t) = \frac{1}{\tau_{1,s} - \tau_{2,s}} [e^{(-t/\tau_{1,s})} - e^{(-t/\tau_{2,s})}] \tag{9.5}$$

- Plasticity
  Neurons and neuronal connections in the brain evolve throughout life. These changes are characterized by a decrease of the number of the neurons and an increase of the density of the neural connections. According to Hebb's postulate, the most often used connections are strengthened, while the weakest ones atrophy.

  Recently, several researchers described a mechanism of spike-timing-dependent plasticity (STDP) [9,42] (cf. Chap. 2). We used it to modify the weight of the coupling between pairs of neurons. The amount of synaptic modification depends on the exact time difference $\Delta t$ between postsynaptic $t_i$ and presynaptic $t_j$ spike arrival (see (9.6, 9.7)). If the presynaptic neuron $j$ fires first ($\Delta t > 0$), long term potentiation (LTP) is induced and the synapse is strengthened. In the opposite case, the synapse is weakened (long term depression, LTD).

$$\Delta t = t_i - t_j \tag{9.6}$$

$$g_{ij}(\Delta t) = \begin{cases} A_+ \exp(-\Delta t/\tau_+) & \text{if } \Delta t > 0 \text{ ;} \\ -A_- \exp(-\Delta t/\tau_-) & \text{if } \Delta t < 0 \text{ .} \end{cases} \tag{9.7}$$

The parameters $A_\pm$ correspond to the maximum number of synaptic changes (when $\Delta t \to 0$). The time parameters $\tau_\pm$ determine the range of the temporal window for synaptic strengthening and weakening. Here, we set the values: $A_+ = 0.01$, $A_- = -0.012$ and $\tau_+ = 20.0$, $\tau_- = 20.0$. For more details, see [40, 42].

- Time Delay
  Spikes require some time to propagate within the network; this time can be determined from the axonal conduction velocities, which depend on the length and diameter of axonal fiber [43]. In our model, we have omitted all spatial properties of the neuron but transduction delays $t_{\text{del}}$ have been considered to capture the time scale of neuronal communication. They play a crucial role in the neuronal dynamics [40, 44], e.g. neuronal synchrony can be enhanced. The typical time delay between neurons varies between 0.1–20 ms corresponding to axon conduction velocities around 1–20 m/s [45].
  The time delay $t_{\text{del},1}$ of the range of 1–10 ms was initially set up for local neurons within one area. For the inter-areal delay, we considered values $t_{\text{del},2}$ of 10–30 ms.

Chapter 2 offers more information and details of the synaptic properties of neuronal connections, their models and the mechanism of spike-timing-dependent plasticity. Here, we finally summarize the parameters of the neurons defined in this section, whose alteration provides freedom in the exploration of the neuronal dynamics.

## 9.4 Parallel Implementation of the Code

We have presented a large-scale network model of the cortex that accounts for several biological features at different scales. From our previous experience, we know that even simple neural dynamics, omitting properties like time delay, synaptic plasticity, neural response etc., demand large computational power, see Chap. 5 and [46, 47]. The inclusion of these omitted components causes the simulation of the system to be computationally infeasible to run on single CPUs. To reduce computational time and to improve the efficacy of the code, the parallelization of the code was the only possibility to perform simulations on a reasonable time scale [48] (cf. Chap. 10). For the parallel communication, we chose the message-passing interface (MPI) [49]. The main idea is based on the exchange of packages between different CPUs, i.e. sending and receiving messages. All details of the code and process of parallelization are described in Chap. 11.

The flexibility of the program allows one to replace various parts with new ones or redefined properties. The groups of students had free access to this parallel code and used it for their own simulations. Students chose and implemented the neural models, in some cases including their own modules (Chaps. 13 and 14).

## 9.5 Summary

In this chapter, we introduced the general concepts of neuronal modeling, especially the construction of a large-scale computational model of cortical neuronal network. First, we reviewed the main neurophysiological properties that should be included in such a complicated model. The general idea of the connectome was introduced and the structural properties of neural networks discussed. Second, we described the general dynamical features of single neurons and interactions between them. All these individual structural and dynamical properties are explained in more depth and summarized in the following chapters, which give basic information about the different biological and physical phenomena occurring in the brain, e.g. dynamics of individual neurons and populations (Chaps. 1, 2, 7, and 8), structure and its relation to the dynamics (Chaps. 3, 5, and 7). Additionally, we discussed the need of the parallelization of the code. More details and parallel implementation are given in Chaps. 10 and 11. This general framework we created was later used by the groups of students during the summer school.

Although, we have previously described our approach as bottom-up modeling (where we go from single neuron dynamics to the dynamics of cortical areas by averaging), the model also exhibits features of a top-down modeling scheme. We started from the systems level, cortical areas, connected according to the cat map. The internal structure of each area was later expressed as the local network of the neurons. From the structural point of the view, we have omitted the complex character of the connections on the cellular level, e.g. layered structure, morphology of the different types of neurons etc. Future possible improvements could include hierarchical organization of neurons into different layers together with substructures like cortical minicolumns and columns. But such a detailed approach would increase the number of parameters and demand even higher computational power. The improved model follows the same goal as an ambitious project, the Blue Brain Project (BBP), attempting to create a computational model of the mammalian brain [6]. The current effort of BBP concentrates on an accurate computational replica of the neocortical column using one of the fastest supercomputers in the world. Later, simulations of the whole brain with detailed anatomical structure and dynamical properties are planned to discover the secrets of dynamical processes in the brain.

# Acknowledgments

We would like to thank G. Zamora and J. Ong for discussion and helpful suggestions.

# References

1. E. R. Kandel, J. H. Schwartz, and T. M. Jessell, Principles of Neural Science, 2000, 4th ed., New York, McGraw Hill.
2. O. Sporns, G. Tononi, and R. Kötter, The human connectome: a structural description of the human brain, PLoS Comput. Biol. 1(4) (2005) 0245–0251.
3. J. J. Wright, C. J. Rennie, G. J. Lees, P. A. Robinson, P. D. Bourke, C. L. Chapman, E. Gordon, and D. L. Rowe, Simulated electrocortical activity at microscopic, macroscopic and global scales, Neuropsychopharmacology 28 (2003) S80–S93.
4. N. Rulkov, I. Timofeev, and M. Bazhenov, Oscillations in large-scale cortical networks: map-based model, J. Comput. Neurosci. 17 (2004) 203–223.
5. E. M. Izhikevich, J. A. Gally, and G. M. Edelman, Spike-timing dynamics of neuronal groups, Cereb. Cortex 14 (2004) 933–944.
6. H. Markram, The blue brain project, Nat. Rev. Neurosci. 7 (2006) 153–160.
7. V. Braitenberg and A. Schüz, Anatomy of the Cortex: Statistics and Geometry, 1991, Springer, Berlin.

8.  D. Rodney and M. Kevan, Neocortex, 459–509, in Synaptical Organisation of the Brain, G. Shepherd, 1991, Springer, New York.
9.  G.-q. Bi and M.-m. Poo, Synaptic modification by correlated activity: Hebb's postulate revisited, Annu. Rev. Neurosci. 24 (2001) 139–166.
10. N. Brunel, Dynamics of sparsely connected networks of excitatory and inhibitory spiking neurons, J. Comput. Neurosci. 8 (2000) 183–208.
11. P. Kudela, P. J. Franaszczuk, and G. K. Bergey, Changing excitation and inhibition in simulated neural networks: effects on induced bursting behavior, Biol. Cybern. 88 (2003) 276–285.
12. V. P. Zhigulin, Dynamical motifs: building blocks of complex dynamics in sparsely connected random networks, Phys. Rev. Lett. 92, 23 (2004) 238701.
13. N. Masuda and K. Aihara, Global and local synchrony of coupled neurons in small-world networks, Biol. Cybern. 90 (2004) 302–309.
14. L. Lago-Fernández, R. Huerta, F. Corbacho, and J. A. Sigüenza, Fast response and temporal coherent oscillations in small-world networks, Phys. Rev. Lett. 84 (2000) 2758–2761.
15. D. Hansel and G. Mato, Existence and stability of persistent states in large neuronal networks, Phys. Rev. Lett. 86 (2001) 4175–4178.
16. J. Ito and K. Kaneko, Spontaneous structure formation in a network of chaotic units with variable connection strengths, Phys. Rev. Lett. 88 (2001) 028701.
17. I. Belykh, E. de Lange, and M. Hasler, Synchronization of bursting neurons: what matters in the network topology, Phys. Rev. Lett. 94 (2005) 188101.
18. M. Kaiser and C. C. Hilgetag, Spatial growth of real-world networks, Phys. Rev. E 69 (2004) 036103.
19. D. Buxhoeveden and M. F. Casanova, The minicolumn hypothesis in neuroscience, Brain 125 (2002) 935–951.
20. J. W. Scannell, C. Blakemore, and M. P. Young, Analysis of connectivity in the cat cerebral cortex, J. Neurosci. 15 (1995) 1463–1483.
21. J. W. Scannell, G. A. P. C. Burns, C. C. Hilgetag, M. A. O'Neill, and M. P. Young, The connectional organization of the cortico-thalamic system of the cat, Cereb. Cortex 9 (1999) 277–299.
22. D. J. Felleman and D. C. Van Essen, Distributed hierarchical processing in the primate cerebral cortex, Cereb. Cortex 1 (1991) 1–47.
23. Z. Ding, J. C. Gore, and A. W. Anderson, Classification and quantification of neuronal fiber pathways using diffusion tensor MRI, Magn. Reson. Med. 49 (2003) 716–721.
24. Neocortical Microcircuit Database, Copyright 2003 Brain & Mind Institute, EPFL, Lausanne, Switzerland, http://microcircuit.epfl.ch/, (2006)
25. WormAtlas. Z. F. Altun and D. H. Hall (eds.), 2002–2006, http://www.wormatlas.org/, (2006)
26. Cortical Connectivity in Macaque, http://cocomac.org/ (2006)
27. Brain Maps, copyright UC Regents Davis campus, 2005–2006, http://brainmaps.org/, (2006)
28. C. C. Hilgetag, G. A. P. C. Burns, M. A. O'Neill, J. W. Scannell, and M. P. Young, Anatomical connectivity defines the organization of clusters of cortical areas in macaque monkey and cat, Phil. Trans. R. Soc. Lond. B 355 (2000) 91–110.
29. C. C. Hilgetag and M. Kaiser, Clustered organization of cortical connectivity, Neuroinformatics 2 (2004) 353–360.

30. O. Sporns, D. R. Chialvo, M. Kaiser, and C. C. Hilgetag, Organization, development and function of complex brain networks, Trends Cogn. Sci 8 (2004) 418–425.

31. O. Sporns, Network analysis, complexity and brain function, Complexity 8 (2003) 56–60.

32. D. J. Watts and S. H. Strogatz, Collective dynamics of 'small-world' networks, Nature 393 (1998) 440–442.

33. V. M. Eguíluz, D. R. Chialvo, G. Cecchi, M. Baliki, and A. Vania Apkarian, Scale-free brain functional networks, Phys. Rev. Lett. 94 (2005) 018102.

34. M. P. Young, The architecture of visual cortex and inferential processes in vision, Spatial Vis. 13 (2000) 137–146.

35. M. I. Rabinovich, P. Varona, A. I. Selverston, and H. D. I. Abarbanel, Dynamical principles in neuroscience, Rev. Mod. Phys. 78 (2006) 1213–1265.

36. E. M. Izhikevich, Which model to use for cortical spiking neurons? IEEE Trans. Neural Netw. 15 (2004) 1063–1070.

37. J. Rinzel and G. B. Ermentrout, Analysis of neural excitability and oscillations, methods in neuronal modeling: From synapses to networks, 135–169, C. Koch and I. Segev, 1989, The MIT Press, Cambridge, MA.

38. M. St-Hilaire and A. Longtin, Comparison of coding capabilities of type I and type II neurons, J. Comput. Neurosci 16 (2004) 299–313.

39. W. J. Freeman, Characteristics of the synchronization of brain activity imposed by finite conduction velocities of axons, Int. J. Bifurcation Chaos Appl. Sci. Eng. 10 (2000) 2307–2322.

40. E. M. Izhikevich, Polychronization: computation with spikes, Neural Comput. 18 (2006) 245–282.

41. M. V. Ivanchenko, G. V. Osipov, V. D. Shalfeev, and J. Kurths, Synchronized bursts following instability of synchronous spiking in chaotic neural networks, arXiv.org:nlin/0601023 (2006).

42. S. Song, K. D. Miller, and L. F. Abbot, Competitive Hebbian learning through spike-timing-dependent synaptic plasticity, Nat. Neurosc. 3, 9 (2000) 919–926.

43. C. W. Eurich, K. Pawelzik, U. Ernst, J. D. Cowan, and J. G. Milton, Dynamics of self-organized delay adaptation, Phys. Rev. Lett. 82 (1999) 1594–1597.

44. M. Dhamala, V. K. Jirsa, and M. Ding, Enhancement of neural synchrony by time delay, Phys. Rev. Lett. 92 (2004) 074104.

45. N. Kopell, G. B. Ermentrout, M. A. Whittington, and R. D. Traub, Gamma rhythms and beta rhythms have different synchronization properties, Proc. Natl. Acad. Sci. USA 97 (2000) 1867–1872.

46. L. Zemanová, C. Zhou, and J. Kurths, Structural and functional clusters of complex brain networks, Physica D 224 (2006) 202–212.

47. C. Zhou, L. Zemanová, G. Zamora, C. C. Hilgetag, and J. Kurths, Hierarchical organization unveiled by functional connectivity in complex brain networks, Phys. Rev. Lett. 97 (2006) 238103.

48. A. Morrison, C. Mehring, T. Geisel, A. Aertsen, and M. Diesmann, Advancing the boundaries of high-connectivity network simulation with distributed computing, Neural Comput. 17 (2005) 1776–1801.

49. W. Gropp, E. Lusk, and A. Skjellum, Using MPI: Portable parallel programming with the message-passing interface, 1999, 2nd ed., The MIT Press, Cambridge, MA.

# 10

# Maintaining Causality in Discrete Time Neuronal Network Simulations

Abigail Morrison and Markus Diesmann

Brain Science Institute, RIKEN, Wako, Japan

**Summary.** When designing a discrete time simulation tool for neuronal networks, conceptual difficulties are often encountered in defining the interaction between the continuous dynamics of the neurons and the point events (spikes) they exchange. These problems increase significantly when the tool is designed to be distributed over many computers. In this chapter, we bring together the methods that have been developed over the last years to handle these difficulties. We describe a framework in which the temporal order of events within a simulation remains consistent. It is applicable to networks of neurons with arbitrary subthreshold dynamics, both with and without delays, exchanging point events either constrained to a discrete time grid or in continuous time, and is compatible with distributed computing.

## 10.1 Introduction

Neural network simulations are crucial for the advancement of computational neuroscience, as the nonlinear activity dynamics is only partially accessible by purely analytical methods and experimental techniques are still severely limited in their ability to observe and manipulate large numbers of neurons. The brain is an unusual physical system, as it consists of elements (neurons) which can best be described by a set of differential equations, yet the interaction between these elements is mediated by point-like events (action potentials or spikes). It is, moreover, a very complex system — for example, each neuron in the cortex receives in the order of $10^4$ connections from other neurons, both within its immediate area and from more remote parts of the brain. Simulating networks with this degree of complexity naturally suggests the use of distributed computing techniques. However, the meshing of continuous-time dynamics and discrete-time communication makes it notoriously difficult to define a consistent and sufficiently general framework for the integration of the dynamics.

There are two classical approaches to simulation: time-driven and event-driven. In the former, a computational time step $h$ is defined. One iteration of a simulation involves each neuron advancing its dynamics over one time step. If its conditions for generating an action potential are met, a spike is delivered to each of the neurons to which it projects. After all neurons have been updated, the next iteration begins. In the latter approach, an event queue manages the order in which spikes are delivered. Each neuron is only updated when it receives an event. If its conditions for generating an action potential are met, the new event is inserted into the queue. This algorithm can be defined very simply and can be very efficient if the neuronal dynamics is invertible — for example, if the arrival of a spike causes an immediate jump in the membrane potential which then decays exponentially. In this case, the neuron can only fire at the arrival of an incoming event, so the behavior of the neuron between the arrival of one event and the next is not relevant for the correct integration of the network. For neuron models with non-invertible dynamics, such as those where the maximum excursion of the membrane potential occurs some time after the arrival of a spike, it is much harder to define an event-driven algorithm. More sophisticated mechanisms are needed: for example, neurons might place provisional events in the queue if they are close to their firing conditions, but may have to revise their predicted spike times upon the arrival of further events [1, 2]. In the following, we concentrate on the time-driven approach as defined above, which can incorporate any kind of subthreshold dynamics without changes being made to the updating and spike delivery algorithm, and has been shown to have good performance in simulating large-scale neuronal networks and to scale excellently when distributed [3].

Here, we present a framework which defines the interactions between the neurons without damaging causality, i.e. such that the order in which neurons are updated does not affect the outcome of a simulation. The framework is suitable for distributed computing. In Sect. 10.2, we cover the basics of point event interaction between continuous-time neuronal elements. We first discuss the historically important concept of neuronal networks with no propagation delay and describe an updating scheme ensuring that the simulation results are independent of the order in which neurons are updated (Sect. 10.2.1). We then demonstrate how this scheme needs to be adapted to incorporate delays that are multiples of the computational time step $h$ (Sect. 10.2.2). Such networks have traditionally constrained spike times to the discrete time grid. However, for networks with propagation delays greater or equal to the computational time step, this constraint can be relaxed. In Sect. 10.3 we show how the scheme can be extended to permit neurons to generate and receive off-grid point events. Finally, we discuss how the propagation delays between neurons can be exploited to optimize communication efficiency between machines in a distributed environment (Sect. 10.4).

## 10.2 Networks with Discrete Spike Times

In the following, we will assume that communication between neurons is mediated by synapses. When a neuron spikes, all of its outgoing synapses send a discrete event to their respective postsynaptic neurons. The event is parameterized with a weight $w$, which is interpreted by the postsynaptic neuron with respect to the postsynaptic dynamics it implements. For example, the postsynaptic neuron may interpret $w$ as the size of an instantaneous jump in its membrane potential, or as the maximum amplitude of a postsynaptic current implemented as an alpha function (see Chap. 1). In Sect. 10.2.2, the event is further parameterized by an integer delay $k$, which expresses the propagation delay $d$ between the neurons in units of the computational time step $h$, i.e. $d = k \cdot h$.

### 10.2.1 Networks without Propagation Delays

Consider the following situation: neuron $i$ and neuron $j$ have a strong reciprocal inhibitory connection, such that a spike causes an instantaneous reduction in the membrane potential of the postsynaptic neuron. Each neuron is receiving enough input to drive it to spike at time $t$. If neuron $i$ is updated to time $t$ first, the spike is instantaneously delivered to neuron $j$. When $j$ is updated, the strong inhibition prevents the membrane potential from passing the threshold, and so it does not itself generate a spike. Conversely, if neuron $j$ is updated first, neuron $j$ spikes at time $t$ and neuron $i$ is inhibited.

The order dependence in the above example is extremely undesirable. However, with a small conceptual adjustment, the simulation can be made internally consistent. The convention is to define that the generation of a spike may only be influenced by spikes which precede it. This is depicted in the flowchart in Fig. 10.1(a). When the neuron is updated from $t$ to $t + h$, it first modifies its state according to the new spikes from its upstream neurons which fired at time $t$ (operator $G$), for example incrementing the membrane potential or postsynaptic current. Then the subthreshold dynamics is carried out to propagate the modified neuron state, including the new events, to $t + h$ (operator $F_h$). At this point, the spiking criteria are applied; if they are fulfilled, the neuron emits a spike. Thus, the effects of the spike on the neuron state are consistent with receiving a spike at $t$, as can be seen in the membrane potential of neuron *post* in Fig. 10.1(b), but the earliest time the neuron can emit a spike as a result of receiving that spike is $t + h$. This is the equivalent of considering spikes to have an infinitesimal $\epsilon$-delay, and has the effect of making simulations consistent, in that the order of updates does not affect the outcome. In our previous example of two neurons with mutual inhibition, both neurons would fire at time $t$. Assuming no refractory period, the effect of the mutual inhibition would result in a hyperpolarization of both neurons at time $t + h$.

### 10.2.2 Networks with Propagation Delays

*Minimal Delay*

It is particularly simple to alter the algorithm described in Sect. 10.2.1 to one in which all propagation delays in the network are equal to the computation

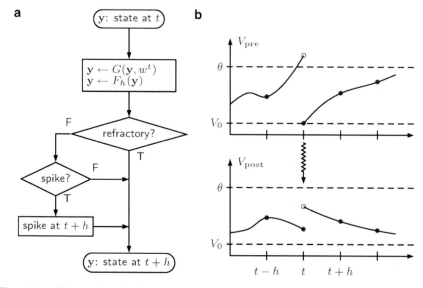

**Fig. 10.1.** Schematic of a discrete time simulation with no propagation delay: (**a**) Neuron update algorithm. The flowchart depicts the order of operations required to propagate the state $y$ of an individual neuron by one time step $h$. Operator $G$ modifies the state according to the incoming events and operator $F_h$ performs the subthreshold dynamics; (**b**) Spike transmission and its effect on the postsynaptic neuron. The membrane potential $V_{\mathrm{pre}}$ of neuron *pre* crosses the threshold $\theta$ in the time step $(t-h, t]$, so a spike is emitted at time $t$ and the membrane potential is reset to $V_0$. The spike arrives at neuron *post* with no delay (*zig-zag arrow*). Filled circles denote the values of the membrane potential that can be reported by the neuron at the end of a time step. Intermediate (non-observable) values of the membrane potential are shown as unfilled circles

time step $h$. In fact, all it amounts to is changing the order of the two operators $F_h$ and $G$, see Fig. 10.1 and Fig. 10.2. If neuron *pre* spikes at time $t$, this spike is delivered immediately to neuron *post* (Fig. 10.2(b)). When neuron *post* is being updated from $t$ to $t + h$, first the subthreshold dynamics are performed to propagate the neuron by a step of $h$ (operator $F_h$), then the neuron state is modified to include the new events visible at $t + h$, including the spike sent by neuron *pre*. Note that for this case and the case where no propagation delay is assumed, the infrastructure of the simulation is the same. A spike produced at time $t$ is delivered immediately to its target, but due to the different order of operations, the effect of the spike is instantaneous in the first case, but delayed by $h$ in the second.

The data structure used to store the pending events can be very simple. If all the synapses have the same dynamics, varying only in amplitude, then each neuron only requires one buffer to store incoming events. Examples of neuron models that only require one buffer are those in which synaptic interactions

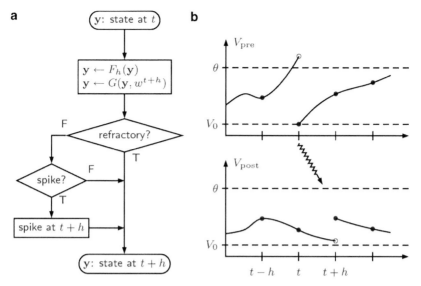

**Fig. 10.2.** Schematic of a discrete time simulation with propagation delays: (**a**) Neuron update algorithm. The flowchart depicts the order of operations required to propagate the state $\mathbf{y}$ of an individual neuron by one time step $h$. Operator $F_h$ applies the subthreshold dynamics and operator $G$ modifies the state according to the incoming events. Note that the order of these two operations is the reverse of the order shown in Fig. 10.1; (**b**) Spike transmission and its effect on the postsynaptic neuron. As in Fig. 10.1(b), neuron *pre* emits a spike at time $t$. The spike arrives at neuron *post* with a minimal delay of $h$ (*zig-zag arrow*). Filled circles denote the values of the membrane potential that can be reported by the neuron at the end of a time step. Intermediate (non-observable) values of the membrane potential are shown as unfilled circles

cause an instantaneous increment to the membrane potential, or induce exponential postsynaptic currents. Other neuron models may have more than one set of synaptic dynamics, such as a longer time constant for inhibitory than for excitatory interactions. Clearly, in this case, one buffer per time constant would be required. However, for the sake of simplicity, we will focus on neuron models with only one set of synaptic dynamics.

Depending on the implementation, either a one-element or a two-element buffer is sufficient to maintain causality in the system. If the global scheduling algorithm iterates through all the neurons twice — once to advance the dynamics, the second time to apply the spiking criteria and deliver any generated events — then a one-element buffer is sufficient, as the 'read' and 'write' phases are cleanly separated. However, for reasons of cache effectiveness it may be preferable to iterate through all the neurons only once — i.e. for each neuron, advance its dynamics, apply its spiking criteria and deliver the new events if it spikes. In this case, the 'read' and 'write' phases are no longer cleanly separated, and a two-element buffer is required.

A two-element buffer is depicted in Fig. 10.3. One side of the buffer can be considered as the 'read' side, the other as the 'write' side. When neuron $i$ is modifying its state to incorporate the new events (operator $G$ in Fig. 10.2), it collects the summed weights becoming visible at $t + h$ from the 'read' side. The act of reading clears that side of the buffer. If any neurons that project to $i$ emit a spike at $t + h$, the weight of this event is added to the 'write' side. After all the neurons have been updated, all their buffers are toggled so that the empty 'read' sides are now 'write' sides, and the 'write' sides, containing those events which become visible at $t + 2h$, are now 'read' sides. Thus, the order in which the neurons are updated does not affect the outcome, as events generated in one time step are always cleanly separated from those generated in the next.

Exactly the same structure can be used for networks with no propagation delay, except the assignation of times to buffer elements is shifted by $h$: in Fig. 10.3(a), the left side receives events for the time step $t + h$ while the summed weight of events becoming visible in time step $t$ is read out of the right side.

### General Delay

A system to simulate a network with a minimal propagation delay $h$ can be converted into one encompassing many different delays, as long as they are all integer multiples of $h$, by replacing the simple two-element buffer with a ring buffer. A traditional ring buffer is an implementation of a queue. The data is represented in a contiguous series of segments. New elements are appended to one end of the series, and the oldest elements are popped off the other end.

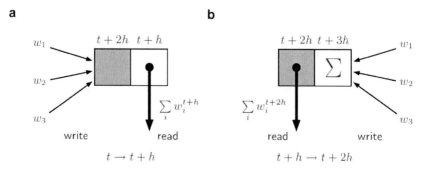

**Fig. 10.3.** A two-element buffer, suitable for use in networks with a delay of $h$: (a) In the time step $(t, t + h]$, the gray side of the buffer is the 'write' side, and it sums the weights of events generated in this time step that are to become visible in the next step. The white side of the buffer is the 'read' side, containing the summed weights of all the events received in the previous time step, $\sum_i w_i^{t+h}$. Once the neuron has read out the buffer, the 'read' side is emptied; (b) After all neurons have been updated, the neuron buffers are toggled. Now the empty white side receives new events, and the gray side is read by the neuron as it updates from $t + h$ to $t + 2h$

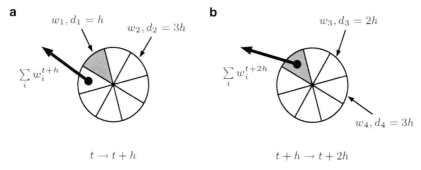

**Fig. 10.4.** A random access ring buffer, suitable for use in networks with delays that are integer multiples of $h$. The weights $w_i$ of incoming events are written to the segments corresponding to their delays $d_i$, such that a delay $d = k \cdot h$ corresponds to the segment $k$ along from the current read position: (**a**) Time step $(t, t + h]$: the read position of the buffer is the segment containing the summed weights of all the events which are to become visible at $t + h$; (**b**) Time step $(t + h, t + 2h]$: The read position has moved to the next segment (*gray*), containing the summed weights of all the events which are to become visible at $t + 2h$

Thus, the arc containing the data can be imagined as rotating around the ring as data is added and removed. In this way, a queue can be implemented without continually having to allocate fresh memory. For our purposes, we need something more like a random access ring buffer, as shown in Fig. 10.4 (see also [3]). Each segment of the ring corresponds to one time step. When the neuron is updating from $t$ to $t + h$, it reads from the segment containing the summed weights of the events that become visible at $t + h$, and then clears this segment. Incoming events are sorted into the other segments depending on their delays: a spike with a delay of $k \cdot h$ would be sorted into the $k$th segment along from the current read position. After all the neurons have been updated, the read positions for all buffers are moved around one segment. The ring buffer needs to be appropriately sized if the correct order of events is to be maintained — it must be large enough to accommodate the largest propagation delay between neurons in the simulated system, $d_{\max} = k_{\max} \cdot h$, without 'wrapping'. Therefore the optimal size for the buffer is $k_{\max} + 1$. Depending on implementation, $d_{\max}$ may either be specified before the creation of the network, or, more elegantly, determined dynamically whilst the network is created.

## 10.3 Networks with Continuous Spike Times

In the systems discussed above, spike times were constrained to the discrete time grid. However, Hansel [4] showed that forcing spikes onto the grid can significantly distort the synchronization dynamics of certain networks. The integration error decreases only linearly with the computational step size, so a very small $h$ is necessary to capture the dynamics accurately. An alternative solution is to interpolate the membrane potential between grid points and evaluate the effect of incoming spikes on the neuronal grid in continuous time [4, 5]. This concept was extended in [6] by combining it with exact integration of the subthreshold dynamics (see [7]). Here, we discuss how the scheme described in Sect. 10.2.2 can incorporate off-grid spike times.

Continuous spike times can easily be incorporated into discrete time simulations without having to implement a central queuing structure if the minimum propagation delay is greater than or equal to the computational step size $h$ and an appropriate representation of time is used. In the networks discussed in Sect. 10.2.2, first the subthreshold dynamics is advanced by one time step from $t$ to $t + h$, and then the spiking criteria are applied. If the neuron state passes the criteria (for example, by having a membrane potential above a threshold), the neuron emits a spike which is assigned to the time $t + h$. Now, let us assume that the actual spike time can be determined more precisely, either by interpolation of the membrane potential or by inverting the dynamics or by any other method, such that $t < t_{\text{spike}} \leq t + h$. If the propagation delay to the neuron's postsynaptic target is $k \cdot h$, the event should become visible at time $t_{\text{spike}} + k \cdot h$, which is in the update interval $(t + k \cdot h, t + (k + 1) \cdot h]$. An appropriate representation of the spike time therefore consists of an integer time stamp $t + h$ and a floating point offset $\delta = t_{\text{spike}} - t$. By definition, $\delta$ is in the interval $(0, h]$. This choice of representation allows the infrastructure described in Sect. 10.2.2 to be kept with only minimal changes. The propagation delay $k$ can still be used to sort the event into the segment that will be read in the step $(t + k \cdot h, t + (k + 1) \cdot h]$, but the ring buffer is adapted to hold a vector of events in each segment instead of a single value. The weight $w$ and offset $\delta$ of the event are appended to the vector. When the neuron performs this update step, the vector is first sorted in order of increasing $\delta$. Note that the vector is just the simplest possible implementation and could be replaced by a more sophisticated data structure such as a calendar queue [8].

The subthreshold dynamics is then advanced from the beginning of the time step to the arrival time of the first event, at which point the neuron state is modified to take account of the first event. Then the dynamics is advanced between the arrival time of the first event and the arrival time of the second event, at which point the neuron state is modified to take account of the second event, and so on until all the events for that update step have been processed. Finally the dynamics is advanced from the arrival time of the final event to the end of the time step. Thus all incoming events have been processed in the correct temporal order.

The scheme described above is very general and can be applied to any kind of subthreshold dynamics, allowing the processing and generation of spikes in continuous time within a discrete time algorithm. No global queuing of events is required, as each neuron queues its events locally. In the case that the subthreshold dynamics is linear, this can be exploited such that not even local queuing is required. This involves a slightly more complicated mechanism for receiving spikes, see [6].

## 10.4 Distributed Networks

Neuronal network simulations can consume a huge amount of memory, especially if biologically realistic levels of connectivity are assumed. In the cortex, each neuron has of the order of $10^4$ incoming synapses and a connection probability of about 0.1 in its local area [9]. Therefore, a network fulfilling both of these constraints must have at least $10^5$ neurons. This is equivalent to about 1 mm$^3$ of cortical tissue, and represents a threshold network size for simulations, as beyond this point, the number of synapses increases only linearly with the number of neurons, rather than quadratically as is the case for smaller systems. Such networks contain $10^9$ synapses, which, even using an extremely simple synapse representation, require several gigabytes of RAM. This state of affairs has naturally prompted much interest in distributed computing, for example [3, 10, 11]. However, distribution raises new issues about maintaining causality in the simulated system. If neuron *pre* projects to neuron *post* with a delay of $k \cdot h$, a spike produced by neuron *pre* at time $t$ should be visible to neuron *post* at time $t + k \cdot h$, no matter whether the two neurons are located on the same machine.

In [3], it was demonstrated that it is more efficient to distribute a neuronal simulation by placing the synapses on the machines of their postsynaptic neurons than of their presynaptic neuron. This is equivalent to distributing a neuron's axon but keeping its dendrite local. That way, when a neuron fires, only its index must be sent across the computer network, rather than a weight and delay for every one of its postsynaptic targets. For neuronal networks with biologically realistic levels of connectivity, this can represent a difference of several orders of magnitude in the amount of information being communicated. One way of ensuring that spikes are always delivered on time is to communicate in each time step, after all the neurons have been updated but before the read positions of their buffers has been incremented (see Sect. 10.2.2). However, this approach is sub-optimal with respect to communication efficiency. Communication between machines has an overhead, so it is more efficient to send one message of $N$ bytes than $N$ messages of one byte each. Fortunately, it is generally possible to communicate less often and still deliver the events correctly. For this, it is necessary to determine the minimum propagation delay between neurons in the simulated system, $d_{\min} = k_{\min} \cdot h$. As in the case of $d_{\max}$, described in Sect. 10.2.2, depending on implementation, $d_{\min}$ could either be specified before the creation of the neuronal network, or determined dynamically whilst the neurons are connected. By definition, a spike cannot have an effect on a postsynaptic neuron earlier than $k_{\min}$ time steps after generation. Therefore, as long as the temporal order of spikes is preserved, it is possible to communicate in intervals of $k_{\min}$ time steps. This can be a significant improvement, as the minimum delay can be considerably larger than the computational time step.

Preserving the temporal order of spikes has two parts: correct storage before communication, and correct delivery after communication. If spikes are constrained to the discrete time grid, for correct storage it is sufficient for each machine to store the indices of spiking neurons in a buffer, with tokens separating the indices of neurons which spiked in one time step from those which spiked in the previous or next step. Thus, at the end of $k_{min}$ time steps, the buffer contains $k_{min}$ blocks neuron indices separated by $k_{min} - 1$ tokens. Then the machine sends this buffer of indices to all other machines, and receives buffers in turn. If the spikes are not constrained to the grid, as described in Sect. 10.3, in addition to the index of a spiking neuron, its spike offset $\delta$ must be buffered and communicated as well. For correct delivery, it is necessary to activate the synapses of the neurons registered in these buffers whilst taking the temporal order into consideration. The position of an index in the buffer represents the communication lag, $k_{lag}$, in delivering the information that a neuron has fired, i.e. $k_{lag} = k_{min}$ for an index in the first block of data, $k_{lag} = k_{min} - 1$ for an index in the second block of data and so on until $k_{lag} = 1$ for indices in the last block of data. Note that this assumes that the read positions of all the ring buffers had been incremented before exchange of spike data (see Sect. 10.2.2). If the order is exchanged, then the communication lag ranges from $k_{min} - 1$ for the first block to 1 for the last block. If the communication lag is subtracted from the propagation delay encoded in a synapse, then the event will become visible to the postsynaptic neuron at exactly the same time as it would in a serial simulation. For example, consider a synapse from neuron *pre* to neuron *post* with a weight $w$ and delay $k \cdot h$. In a serial simulation, if neuron *pre* emitted a spike at time $t$, $w$ would be added to the ring buffer in the $k$th segment along from the current read position. In a distributed simulation, $w$ would be added to neuron *post*'s ring buffer $k - k_{lag}$ segments along from the current read position, with $k_{lag}$ determined by the position of the index of neuron *pre* in the received index buffer. A slightly more sophisticated version of this approach is discussed in [3].

## 10.5 Conclusions and Perspectives

We have shown how relatively simple methods and data structures can be used to simulate networks of spiking neurons in discrete time whilst reliably maintaining the temporal order of events. If spike times are constrained to the discrete time grid, the framework is applicable to networks with no propagation delay and to networks with arbitrary delays that are integer multiples of the computation step size $h$. If spike times are not constrained to the grid, the framework is only applicable to networks with propagation delays greater than or equal to the computational step size $h$. In this case, the constraint that delays must be an integer multiple of $h$ could be relaxed, because floating point offsets and delays can always be recombined on the fly to produce integer delays and floating point offsets. All these networks can be implemented in a distributed environment in a symmetrical fashion, i.e. the architecture is peer-to-peer rather than master-slave.

These networks can be simulated efficiently because the delivery time of an event in the simulated system has been decoupled from the arrival time at the postsynaptic neuron and the temporal resolution of the simulation. This is the concept that underlies both the ring buffers and the minimum delay intervals for communication. However, if delivery and arrival times are decoupled, this can be problematic for synaptic processes that depend on the state of the postsynaptic neuron, for example spike-timing-dependent plasticity [12, 13] (see also Chaps. 2 and 9). An algorithm has been developed that maintains the correct relationships if propagation delays are assumed to be predominantly dendritic [14]. However, if the propagation delays are predominantly axonal, the framework presented here is not sufficient and will have to be adapted.

## Acknowledgements

We acknowledge constructive discussions with the members of the NEST initiative (http://www.nest-initiative.org). Partially funded by DIP F1.2, BMBF Grant 01GQ0420 to the Bernstein Center for Computational Neuroscience Freiburg and EU Grant 15879 (FACETS).

## References

1. Brette, R. (2006). Exact simulation of integrate-and-fire models with synaptic conductances. *Neural Comput. 18*(8), 2004–2027.
2. Lytton, W. W., & Hines, M. L. (2005). Independent variable time-step integration of individual neurons for network simulations. *Neural Comput. 17*, 903–921.
3. Morrison, A., Mehring, C., Geisel, T., Aertsen, A., & Diesmann, M. (2005). Advancing the boundaries of high connectivity network simulation with distributed computing. *Neural Comput. 17*(8), 1776–1801.
4. Hansel, D., Mato, G., Meunier, C., & Neltner, L. (1998). On numerical simulations of integrate-and-fire neural networks. *Neural Comput. 10*(2), 467–483.
5. Shelley, M. J., & Tao, L. (2001). Efficient and accurate time-stepping schemes for integrate-and-fire neuronal networks. *J. Comput. Neurosci. 11*(2), 111–119.
6. Morrison, A., Straube, S., Plesser, H. E., & Diesmann, M. (2006). Exact subthreshold integration with continuous spike times in discrete time neural network simulations. *Neural Comput. 19*, 47–79.
7. Rotter, S., & Diesmann, M. (1999). Exact digital simulation of time-invariant linear systems with applications to neuronal modeling. *Biol. Cybern. 81*(5/6), 381–402.

8. Brown, R. (1988). Calendar queues: a fast O(1) priority queue implementation for the simulation event set problem. *Communications of the ACM 31*(10), 1220–1227.

9. Braitenberg, V., & Schüz, A. (1991). *Anatomy of the Cortex: Statistics and Geometry*. Berlin, Heidelberg, New York: Springer-Verlag.

10. Hammarlund, P., & Ekeberg, O. (1998). Large neural network simulations on multiple hardware platforms. *J. Comput. Neurosci. 5*(4), 443–459.

11. Harris, J., Baurick, J., Frye, J., King, J., Ballew, M., Goodman, P., & Drewes, R. (2003). A novel parallel hardware and software solution for a large-scale biologically realistic cortical simulation. Technical report, University of Nevada.

12. Bi, G.-q., & Poo, M.-m. (1998). Synaptic modifications in cultured hippocampal neurons: Dependence on spike timing, synaptic strength, and postsynaptic cell type. *J. Neurosci. 18*, 10464–10472.

13. Markram, H., Lübke, J., Frotscher, M., & Sakmann, B. (1997). Regulation of synaptic efficacy by coincidence of postsynaptic APs and EPSPs. *Science 275*, 213–215.

14. Morrison, A., Aertsen, A., & Diesmann, M. (2007). Spike-timing dependent plasticity in balanced random networks. *Neural Comp. 19*, 1437–1467.

# Sequential and Parallel Implementation of Networks

Werner von Bloh

Potsdam Institute for Climate Impact Research, PO Box 601203, 14412 Potsdam, Germany
bloh@pik-potsdam.de

## 11.1 Implementation of Diffusive Coupled Networks

A diffusive coupled network of $n$ neurons can be described by a state vector $N_i(t)$:

$$\frac{\mathrm{d}N_i(t)}{\mathrm{d}t} = f(N_i(t)) + \sum_{j=1}^{n} w_{ij} \times (N_j(t) - N_i(t)), \qquad (11.1)$$

where the function $f$ describes the evolution of the neuron and the weight matrix $w_{ij}$ the coupling strength between the neurons. For simplicity, we assume that the state of the neuron can be described by a single scalar. The matrix $w_{ij}$ has the following properties:

$$w_{ij} \begin{cases} > 0 \text{ excitatory coupling} \\ = 0 \text{ no coupling} \\ < 0 \text{ inhibitory coupling} \end{cases} \qquad (11.2)$$

The matrix can be implemented by a two-dimensional array. The size of this array scales with $O(n^2)$ independent of the number of couplings (equivalent to non-zero elements in $w_{ij}$). This is in particular inefficient for sparsely coupled neurons. In this case, it is better to use the list structure (see Fig. 11.1) [1].

From now, on we will apply the Morris-Lecar (ML) neuron model [2] (see Chaps. 1, 9, 13, 14). The dynamics can be described by two coupled ordinary differential equations

$$c\frac{\mathrm{d}v}{\mathrm{d}t} = f_v(v, w) = I - g_L(v - v_L) - g_K w(v - v_K)$$
$$\qquad\qquad -g_{\mathrm{Ca}} m_v(v)(v - v_{\mathrm{Ca}}) \qquad (11.3)$$

$$\frac{\mathrm{d}w}{\mathrm{d}t} = f_w(v, w) = \lambda(v)(w_v(v) - w), \qquad (11.4)$$

$$m_v(v) = \frac{1}{2}(1 + \tanh((v - v_1)/v_2)),$$

(a)

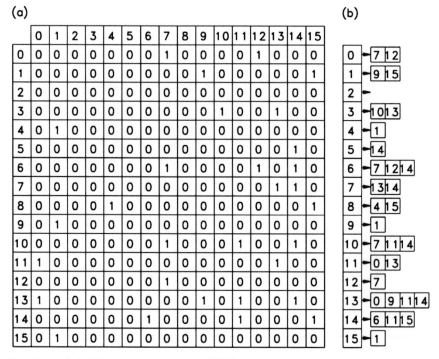

| | 0 | 1 | 2 | 3 | 4 | 5 | 6 | 7 | 8 | 9 | 10 | 11 | 12 | 13 | 14 | 15 |
|---|---|---|---|---|---|---|---|---|---|---|---|---|---|---|---|---|
| 0 | 0 | 0 | 0 | 0 | 0 | 0 | 0 | 1 | 0 | 0 | 0 | 0 | 1 | 0 | 0 | 0 |
| 1 | 0 | 0 | 0 | 0 | 0 | 0 | 0 | 0 | 0 | 1 | 0 | 0 | 0 | 0 | 0 | 1 |
| 2 | 0 | 0 | 0 | 0 | 0 | 0 | 0 | 0 | 0 | 0 | 0 | 0 | 0 | 0 | 0 | 0 |
| 3 | 0 | 0 | 0 | 0 | 0 | 0 | 0 | 0 | 0 | 0 | 1 | 0 | 0 | 1 | 0 | 0 |
| 4 | 0 | 1 | 0 | 0 | 0 | 0 | 0 | 0 | 0 | 0 | 0 | 0 | 0 | 0 | 0 | 0 |
| 5 | 0 | 0 | 0 | 0 | 0 | 0 | 0 | 0 | 0 | 0 | 0 | 0 | 0 | 0 | 1 | 0 |
| 6 | 0 | 0 | 0 | 0 | 0 | 0 | 0 | 1 | 0 | 0 | 0 | 0 | 1 | 0 | 1 | 0 |
| 7 | 0 | 0 | 0 | 0 | 0 | 0 | 0 | 0 | 0 | 0 | 0 | 0 | 0 | 1 | 1 | 0 |
| 8 | 0 | 0 | 0 | 0 | 1 | 0 | 0 | 0 | 0 | 0 | 0 | 0 | 0 | 0 | 0 | 1 |
| 9 | 0 | 1 | 0 | 0 | 0 | 0 | 0 | 0 | 0 | 0 | 0 | 0 | 0 | 0 | 0 | 0 |
| 10 | 0 | 0 | 0 | 0 | 0 | 0 | 0 | 1 | 0 | 0 | 0 | 1 | 0 | 0 | 1 | 0 |
| 11 | 1 | 0 | 0 | 0 | 0 | 0 | 0 | 0 | 0 | 0 | 0 | 0 | 0 | 1 | 0 | 0 |
| 12 | 0 | 0 | 0 | 0 | 0 | 0 | 0 | 1 | 0 | 0 | 0 | 0 | 0 | 0 | 0 | 0 |
| 13 | 1 | 0 | 0 | 0 | 0 | 0 | 0 | 0 | 0 | 1 | 0 | 1 | 0 | 0 | 1 | 0 |
| 14 | 0 | 0 | 0 | 0 | 0 | 0 | 1 | 0 | 0 | 0 | 0 | 1 | 0 | 0 | 0 | 1 |
| 15 | 0 | 1 | 0 | 0 | 0 | 0 | 0 | 0 | 0 | 0 | 0 | 0 | 0 | 0 | 0 | 0 |

(b)

| | |
|---|---|
| 0 | → 7 12 |
| 1 | → 9 15 |
| 2 | → |
| 3 | → 10 13 |
| 4 | → 1 |
| 5 | → 14 |
| 6 | → 7 12 14 |
| 7 | → 13 14 |
| 8 | → 4 15 |
| 9 | → 1 |
| 10 | → 7 11 14 |
| 11 | → 0 13 |
| 12 | → 7 |
| 13 | → 0 9 11 14 |
| 14 | → 6 11 15 |
| 15 | → 1 |

**Fig. 11.1.** (a) Adjacency matrix and; (b) list representation of a random network of 16 neurons with connection probability of $p = 0.15$

$$m_w(v) = \frac{1}{2}(1 + \tanh((v - v_3)/v_4)),$$
$$\lambda(v) = \theta \cosh((v - v_3)/(2v_4)).$$

The diffusive coupling (corresponding to electrical synapses; cf. Chaps. 2 and 9) is achieved with the component $v$, i.e

$$c\frac{dv_i}{dt} = f_v(v_i, w_i) + \sum_{j=1}^{n} w_{ij} \times (v_j - v_i). \tag{11.5}$$

### 11.1.1 Sequential Code

The implementation of the list and the dynamics of the network is performed with the help of the programming language C. It allows the definition of abstract datatypes, where definition and implementation are separate [3]. An introduction to the programming language C is given in [4]. A neuron can be declared as a datatype Neuron in the following way (in the declaration header file neuron.h):

```
typedef struct
{
   float v[2];    /* state of ML neuron */
   int inhib;     /* inhibitory (TRUE) or not (FALSE) */
   List connect; /* list of connections */
} Neuron;
```

The datatype of lists List used by Neuron is declared in the header file list.h:

```
typedef struct
{
   int *index; /* pointer to list of indices itself */
   int len;    /* length of list */
} List;

/* Declaration of functions */

/* initialize empty list to  */
extern void initlist(List *);
/* add connection to list */
extern int  addlistitem(List *,int);
```

The header file contains the declaration of the structure List and the prototypes of the functions for initialization and adding a connection to the network. Lists can be implemented in two ways: (1) as a variable sized array, and (2) with linked pointers. The advantage of a implementation using arrays is the better storage efficiency. A pointer implementation uses for every element an additional pointer causing some memory overhead. Adding an element to an array, however, is performed by a copying process of the full list not necessary in the pointer implementation. If the network is not changed often during runtime, an array implementation is usually more efficient. The initialization function sets the length of the list to zero and initializes the index array to the NULL pointer.

```
void initlist(List *list)
{
   list->len=0;
   list->index=NULL;
} /* of 'newlist' */
```

An item can be added to the list by a call to the addlistitem function. The updated length of the list is returned by this function. The memory allocation can be done by the realloc function of the standard C library. It increases the allocated memory by a certain number of bytes.

```
int addlistitem(List *list,int item)
{
   /* add item to index vector */
```

```
list->index=(int *)realloc(list->index,
                            sizeof(int)*(list->len+1));
list->index[list->len]=item;
list->len++;
return list->len;
} /* of 'addlistitem' */
```

A random network can be built up with the function `randomnet`. For each element of the neuron array, the list is first initialized and then random connections are added to the list with a probability $p_{conn}$. The neuron is marked as an inhibitory neuron with a probability $p_{inhib}$.

```
void randomnet(Neuron net[], /* array of neurons */
               int n,        /* number of neurons */
               float p_conn,  /* probability of establishing
                                 connection */
               float p_inhib /* probability of a inhibitory
                                neuron */
               )
{
  int i,j;
  for(i=0;i<n;i++) /* iterate over all neurons */
  {
   initlist(&net[i].connect);
   net[i].inhib=(drand48()<p_inhib);
   for(j=0;j<n;j++)
     if(i!=j &&  /* avoid self connections */
        drand48()<p_conn)
       addlistitem(&net[i].connect,j);
  }
} /* of 'randomnet' */
```

The `drand48` function is a generator of uniformly distributed pseudo-random numbers defined in `stdlib.h`. The update of the ML model can be implemented by the following `updateml` function:

```
/* constants for ML model */
#define    c 1.0
#define    gL 0.5
#define    gK 2.0
#define    gCa 1.0
#define    vL (-0.5)
#define    vK (-0.7)
#define    vCa 1.0
#define    v1 (-0.01)
#define    v2 0.15
#define    v3 0.1
```

```
#define    v4 0.145
#define    theta (1.0/3.0)
void updateml(float dv[2], /* derivatives dv/dt  */
              float v[2],  /* state vector v, v[0]=v,
                              v[1}=w */
              float I      /* applied current */
             )
{
  float mv,wv,lambda;
  mv=0.5*(1+tanh((v[0]-v1)/v2));
  wv=0.5*(1+tanh((v[0]-v3)/v4));
  lambda=theta*cosh((v[0]-v3)/(2*v4));
  dv[0]=I-gL*(v[0]-vL)-gK*v[1]*(v[0]-vK)-gCa*mv*(v[0]-vCa);
  dv[1]=lambda*(wv-v[1]);
} /* of 'updateml' */
```

Then the update of the state of all coupled neurons can be performed by the update function. A simple explicit Euler scheme is used to solve the ordinary differential equations:

$$v_i(t + \Delta t) = v_i(t) + \frac{\Delta t}{c} \times f_v(v_i(t), w_i(t)),$$
$$w_i(t + \Delta t) = w_i(t) + \Delta t \times f_w(v_i(t), w_i(t)). \tag{11.6}$$

```
void update(Neuron net_new[], /* updated array of neurons */
            Neuron net[],      /* array of neurons */
            int n,             /* number of neurons */
            float I,           /* applied current */
            float w_in,        /* inhibitory coupling
                                  strength */
            float w_ex,        /* excitatory coupling
                                  strength */
            float h            /* time step */
           )
{
  int i,j,index;
  float sum,dv[2];
  for(i=0;i<n;i++) /* calculate coupling */
  {
    sum=0;
    for(j=0;j<net[i].connect.len;j++)
    {
      index=net[i].connect.index[j];
      if(net[index].inhib)
        /* inhibitory neuron */
        sum-=w_in*(net[index].v[0]-net[i].v[0]);
```

```
    else
      /* excitatory neuron */
      sum+=w_ex*(net[index].v[0]-net[i].v[0]);
  }
  updateml(dv,net[i].v,I);
  /* apply simple Euler scheme */
  net_new[i].v[0]=net[i].v[0]+(h/c)*(dv[0]+sum);
  net_new[i].v[1]=net[i].v[1]+h*dv[1];
} /* of 'update' */
```

## 11.1.2 Parallel Code

The sequential code is limited by the speed and storage capacity of a single workstation. In particular, a large network of neurons uses $O(n^2)$ memory for the connections, quite easily exceeding the storage of a typical workstation. Instead of running the code on a faster computer with more memory, it is usually more cost efficient to run the code in parallel on a network of computers. There are two different parallel computational models: shared memory and distributed memory. In the shared memory model, the processors share the same memory and address space. The memory bandwidth, however, limits the achievable maximum number of processors. In the distributed memory model, each processor has its own local memory. Data exchange is performed via a communication network. This approach allows parallel computers consisting of several thousands of processors.

*Message-passing Paradigm*

The parallelization of the code uses the message-passing paradigm based on the distributed memory model. The message-passing model consists of a set of processes or tasks that only have local memory but are able to communicate with other tasks by sending and receiving messages. The data transfer from the local memory of one task to the local memory of another task requires operations to be performed on both processes. A portable implementation running on different platforms and architectures is provided by the message-passing interface (MPI). An introduction to MPI is given in [5]. The basic functions of MPI are listed in Table 11.1. The datatypes and message-passing functions are declared in the header mpi.h: A simple *hello world* program in MPI looks like:

```
#include <stdio.h>
#include <mpi.h>
int main(int argc,char **argv)
{
  int mytask,ntask;
```

**Table 11.1.** The basic six-function version of MPI

| Routine | Description |
| --- | --- |
| MPI_Init | Initialize MPI |
| MPI_Comm_size | Find out how many tasks there are |
| MPI_Comm_rank | Find out which task I am |
| MPI_Finalize | Finish MPI |
| MPI_Send | Send a message |
| MPI_Recv | Receive a message |

```
MPI_Init(&argc,&argv);
MPI_Comm_size(MPI_COMM_WORLD,&ntask);
MPI_Comm_rank(MPI_COMM_WORLD,&mytask);
printf("Hello! I am task %d out of %d tasks\n",
       mytask,ntask);
MPI_Finalize();
return 0;
}   /* of 'main' */
```

This code can be run in parallel on a arbitrary number of processors. On four processors, it will produce the following output:

```
% mpirun -np hello
Hello! I am task 0 out of 4 tasks
Hello! I am task 2 out of 4 tasks
Hello! I am task 1 out of 4 tasks
Hello! I am task 3 out of 4 tasks
```

The task belonging to a group of $n$ tasks is identified by a number ranging from 0 to $n-1$. The default group containing all tasks is named MPI_COMM_WORLD. There are two types of communication routines. The first class are point to point routines, like MPI_Recv and MPI_Send in order to send to/receive from a specified task (Fig 11.2). The basic send operation in MPI is declared as

```
MPI_Send(address,count,datatype,destination,tag,comm),
```

where

- address, count, datatype describe count occurrences of items of the form datatype starting at address.
- destination is the task identifier of the destination in the group associated with the communicator comm.
- tag is an integer used for message matching.
- comm identifies a group of tasks and a communication context.

The corresponding receive is

```
MPI_Recv(address,maxcount,datatype,source,tag,comm,status)
```

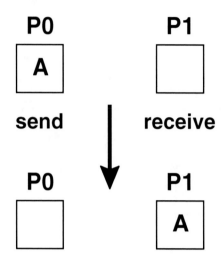

**Fig. 11.2.** Basic message-passing function of sending information from task P0 to task P1

where

- `address`, `maxcount`, `datatype` describe the receive buffer as they do in the case of `MPI_Send`.
- `source` is the task identifier of the source of the message in the group associated with the communicator `comm`.
- `status` holds information about the actual message size, source and tag.

MPI has predefined datatypes of the objects sent to or received from remote tasks. They are listed in Table 11.2. The size of the MPI datatypes can be calculated by a call of the `MPI_Type_extent` routine.

The second class of MPI functions are collective operations that are called by all processors simultaneously. The most important collective routines are summarized in Table 11.3. Their communication patterns are graphically represented in Fig. 11.3. Finally, the C bindings of all MPI routines used are given in Table 11.4.

**Table 11.2.** Subset of basic (predefined) MPI datatypes in C

| MPI Datatype | C Datatype |
|---|---|
| MPI_BYTE | signed char |
| MPI_DOUBLE | double |
| MPI_FLOAT | float |
| MPI_INT | int |

**Table 11.3.** Collective operations of the MPI

| Routine | Description |
|---------|-------------|
| MPI_Bcast | Broadcast data to all tasks |
| MPI_Gather | Gather all data to a single task |
| MPI_Reduce | Reduce data to one task |
| MPI_Alltoall | All to all communication |
| MPI_Alltoallv | All to all communication with variable size |
| MPI_Scatter | Scatter data to all tasks |

*Distributing the Network*

In a parallel application, the neurons have to be distributed evenly on all tasks. The parallel algorithm is described in [6] in detail (cf. Chap. 10). They use basic send/receive routines, while our implementation is based on collective MPI operations. An algorithm based on send/receive must use a complete pairwise exchange algorithm [7] in order to prevent deadlocks.

In a distributed network, the connection list contains entries to remote neurons. Then, the state of the neuron has to be transferred to the remote

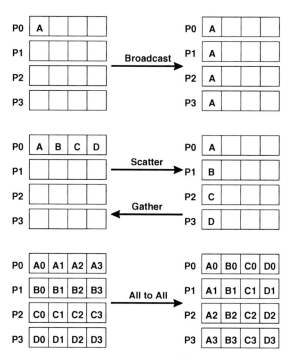

**Fig. 11.3.** Communication patterns for the MPI collective operations MPI_Bcast, MPI_Gather/MPI_Scatter, and MPI_Alltoall on four tasks P0–P3

neuron stored in a different task. The algorithm for setting up the communication structure works in the following way:

(i) For each task, a sorted list of neuron indices to which a connection exists has to be created. Sorting is necessary in order to delete duplicate entries.
(ii) It has to be determined how many neuron states have to be sent to remote neurons of all tasks. This defines the length of the output buffers.
(iii) This information is distributed to all tasks via a `MPI_Alltoall` collective operation call.
(iv) The indices of the neurons that have to be sent to a specific task must be distributed from all tasks to all tasks via a `MPI_Alltoallv` call. The length of the packets has been determined by step 2 and 3.
(v) Two index vectors are built up containing the mapping from the input buffer to the connection lists and the neuron indices to the output buffer.

Then, the exchange of the neuron states can be performed in three steps:

(i) Copy the state of the neurons to be sent to other tasks to the output buffer.
(ii) Distribute the information with a call of `MPI_Alltoallv`. The `MPI_Alltoallv` collective operation transports the output buffers to the corresponding input buffers.
(iii) The input buffer must be mapped to the connection list.

**Table 11.4.** C bindings for the MPI functions used in the parallelization of networks

| |
|---|
| int **MPI_Alltoall**( void *sendbuf,int sendcount,MPI_Datatype sendtype, void *recvbuf,int recvcount,MPI_Datatype recvtype, MPI_Comm comm) |
| int **MPI_Alltoallv**( void *sendbuf,int *sendcounts,int *sdispls, MPI_Datatype sendtype,void *recvbuf,int *recvcounts, int *rdispls,MPI_Datatype recvtype,MPI_Comm comm) |
| int **MPI_Comm_size**(MPI_Comm comm,int *size) |
| int **MPI_Comm_rank**(MPI_Comm comm,int *rank) |
| int **MPI_Finalize**() |
| int **MPI_Gather**( void *sendbuf,int sendcount,MPI_Datatype sendtype, void *recvbuf,int recvcount,MPI_Datatype recvtype,int root, MPI_Comm comm) |
| int **MPI_Init**(int *argc,char ***argv) |
| int **MPI_Reduce**( void *sendbuf,void *recvbuf,int count,MPI_Op op,int root, MPI_Comm comm) |
| int **MPI_Recv**( void *buf,int count,MPI_Datatype datatype,int source,int tag, MPI_Comm comm) |
| int **MPI_Send**( void *buf,int count,MPI_Datatype datatype,int source,int tag, MPI_Comm comm,MPI_Status *status) |
| int **MPI_Type_extent**(MPI_datatype datatype,MPI_Aint *extent) |

The datatype of a distributed network `Pnet` can be defined by the following structure:

```
typedef struct {
  int n;   /* total number of neurons */
  int lo; /* lower bound of subarray */
  int hi; /* upper bound of subarray */
  int ntask; /* number of tasks */
  int taskid;  /* my task identifier */
  int outsize; /* size of output buffer */
  int insize; /* size of input buffer */
  int *outdisp; /* displacement vector  for output */
  int *indisp; /* displacement vector for input */
  int *inlen,*outlen; /* vector length for input/output */
  MPI_Datatype datatype; /* datatype  of input/output
                                 buffer */
  void *outbuffer,*inbuffer; /* input/output buffer of
                                 /* generic type void */
  int *outindex; /* index vector of output */
  List *connect;  /* list of connections */
} Pnet;
```

The topology of the network is now part of this datatype and not part of `neuron`. The basic functions for the datatype `Pnet` are:

```
/* Initialization of datatype */
extern Pnet *pnet_init(MPI_Datatype,int);
/* Random network setup */
extern void pnet_random(Pnet *,float);
/* Creating communication structure */
extern void pnet_setup(Pnet *);
/* Exchange information */
extern void pnet_exchg(Pnet *);
/* Macros for convenience */
/* iterator of subarray */
#define pnet_foreach(pnet,i) for(i=(pnet)->lo;
        i<=(pnet)->hi;i++)
/* allocating an array ar[lo:hi] of datatype type */
#define newvec(type,lo,hi) \
    (type *)malloc(sizeof(type)*(hi-(lo)+1)-(lo))
#define freevec(ptr,lo) free(ptr+(lo))
```

The datatype is initialized by a call to `pnet_init`. The function has two arguments. The first argument defines the datatype of the data to be distributed between the different tasks. The second argument defines the total number of neurons. In the first part of the function, the necessary arrays are allocated, the number of tasks and the task identifier are determined. In the next part, the lower and upper bounds of the neuron array are calculated. In particular, it must be considered that the total number of neurons cannot be divided by the total number of tasks. Finally, an array for the neuron connections is allocated and initialized:

```
Pnet *pnet_init(MPI_Datatype datatype, /* MPI datatype */
                int n /* total number of neurons */
                )       /* returns allocated struct */
{
  int slice,rem,i;
  Pnet *pnet;
  pnet=(Pnet *)malloc(sizeof(Pnet));
  pnet->n=n;
  pnet->datatype=datatype;
  MPI_Comm_size(MPI_COMM_WORLD,&pnet->ntask);
  MPI_Comm_rank(MPI_COMM_WORLD,&pnet->taskid);
  /* calculate lower and upper bound of subarray */
  slice=pnet->n/pnet->ntask;
  pnet->lo=pnet->taskid*slice;
  pnet->hi=(pnet->taskid+1)*slice-1;
  rem=pnet->n % pnet->ntask;
  /* distribute the remainder evenly on all tasks */
  if(pnet->taskid<rem)
  {
    pnet->lo+=pnet->taskid;
    pnet->hi+=pnet->taskid+1;
  }
  else
  {
    pnet->lo+=rem;
    pnet->hi+=rem;
  }
  /* allocate arrays */
  pnet->outdisp=(int *)malloc(sizeof(int)*pnet->ntask);
  pnet->indisp=(int *)malloc(sizeof(int)*pnet->ntask);
  pnet->outlen=(int *)malloc(sizeof(int)*pnet->ntask);
  pnet->inlen=(int *)malloc(sizeof(int)*pnet->ntask);
  pnet->connect=newvec(List,pnet->lo,pnet->hi);
  pnet_foreach(pnet,i)
    initlist(pnet->connect+i);
  return pnet;
} /* of 'pnet_init' */
```

A random network is set up by calling pnet_random. The function uses the macro pnet_foreach, ensuring that only the local subarray of each task is accessed.

```
void pnet_random(Pnet *pnet,
                 float p_conn /* connection probability */
                 )
{
  int i,j;
  pnet_foreach(pnet,i)
    for(j=0;j<pnet->n;j++)
      if(i!=j && drand48()<p_conn)
        addlistitem(pnet->connect+i,j);
} /* of 'pnet_random' */
```

The setup of the necessary communication patterns between the different tasks is performed by the pnet_setup function. Each task has to know the upper and lower bounds of the subarrays of all other tasks. This information is stored in the arrays lo and hi:

```
void pnet_setup(Pnet *pnet)
{
  int *lo,*hi;
  int i,j,k,*in,size,slice,rem,task;
  slice=pnet->n/pnet->ntask;
  rem=pnet->n % pnet->ntask;
  lo=newvec(int,0,pnet->ntask-1);
  hi=newvec(int,0,pnet->ntask-1);
  for(i=0;i<pnet->ntask;i++)
  /* calculate boundaries of all tasks for n mod ntask<>0 */
  {
    lo[i]=i*slice;
    hi[i]=(i+1)*slice-1;
    if(i<rem)
    {
      lo[i]+=i;
      hi[i]+=i+1;
    }
    else
    {
      lo[i]+=rem;
      hi[i]+=rem;
    }
  }
```

Then, the total number of connections and their indices are calculated. The array in stores the neuron indices of all connections (Fig. 11.4).

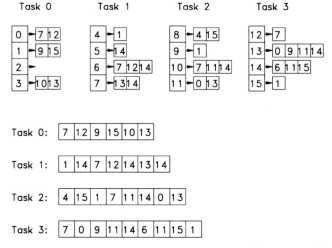

**Fig. 11.4.** Network of 16 neurons distributed to 4 tasks (*upper figure*). These are the contents of `pnet->connect` before the call of `pnet_setup`. The lower figure shows the combined lists of network connections for each task 0–3

```
for(i=0;i<pnet->ntask;i++)
  pnet->outlen[i]=pnet->inlen[i]=0;
/* calculating total length of connection list */
size=0;
pnet_foreach(pnet,i)
  size+=pnet->connect[i].len;
in=(int *)malloc(sizeof(int)*size);
k=0; /* concatenating connection lists */
pnet_foreach(pnet,i)
  for(j=0;j<pnet->connect[i].len;j++)
    in[k++]=pnet->connect[i].index[j];
```

This array is sorted and duplicated entries are deleted (see Fig. 11.5).

```
/* sort connection list */
qsort(in,size,sizeof(int),
      (int (*)(const void *,const void *))compare);
pnet->insize=1; /* delete duplicate entries */
for(i=1;i<size;i++)
  if(in[i]!=in[i-1]) /* same indices? */
  {
    /* no, increase insize by one */
    in[pnet->insize]=in[i];
    pnet->insize++;
  }
```

1 2 3
Task 0: 7 9 10 12 13 15
0 1 3
Task 1: 1 7 12 13 14
0 1 2 3
Task 2: 0 1 4 7 11 13 14 15
0 1 2 3
Task 3: 0 1 6 7 9 11 14 15

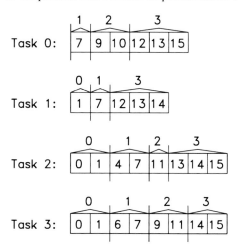

**Fig. 11.5.** List of network connections for each task after sorting and deleting duplicate entries. The numbers above the lists are the corresponding task identifiers to which the information must be sent

After this step, the length of the vector sent to all tasks has to be calculated. This can be performed by counting all neuron indices inside the lower and upper bound for each task (Fig. 11.6).

```
task=0;
pnet->inlen[task]=0;
/* calculating inlen vector */
for(i=0;i<pnet->insize;)
  if(in[i]<=hi[task]) /* inside the boundaries of task? */
  {
    /* yes, increment inlen by one */
    pnet->inlen[task]++;
    i++;
  }
```

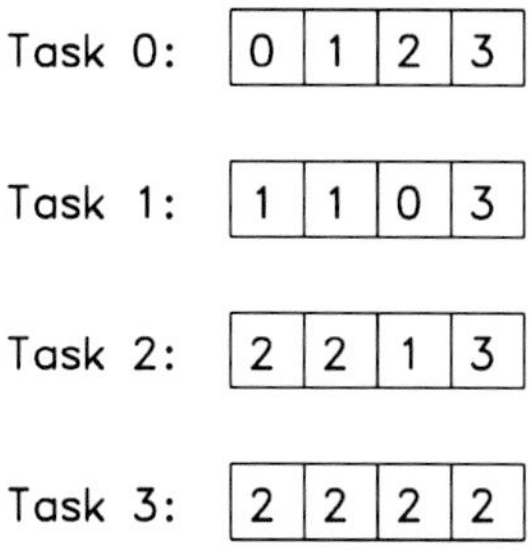

Task 0: 0 1 2 3
Task 1: 1 1 0 3
Task 2: 2 2 1 3
Task 3: 2 2 2 2

**Fig. 11.6.** The length of the communication packets needed by `MPI_Alltoallv`. The information has to be distributed by `MPI_Alltoall` to all tasks

```
else
{
    /* no, goto next task, set new inlen to zero */
    task++;
    pnet->inlen[task]=0;
}
```

It is important to map the connection from the input buffer to the connection list of the local neurons. This information is stored again in the array of lists pnet->connect, because the original connection lists are not needed any more (Fig. 11.7). The current implementation uses a linear search algorithm to find the indices in the connection list leading to a runtime characteristic of $O(n^2)$. By using a better algorithm (e.g. binary search), the runtime can be significantly reduced. The setup function, however, is called only once during initialization of the network.

```
/* calculating mapping from input buffer to connection */
pnet_foreach(pnet,i)
{
    for(j=0;j<pnet->connect[i].len;j++)
        /* search for index in array in */
        for(k=0;k<pnet->insize;k++)
            if(in[k]==pnet->connect[i].index[j])
            {
                /* index found and stop searching */
                pnet->connect[i].index[i]=k;
                break;
            }
} /* of pnet_foreach */
```

The information of the lengths of the communication packets to be received from other tasks has to be distributed by a call to MPI_Alltoall. Then, the pnet->outlen array contains the length of the outgoing packets. This information is needed by the MPI_Alltoallv function. MPI_Alltoallv sends a distinct message from each task to every task, where the messages can

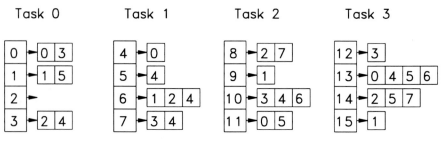

**Fig. 11.7.** Mapping of the input buffer to the connection list. These are the contents of pnet->connect after the call of pnet_setup

have different sizes and displacements. The displacement is the offset from the first element of the array to the first element of the message and can be simply calculated by summing up the `pnet->inlen` array. After calling `MPI_Alltoallv`, the array `pnet->outindex` contains the indices of neurons that must be sent to other tasks (see Fig. 11.8).

```
MPI_Alltoall(pnet->inlen,1,MPI_INT,pnet->outlen,1,
             MPI_INT,MPI_COMM_WORLD);
/* calculating displacements needed by MPI_Alltoallv */
pnet->indisp[0]=pnet->outdisp[0]=0;
for(k=1;k<pnet->ntask;k++)
{
   pnet->indisp[k]=pnet->inlen[k-1]+pnet->indisp[k-1];
   pnet->outdisp[k]=pnet->outlen[k-1]+pnet->outdisp[k-1];
}
pnet->outsize=0;
for(i=0;i<pnet->ntask;i++)
   pnet->outsize+=pnet->outlen[i];
/* allocating outindex */
pnet->outindex=(int *)malloc(sizeof(int)*pnet->outsize);
/* information is moved from in to pnet->outindex calling
   the collective operation MPI_Alltoallv */
MPI_Alltoallv(in,pnet->inlen,pnet->indisp,MPI_INT,
              pnet->outindex,pnet->outlen,
              pnet->outdisp,MPI_INT,MPI_COMM_WORLD);
```

Finally, we allocate the input and output buffers of type stored in `pnet->datatype`. The function `MPI_Type_extent` returns the number of bytes for this datatype. These buffers of generic type `void` have to be cast by the user to the appropriate type.

```
/* allocating input and output buffer */
MPI_Type_extent(pnet->datatype,&size);
pnet->outbuffer=malloc(pnet->outsize*size);
```

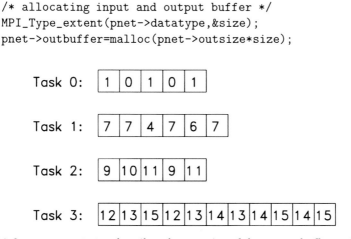

**Fig. 11.8.** `pnet->outindex` describes the mapping of the output buffer to the neuron indices

```
    pnet->inbuffer=malloc(pnet->insize*size);
    /* free auxiliary storage */
    free(in);
    free(lo);
    free(hi);
} /* of 'pnet_setup' */
```

The compare function is needed for the quicksort C library function qsort. It sorts integer values in ascending order:

```
static int compare(const int *a,const int *b)
{
    return *a-*b;
} /* of 'compare' */
```

Then, the exchange is done by the pnet_exchg function calling only the MPI_Alltoallv function. Information is copied from the output buffers to the input buffers.

```
void pnet_exchg(Pnet *pnet)
{
    MPI_Alltoallv(pnet->outbuffer,
                  pnet->outlen,pnet->outdisp,
                  pnet->datatype,pnet->inbuffer,
                  pnet->inlen,pnet->indisp,
                  pnet->datatype,MPI_COMM_WORLD);
} /* of 'pnet_exchg' */
```

The parallel initialization of the network is performed by init:

```
Neuron *init(Pnet **pnet,
             int n, /* number of neurons */
             float p_conn /* connection probability */
             ) /* returns allocated array of neurons */
{
    Neuron *net;
    int i;
    *pnet=pnet_init(MPI_FLOAT,n);
    pnet_random(*pnet,p_conn);
    pnet_setup(*pnet);
    net=newvec(Neuron,(*pnet)->lo,(*pnet)->hi);
    /* random initial state of ML neurons */
    pnet_foreach(*pnet,i)
    {
        net[i].v[0]=-0.02*0.01*drand48();
        net[i].v[1]=0.05+0.20*drand48();
    }
    return net;
} /* of 'init' */
```

Then, the parallel update for diffusive coupling can be done in the following way. For simplicity, only excitatory coupling is used. First, the state of the neuron has to be copied to the output buffers. After calling `pnet_exchg`, the information is moved to the input buffer. The list `pnet->connect` provides the information about the mapping of the input buffer to the connections.

```
void update(Pnet *pnet,
            Neuron net[], /* subarray of neurons */
            float I,      /* applied current */
            float w,      /* coupling strength */
            float h       /* time step */
            )
{
  int i;
  float sum,*buffer,dv[2];
  /* cast outbuffer to float pointer */
  buffer=(float *)pnet->outbuffer;
  /* copy state to output buffer */
  for(i=0;i<pnet->outsize;i++)
    buffer[i]=net[pnet->outindex[i]].v[0];
  /* Exchange of necessary information to all tasks */
  pnet_exchg(pnet);
  /* cast inbuffer to float pointer */
  buffer=(float *)pnet->inbuffer;
  pnet_foreach(pnet,i)
  {
    sum=0;
    for(j=0;j<pnet->connect[i].len;j++)
      sum+=buffer[pnet->connect[i].index[j]]-net[i].v[0];
    updateml(dv,net[i].v,I);
    /* performing Euler step */
    net[i].v[0]+=(h/c)*(dv[0]+w*sum);
    net[i].v[1]+=h*dv[1];
  }
} /* of 'update' */
```

*Efficiency of Parallelization*

Communication is necessary after every time step in the case of diffusive coupling. This limits the efficiency of the parallel code, in particular for large networks with dense couplings. Efficiency $E$ is defined by

$$E(p) := \frac{T(1)}{p \times T(p)}, \tag{11.7}$$

where $T(p)$ denotes the computation time running on $p$ parallel tasks. The computation time can be divided into two parts:

$$T = T_{\text{calc}} + T_{\text{comm}}, \tag{11.8}$$

where $T_{\text{calc}}$ denotes the computation part and $T_{\text{comm}}$ the communication part. In the case of a fully coupled network of $n$ neurons, $T_{\text{comm}}$ scales as $O(n^2)$, while the computation part scales as $O(n^2/p)$:

$$T(p) \sim \frac{n^2}{p} T'_{\text{calc}} + n^2 T'_{\text{comm}}, \tag{11.9}$$

Thus,

$$E(p) = \frac{T'_{\text{calc}}}{T'_{\text{calc}} + p T'_{\text{comm}}} \tag{11.10}$$

resembling Amdahl's law [8]. For large $p$, the total runtime is dominated by the communication time and the efficiency tends to zero.

## 11.2 Non-diffusive Coupling

The coupling of neurons can also be done in a different way: If the integrated postsynaptic potentials (PSP) reach a certain threshold, a spike train is sent to remote neurons with a delay time $t_{\text{delay}}$. The PSP evoked by one single spike can be parameterized by a gain function $g(t)$ with a rise time $\tau_1$ and decay time $\tau_2$ (cf. Chap. 1):

$$g(t) = \frac{\exp(-t/\tau_1) - \exp(-t/\tau_2)}{\tau_1 - \tau_2} \tag{11.11}$$

For $t > t_{\max} \approx 15\text{ms}$ and the chosen parameter $\tau_1 = 1$ ms and $\tau_2 = 2$ ms, the PSP gain function $g$ is nearly zero (Fig. 11.9).

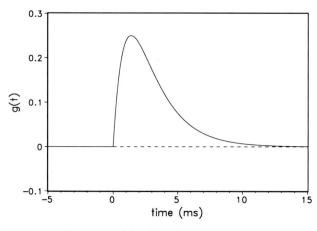

**Fig. 11.9.** PSP gain function $g(t)$ with rise time $\tau_1 = 1$ ms and decay time $\tau_2 = 2$ ms

A spike is generated if $v$ exceeds the critical threshold of zero at time $t_{\text{spike}}$. The coupled model of $n$ neurons is defined by

$$c\frac{\mathrm{d}v_i}{\mathrm{d}t} = f_v(v_i, w_i) + \sum_{j=1}^{n}\sum_{k=1}^{m_j} w_{\text{in,ex}}g(t - t_{\text{spike},j,k} - t_{\text{delay},i,j})(v_i - v_{\text{in,ex}}),$$

$$\frac{\mathrm{d}w_i}{\mathrm{d}t} = f_w(v_i, w_i) + \sigma\xi, \qquad\qquad (11.12)$$

$$i = 1, \ldots, n,$$

where $t_{\text{spike},j,k}, k = 1, \ldots, m_j$ are the $m_j$ spike times of neuron $j$, $w_{\text{in,ex}}, v_{\text{in,ex}}$ are weights for inhibitory/excitatory coupling, and $\xi$ is additional Gaussian white noise with amplitude $\sigma$. The numerical discretization using an Euler scheme for a stochastic differential equation is:

$$v_i(t + \Delta t) = v_i(t) + \frac{\Delta t}{c} \times f_v(v_i(t), w_i(t)),$$

$$w_i(t + \Delta t) = w_i(t) + \Delta t \times f_w(v_i(t), w_i(t)) + \sqrt{\Delta t} \times \xi. \qquad (11.13)$$

### 11.2.1 Sequential Version

In order to speed up the evaluation of the gain function $g(t)$, a lookup table is used created by the function getg:

```
#define tau_1 1.0 /* rise time (ms) */
#define tau_2 2.0 /* decay time (ms) */

float *getg(int tmax, /* size of lookup table */
            float h   /* time step (ms) */
            )         /* returns calculated lookup table */
{
  float *g,gmax;
  int t;
  g=(float *)malloc(sizeof(float)*tmax);
  gmax=0;
  for(t=0;t<tmax;t++)
  {
    g[t]=1/(tau_1-tau_2)*(exp(-t*h/tau_1)-exp(-t*h/tau_2));
    if(g[t]>gmax)
      gmax=g[t];
  }
  /* normalize g */
  for(t=0;t<tmax;t++)
    g[t]/=gmax;
  return g;
} /* of 'getg' */
```

An efficient implementation of such coupling uses a finite event buffer where the time of spike events are stored (Chap. 10). It is assumed that the maximum number of overlapping spikes is limited by the size of the buffer. The datatype Rbuffer can be defined by using an array that stores a limited number of events. After the maximum number of spikes is reached, the oldest event is overwritten. This can be implemented by a ring topology using the modulo operator. A graphical representation of datatype Rbuffer is shown in Fig. 11.10.

```
typedef struct
{
   int len;   /* length of ring buffer */
   int top;   /* index of first element in ring buffer */
   int size;  /* maximum length of ring buffer */
   int *events; /* stores up to size latest events */
} Rbuffer;
void initrbuffer(Rbuffer *,int);
void addrbuffer(Rbuffer *,int);
```

A macro is defined in order to access an event at position pos from the top of the ring buffer:

```
#define getrbuffer(rbuf,pos) \
   (rbuf)->events[((rbuf)->top-1-(pos)+(rbuf)->size) % (rbuf)
   ->size]
```

The datatype is initialized by a call to initrbuffer. The parameter size sets the maximum number of events that can be stored.

```
void initrbuffer(Rbuffer *rbuf,int size)
{
   rbuf->len=rbuf->top=0;
   rbuf->events=(int *)malloc(sizeof(int)*size);
   rbuf->size=size;
} /* of 'initrbuffer' */
```

The implementation for adding an event to the ring buffer is straightforward. The modulo operator is used after incrementing the index to the first element in the buffer top defining a ring topology.

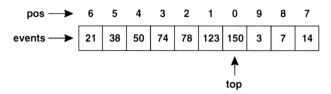

**Fig. 11.10.** Graphical representation of datatype Rbuffer with size equal to 10

```
void addrbuffer(Rbuffer *rbuf,int event)
{
  /* have we reached the maximum number of events? */
  if(rbuf->len<rbuf->size)
    rbuf->len++; /* no, increase len by one */
  rbuf->events[rbuf->top]=event;
  /* increase top by one, then perform  modulo operator */
  rbuf->top=(rbuf->top+1) % rbuf->size;
} /* of 'addrbuffer' */
```

For the ML model, the datatype Neuron has to be modified:

```
typedef struct
{
  float v[2];    /* state of neuron (ML) */
  Rbuffer spikes;  /* times of last spikes */
  List connect;  /* list of connections */
  int inhib; /* inhibitory (TRUE) or excitatory (FALSE) */
} Neuron;
```

The neurons are initialized and the network is established by the function init:

```
#define RBUFF_LEN 10 /* maximum length of ring buffer */

void init(Neuron net[], /* array of neurons */
          int n,        /* total number of neurons */
          float p_conn, /* probability of establishing
                            connection */
          float p_inhib /* probability of a inhibitory
                            coupling */
         )
{
  int i,j;
  /* Setup of random network */
  randomnet(net,n,p_conn,p_inhib);
  for(i=0;i<n;i++)
  {
    /* Set initial conditions of neurons */
    net[i].v[0]=-0.02*0.01*drand48();
    net[i].v[1]=0.05+0.20*drand48();
    /* Initializing event ring buffer */
    initrbuffer(&net[i].spikes,RBUFF_LEN);
  }
} /* of 'init' */
```

The update function uses the datatype Rbuffer and List. The Gaussian white noise is generated by a call of the gasdev function as part of the numerical recipes library [9]. The function has been slightly modified and uses the `drand48` random number generator.

```
/* constants for ML model */
#define    c 1.0
/* coupling constants */
#define V_in (-0.55)
#define V_ex 0.05

void update(FILE *file, /* output file for spikes */
            Neuron net[], /* array of neurons */
            int n, /* number of neurons */
            int time, /* integer time */
            float h, /* time step */
            float *g, /* lookup table */
            int t_max, /* size of lookup table */
            float w_in,float w_ex,int delay,
            float I, /* applied current */
            float sigma /* amplitude of Gaussian white
                            noise */
            )
{
  int i,j,k,index;
  float sum,v_new[2],dv[2];
  int spike;
  for(i=0;i<n;i++)
  {
    sum=0; /* calculate interactions */
    for(j=0;j<net[i].connect.size;j++)
    {
      index=net[i].connect.index[j];
      /* iterating over the ring buffer of neuron index */
      for(k=0;k<net[index].spikes.len;k++)
      {
        spike=getrbuffer(&net[index].spikes,k)+delay;
        /* testing whether spike is active */
        if(time>=spike && time<spike+t_max)
        {
          /* yes */
          if(net[index].inhib)
            /* inhibitory coupling */
            sum-=w_in*g[time-spike]*(net[i].v[0]-V_in);
          else
            /* excitatory coupling */
            sum-=w_ex*g[time-spike]*(net[i].v[0]-V_ex);
```

```
      }
      else
        /* no, break if all spikes are not active
           anymore */
        if(spike+t_max<time)
          break;
    } /* of for(k=...) */
  } /* of for(j=...) */
  /* update of ML */
  updateml(dv,net[i].v,I);
  v_new[0]=net[i].v[0]+(h/c)*(dv[0]+sum);
  v_new[1]=net[i].v[1]+h*dv[1]+sqrt(h)*sigma*gasdev();
  /* critical threshold of zero reached for v? */
  if(net[i].v[0]<0 && v_new[0]>0)
  {
    /* yes, we have a spike event, add to my ring buffer */
    addrbuffer(&net[i].spikes,time);
    fprintf(file,"%g %d\n",time*h,i);
  }
  net[i].v[0]=v_new[0];
  net[i].v[1]=v_new[1];
} /* of for(i=...) */
} /* of 'update' */
```

The mean value of $v_i$ over the $n$ neurons as a diagnostic variable, $\bar{v} = \frac{1}{n}\sum_{i=1}^{n} v_i$, can be calculated using the **mean** function:

```
float vmean(const Neuron net[],int n)
{
  int i;
  float sum;
  sum=0;
  for(i=0;i<n;i++)
    sum+=net[i].v[0];
  return sum/n;
} /* of 'vmean' */
```

The main program calculating the dynamics of the network is as follows:

```
#include "list.h"
#include "neuron.h"
#define n 100 /* number of neurons */
#define I 0.1 /* input current */
#define sigma 0.0 /* amplitude of Gaussian white noise */
#define t_max 20.0 /* duration of spike (ms) */
#define h 0.01  /* time step  (ms) */
#define p_conn 1.0 /* fully coupled network */
```

```
#define p_inhib 0.0 /* no inhibitory coupling */
#define t_end 100.0 /* simulation time (ms) */
#define delay 0 /* no delay */

int main(int argc,char **argv)
{
  Neuron net[n];
  float *g,g1,w_in,w_ex;
  int t,nstep,ostep,tmax;
  FILE *file,*log;
  init(net,n,p_conn,p_inhib);
  w_in=0.1/((n-1)*p);
  w_ex=0.1/((n-1)*p);
  nstep=t_end/h;
  ostep=nstep/100;
  tmax=t_max/h;
  g=getg(tmax,h); /* creating lookup table for g */
  file=fopen("neuron.spike","wb");
  log=fopen("neuron.mean","w");
  for(t=0;t<nstep;t++) /* time loop */
  {
    if(t % ostep==0)
      /* write to output file every ostep time steps */
      fprintf(log,"%g %g\n",t*h,vmean(net,n));
    update(file,net,n,t,h,g,t_max,w_in,w_ex,delay,I,sigma);
  } /* of   time loop */
  fclose(file);
  fclose(log);
  return 0;
} /* of 'main' */
```

### 11.2.2 Parallel Version

The parallelization is based on Pnet. The data to be exchanged now have the type MPI_INT. In order to incorporate the ring buffer storing the spike events, a new datatype Rnet has to defined. inhib is not part of datatype neuron, because information about remote neurons is also needed.

```
typedef struct
{
  Pnet *pnet;    /* parallel network datatype */
  Rbuffer *rbuffer; /* event ring buffer */
  int *inhib;    /* inhibitory list */
} Rnet;
```

The datatype `Neuron` now contains only the last spike event of the neuron itself:

```
typedef struct
{
   float v[2];   /* state of neuron (ML) */
   int spike;    /* time of last spike */
} Neuron;
```

Both the initialization of the neuron and the parallel communication pattern are organized by the function `init`. The information about whether a remote neuron is inhibitory or not is distributed by a call to `pnet_exchg`. The output buffer contains boolean values.

```
#define NOFIRE -1
Neuron *init(Rnet *rn,
                int n, /* total number of neurons */
                float p_conn, /* probability of establishing
                                   connection */
                float p_inhib /* probability of a inhibitory
                                   neuron */
                ) /* returns allocated subarray of neurons */
{
   Neuron *net;
   int i,*buffer;
   int *inhib;
   rn->pnet=pnet_init(MPI_INT,n);
   /* setup of random network */
   pnet_random(rn->pnet,p_conn);
   /* allocate subarray of neurons and temp. inhibitory
      array*/
   net=newvec(Neuron,rn->pnet->lo,rn->pnet->hi);
   inhib=newvec(int,rn->pnet->lo,rn->pnet->hi);
   pnet_foreach(rn->pnet,i)
   {
      net[i].spike=NOFIRE;
      /* Set initial condition of neuron */
      net[i].v[0]=-0.02*0.01*drand48();
      net[i].v[1]=0.05+0.20*drand48();
      inhib[i]=(drand48()<p_inhib);
   }
   pnet_setup(rn->pnet);
   /* initialization of ring buffer  */
   rn->rbuffer=(Rbuffer *)malloc(sizeof(Rbuffer)*rn->pnet->
       insize);
   for(i=0;i<rn->pnet->insize;i++)
      initrbuffer(rn->rbuffer+i,RBUFF_LEN);
```

```
    /* allocating information about inhibitory neurons */
    rn->inhib=(int *)malloc(sizeof(int)*rn->pnet->insize);
    /* mapping inhibitory array to output buffer; */
    buffer=(int *)rn->pnet->outbuffer;
    for(i=0;i<rn->pnet->outsize;i++)
      buffer[i]=inhib[rn->pnet->outindex[i]];
  /* distributing inhibitory vector */
    pnet_exchg(rn->pnet);
    buffer=(int *)rn->pnet->inbuffer;
    for(i=0;i<rn->pnet->insize;i++)
      rn->inhib[i]=buffer[i];
    /* free temporary storage */
    freevec(inhib,rn->pnet->lo);
    return net;
  } /* of 'init' */
```

In order to write out the timing and the corresponding index of the firing neuron in a sequential way, a datatype Slist has to be defined:

```
typedef struct
{
  int time,neuron;
} Spike;
typedef struct
{
  Spike *index;
  int len;  /* length of list */
} Slist;

/* Declaration of functions */

/* Initialize empty list */
extern void initspikelist(Slist *);
/* Add spike to list */
extern int  addspikelistitem(Slist *,Spike);
/* Empty list */
extern void emptyspikelist(Slist *);
```

Implementation of Slist is identical to List. If neuron fires, then the index together with the timing of the event is stored in the Spike structure and added to the spike list. The parallel update function can be written as:

```
void update(Slist *spikelist, /* Spike list */
            Rnet *rn,
            Neuron net[], /* array of neurons */
            int time, /* integer time */
            float h,  /* time step (ms) */
```

```
                  float *g, /* lookup table */
                  int t_max, /* size of lookup table */
                  float w_in,float w_ex,int delay,
                  float I, /* applied current */
                  float sigma /* amplitude of Gaussian white
                                  noise */
                )
{
  int i,j,k,index;
  float sum,v_new[2],dv[2];
  int spike;
  Spike event;
  pnet_foreach(rn->pnet,i)
  {
    sum=0; /* calculate interactions */
    for(j=0;j<rn->pnet->connect[i].len;j++)
    {
      index=rn->pnet->connect[i].index[i];
      for(k=0;k<rn->rbuffer[index].len;k++)
      {
        spike=getrbuffer(rn->rbuffer+index,k)+delay;
        if(time>=spike && time<spike+t_max)
        {
          if(rn->inhib[index])
            sum-=w_in*g[time-spike]*(net[i].v[0]-V_in);
          else
            sum-=w_ex*g[time-spike]*(net[i].v[0]-V_ex);
        }
        else if(spike+t_max<time)
          break;

      }
    } /* of for(j=..) */
    updateml(dv,net[i].v,I);
    v_new[0]=net[i].v[0]+(h/c)*(dv[0]+sum);
    v_new[1]=net[i].v[1]+h*dv[1]+sqrt(h)*sigma*gasdev();
    /* critical threshold of zero reached for v? */
    if(net[i].v[0]<0 && v_new[0]>0)
    {
      /* yes, store time */
      net[i].spike=time;
      event.time=time;
      event.neuron=i;
      addspikelist(spikelist,event); /* add spike to list */
    }
```

```
      net[i].v[0]=v_new[0];
      net[i].v[1]=v_new[1];
    } /* of for(i=..) */
  } /* of 'update' */
```

The exchange of spike timings is performed by a call of the exchg function:

```
void exch(Rnet *rn,Neuron net[])
{
  int i,*buffer;
  /* write time of last spike to output buffer */
  buffer=(int *)rn->pnet->outbuffer;
  for(i=0;i<rn->pnet->outsize;i++)
    buffer[i]=net[rn->pnet->outindex[i]].spike;
  pnet_exchg(rn->pnet); /* Communication */
  /* add times of last spike to the corresponding ring
     buffer */
  buffer=(int *)rn->pnet->inbuffer;
  for(i=0;i<rn->pnet->insize;i++)
    if(buffer[i]!=NOFIRE) /* spike occured */
      if(rn->rbuffer[i].len==0 ||
         rn->pnet->inbuffer[i]!=getrbuffer(rn->rbuffer+i,0))
        /* we have a new spike */
        addrbuffer(rn->rbuffer+i,buffer[i]);
} /* of 'exch' */
```

The exchange is only necessary after time $t_{delay}$. Therefore, communication is significantly reduced in comparison to diffusive coupled neurons.

Each task contains a list of all spike events occured stored locally in the Slist datatype. The serialized output of the spike events collected from all tasks is achieved by the fwritespikes function. All tasks initially send the number of events recorded via the collective MPI_Gather operation to task zero. Then the content of the spike list is sent via MPI_Send to task zero.

```
#define MSG_TIME 99 /* message tag used by send/recv */

void fwritespikes(FILE *file,Pnet *pnet,Slist *list)
{
  int len,i,*list_len;
  Spike *vec;
  MPI_Status status; /* needed by MPI_Recv */
  len=list->len;
  list_len=newvec(int,0,pnet->ntask-1);
  /* Gather number of spikes from all tasks */
  MPI_Gather(&len,1,MPI_INT,list_len,1,MPI_INT,
             0,MPI_COMM_WORLD);
  if(pnet->taskid==0)
```

```
{
  /* write spike events of task 0 */
  fwrite(list->index,sizeof(Spike),len,file);
  /* collect spike events from all other tasks */
  for(i=1;i<pnet->ntask;i++)
    if(list_len[i]>0) /* spike occured in task i? */
    {
      /* yes, allocate temporal storage */
      vec=newvec(Spike,0,list_len[i]-1);
      /* receive spike list from task i */
      MPI_Recv(vec,sizeof(Spike)*list_len[i],
               MPI_BYTE,i,MSG_TIME,
               MPI_COMM_WORLD,&status);
      fwrite(vec,sizeof(Spike),list_len[i],file);
      free(vec);
    }
}
else if (len>0) /* spike occured in my task */
  /* send to task zero */
  MPI_Send(list->index,sizeof(Spike)*len,MPI_BYTE,
           0,MSG_TIME,MPI_COMM_WORLD);
free(list_len);
} /* of 'fwritespikes' */
```

The function for calculating the mean value $\bar{v}$ in parallel uses the global reduction function MPI_Reduce. The reduction function MPI_SUM adds the values of all tasks. The function returns in task zero the global sum:

```
float vmean(Rnet *rn,
            const Neuron net[] /* subarray of neurons */
            ) /* returns mean value of v on task zero */
{
  int i;
  float sum,globalsum;
  sum=0;
  pnet_foreach(rn->pnet,i)
    sum+=net[i].v[0];
  /* global reduction of sum o globlasum on task zero
     using add operator */
  MPI_Reduce(&sum,&globalsum,1,MPI_FLOAT,MPI_SUM,
             0,MPI_COMM_WORLD);
  return globalsum/rn->pnet->n;
} /* of 'vmean' */
```

The main program of the parallel version is:

```
#include <stdlib.h>
```

```c
#include <stdio.h>
#include <mpi.h> /* MPI prototypes */
#include "list,h" /* list datatype */
#include "pnet.h"
#include "rnet.h"
#include "neuron.h"
int main(int argc,char **argv)
{
  Neuron *net;
  float *g,g1,w_in,w_ex,v;
  int t,nstep,ostep,tmax;
  FILE *file,*log;
  Rnet rnet;
  Slist spikelist;
  MPI_Init(&argc,&argv); /* initialize MPI */
  net=init(&rnet,n,p_conn,p_inhib);
  /* set random seeds differently for each task */
  srand48(22892+38*rnet.pnet->taskid);
  /* setting inhibitory and excitatory coupling strength */
  w_in=0.1/((n-1)*p_conn);
  w_ex=0.1/((n-1)*p_conn);
  nstep=t_end/h;
  ostep=nstep/100;
  tmax=t_max/h;
  g=getg(tmax,h);
  if(rnet.pnet->taskid==0)
  {
    /* opening output files on task 0 */
    file=fopen("neuron.spike","wb");
    log=fopen("neuron.mean","w");
  }
  initspikelist(&spikelist);
  for(t=0;t<nstep;t++) /* time loop */
  {
    if(t % ostep==0)
    {
      /* write mean value of v to output file every
         ostep time steps */
      v=vmean(&rnet,net);
      if(rnet.pnet->taskid==0)
        fprintf(log,"%g %g\n",t*h,v);
    }
    update(&spikelist,&rnet,net,t,h,g,t_max,
           w_in,w_ex,delay,I,sigma);
    if(delay==0 || t % (delay-1)==0)
```

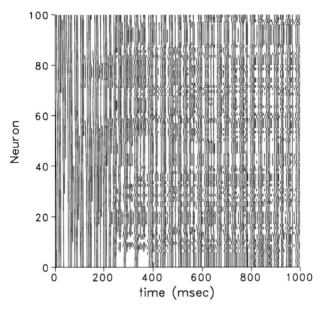

**Fig. 11.11.** Spike pattern for a network of 100 neurons derived from file `neuron.spike`

```
      { /* exchange necessary every delay time steps */
        exch(&rnet,net);
        /* write out spike timings */
        fwritespikes(file,rnet.pnet,&spikelist);
        emptyspikelist(&spikelist);
      }
    } /* of time loop */
    if(rnet.pnet->taskid==0)
    {
      fclose(file);
      fclose(log);
    }
    MPI_Finalize(); /* end MPI */
    return 0;
  } /* of 'main' */
```

Sample output of the spiking times of a ML network is shown in Fig. 11.11.

## 11.3 Connection Dependent Coupling Strengths and Delays

For the sake of simplicity, we have up to now only considered globally uniform values for the coupling strength $w_{\mathrm{in,ex}}$ and delays $t_{\mathrm{delay}}$. In general, these are connection dependent values. This can be implemented by defining a new datatype for connections:

```
typedef struct
{
  int index; /* index of neuron */
  float w;   /* connection dependent weight */
  int delay; /* delay */
} Conn;

typedef struct
{
  Conn *conns; /* array of connections */
  int size;    /* number of connections */
} Connlist;
```

Then, the datatype neuron is defined as:

```
typedef struct
{
  float v[2];        /* state of neuron (ML) */
  Rbuffer spikes;    /* times of last spikes */
  int inhib;         /* inhibitory or not */
  Connlist connect;  /* connection list */
} Neuron;
```

Using this data structure, it is possible to model spike-timing-dependent plasticity (STDP), i.e. the weights are modified differently, dependent on the times of the pre- and postsynaptic spike arrival times $t_i$ and $t_j$ (Chaps. 2 and 9). The weight $w_{ij}$ of a connection is increased or decreased by $\Delta w_{ij}$ according to:

$$\Delta w_{ij} = \begin{cases} A_+ \exp(\Delta t/t_+) \text{ for } \Delta t > 0 \\ A_- \exp(\Delta t/t_-) \text{ for } \Delta t < 0 \end{cases}, \qquad (11.14)$$

where $A_+ > 0$, $A_- < 0$ and $\Delta t = t_i - t_j$. The function $\Delta w_{ij}$ (Fig. 11.12) is implemented in the following way:

```
#define A_plus 0.01
#define A_minus (-0.012)
#define t_plus 20.0
#define t_minus 20.0

float deltaw(float deltat)
{
  return (deltat>0) ? A_plus*exp(-deltat/t_plus)
                    : A_minus*exp(deltat/t_minus);
} /* of 'deltaw' */
```

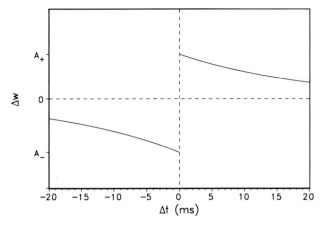

**Fig. 11.12.** Plasticity $\Delta w$ as a function of difference between pre- and postsynaptic spike arrival $\Delta t$

Then the sequential update function with plasticity is:

```
void update(FILE *file, /* output file for spikes */
            Neuron net[], /* array of neurons */
            int n, /* number of neurons */
            int time, /* integer time */
            float h, /* time step */
            float *g, /* lookup table */
            int t_max, /* size of lookup table */
            float I, /* applied current */
            float sigma /* amplitude of Gaussian white
                              noise */
            )
{
  int i,j,k,index,last;
  float sum,v_new[2],dv[2];
  int spike;
  for(i=0;i<n;i++)
  {
    sum=0; /* calculate interactions */
    for(j=0;j<net[i].connect.size;j++)
    {
      index=net[i].connect.conns[j].index;
      /* iterating over the ring buffer of neuron index */
      for(k=0;k<net[index].spikes.len;k++)
      {
        spike=getrbuffer(&net[index].spikes,k)+
              net[i].connect.conns[j].delay;
        /* testing whether spike is active */
        if(time>=spike && time<spike+t_max)
```

```
        {
          /* yes */
          if(net[index].inhib)
            /* inhibitory coupling */
            sum-=net[i].connect.conns[j].w*
                g[time-spike]*(net[i].v[0]-V_in);
          else
            /* excitatory coupling */
            sum-=net[i].connect.conns[j].w*
                g[time-spike]*(net[i].v[0]-V_ex);
          /* plasticity */
          if(time==spike)
          {
            /* previous spike occured? */
            if(net[i].spikes.len>0)
            {
              /* yes, change weight of connection, delta
                 t<0 */
              last=getqueue(net[i].spikes,0);
              net[i].connect.conns[j].w+=
                  deltaw(-(spike-last)*h);
            }
          }
          else
            /* no, break if all spikes are not active
               anymore */
            if(spike+t_max<time)
              break;
      } /* of for(k=...) */
    } /* of for(j=...) */
    /* update of ML */
    updateml(dv,net[i].v,I);
    v_new[0]=net[i].v[0]+(h/c)*(dv[0]+sum);
    v_new[1]=net[i].v[1]+h*dv[1]+sqrt(h)*sigma*gasdev();
    /* critical threshold of zero reached for v? */
    if(net[i].v[0]<0 && v_new[0]>0)
    {
      /* yes, we have a spike event, add to my ring buffer */
      addrbuffer(&net[i].spikes,time);
      fprintf(file,"%g %d\n",time*h,i);
      /* plasticity */
      for(j=0;j<net[i].connect.len;j++)
      {
        index=net[i].connect.conns[j].index;
        for(k=0;k<net[index].spikes.len;k++)
```

```
    {
      spike=getqueue(net[index].spikes,k)+
          net[i].connect.conns[j].delay;
      if(spike<time)
      {
        /*change weight of connection, delta t>0 */
        net[i].connect.conns[j].w+=deltaw((time-spike)
        *h);
        break;
      }
    } /* of for(k=...) */
  } /* of for(j=...) */
}
net[i].v[0]=v_new[0];
net[i].v[1]=v_new[1];
} /* of for(i=...) */
} /* of 'update' */
```

The initialization of the network has to include setting up the connection-dependent weights and delays. The parallel version of the code can be implemented analogously. The parallel exchange performed by **pnet_exch** is only necessary every $t_{\min} \times \Delta t$ time steps, where $t_{\min}$ is defined as

$$t_{\min} = \min_{i,j=1...n} t_{\mathrm{delay},i,j}. \tag{11.15}$$

# References

1. R. Sedgewick: *Algorithms in C, Part 5: Graph algorithms*, 3rd edn (Addison Wesley Reading MA 2001)
2. C. Morris, H. Lecar: Biophys. J., **35**, 193 (1981)
3. T. H. Cormen, C. E. Leiserson, R. L. Rivest, C. Stein: *Introduction to algorithms*, 2nd edn (MIT Press Cambridge MA 2001)
4. B W. Kernighan, D. Ritchie: *The C programming language* (Prentice Hall 1988)
5. W. Gropp, E. Lusk, A. Skjellum: *Using MPI: Portable parallel programming with the message-passing interface*, 2nd edn (MIT Press Cambridge MA 1999)
6. A. Morrison, C. Mehring, T. Geisel, A. Aertsen, M. Diesmann: Neural Comput., **17**, 1776 (2005)
7. A. Tam, C. Wang: Efficient scheduling of complete exchange on clusters. In: *13th international conference in parallel and distributed computing systems* (PCDS Las Vegas 2000)
8. G. M. Amdahl: Validity of the single-processor approach to achieving large scale computing capabilities. In: *AFIPS Conference Proceedings*, Vol. 30, (AFIPS Press Reston VA 1967) pp. 483–485.
9. W. H. Press, S. A. Teukolsky, W. T. Vetterling, B. P. Flannery: *Numerical recipes in C: The art of scientific computing*, 2nd edn (Cambridge University Press Cambridge MA 1992)

# Part V

## Applications

# Parametric Studies on Networks of Morris-Lecar Neurons

Steffen Tietsche[1], Francesca Sapuppo[2] and Petra Sinn[1]

[1] Institute of Physics, Potsdam University, Am Neuen Palais 10, 14469 Potsdam, Germany
[2] Dipartimento di Ingegneria Elettrica, Elettronica e dei Sistemi, Universita' degli Studi di Catania, V.le A.Doria 6, 95125, Catania, Italy
fsapuppo@diees.unict.it

## 12.1 Introduction

The properties of a network are determined by the network topology, including the connectivity and the coupling strength. Our network consists of neurons modeled by the Morris-Lecar equations. We study the influence of some important parameters on the network dynamics. The parameters we vary are the network topology, the global coupling strengths for excitatory and inhibitory neurons $g_{ex}$ and $g_{in}$ and the variance of an internal noise term.

In Sect. 12.2, the mean spike rate is taken as a measure for network activity, and scans through the $g_{ex}$–$g_{in}$ parameter plane for different noise strengths reveal how the network activity typically depends on the coupling strength. Most of the study is concerned with ER random networks, but in Sect. 12.2.3, it is shown that the typical dependencies also hold for small-world networks.

In Sect.12.3, the network behavior near the lower critical coupling strength, where network activity sets in, is inspected more closely. Raster plots of random networks as well as small-world networks show the time series of spiking activity.

### 12.1.1 The Morris-Lecar Neuron Model

To simulate our neural network, we use the Morris-Lecar model (see Chaps. 1, 9, 11, 14). This model represents an electrical circuit similar to a cellular membrane. It consists of three general synaptic currents and an additional external current $I_{ext}$. Equation (12.1) describes the development of the membrane potential $V$, with instantaneous activation of the inward $Ca^{2+}$ current, slower activation of the outward $K^+$ current and a leakage current.

$$\dot{V} = -g_{\text{Ca}}n_\infty(V - V_{\text{Ca}}) - g_{\text{K}}w(V - V_{\text{K}}) - g_{\text{leak}}(V - V_{\text{leak}}) + I_{\text{ext}} \quad (12.1)$$

$$\text{with } n_\infty = 0.5 \left[ 1 + \tanh\left(\frac{V - V_1}{V_2}\right) \right]$$

$$\dot{w} = \phi\cosh\left(\frac{V - V_3}{2V_4}\right)(w_\infty - w) + \sigma\eta(t)\sqrt{h} \quad (12.2)$$

$$\text{with } w_\infty = 0.5 \left[ 1 + \tanh\left(\frac{V - V_3}{V_4}\right) \right]$$

This is the normalized form of the Morris-Lecar equations used by Rinzel and Ermentrout [1]. The parameters $g_{\text{Ca}}$, $g_{\text{K}}$ and $g_{\text{leak}}$ are the maximal conductances for each synaptic current and $V_{\text{Ca}}$, $V_{\text{K}}$ and $V_{\text{leak}}$ are the corresponding reversal potentials. The number of open $Ca^{2+}$ channels is $n_\infty$. The number of activated $K^+$ channels is $w$, its time dependence being described in (12.2). Equation (12.2) contains an internal noise term $\sigma\eta(t)\sqrt{h}$, which is additive Gaussian white noise with variance $\sigma^2$. The external current $I_{\text{ext}}$ is given by

$$I_{\text{ext}} = I + I_{\text{syn}},$$

where $I$ is a constant external current and $I_{\text{syn}}$ the synaptic current coming from connected neurons (see below). For a discussion of the equations and their parameters, refer to [2,4], or Chap. 1.[3]

### 12.1.2 Network Setup

The statistical properties of the networks (e.g. connection probability) are fixed, but the network behavior may vary for different realizations of the setup. We set up a new network for every simulation run, thereby ensuring that most of our data points are from typical realizations of the network setup.

The coupling of the neurons enters the Morris-Lecar equations through the synaptic current $I_{\text{syn}}$, which is defined as

$$I_{\text{syn}} = -\sum_{j=1}^{N}\sum_{k=1}^{\infty} g(t - t_{\text{spike}}(j, k))(V - V_j) \quad (12.3)$$

with

$$g(t) = \frac{g_j}{\tau_1 - \tau_2}(e^{-t/\tau_1} - e^{-t/\tau_2}) \quad (12.4)$$

and the replacement rule

---

[3] The parameter values were fixed to: $V_{\text{Ca}} = 1.0$, $V_K = -0.7$, $V_{\text{leak}} = -0.5$, $V_1 = -0.01$, $V_2 = 0.15$, $V_3 = 0.1$, $V_4 = 0.145$, $V_{\text{inh}} = -0.55$, $V_{\text{ex}} = 0.05$, $I = 0.08$, $\phi = 0.33$, $g_{\text{Ca}} = 1.0$, $g_K = 2.0$, $g_{\text{leak}} = 0.5$, $h = 0.01$ ms, $\tau_1 = 1.0$ ms, $\tau_2 = 2.0$ ms, $N = 400$.

$$\text{neuron } j \text{ is excitatory} \quad \Longleftrightarrow \quad V_j = V_{\text{ex}}, \quad g_j = g_{\text{ex}},$$
$$\text{neuron } j \text{ is inhibitory} \quad \Longleftrightarrow \quad V_j = V_{\text{in}}, \quad g_j = g_{\text{in}}. \tag{12.5}$$

In (12.3), $N$ is the number of neurons in the network. The function $g(t)$ modulates the weight of a spike depending on the time that has passed. $V_{\text{ex}} > 0$ is the resting potential for excitatory neurons, $V_{\text{in}} < 0$ the resting potential for inhibitory neurons.

In addition to the spikes produced by the Morris-Lecar equations, we feed spikes to each neuron independently at random times with a mean frequency of 3 Hz. This Poisson process serves as an external forcing of the network and stimulates activity [3].

In our study, the network consists of 400 neurons. 90% of the neurons are excitatory and 10% are inhibitory. The connections between neurons are bidirectional and the connection strengths are uniform. Excitatory neurons are connected with a connection strength $g_{\text{ex}}$, and inhibitory neurons have connection strength $g_{\text{in}}$. The following two network topologies are considered:

(i) An ER random network, in which each neuron is connected to every other neuron with a probability of $p = 0.2$, leading to a connectivity of 20%.
(ii) A small-world network, which is set up by first forming a ring where every neuron is connected to the next 40 neighbors to each side and then replacing 5% of those connections by connections to random neurons. The resulting connectivity is also 20%.

Both topologies have properties that are important for cortical networks: the random networks has a short pathlength but low clustering, and the small-world network has short pathlengths and high clustering.

## 12.2 Influence of Coupling Strengths and Noise on the Network Activity

The dynamics of networks of neurons is affected by excitatory and inhibitory coupling strengths ($g_{\text{ex}}$ and $g_{\text{in}}$, respectively). We take the spike rate as a measure for the global behavior of the network.

For any pair of $g_{\text{ex}}$ and $g_{\text{in}}$, the spike rate per neuron is computed by running the simulation, counting the total number of spikes, and dividing by simulation time (typically 3 s) and the number of neurons.

We take the code from Chap. 11, make it parallel using MPI and run it on a cluster of 16 nodes. The parameter space ($g_{\text{ex}}, g_{\text{in}}$) is divided into subareas, with each area being assigned to one processor. The parameter $g_{\text{ex}}$ is varied between 0 and 1 with a step size of 0.008, while $g_{\text{in}}$ is varied between 0 and 0.4 with a step size of 0.01. We use a smaller incremental step for $g_{\text{ex}}$ than for $g_{\text{in}}$, since we expect that the network dynamics depends more strongly on the excitatory coupling strength than on the inhibitory one, because only 10% of the neurons are inhibitory.

## 12.2.1 Characteristic Features of the Activity Function

We study a network of $N = 400$ neurons, which are randomly connected with a probability $p = 0.2$. In this first study, the noise level in (12.2) is set to $\sigma = 0$. This initial choice allows us to study the network behavior in the case of deterministic neuron equations.

Figure 12.1 shows the 3-D plot of the spike rate as a function of the coupling strengths $g_{in}$ and $g_{ex}$. For small $g_{ex}$, the spike rate is rather low (3 Hz). This is the rate of spikes that are externally induced by the Poisson process. As expected, there is no self-sustained spiking activity above the input level if the coupling between the neurons is too weak.

At $g_{ex} \approx 0.05$, there is an abrupt increase in network activity. The value of the critical threshold only weakly depends on the inhibitory coupling strength, as Fig. 12.2 shows. As the excitatory coupling strength increases further, the spike rate reaches a maximum of about 130 Hz at $g_{ex} \approx 0.15$ and then decreases again. The most obvious difference when changing $g_{in}$ is the behavior for large $g_{ex}$. For small $g_{in}$, the spike rate quickly drops to a constant value of 50 Hz, whereas for larger $g_{in}$, the decrease is slower, and saturation is not yet reached for the highest $g_{ex}$ in our data.

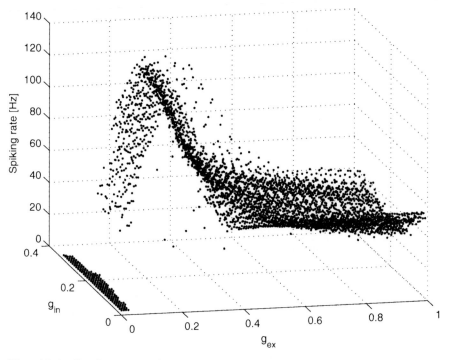

**Fig. 12.1.** Random network: spike rate per neuron as a function of coupling strengths $g_{in}$, $g_{ex}$

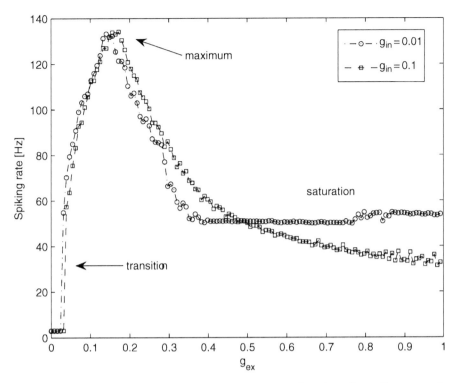

**Fig. 12.2.** Random network: spike rate per neuron as a function of coupling strength $g_{ex}$

### 12.2.2 Effects of Additional Gaussian Noise

Next, we study the effects of additional noise on the global behavior of a random network. The crucial parameter is the noise strength $\sigma$ in the Morris-Lecar equation (12.2). In Fig. 12.3, the activity for $\sigma = 0.05$ is compared to the case where $\sigma = 0$.

The $\sigma = 0.05$ curve does not present the sharp increase in spike rate at a certain excitatory coupling strength like the one we observe for $\sigma = 0$. We speculate that the reason for this is the disturbance of collective behavior by the independent random signals fed to each neuron.

Furthermore, the peak of maximal activity moves towards higher coupling strengths. We compare the spike rates at a fixed value of $g_{in} = 0.1$: for $\sigma = 0.05$ the maximum is at $g_{ex} = 0.25$, and for $\sigma = 0$ it is at $g_{ex} = 0.16$. Similarly to $\sigma = 0$, the range of $g_{ex}$ considered is too small to observe saturation.

### 12.2.3 Activity Function for a Small-world Network

The study of the dynamics of a small-world network with $N = 400$ neurons, probability of long range connection $p = 0.05$ is performed with the same

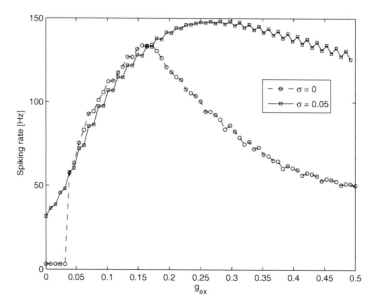

**Fig. 12.3.** Random network: spike rates per neuron for noisy Morris-Lecar neurons compared to the curve $\sigma = 0$. For both curves $g_{in} = 0.1$

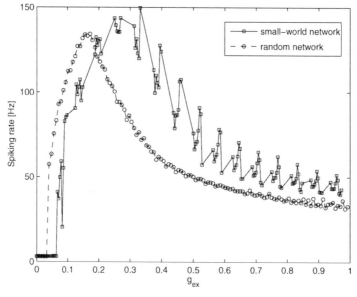

**Fig. 12.4.** Comparison between spike rates for small-world and random networks. For the random network, $g_{in} = 0.15$, for the small-world network, $g_{in} = 0.4$

global connectivity of 20% and the same Poissonian input process. Fig. 12.4 shows the spike rate curves for a random as well as small-world network. For both network types, there is a transition point at some value of the excitatory coupling strength $g_{ex}$, where a sharp increase in network activity occurs.

## 12.3 Network Dynamics Near the Critical Coupling Strength

As shown in the previous section, for both network types, there is a sharp increase in network activity at a certain coupling strength $g_{ex}$. This section will show that close to the critical coupling strength, the network dynamics develops complex features on the time scale of a few seconds (cf. Fig. 12.5).

To analyze the network behavior, the spike events are presented in *raster plots*. This type of plot has time on the horizontal axis, neuron index on the vertical axis, and contains a dot for every spike event. From this, temporal

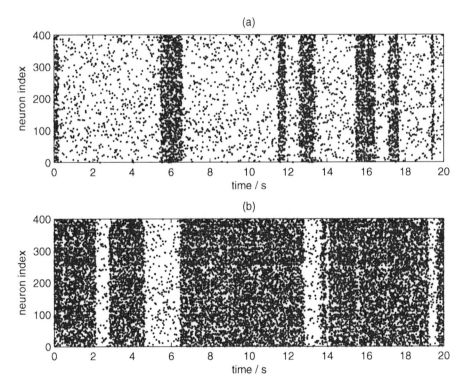

**Fig. 12.5.** Raster plots of a random network of Morris-Lecar neurons for two different coupling strengths that are in the critical range: (**a**) $g_{ex} = 0.062$; (**b**) $g_{ex} = 0.065$. $g_{in} = 0.4$ for both

changes in network activity as well as spatial inhomogeneities can be readily recognized.

When there is Gaussian white noise added to the Morris-Lecar equation (12.2), the sharp transition in activity as well as the complex behavior vanish. This is to be expected, since noise with variance $\sigma^2$ is added to each neuron independently and should disturb correlation and coherence between them. All considerations in the following part are therefore from data with $\sigma = 0$.

### 12.3.1 ER Random Network

In our network configuration (cf. Sect. 12.1) and with $g_{in} = 0.4$, the transition occurs between $g_{ex} = 0.06$ and $g_{ex} = 0.07$. Figure 12.5 shows two simulations with $g_{ex} = 0.062$ and $g_{ex} = 0.065$.

For these parameter values, there are two distinct states of the network. The state with low activity has a spike rate of approximately 3 Hz. This is the main frequency of the external forcing (cf. Sect. 12.1), so there is almost no self-sustained activity. The high-activity state has a spike rate of approximately 26 Hz. In this state, the neurons seem to be synchronized (cf. Fig. 12.6), which suggests that the interaction between the neurons drives the network.

Within one simulation run, the network switches between the two states irregularly. With increasing $g_{ex}$, the network is in the high-activity state for longer time intervals (Fig. 12.5). Interestingly, the spike rate changes rather abruptly within 50 ms, whereas between the switches it remains constant for typically a few seconds.

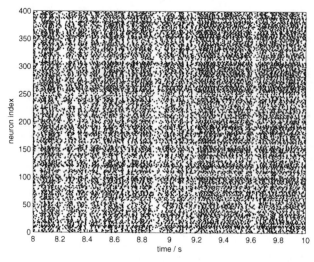

**Fig. 12.6.** Random network near critical coupling strength during high-activity phase. Enlargement of Fig. 12.5(b)

Figure 12.6 shows an enlargement of a high-activity phase of Fig. 12.5(b). There are irregular small-amplitude variations of activity on a time scale of a few milliseconds. The horizontal stripes are due to the rather strong activity variations between neurons: the average spike rate for a single neuron is 23 spikes per second, with a standard deviation of 9 spikes per second. We think that the deviations in activity are due to differences in the number of connections. Neurons with fewer connections have less input and therefore should be less active on average.

### 12.3.2 Small-world Network

Raster plots for small-world networks also show complex behavior near the critical coupling strength, but there are differences to the behavior of random networks, which are related to network topology. In a small-world network, neurons with neighboring indices are strongly connected, while there are only a few random connections between distant neurons (cf. Sect. 12.1).

The small-world topology influences the dynamics of the network in two ways:

(i) By the connection of neurons with neighboring indices, the concepts of neighborhood and distance between neurons are introduced. As a consequence, the spreading of activity is represented in the raster plots as non-orthogonal features, which almost certainly do not occur in random networks (compare Figs. 12.5 and 12.7).

(ii) Neighboring neurons can group into clusters that are strongly interconnected, but weakly connected to the rest of the network. The existence of inhibitory neurons supports this separation into clusters. In a raster plot, the clustering is indicated by broad horizontal stripes with distinct spiking behavior.

Figure 12.7 shows two examples of small-world networks with a close-to-critical coupling strength $g_{ex}$. As in the case of random networks, there is a state in which the network activity is only driven by the 3 Hz Poissonian input noise, and a state in which the coupling leads to a rather strong activity. For small-world networks, the whole network is not all in one single state; rather, the different clusters can have different states. For example, the cluster around neuron 300 in Fig. 12.7(a) is silent all the time, while the state of the neurons 100 to 250 changes irregularly.

Generally, with increasing $g_{ex}$, the high-activity state dominates the low-activity state, which is similar to the case of random networks. But the simulation run shown in Fig. 12.7(a) has a higher activity than the one shown in Fig. 12.7(b), although its coupling strength is lower. Bearing in mind that for each simulation run a new realization of the network setup was used, this gives a hint that the dynamics of near-critical small-world networks is more sensitive to the details of the network setup.

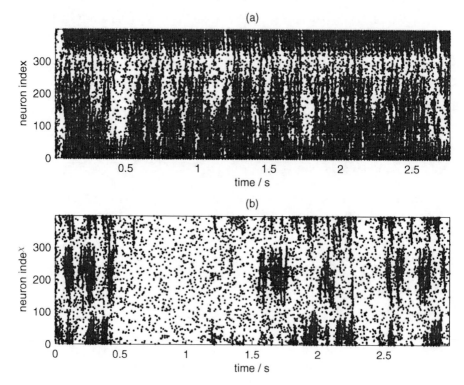

**Fig. 12.7.** Raster plots of small-world networks of Morris-Lecar neurons for two different coupling strengths that are in the critical range: (**a**) $g_{ex} = 0.065$; (**b**) $g_{ex} = 0.071$. $g_{in} = 0.4$ for both

In Fig. 12.8, which is an enlargement of Fig. 12.7(b), it can be seen that during high-activity phases, the neurons are strongly synchronized. They spike together after regular time intervals of approximately 20 ms. Another interesting feature is the spreading of activity that takes place between $t = 0.5$ s and $t = 0.6$ s; initially only neurons 0 to 50 are active, but then activity spreads up to neuron 250, where the propagation stops at the silent cluster around neuron 300.

## 12.4 Conclusion

Our study of networks of Morris-Lecar neurons shows that the network activity, measured as the average spike rate per neuron, strongly depends on the coupling strength between the excitatory neurons. There is a lower threshold coupling strength at which the activity abruptly rises, a maximal activity for intermediate coupling strengths and a saturation effect for high coupling strengths.

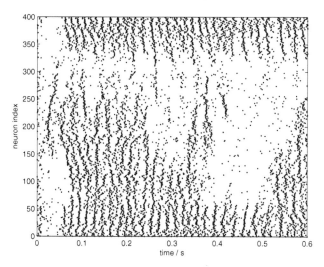

**Fig. 12.8.** Small-world network near critical coupling strength. Enlargement of Fig. 12.7(a)

Additive noise in the Morris-Lecar equations changes the shape of the activity function: for small coupling strengths, the activity increases smoothly — there is no threshold coupling strength. The maximum of activity is at higher coupling strengths. Our data is not sufficient to establish a saturation effect for the case of additive noise. Future studies should be carried out for larger parameter intervals.

For coupling strengths that are close to the lower threshold, complex dynamical patterns are possible. These patterns, visualized in raster plots, depend on the network topology and the coupling strength. Internal noise seems to inhibit the patterns. We suspect that the reason for this is the destruction of coherence between the neurons by the independent random signals fed to each neuron.

Our results suggest that the dependence of activity on coupling strength is similar for small-world networks and random networks. Nevertheless, the dynamical patterns inferred from the raster plots are rather different. Furthermore, the behavior of small-world networks seems to be more sensitive to connection set up. However, our results for small-world networks are only preliminary and require validation. Further studies should also try to systematically investigate the influence of the internal noise strength.

Two important questions remain open. First, how can one quantify the network behavior by different measures than the spike rate? For example, the detection of phase synchronization of the neurons would certainly give deeper insight into the network dynamics (compare also Chap. 6). Second, the character of the transition at the lower threshold of coupling strength should be explored. This regime may be particularly interesting for brain dynamics, since the intermediate degree of synchronisation allows for coherent but complex reaction to external stimulation.

# References

1. J. Rinzel and B. Ermentrout: Analysis of neural excitability and oscillations. In: *Methods in Neuronal Modeling* ed. by C. Koch, I. Segev (MIT Press, Cambridge 1991).
2. M. St-Hilaire and A. Longtin: Journal of Computational Neuroscience **16**, 299–313 (2004).
3. N. Montejo, M. N. Lorenzo, V. Pérez-Villar, and V. Pérez-Muñuzuri: Physical Review E **72**, 011902 (2005).
4. E. Izhikevich, IEEE Transactions on Neural Networks, **15**, 1063–1070 (2004).

# Traversing Scales:
# Large Scale Simulation of the Cat Cortex
# Using Single Neuron Models

Martin Vejmelka[1], Ingo Fründ[2] and Ajay Pillai[3]

[1] Department of Cybernetics, Faculty of Electrical Engineering, Czech Technical
University, Karlovo náměstí 13, 121 35 Praha 2, Czech Republic
Institute of Computer Science, Czech Academy of Sciences, Pod Vodárenskou
věží 4, Praha 8, Czech Republic
vejmelka@cs.cas.cz
[2] Department for Biological Psychology, Otto-von-Guericke University, Magdeburg
PO-Box 4120
ingo.fruend@nat.uni-Magdeburg.de
[3] Theoretical Neuroscience Group, Center for Complex Systems & Brain Sciences,
Florida Atlantic University, Boca Raton, Florida, USA
pillai@ccs.fau.edu

## 13.1 Introduction

The average adult human brain has about 100 billion neurons. Taken together,
these neurons form several trillions [1] of connections with each other. Though
we understand the functioning of single neurons in quite some detail, the same
cannot be said about large-scale neural network dynamics and the mechanisms
that generate them. Issues pertaining to large-scale neural interactions remain
open questions in the neurophysiological community. Even amongst computational neuroscientists who use neural modeling to investigate brain dynamics,
scale has remained a challenging issue.

To date, there are two major lines in modeling neural activity. On the one
hand, individual neurons are modeled in more or less detail [2] (cf. Chaps.1
and 2) and on the other hand, field equations that do not explicitly contain
individual neurons are derived for the propagation of activity in neural tissue
[3](cf. Chaps.1 and 8). Both these approaches have advantages but at the
same time they have serious shortcomings.

When modeling individual neurons, one has access not only to spike rates
but also to spike timings and the relations between the timings of individual spikes. Such timing aspects seem to play important roles for neural processing and coding [4–6]. It has been argued that only codes that rely on
spike timing could account for certain types of experimentally observed neuronal responses [7]. However, modeling individual neurons is usually limited

to relatively small numbers of neurons, typically several hundred, e.g. [8]. To overcome this limitation, several models that describe the propagation of statistical properties (such as spike rate or average postsynaptic potential) of neural tissue, so-called neural "field equations", have been developed [3,9,10]. In these types of models, individual cells are not modeled explicitly but instead the neural tissue by virtue of its high density of neurons is modeled as a continuum. Although these models capture dynamic aspects of large numbers of neurons, they do not capture all the details like an individual neuron's spikes and hence cannot provide information such as spike timing.

In this chapter, we explore the possibility of bringing the best of both worlds into one approach. We modeled a large-scale hierarchically organized neuronal network made up of individual neurons with a connection topology based on real physiological data. The connectivity of this large network mimicked the connectivity patterns found in the cat cortex by Scannel et al. [11] previously.

Simulating hundreds of thousands of neurons requires far more computing power than what can be supplied by even the most powerful desktop computers of today. In addition, it also presents some difficult numerical and computational challenges. We employed clusters of PCs operating in parallel to simulate our very large and detailed neuronal network. Thus, we were able to capture details of the network like spike timing and at the same time have the benefit of a large spatial scale and realistic connectivity. Using this approach, one set of results may be analyzed on multiple levels of detail at the same time.

## 13.2 Materials and Methods

We have modeled the distributed large-scale cortical activity of the cat using a neural network model with three hierarchically organized scales. On the lowest level of hierarchy, single neuron dynamics was modeled using point neuron models (cf. Chap. 1). These neuron models were connected to form local networks with a random topology (cf. Chap.3). The local networks were subsequently connected together to form a global network using a connection scheme respecting physiologically known information about long-range connectivity in the mammalian brain [11] (cf. Chaps. 4 and 9).

### 13.2.1 Neuron Model

Single neurons were modeled using the simple neuronal model proposed by Izhikevich [12] (cf. Chap. 1). This model consists of two state equations (13.2), representing the fast dynamics of the membrane potential and slow recovery effects due to activation of $K^+$ and inactivation of $Na^+$ currents.

$$\frac{dv}{dt} = 0.04v^2 + 5v + 140 - u + I \tag{13.1}$$

$$\frac{du}{dt} = a(bv - u),$$

where $u$ is a variable representing the $Na^+$ channel activation and $K^+$ channel inactivation and $v$ is the membrane potential. If the membrane potential exceeds $30\,mV$, the model is reset to $v = c$   and   $u = u + d$ and a spike is emitted. The model can be tuned using the four parameters $a, b, c, d$ to mimic the dynamical features of a wide variety of cortical and subcortical neurons. The variable $I$ represents the total input current into the neuron. Motivated by the anatomy of the mammalian cortex [13], we chose the ratio of excitatory to inhibitory neurons to be 4 to 1. Excitatory neurons were chosen from a distribution containing the dynamical features of regular spiking, intrinsically bursting and chattering cells [14, 15]; typical activity of each type is shown in Fig. 13.1. For every excitatory neuron, the parameters were set to $(a, b) = (0.02, 0.2)$ and $(c, d) = (-65, 8) + (15, -6)r^2$ where $r \in \langle 0, 1 \rangle$ is a uniformly distributed variable. As regular spiking cells are more frequent in the cortex, the choice of excitatory neurons was biased towards this cell type. The term $r^2$ serves to bias the distribution.

Inhibitory neurons were chosen from a distribution containing fast spiking and low threshold firing cells, the parameters given by $(a, b) = (0.02, 0.25) + (0.08, -0.05)r$ and $(c, d) = (-65, 2)$, where again $r \in \langle 0, 1 \rangle$ is uniformly

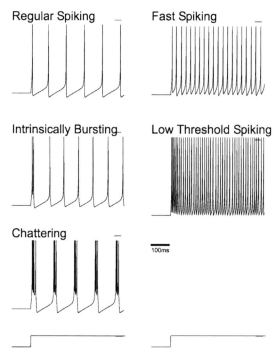

**Fig. 13.1.** Membrane potential evolution of samples from the distribution of excitatory (*left*) and inhibitory (*right*) cells in response to injected current (*bottom traces*)

distributed. Samples of the activity of both types are also in Fig. 13.1. Low threshold firing cells are characterized by strong spike frequency adaption during tonic stimulation [12].

## 13.2.2 Local Networks

We have decided to use a random connection topology for local connections because we found the current anatomical and physiological knowledge insufficient to support the adoption of a more specific topology. In order to mimic the sparsity of neural connections, only 10% of all possible connections were generated. Connections were randomly selected from the set of all possible connections. In total, 53 local networks were modeled, one for each area of the cortex as mapped by Scannel et al. [11]. If one neuron was connected to another neuron, spikes of the first neuron triggered postsynaptic potentials in the second neuron with a certain conduction delay. Postsynaptic potentials were modeled by the function

$$V_{\text{PSP}}(\tau) = \frac{\sigma}{\tau_1 - \tau_2}\Big(\exp(-\tau/\tau_1) - \exp(-\tau/\tau_2)\Big) \tag{13.2}$$

with a rising time constant $\tau_1 = 1\,\text{ms}$ and a falling time constant of $\tau_2 = 3\,\text{ms}$ added to the membrane potential of the postsynaptic cell, and where $\sigma$ is the peak value of the exponential. The time constants of the postsynaptic potential were chosen to mimic the dynamics of AMPA and $\text{GABA}_A$ receptors of excitatory and inhibitory synapses respectively [13]. Cortical conduction delays vary over a broad range between 0.1 ms to delays as long as 44 ms. Data by Swadlow [16] show two peaks in the distribution of conduction delays. One of these peaks is in the range of delays clearly below 10 ms, the other peak is in a range between 10 ms and slightly above 20 ms. We modeled local conduction delays in the range between 0.5 and 4 ms, which roughly corresponds to the first peak in the conduction delay distribution. In this way, conduction delays for long range connections could be modeled using conduction delays from the second peak. The strengths of the connections in the local networks were assigned to the lowest value for which the stimulation of 10% of the neurons in the network still lead to an overall response of that area.

## 13.2.3 Long Range Networks

The local networks were connected to form one single large-scale network. Very little is known about the exact strengths of long range corticocortical connections. However, Scannel et al. [11] rated connectivity data from the cat cortex according to whether a connection was strong, intermediate, weak or absent. Figure 13.2 shows the connections between areas in matrix form. See [11] for assignment of the numbers to anatomical names. A total of 826 connections have been identified.

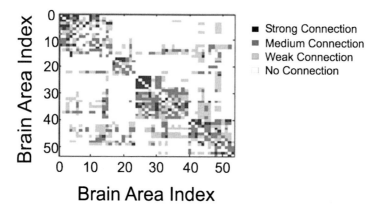

**Fig. 13.2.** The connectivity map used to set up the connections in the model. Adopted from [11]

We simulated long range connectivity by associating each of the 53 local networks with one cortex area. If the data by Scannel et al. [11] reported a connection between two areas, we randomly selected 5% of the excitatory cells from the first area and connected these cells to a randomly selected 5% cells from the other area. Connections were modeled as described in Sect. 13.2.2 (see Chaps. 3, 9). The alpha functions modeling the postsynaptic potential shape of these connections were scaled by a factor depending on the connection strengths reported by [11]. If a strong connection was reported, the alpha function was scaled by a factor of $\sigma = 3$, if an intermediate connection was reported the alpha function was scaled by $\sigma = 2$. For weak connections, the alpha function was not scaled, $\sigma = 1$. We have thus used a linear scale for the connection strengths. Other options include polynomial (e.g. quadratic) scaling or exponential scaling ($10^1, 10^2, 10^3$).

Conduction delays of long range connections were randomly assigned to a value between 10 and 20 ms in accordance with data provided by Swadlow [16]. Connection strengths between local networks were assigned the lowest value for which the stimulation of one local network still visibly spread to other local networks.

### 13.2.4 Stimulation Paradigms

During the simulation, the network was exposed to three different types of input.

(i) Unspecific thalamic input which was modeled by adding Gaussian white noise to the membrane potential of all cells.

(ii) Specific thalamic input, which was modeled by the injection of direct current into some of the cells belonging to the primary visual cortex (Brodmann's area 17).

(iii) Combined specific and unspecific thalamic input.

### 13.2.5 Rastergram Analysis

To find out whether the signal was propagating in an orderly fashion primarily along the paths suggested by the anatomical data, we analyzed the resulting spike time series to recover propagation delays of the signals from Brodmann's area 17 and related this to the shortest pathlength to all other areas.

Two different detection methods were used to estimate the propagation delay. The neural spike traces were preprocessed to generate one time series per local network, resulting in 53 different time traces. The preprocessing algorithm computed the number of spikes per time step in each local network. The first method was a threshold algorithm, which triggered a detection event when spiking activity in a local network crossed a minimum threshold. The propagation delay was defined as the difference of the time instance when activity was detected and the beginning of DC current injection. The second method was cross-correlation, where the propagation delay was derived from the lag at which the cross-correlation exhibited a maximum, zero lag coinciding with the time instant when the DC current injection started.

To provide further evidence that the synaptic connections are well adjusted and excitation is spreading primarily along the long range connectivity paths, we analyzed the local network cumulative spike traces resulting from the simulation to detect dominant connectivity patterns. Again, two different methods were used, cross-correlation and mutual information. The algorithm operated with prior information on how many connections were in the original input matrix. The problem of selecting a suitable detection threshold was thus circumvented. The question was, given the correct number of connections, will the detected connections between the area be similar to the anatomical data used as input to the simulation? Mutual information was computed for each pair of the time series. The computed values were sorted by magnitude from largest to smallest and only the 826 largest values (the number of connections in the data given by Scannel et al.) were considered to be detected connections.

## 13.3 Results

The model was run on 16 nodes of the Linux cluster "Peyote" of the Max Planck Institute for Gravitational Physics, Potsdam, Germany (cf. Chap. 11). The cluster is populated by 128 nodes with 2 CPUs (Intel Dual Xeon) and 2GB RAM each. The model was run under the three above mentioned conditions and a 10 second spike rastergram was obtained from each simulation. Here, we show results from simulations performed with 4096 neurons in each local network, that is, 217,088 neurons total. The rastergrams have simulation time on the horizontal axis and neuron index on the vertical axis (cf. Chaps. 12 and 14). When a neuron emits a spike, a dot is placed on the rastergram at the point corresponding to the time instant and neuron index. The network was simulated using Gaussian integration with a step size of 0.5 ms.

In the first experiments (specific thalamic input), we injected 50 nA of direct current into 10% of the neurons in Brodmann's area 17 (primary visual cortex) periodically every second for 100 ms. The rest of the network had no external excitatory input except incoming postsynaptic potentials from other neurons. The rastergram in Fig. 13.3 shows a one second segment of the simulation.

Figure 13.4 clearly shows that the propagation delays from Brodmann's area 17 to other local networks are well correlated with the minimum path-length between the local network and Brodmann's area 17. Both detection methods show similar results.

The plot showing connections detected using mutual information is shown in Fig. 13.5.

The computed connectivity matrix exhibits a reasonable degree of similarity to the original matrix.

In the second experiment (unspecific thalamic input), Gaussian white noise with mean 0 and standard deviation 5 nA was added to the membrane potential, $v$, in (13.2) of each neuron in each time step. The model network was run for 10 seconds and a trace of the spikes of each neuron was captured. Figure 13.6 shows the resulting trace image of the first two seconds. The network exhibits a synchronous rhythm with a frequency of approximately 5 Hz.

To simulate more realistic conditions, we applied the direct current injection in the presence of unspecific thalamic input. The resulting trace is shown in Fig. 13.7.

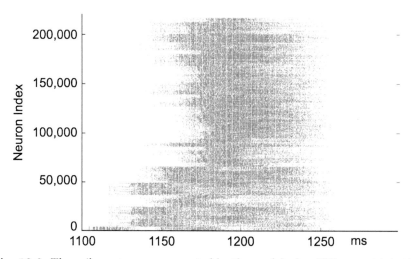

**Fig. 13.3.** The spike rastergram generated by the model when DC current injection was applied to 10% of the cells of Brodmann's area 17

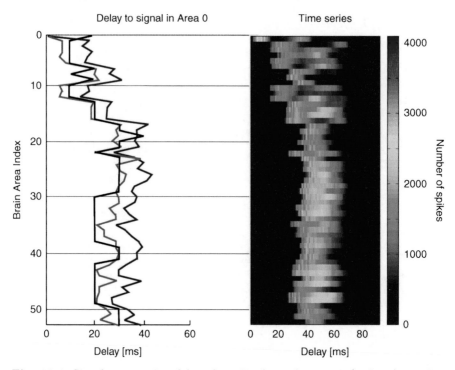

**Fig. 13.4.** Signal propagation delays from Brodmann's area 17 (*index 0*) to other areas. Black line is the delay corresponding to the minimum pathlength from area 17 to each area in turn. The blue line is the propagation delay computed by cross-correlation and the red line is the propagation delay computed by activity detection

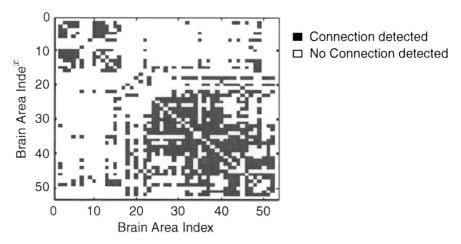

**Fig. 13.5.** Connectivity matrix computed from the time series generated by the model

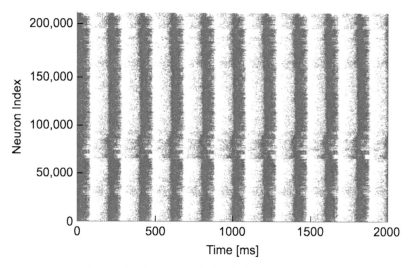

**Fig. 13.6.** Unspecific thalamic input rastergram

Qualitatively, the model exhibited the same behavior when unspecific thalamic input was present in addition to the direct currentinjection and again
generated a 5 Hz rhythm. This could be explained by the fact that we only applied weak stimulation, which we verified would propagate from local network
to local network, but was itself not strong enough to significantly alter the
dynamics of the entire network.

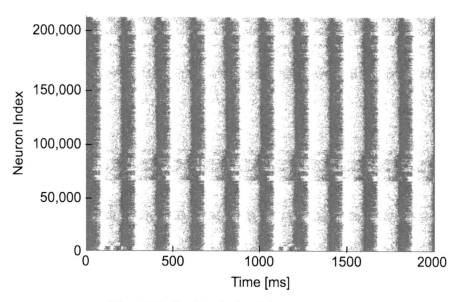

**Fig. 13.7.** Combined stimulation rastergram

## 13.4 Discussion

In the current investigation, we observed highly synchronized spike waves when the network was exposed to global unspecific noise. If one area of the network was directly activated by adding a fixed current to 10% of the neurons in that area, activity spread from this area to other areas.

We observed 5 Hz oscillations that were highly synchronized across the whole network. What is the origin of these oscillations? We can distinguish three different reasons that our network might show these oscillations:

(i) If excited by a constant current, most excitatory cells in our model spike with an interspike interval of roughly 200 ms [12]. It might thus be that an overall 5 Hz rhythm is due to the fact that every single neuron fires spikes with a frequency of 5 Hz.

(ii) If the overall 5 Hz rhythm can not be explained by the properties of the single neurons, this rhythm might evolve from the dynamics of the local randomly connected networks.

(iii) A third possible reason for the emergence of 5 Hz activity could be the large-scale connection properties that were adapted from Scannel et al. [11].

From these three mechanisms that might underlie the 5 Hz oscillations we observed in our simulation, the first mechanism is easily ruled out. In a simulation without any connections, only the properties of the single cells would play a role. In such a simulation, the first spike wave might be synchronized. However, subsequent spike waves will be increasingly scattered due to the noisy input. As long as there is no mechanism that counteracts the scattering of spike times due to noise, an overall rhythm will not emerge as a rule. The second and the third mechanisms are much harder to differentiate. Studies of networks with one level of hierarchy (i.e. local networks, in our simulation) have however not shown any such oscillations. In such simulations, usually much higher network frequencies ($> 25$ Hz) than 5 Hz are observed [2, 8]. Some authors also described lower frequency components in the alpha range (8–12 Hz, [12]). Therefore, we believe that our results can be accounted for by the large-scale connectivity that was adapted from Scannel et al. [11].

The frequency of the synchronized activity was an unexpected result. In general, the main frequency components of the brain can be found around 10 Hz and 40 Hz [17]. However, it could be shown in field equation simulations that under conditions with either very low external input [18] or weak connections [19], oscillation periods can also be in the range that was observed in the current investigation. Both these conditions might be present in the current simulation. However, the data from Fig. 13.4 demonstrate that the connections are at least strong enough to ensure a reliable propagation of activity between areas. A more rigorous investigation of this issue is required to clearly separate these two points.

Under what conditions do "real" brains show such highly synchronized low frequency activity? The electroencephalogram displays synchronized oscillations in the so-called theta range (3–7 Hz) during sleep stage 1 (drowsiness) [20] and most prominent during deep meditation [21, 22]. Interestingly, these brain states can be associated with reduced sensory input. During sleep stage 1, the eyes are closed and sounds are perceived to be damped. Persons performing deep meditation report that sensory stimulation is attenuated, which is also confirmed by event related potentials (Coromaldi, pers. comm.). Thus, the emergence of 5 Hz rhythmic activity during very weak stimulation seems to have a psycho-physiological counterpart.

If unspecific input to all areas was combined with direct excitation of one area, the well defined propagation pattern observed in Figs. 13.3 and 13.4 was lost. A plausible reason for this might be that the specific stimulus was too weak to trigger an overall change of the dynamics of the network. However, as we did not analyze the spike patterns at the level of single neurons, we cannot exclude the possibility that there were stimulus specific patterns in the spike responses of the single neurons. The importance of such patterns has been highlighted by several authors [2, 4, 7, 23].

## 13.5 Conclusion

The current results indicate that spiking neuron models can display dynamics that are comparable to those obtained from neural field equations. If the number of simulated spiking neurons is sufficiently large, weak noisy input can drive activity patterns that are observed in field equations and neural mass recordings. Further investigations are required to identify the precise relations between spike timing and spike rates in such large scale simulations.

## Acknowledgments

We would like to extend our grateful thanks to the scientific directors of the Helmholtz International Summer School Dr. Changsong Zhou, Dr. Marco Thiel and Dr. Peter beim Graben for giving us the chance to work in an exciting area of today's research and for providing us with state-of-the-art hardware to run the simulations. We would also like to thank the lecturers for their engaging talks and interesting discussions. Special thanks are due to Werner von Bloh for providing us with the original version of the simulation code and for helping with the modifications. Finally, we would like to thank all the students who participated in the Summer School for the stimulating discussions and inspiring time we all had in Potsdam in September 2005.

## References

1. Christof Koch. *Biophysics of Computation: Information Processing in Single Neurons*. Oxford University Press, New York, 1999.
2. E M Izhikevich. Polychronization: computation with spikes. *Neural Computation*, 18:245–282, 2006.

3. V K Jirsa and H Haken. A derivation of a macroscopic field theory of the brain from the quasi-microscopic neural dynamics. *Physica D*, 99:503–526, 1997.
4. M Abeles. Time is precious. *Science*, 304:523–524, 2004.
5. E Ahissar and A Arieli. Figuring space by time. *Neuron*, 32:185–201, 2001.
6. E M Izhikevich, N S Desai, E C Walcott, and F C Hopensteadt. Bursts as a unit of neural informations: selective communication via resonance. *Trends in Neuroscience*, 26(3):161–167, 2003.
7. S Thorpe, A Delorme, and R Van Rullen. Spike-based strategies for rapid processing. *Neural Networks*, 6–7(14):715–725, 2001.
8. N Kopell, G B Ermentrout, M A Whittington, and R D Traub. Gamma rhythms and beta rhythms have different synchronization properties. *Proc Natl Acad Sci USA*, 97(4):1867–1872, 2000.
9. W J Freeman. *Mass Action in the Nervous System. Examination of the Neurophysiological Basis of Adaptive Behavior through the EEG*. Academic Press, New York, 1975.
10. H R Wilson and J D Cowan. Excitatory and inhibitory interactions in localized populations of model neurons. *Biophys J*, 12(1):1–24, 1972.
11. J W Scannel, G A P C Burns, C C Hilgetag, M A O'Neil, and M P Young. The connectional organization of the cortico-thalamic system of the cat. *Cerebral Cortex*, 9(3):277–299, 1999.
12. E M Izhikevich. Simple model of spiking neurons. *IEEE Transactions on neural networks*, 14(6):1569–1572, 2003.
13. E M Izhikevich, J A Gally, and G M Edelmann. Spike-timing dynamics of neuronal groups. *Cereb Cortex*, 14(8):933–944, 2004.
14. B W Connors and M J Gutnick. Intrinsic firing patterns of diverse neocortical neurons. *Trends in Neuroscience*, 13(9):365–366, 1990.
15. C M Gray and D A McCormick. Chattering cells: superficial pyramidal neurons contributing to the generation of synchronous osillations in the visual cortex. *Science*, 274(5284):109–113, 1996.
16. H A Swadlow. Physiological properties of individual cerebral axons studied in vivo for as long as one year. *J Neurophysiol*, 54(5):1346–1362, 1985.
17. O David and K J Friston. A neural mass model for MEG/EEG: coupling and neuronal dynamics. *NeuroImage*, 20:1743–1755, 2003.
18. L H A Monteirony, M A Bussaby, and J G Chaui Berlinckz. Analytical results on a Wilson-Cowan neuronal network modified model. *J Theor Biol*, 219:83–91, 2002.
19. R M Borisyuk and A B. Kirillov Bifurcation analysis of a neural network model. *Biol Cybern*, 66(4):319–325, 1992.
20. J S Barlow. *The Electroencephalogram. Its Patterns and Origins*. MIT Press, 1993.
21. L I Aftanas and S A Golocheikine. Human anterior and frontal midline theta and lower alpha reflect emotionally positive state and internalized attention: High-resolution EEG investigation of meditation. *Neurosci Lett*, 310:57–60, 2001.
22. E Coromaldi, C Başar-Eroglu, and M A Stadler. Langanhaltende Theta-Aktivität während tiefer Meditation: Eine Einzelfallstudie mit einem Zen-Meister. *Hypnose und Kognition*, 21:61–77, 2004.
23. E Körner, M-O Gewaltig, U Körner, A Richter, and T Rodemann. A model of computation in neocortical architecture. *Neural Networks*, 12:989–1005, 1999.

# 14

# Parallel Computation of Large Neuronal Networks with Structured Connectivity

Marconi Barbosa[1], Karl Dockendorf[2], Miguel Escalona[3], and Borja Ibarz[4]
Aris Miliotis[2], Irene Sendiña-Nadal[4], Gorka Zamora-López[5]
and Lucia Zemanová[5]

[1] University of São Paulo, Brazil
   `marconi@if.sc.usp.br`
[2] University of Florida, USA
   `am397@ufl.edu`
[3] Universidad de Los Andes, Venezuela
   `angele@ula.ve`
[4] Universidad Rey Juan Carlos, Spain
   `borja.ibarz@urjc.es`
[5] University of Potsdam, Germany
   `gorka@agnld.uni-potsdam.de`

## 14.1 Introduction

One does not need to delve into complex modern physical phenomena to realize that laws of physical nature are vastly employed and exploited by nature. We can see in an object as ubiquitous as a flower, among other striking properties related to its form, that the anther is isolated and the stigma is grounded, readily providing an electrostatic mechanism for charged bees to carry the pollen [1]. This phenomenon is so basic, yet shows what years of evolution manage with one charged particle, its surplus and absence.

The quest to understand and somehow control biological systems, using fundamentals and even by-products from physics and mathematics, has a long history. Only to mention a very few examples and recent observations: an original model of rhythmic waves coordinated by a central pattern generator in multi-legged animal locomotion [2, 3], models dealing with the mechanism of pattern formation (a perspective in [4]); a description of the formation of sunflowers' spirals and their relationship to Fibonacci series [5]; and new phenomena and nonlinear mechanical models of hearing [6, 7]. Most of these works require multidisciplinary thinking and address a larger community at various levels of mathematical sophistication, see [8] for a recent selection of biologically related material.

Among biological systems in general, the mammalian nervous system is arguably the richest example of the interplay between biology, physics, chemistry and geometry. In the brain, vastly complex physical phenomena occur, but not only because of the raw number of interacting units and their hierarchical organization. The units themselves [9–12] as well as their effective connectivity [13–15] are still a major challenge to comprehension.

Various reports attempting to model aspects of neuronal activity with tools from nonlinear dynamics have aroused a growing acceptance of the fact that the diverse biological reactions of a cell to a stimulus can be explained by bifurcation theory. While many types of receptors and channels (which need to be taken into account in a description of the cell's intrinsic properties) might be present in a specific cell, they only determine the type of bifurcation of the neural dynamical behavior [16].

Of more interest to us is the thriving activity in modeling networks of neurons with diverse levels of biological realism, focused on explaining a few features of the collective behavior. Using tools related to nonlinear dynamics, such models come in various forms like traditional dynamical systems [17–19], continuous media [20–22], mean field approximations [23], maps [24,25]; models incorporating morphology and structure [26–31]; models with competition both at the network level [19] or at the synaptic level via plasticity [15] (see Chap. 1 for a survey). As usual, a trade-off is necessary to balance the level of detail and the scope of any attempt to come out with a useful model, i.e. to predict behavior or a particular trait accurately and reproducibly.

In [16], a thorough exposition of the basic mechanisms by which neuronal dynamical features can be understood is presented and the common idea of the existence of a threshold is challenged. In [32], a review of the state of affairs of neuronal network modeling surveying a wide range of techniques and information of the present state of phenomenological advances is available. The frontier between complex network structures and associated dynamics is extensively detailed in [33]. A bold perspective appearing in [34] argues that the level of knowledge of the biological intricacies of brain areas and layers has reached such a mature level that time is ripe to attempt a larger scale brain simulation of a microcolumn, calibrated to be indistinguishable from a real one.

The real lack of detailed knowledge of connectivities in the mammalian brain suggests that a wide range of possibilities should be tried when simulating networks of many neurons. This poses a computational challenge. The main goal of this chapter is to show the simulation work we produced during the Summer School using and modifying the code presented in Chap. 11 using the general scheme of Chap. 9 that tries to take on account neural structures and connectivities. It is important to remember that other efforts to simulate computationally large communities of neuronal cells have been done with different objectives in mind, for example [35,36] or the various references in [34]. Although interesting and efficient, those initiatives do not permit a straightforward extension to allow for structured areas with their natural connectivity patterns.

This chapter reports on a number of activities developed at the 5th summer school where we implemented a working framework for the simulation of large populations of neurons. Those neurons were treated as dynamical systems near a bifurcation so that variations of critical parameters would convert an otherwise quiescent state into a state of firing/bursting activity. Keeping track of all the information describing the network state in files would of course become prohibitive, so strategies for dealing with the data also were envisaged.

We chose the Morris-Lecar model as the unit in our simulations. This choice is based on the fact that its dynamics is well understood for the sake of the computational efficiency needed to produce data on a fine time scale and long simulation range. Its efficiency could be one order of magnitude better than the full Hodgkin-Huxley prototype neuron. Morris-Lecar is a conductance-based model like Hodgkin-Huxley but with only two persistent channels, one fast ($Ca^{2+}$) and one slow ($K^+$). Another group of participants dealing with the cat map (Chap. 13) implemented the Izhikevich model, see [17]. Any other dynamical model can be easily implemented as an additional module to the code (cf. Chap. 1).

The choice of the Morris-Lecar model could seem odd to a biologist, for this model was originally proposed to simulate features of a muscle cell, but the trade-off of biological realism for the possibilities of investigating a longer simulation and actually observing size dependent phenomena has paid off, especially when considering that many other biological facts were already left aside or oversimplified (e.g. synaptic dynamics, morphology, delay, etc.) for this particular set of studies. This is in full accordance with the original proposal of building a simple framework from scratch, avoiding black boxes, in a bid to better understand the role played by connectivity in large networked dynamical systems.

The idea of building a framework as general as we implemented is to start probing the functioning of the cat's brain. A vast amount of knowledge of its structures and connectivities has been collected during the last few years [37]. An important feature of the cat's brain is its subdivision into 53 functioning areas. The connection strength of these areas is assumed to be proportional to the thickness of nerves connecting them and is implemented in our code as well. One question that comes to mind is to what extent stimulating one area would spread activity to other areas. This main task is what we set about to address after polishing the code. Our results, while still too preliminary, look promising.

The chapter begins with a brief description, in Sect. 14.2, of the Morris-Lecar prototype neuron dynamics and its coupling to other neurons and the noisy environment. In Sect. 14.3, the idea of one neuronal area is developed and the conjecture of small world connectivity is implemented. This section also addresses the tuning of the network to a natural baseline behavior through parameter search. In Sect. 14.4, the procedure of tuning parameters to baseline behavior is re-introduced and preliminary results of our simulations, mainly the stimulation/ablation of areas and the effect of the size of the network, are presented. Section 14.5 describes the relation of the known connectivity structure of cortex with the observed activity in the simulations. Last, Sect. 14.6 provides discussions on our results and poses important perspectives to be considered in the future using the general framework presented in Chap. 9.

## 14.2 The Model

### 14.2.1 The Morris-Lecar Neuron Model

The neuron model used in our simulations is the Morris-Lecar model. It is a simplified conductance model of the barnacle muscle fiber, with two variables obeying the equations:

$$\dot{v} = -g_{Ca}(v)(v - V_{Ca}) - g_K w(v - V_K) - g_L(v - V_L) + I_{ext} \tag{14.1}$$

$$\dot{w} = \frac{\phi}{\tau_w(v)}(W_\infty(v) - w) \tag{14.2}$$

This is the dimensionless form of the model as presented by Rinzel and Ermentrout in their classic exposition [38]. Voltage $v$ is normalized to the reversal potential of the excitatory ion $Ca^{2+}$, so voltage parameters are $V_{Ca} = 1.0$, $V_K = -0.7$ and $V_L = -0.5$. Conductances have been normalized by a reference conductance $G_{ref} = 4$ mS/cm$^2$ and time by the time constant $\tau = C/G_{ref} = 5$ ms, where $C = 20$ $\mu$F/cm$^2$. Thus we arrive at values $g_K = 2.0$, $g_{Ca} = 1.0$ and $g_L = 0.5$. In order to better match the model to the simulation of mammal cortex rhythms, one (dimensionless) time unit will henceforth be equivalent to 1 ms. Finally, $\phi = \frac{1.0}{3.0}$, and:

$$g_{Ca}(v) = 0.5 \left[ 1 + \tanh \left( \frac{v + 0.01}{0.15} \right) \right] \tag{14.3}$$

$$\tau_w(v) = \frac{1}{\cosh \left( \frac{v - 0.1}{0.145} \right)} \tag{14.4}$$

$$W_\infty(v) = 0.5 \left[ 1 + \tanh \left( \frac{v - 0.1}{0.29} \right) \right] \tag{14.5}$$

This leaves the external current $I_{ext}$ as the only free parameter. For low values ($I_{ext} < I_{SN}$), all orbits of the system are attracted to the unique stable node with most potassium channels closed and a low, polarized voltage (between $V_L$ and $V_K$). As the external current grows beyond $I_{SN} = 0.0833$, the stable node disappears via a saddle-node bifurcation in an invariant circle, giving way to an attracting limit cycle. In this cycle, the voltage jumps towards $V_{Ca}$ and triggers the opening of potassium channels, which in turn pull the voltage back to polarized values; potassium channels then close and the cycle begins again. The result is rhythmic spiking beginning at infinitely low frequencies,

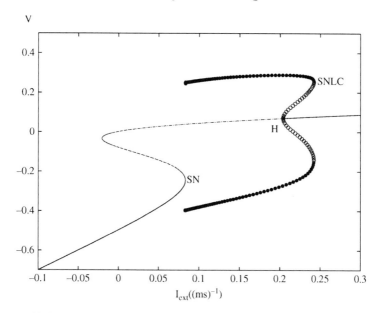

**Fig. 14.1.** Bifurcation diagram of the Morris-Lecar model used in this chapter. *SN*, saddle-node on invariant circle. *H*, Hopf. *SNLC*, saddle-node of limit cycles

and therefore, for the values of parameters chosen, Morris-Lecar is a class I model (for other choices of parameters, it may be a class II model ; see [38]).

Figure 14.1 completes the picture of bifurcations with external current. When $I_{\text{ext}} = I_{\text{SNLC}} = 0.242$, the limit cycle disappears through collision with an unstable limit cycle (born at a subcritical Hopf bifurcation of the hitherto unstable focus of the system) and the neuron becomes silent again, this time at a depolarized value of voltage. In this state, the external current is so strong that it effectively offsets the injection of potassium ions into the cell, preventing the firing of action potentials.

The parameter region of interest for our networks is the so-called excitable regime found at values of $I_{\text{ext}}$ just below the threshold $I_{\text{SN}}$ of rhythmic spiking. In this regime, if the voltage is pushed by the arrival of a synaptic impulse or by random noise out of equilibrium and across the unstable manifold of the saddle point, the system will make a long excursion in the phase plane, producing a single spike. This is illustrated in Fig. 14.2.

### 14.2.2 Coupling Between Neurons and External Stimulation

In the previous section, external current $I_{\text{ext}}$ has been treated as a constant parameter. In network simulations, $I_{\text{ext}}$ is a time-varying current coming from three sources:

$$I_{\text{ext}}(t) = I_{\text{bias}} + I_{\text{syn}}(t) + I_{\text{Poiss}}(t)$$

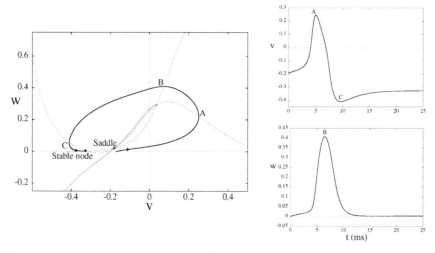

**Fig. 14.2.** Phase plane representation (**left**) and time evolution of variables $v$ and $w$ (**right**) for a single spike in the excitable regime ($I_{\text{ext}} = 0.07$) of the Morris-Lecar model. In the phase plane diagram, the thin continuous line is the unstable manifold of the saddle, dashed lines are the $v$ and $w$ nullclines, and the thick continuous line is the system trajectory. Point $A$, both in the phase plane and in the $v - t$ diagram, is the maximum of the action potential, where the trajectory crosses the $v$-nullcline. At point $B$, where the trajectory intersects the $w$-nullcline, there is maximal opening of potassium channels. Point $C$ is the after-hyperpolarization peak due to remaining open potassium channels when the voltage first returns to the equilibrium level

- $I_{\text{bias}}$ is a constant external bias current which will set the neurons in the excitable regime, as described in the previous section. It is common for all neurons.
- $I_{\text{syn}}$ is the sum of synaptic currents arising from connections with other neurons in the network.
- $I_{\text{Poiss}}$ also comes in the form of a synaptic current, but its origin lies outside the network; the presynaptic spikes that generate this current do not come from neurons in the network, but are instead randomly generated according to a Poisson process. These currents allow us to inject external stimulation.

We now describe the synaptic current $I_{\text{syn}}(t)$. It is a sum of currents due to both excitatory and inhibitory chemical synapses. Indeed, neurons in the network are classified as excitatory or inhibitory. If an excitatory presynaptic neuron fires at time $t_{\text{sp}}$ (i.e. its voltage crosses zero with positive derivative), it adds to the term $I_{\text{syn}}(t)$ of the postsynaptic neurons an ohmic current $I_{\text{syn,exc}}(t)$ with reversal potential $V_{\text{exc}} = 0.05$ and time-varying, alpha function shaped conductance, thus:

$$I_{\text{syn,exc}}(t) = -g_{\text{exc}}\alpha_{\text{exc}}(t - t_{\text{sp}} - t_{\text{del}}) \cdot \Theta(t - t_{\text{sp}} - t_{\text{del}}) \cdot (v_{\text{post}}(t) - V_{\text{exc}}) \quad (14.6)$$

Here, $g_{exc}$ is the strength of connection between pre- and postsynaptic neurons, $t_{del}$ is the time delay of this connection, $\Theta(t)$ is the Heaviside step function, $v_{post}(t)$ is the postsynaptic neuron voltage and $\alpha_{exc}(t)$ is the alpha function:

$$\alpha_{exc}(t) = \frac{1}{\tau_{1,exc} - \tau_{2,exc}}(e^{-\frac{t}{\tau_{1,exc}}} - e^{-\frac{t}{\tau_{2,exc}}})$$

The smaller of the two time constants $\tau_{1,exc}$ and $\tau_{2,exc}$ is the rise time and the larger one is the decay time of the function. If, instead, the presynaptic neuron is inhibitory, the reversal potential is $V_{inh} = -0.50$ (i.e. equal to $V_L$) and the synaptic current added to $I_{syn}(t)$ is similarly:

$$I_{syn,inh}(t) = -g_{inh}\alpha_{inh}(t - t_{sp} - t_{del}) \cdot \Theta(t - t_{sp} - t_{del}) \cdot (v_{post}(t) - V_{inh}),$$

where $\alpha_{inh}(t)$ is now timed according to (possibly different) constants $\tau_{1,inh}$ and $\tau_{2,inh}$.

Finally, the $I_{Poiss}(t)$ term is very similar to $I_{syn}(t)$. The only difference is that it is made up exclusively of excitatory currents of the form of 14.6, and the times $t_{sp}$ do not correspond to spikes of presynaptic neurons but are instead generated by a Poisson process. By varying the rate $\lambda$ of this process, the amount of external stimulation injected into the different areas of our network may be chosen. Baseline values for non-stimulated areas are around $\lambda = 3$ Hz (mean period $T_{Poiss} = 333$ ms).

## 14.3 Setting Proper Parameters

The two-level network described in Chap. 9 gives rise to a moderately large number of parameters (connection numbers and strengths) that have to be tuned if we want our model to mimic cortex behavior. In this section, we describe the tuning procedure in two steps: first for intra-area parameters, and then for the whole 53 area network. Table 14.1 summarizes all the relevant parameters of the model and gives default values for them.

### 14.3.1 Optimal Inhibitory and Excitatory Coupling Strength for one area

In the absence of specific external stimulation, we would like neurons in our network to receive balanced excitatory and inhibitory input, so as to maintain a baseline activity corresponding to the non-specific Poissonian stimulation of 3 Hz. If this balance is achieved, scaling of connection strength with the square root of the degree (see Chap. 9) will ensure that input amplitude is independent of the number of neurons in the network (which is bounded by computational constraints). Balance of excitation and inhibition in one area depends on coupling parameters $g_{1,inh}$ and $g_{1,exc}$. In order to find appropriate values for these parameters, we did the following:

**Table 14.1.** Parameters of the network model and their default values grouped as: neuronal model parameters, network topology, connectivity strength and delays

| Parameter | Description | Default value |
|---|---|---|
| $I_{bias}$ | Constant bias current | 0.08 |
| $V_{exc}$ | Reversal potential for excitatory synapses | 0.05 |
| $V_{inh}$ | Reversal potential for inhibitory synapses | $-0.5$ |
| $n$ | neurons per area | 512 |
| $p_{inh}$ | Ratio of inhibitory neurons | 0.2 |
| $p_{ring}$ | Ratio of connections inside one area | 0.1 |
| $p_{rew}$ | Probability of rewiring | 0.3 |
| $p_3$ | Ratio of neurons receiving synapses from a connected area | 0.05 |
| $p_4$ | Ratio of neurons with synapses towards a connected area | 0.05 |
| $g_{1,exc}$ | Non-normalized strength of intra-area excitatory synapses | 0.075 |
| $g_{1,inh}$ | Non-normalized strength of intra-area inhibitory synapses | 2.5 |
| $g_{2,exc}$ | Non-normalized strength of inter-area excitatory synapses | 0.075 |
| $g_{2,inh}$ | Non-normalized strength of inter-area inhibitory synapses | 0 |
| $g_{ext}$ | Strength of connection for Poissonian currents | 0.1 |
| $\tau_{1,exc}, \tau_{2,exc}$ | Rise and delay times of excitatory synaptic current | 1 ms, 3 ms |
| $\tau_{1,inh}, \tau_{2,inh}$ | Rise and delay times of inhibitory synaptic current | 1 ms, 3 ms |
| $t_{del,1,exc}$ | Delay of intra-area excitatory synapses | 1 ms |
| $t_{del,1,inh}$ | Delay of intra-area inhibitory synapses | 3 ms |
| $t_{del,2,exc}$ | Delay of inter-area excitatory synapses | 3 ms |
| $t_{del,2,inh}$ | Delay of inter-area inhibitory synapses | 9 ms |
| $T_{Poisson}$ | Mean period of Poisson excitation | 333 ms |

- In the absence of inhibition ($g_{1,inh} = 0$), we measured the mean firing rate (MFR) of the network (total number of spikes per second per neuron) as a function of the strength of excitatory synapses $g_{1,exc}$. Poissonian stimulation at a rate of 3 Hz is added to elicit baseline network activity. As is shown in Fig. 14.3(left), for $g_{1,exc} < 0.05$, the MFR is close to the externally imposed Poisson rate. At around $g_{1,exc} \approx 0.05$, activity blows up to a state of higher MFR where spiking is self-sustained and independent of the external input.
- Selecting a value for $g_{1,exc}$ barely above the threshold of sustained MFR ($g_{1,exc} = 0.075$), we increased the strength of inhibitory connections until the activity turned back to the background level as shown in

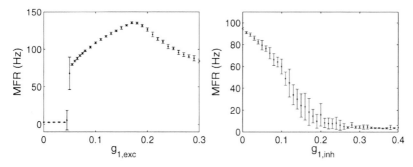

**Fig. 14.3.** (**Left**) Mean firing rate in one area of $n = 512$ neurons as a function of the strength of excitatory synapses for $g_{1,\text{inh}} = 0$. Each point is an average of 25 simulations; (**Right**) Mean firing rate in one area of $n = 512$ neurons as a function of the strength of inhibitory synapses for $g_{1,\text{exc}} = 0.075$. Each point is an average of 25 simulations

Fig. 14.3(right). At this point, excitatory and inhibitory forces within one area are balanced.

In order to check if both forces are balanced, we have measured the MFR as a function of the number of neurons $n$ in one area for two sets of parameters. The first one corresponds to the pair $(g_{1,\text{exc}} = 0.075,\ g_{1,\text{inh}} = 0.25)$, which produces a 3 Hz firing rate (see Fig. 14.3) and the second one, $(g_{1,\text{exc}} = 0.075,\ g_{1,\text{inh}} = 0.1)$, is chosen such that there is more excitation than inhibition. This imbalance is going to be dependent on area size as shown in Fig. 14.4, while for the right selection of the excitatory and inhibitory strengths, the MFR remains constant at background level.

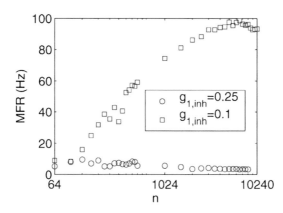

**Fig. 14.4.** Comparison of the MFR as a function of the number of neurons $n$ per area between two sets of coupling strengths. When inhibition is not well balanced, the MFR increases as $n$ becomes larger

### 14.3.2 Tuning Inter-area Parameters

When the 53 areas are coupled with the choice of parameters of the previous section, making $g_{2,exc} = g_{1,exc}$ and $g_{2,inh} = 0$, a firing pattern that we will call "generic" was observed, see Fig. 14.5. All areas show a similar homogeneous pattern of activity, only the firing rate differs. Overexcitation can be determined to be responsible for this result. Therefore, a more extensive search for parameters that yielded spontaneous bursting (the desired physiologic phenomenon) was performed to elucidate suitable ranges and ratios of parameters in the model. In addition, trends were noted on how such parameters affect general behavior, including bursting, spike rates, and propagation through the systems of the simulated cortical structures.

Even though the parameter values chosen for a single area produced activity patterns and firing rates analogous to "natural" activity, the extension to 53 areas, through incorporation of the connectivity matrix, affected the behavior of each area. The primary effect was the introduction of additional activity in each area from all the areas that it is connected to with afferent connections. This extra activity increased the mean firing rate of each area to frequencies higher than desired, higher than the 10–40 Hz range. Since the parameters for excitatory and inhibitory connections within each area have the greatest influence on the mean firing rate, we performed simulations of the whole model with different combinations of these values. Figure 14.6 summarizes the results of all these simulations. From these plots, we chose the parameters $g_{1,inh} = 0.4$ and $g_{2,exc} = 0.075$ as the ones providing the most appropriate firing rates.

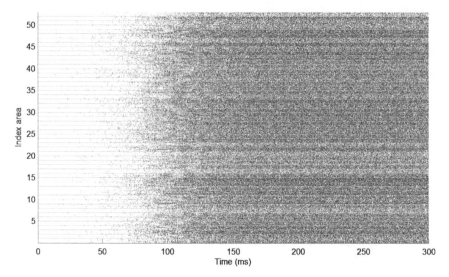

**Fig. 14.5.** Raster plot of the behavior of the whole cortex characterized as "generic". $g_{1,inh} = 0.4$, $g_{2,inh} = 0$, $g_{1,exc} = g_{2,exc} = 0.075$, $p_{ring} = 0.05$, $p_{rew} = 0.2$

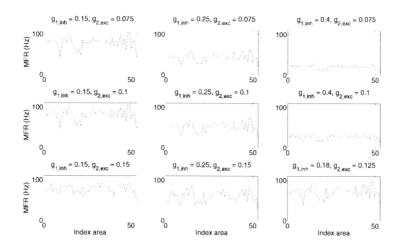

**Fig. 14.6.** Mean firing rate of each area, under nine different combinations of $g_{1,inh}$ and $g_{2,exc}$. $p_{ring} = 0.05$, $p_{rew} = 0.2$. Chosen parameters for further simulations: $g_{1,inh} = 0.4$ and $g_{2,exc} = 0.075$

Despite the fact that our model was now able to produce activity within the desired firing rate range, the overall behavior of areas continued to be rather homogeneous. We interpreted this result as not being "natural" behavior and called it "generic".

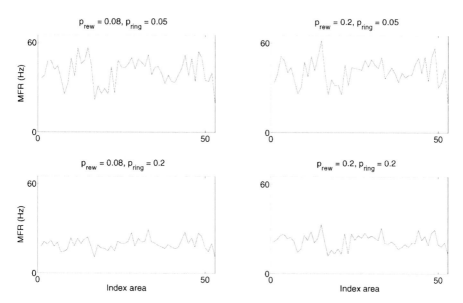

**Fig. 14.7.** Mean firing rate of each area, under four different combinations of $p_{ring}$ and $p_{rew}$. Chosen parameters for later simulations: $p_{ring} = p_{rew} = 0.2$

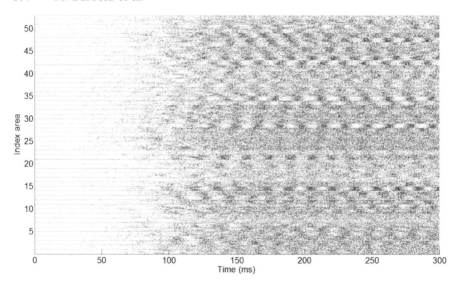

**Fig. 14.8.** Raster plot of the behavior of the whole cortex characterized as "natural". $g_{1,\mathrm{inh}} = 0.4$, $g_{1,\mathrm{exc}} = 0.075$, $p_{\mathrm{ring}} = p_{\mathrm{rew}} = 0.2$

Therefore, a set of further parameter search simulations were carried out where different combinations of the connectivity within each area were investigated, see Fig. 14.7. In this set of simulations, as before, our primary determining factor for "natural" behavior was to conserve the mean firing rate within the desired ranges. The additional condition was to obtain raster plots that presented different patterns of activity. Considering that the highest number of connections between areas is within each system, we desired that the areas of each system behave in a similar manner, and the behavioral pattern of each system be different from each other, hence representing each system's different function. This second condition was satisfied by setting parameters $p_{\mathrm{ring}}$ and $p_{\mathrm{rew}}$ to 0.2 (each area is modeled by a "small-world" subnetwork composed of 512 neurons. The number of connections is controlled by parameter $p_{\mathrm{ring}}$ and the deviation from the initial ring by $p_{\mathrm{rew}}$ (see Chaps. 3 and 9 for detailed descriptions)). Figure 14.8 presents the behavior obtained with these parameters.

## 14.4 Simulation of the Cat Cerebral Cortex

Once we managed to configure our model to behave in the manner of Fig. 14.8, we were more confident that the effects of the connectivity matrix were significant and that the behavior of the model could be described as "natural". We remind the reader that, as discussed in Chap. 3, the given corticocortical network has been found to be divided into four major clusters, corresponding to sensorial systems. Areas indexed 1–16 correspond to *visual cortex*, 17–23 represents the *auditory cortex*, indices 24–39 *somatosensory-motor cortex* and 40−53 *frontolimbic* cortical areas. Next, we aimed to simulate different conditions of the neural network:

- The already achieved natural behavior, under the conditions of background noise (simulated by low frequency Poisson noise). Area coupling propels connected areas into correlated patterns of bursting, behavior remaining similar within each of the 4 cortical systems. Due to the small number of connections between systems, each system is to follow its own independent behavior.
- Stimulation of a single area, and of a whole system. From the sparseness of connection weights in the matrix beyond each system, and the existence of areas operating as "communication hubs", a preferential pathway of inter-system communication is implied. Stimulation of an area, or system, with a simulated external signal was expected to propagate to other systems of the network along this preferential pathway.
- The effects of ablating an area. The existence of only a small number of areas preferential for inter system communication implies a greater significance for these areas for network operation. Removal of these areas from the model should contribute to a great change of the overall activity of the network. To further confirm that our model was adequately representative of cat cerebral cortex and to observe the effects of the removal of such areas, we carried out simulations where such areas were ablated.

We thus proceeded to simulate stimulation of a part of the cortex to observe the propagation of the stimulation and the effects on the non-stimulated parts. Considering that the visual system is one of the most important, if not the primary, stimulus receiving part of the cortex, we simulated our model with stimulation of pulses at 25 Hz on the whole visual system. It is clear from Fig. 14.9, as expected, that the independent behavior of the other systems and areas of the cortex was dominated by the stimulation. Specifically, we can observe that the visual system transfers the introduced activity into the other systems and functions as the driving system for the whole cortex.

Our final simulations considered the effects of damaged tissue. To observe these effects, we simulated our model with stimulation and in addition, we inactivated areas that operated as midpoints in the pathway from the stimulation area to other areas. In these simulations, to represent area "death" or "ablation", we lowered the excitability parameter, $I_{bias}$, of the neurons within the target area slightly below their excitable regime. In addition, since the effects of stimulation of the whole visual cortex were so dominating that subtle changes of the network ("ablation" of an area) were not affecting the overall behavior, we limited ourselves to stimulating only "area 17" (*primary visual cortex*, index: 1) with pulses at 25 Hz. Areas "Ia", "35", and "36" (indexes: 43, 48 and 49; frontolimbic areas) were active (Fig. 14.10) or inactive (Fig. 14.11). Inactivation of these frontolimbic areas can be seen to

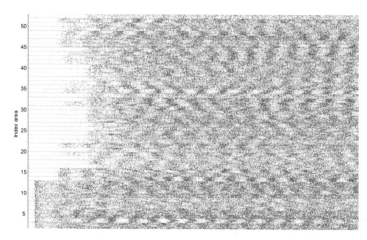

**Fig. 14.9.** Raster plot of the behavior of the whole cortex characterized under stimulation of the whole visual cortex

affect the auditory system (indices: 17–23) lowering its activity and synchronization as expected, since part of the functions of the limbic system is to connect the visual system with the auditory system.

Aside from configuring and simulating our model to observe the above behaviors, we also attempted a simulation with a large number of neurons ($> 10^6$). This was done mostly as a computational task and the parameters for this simulation were not in accordance to the other simulations. Regardless, the behavior exhibited by our model with this large number of neurons is

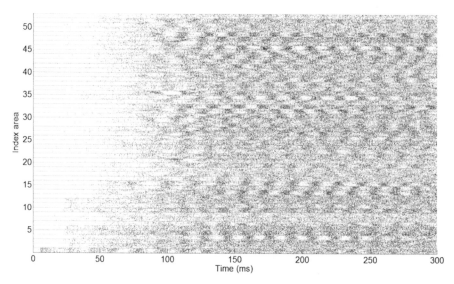

**Fig. 14.10.** Raster plot of stimulation of area "17" (primary visual cortex, index: 0)

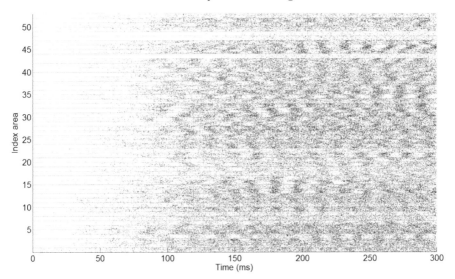

**Fig. 14.11.** Stimulation of area "17", while areas "LA", "35", and "36" (indices: 43, 48 and 49; frontolimbic cortex) are inactive

presented in Fig. 14.12. Considering that the parameters were different from our other simulations, the result is obviously misleading. We believe the strong oscillatory pattern observed to be related to the dynamics of the Morris-Lecar neuronal model. Looking at its bifurcation diagram, shown in Fig. 14.1, we argue that for the given parameters, all neurons must have collectively followed similar changes in their dynamical states. When so many neurons are present

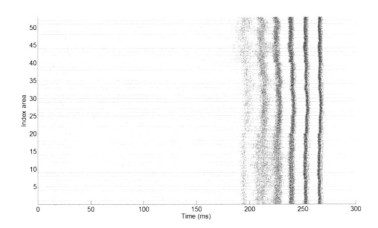

**Fig. 14.12.** Raster plot of the behavior of a cortex of $1,085,440$ neurons, which corresponds to $20,480$ neurons per area. $g_{1,\text{inh}} = 0.4$, $g_{1,\text{exc}} = 0.075$, $g_{2,\text{exc}} = 0.15$, $p_{\text{ring}} = 0.05$, $p_{\text{rew}} = 0.1$

within each cortical area, each neuron receives an extremely large amount of input, which, after a brief period of intense firing, raises their $I_{\text{bias}}$ triggering all neurons beyond the $SNLC$ point into the "silent regime". Once all neurons are silent, only noise is present in the system, allowing neurons to recover their "excitatory state" and start firing again after a brief pause. Unfortunately, we could not perform further simulations of this size due to computer time limitations.

## 14.5 Dependence of MFR on Anatomical Connectivity

In this section, we present a brief attempt to explore the relationship between the observed behavior of the simulated system and the structural properties of the network. We will look for correlations between the characteristic firing rate of each cortical area obtained in the simulations with its degree and intensity. We will also try to find a simple analytical solution to explain the observed dynamics.

In Sect. 14.4, firing rates of individual cortical areas were estimated from the simulations. Frequency is observed to be modulated within about 10–20 ms (look at the fine structure of Fig. 14.8). On the other hand, under stationary conditions it varies significantly from area to area (Fig. 14.13). The variation of $g_{2,\text{exc}}$ (inter-areal excitatory coupling strength) and $g_{1,\text{inh}}$ (intra-areal inhibitory coupling strength) contributes to the absolute value of the mean firing rate. In the chosen parameter range, $g_{2,\text{exc}} \in [0.075, 0.15]$ and

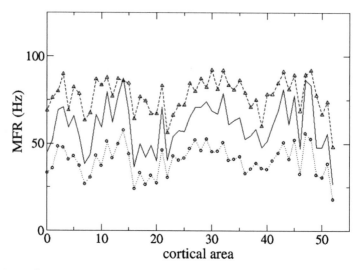

**Fig. 14.13.** Dependence of the firing rate on the internal and external coupling strength. $g_{2,\text{exc}} = 0.15$ in all cases. Dashed line: $g_{1,\text{inh}} = 0.15$, solid: $g_{1,\text{inh}} = 0.25$, dotted: $g_{1,\text{inh}} = 0.4$

$g_{1,\text{inh}} \in [0.15, 0.4]$, the mean firing rate of the areas was found to lie between 25 to 100 Hz. The main curve profile remains unchanged for different values of $g_{1,\text{inh}}$ and $g_{2,\text{exc}}$. Modification of the $g_{1,\text{inh}}$ coupling shifts the curve in the vertical direction, i.e. higher frequencies are achieved with lower inhibition. This fact indicates the importance of inhibitory connections in the modulation of brain dynamics. Indeed, inhibitory coupling has been shown to suppress oscillations induced by the excitatory coupling as is known to happen in pathology such as epileptic seizures caused by excessive synchronization of neuronal activity [39].

### 14.5.1 Correlation to $k^{\text{in}}$ and $S^{\text{in}}$

The input degree of a node $k^{\text{in}}$ refers to the number of connections a node receives. Its natural extension for weighted networks, the input intensity of a node $S^{\text{in}}$, is the sum of the strength of its input connections. Although there are many existing network measures (see Chap. 3 for descriptions of network characterization and properties of the cat cortex), here, we will only explore the relationship between the mean firing rate and $k^{\text{in}}$ and $S^{\text{in}}$.

The average response of a cortical area depends directly on the amount of input signal received, thus we will correlate the MFR to $k^{\text{in}}$ and $S^{\text{in}}$ of the cortical areas. Linear correlation of both measures with the MFR obtained from simulations is depicted on Fig. 14.14. As expected, it is a monotonously increasing function of $k^{\text{in}}$ and $S^{\text{in}}$. The more input a cortical area receives, the more often its neurons will fire. The linear fit is better in the case of intensity, since the relation between the MFR and $k^{\text{in}}$ slightly saturates at high degrees ($k^{\text{in}} \geq 20$). This saturation is more pronounced in other parameter sets (not shown). Figure 14.17 shows that correlation between the MFR and $S^{\text{in}}$ is higher for most of the parameter sets.

### 14.5.2 Analytical Estimation

In the following, our modest effort to model the observed MFRs for each cortical area is presented. A commonly used approach in artificial neural networks

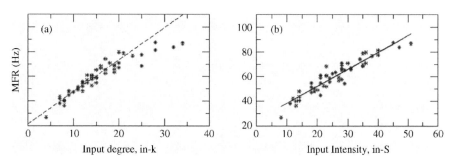

**Fig. 14.14.** Correlation of average simulated Mean Firing Rates per area (MFR) with: **(a)** degree and; **(b)** intensity

is to define an activation function describing the average response of neurons to input received from its neighbors. In a general form, the equations are written as:

$$r_i(t+1) = F(h_i)$$

$$h_i = a \sum_{j=1}^{N} W_{ij} r_j(t) + \xi, \quad i = 1, ..., N,$$

$r_i$ being the activity of the neuron, $W_{ij}$ the weighted adjacency matrix of the network and $\xi$ some external input, i.e. noise. $F(h_i)$ is usually some sigmoidal saturation function and is normalized either to $[0, 1]$ or $[-1, 1]$. Parameter $a$ controls the slope of the saturation function tuning the scale of the response. Such a function sums up all inputs the neuron receives and returns a normalized output representing its average activity response.

Similar approaches have already been used for cortical models of the cat [40, 41]. Here, we simulate cortical areas instead of individual neurons. This approach is reasonable since the mean activation level of a cortical area strongly depends on the amount of input received from its neighbors in a cumulative and smooth manner, which could be highly arguable in the case of individual neurons [42, 43]. As said above, the number of firing neurons scales with the input an area receives. Here, mean activity level will be considered to be equivalent to MFR. The simplest choice is to assume a linear approximation

$$F(h_i) = \alpha h_i + \beta, \text{ where } \beta = 0$$

(the saturation function crosses the origin). Our study is then limited to estimating the slope parameter $a$ and noise level $\xi$ to provide an optimal approximation to the results obtained from the simulations.

In the steady state $r_i(t+1) = r_i(t)$ and after taking $F(h_i)$ to be linear, equations reduce to:

$$r_i = \alpha \left( a \sum_{j}^{N} W_{ij} r_j + \xi \right).$$

After rescaling, we arrive at the following equation in matrix form:

$$\boldsymbol{r'} = (\boldsymbol{I} - a'\boldsymbol{W})^{-1} \boldsymbol{\xi'}, \tag{14.7}$$

where $\boldsymbol{r'}$ is the MFR vector of the 53 areas, $\boldsymbol{I}$ is the identity matrix, $\boldsymbol{W}$ is the adjacency matrix of the cortical network, $a' = \alpha a$, and $\boldsymbol{\xi'}$ is a column vector where all elements are $\alpha\xi$.

Our main purpose is to find the coupling strengths $g_{1,\text{inh}}$ and $g_{1,\text{exc}}$ that produce estimated MFRs as closely correlated to the MFRs from our simulations as possible. We face the problem of setting both slope $a'$ and noise

$\xi'$ parameters satisfactorily. There is a limited range of $a'$s we can examine because the maximum eigenvalue of our dynamical system (given by the adjacency matrix $\boldsymbol{W}$) is $\lambda_{\max} = 29.08$, and thus $a' \approx 1/\lambda_{\max} = 0.034385$. At this value, the system has a singularity and solutions with larger $a'$ are unstable. Correlation is "blind" to this singularity, and thus, in order to look for proper $a'$ and $\xi'$, we will calculate the Euclidean distance between the estimated $\boldsymbol{r'}$ vectors and the vectors of MFRs for different $g_{1,\text{inh}}$ and $g_{2,\text{exc}}$ obtained from our simulations. Figure 14.15a shows how estimated MFRs differ from simulated MFRs for different values of $a'$. In this representation, the singularity is clearly observed as the distance between $\boldsymbol{r'}$ and MFR vectors grows to infinity (Fig. 14.15b).

Importantly, Fig. 14.15(b) also shows the presence of a minimal distance, so in the following, our optimization problem is to find the closest $\boldsymbol{r'}$ solutions to the MFR from simulations. Among the stable solutions ($a' < 0.034385$), the shortest distance depends both on $a'$ and $\xi'$ for each coupling strength combination. After solving (14.7) for different parameters $a'$ and $\xi'$, optimal values were found for each combination of $g_{1,\text{inh}}$ and $g_{2,\text{exc}}$ as summarized in Table 14.2.

We are now ready to look for the optimal coupling strengths. For each pair of $g_{1,\text{inh}}$ and $g_{2,\text{exc}}$, the simulated MFRs and the vector $\boldsymbol{r'}$ estimated for corresponding optimal $a'$ and $\xi'$ (see Table 14.2) are correlated. Results are shown in Fig. 14.16. Note that the best correlation is for the values of the coupling parameters that produce lower MFRs. Interestingly, the properly balanced inhibitory coupling allows us to achieve both "natural behavior" and maximal correlation to the linear approximation. Too low as well as too high inhibition gives rise to a marked decrease of this correlation for all excitatory values.

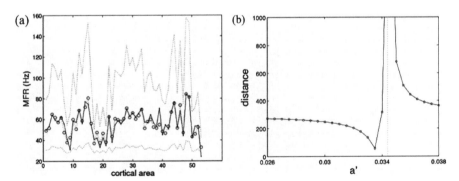

**Fig. 14.15. a)** Simulated MFR of the 53 cortical areas (*solid*) and estimated MFRs with $\xi = 24.5$ and different $a'$ values (*dotted lines*). Distance between simulated and estimated MFRs varies significantly for different parameters; **b)** Setting $\xi' = 1.0$, the singularity of the dynamical system appears as distance going to infinity at $a' = 0.034385$. Larger values of $a'$ represent unstable solutions

**Table 14.2.** Optimal values of slope $a'$ and noise $\xi'$ for different coupling strengths

| $g_{2,exc}$ | $g_{1,inh}$ | opt $a'$ | opt $\xi'$ | $g_{2,exc}$ | $g_{1,inh}$ | opt $a'$ | opt $\xi'$ |
|---|---|---|---|---|---|---|---|
| 0.075 | 0.15 | 0.0145 | 47.0 | 0.125 | 0.15 | 0.0135 | 50.0 |
| | 0.18 | 0.0175 | 33.5 | | 0.18 | 0.0175 | 37.5 |
| | 0.2 | 0.0185 | 28.0 | | 0.2 | 0.0190 | 32.0 |
| | 0.25 | 0.0200 | 19.0 | | 0.25 | 0.0205 | 24.5 |
| | 0.3 | 0.0200 | 14.5 | | 0.3 | 0.0210 | 20.0 |
| | 0.4 | 0.0195 | 9.0 | | 0.4 | 0.0195 | 16.0 |
| 0.1 | 0.15 | 0.0140 | 50.0 | 0.15 | 0.15 | 0.0130 | 50.0 |
| | 0.18 | 0.0175 | 36.0 | | 0.18 | 0.0165 | 39.5 |
| | 0.2 | 0.0190 | 30.0 | | 0.2 | 0.0190 | 32.5 |
| | 0.25 | 0.0205 | 22.0 | | 0.25 | 0.0220 | 23.5 |
| | 0.3 | 0.0200 | 18.5 | | 0.3 | 0.0215 | 21.5 |
| | 0.4 | 0.0210 | 11.5 | | 0.4 | 0.0205 | 17.5 |

Increasing the excitatory coupling $g_{2,exc}$ produces, in general, a monotonous increase of correlation.

Finally, we compare the results from this analytical linear estimation to the effects of degree and intensity distribution. Vectors of input degrees $k^{in}$ and input intensities $S^{in}$ of cortical areas are correlated to the MFRs from the simulation as shown in Fig. 14.17. As expected from the weighted nature of the adjacency matrix, intensities do correlate better than degrees. After all the optimization effort, high correlation values suggest that our model behaves as a linear system for a certain range of coupling strengths (see Fig. 14.17). This is also supported by the high correlation between simulated MFRs and intensities $S^{in}$. Indeed, the analytical estimations show only slightly better correlation than the $S^{in}$.

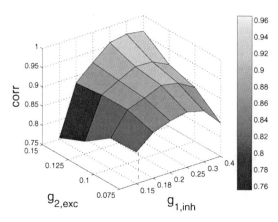

**Fig. 14.16.** Correlation between MFR from simulations and estimated $r'$ using optimal $a'$ and $\xi'$ for each set of $g_{1,inh}$ and $g_{2,exc}$

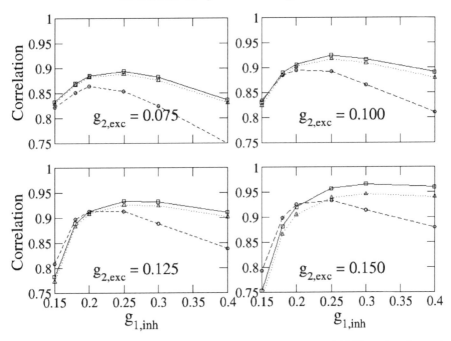

**Fig. 14.17.** Comparison of the correlation between computed MFR and estimated MFR at parameter $a'$ (*solid line*), degree (*dashed*), intensity (*dotted*)

## 14.6 Conclusions and Outlook

During this set of computational exercises, we learned principles of large-scale neuronal simulations using the parallel code initially given to us. We defined a method to look for parameters that would provide a *realistic behavior*. We observed that finding suitable parameters and obtaining robust behavior is difficult. First, for the internal sub-network representing each cortical area, a set of excitatory/inhibitory strengths was found that would provide stable response with increasing size of the sub-network (see Fig. 14.4). The requirement was to keep average firing rate of individual neurons around 3 Hz. Then, after connecting the 53 subnetworks and assuming equal $g_{1,\text{exc}}$ and $g_{2,\text{exc}}$, the network showed too high a MFR, which was re-balanced by increasing the strength of inhibitory connections (Fig. 14.7). But this change happened to break the balance again as seen in the simulation with a million neurons, where the effect of scaling is evident.

However, raster plots displayed rather homogeneous and similar behavior of all cortical areas (Fig. 14.5), a behavior we called "generic". Looking for different, more realistic behaviors, we performed a parameter search on $p_{\text{ring}}$

and $p_{rew}$ in order to change the network structure and hopefully also the behavior. A regime in which cortical areas exhibit bursting and silent epochs was found providing more interesting dynamics that we called "natural behavior" (see Fig. 14.8).

Finally, as discussed in Sect. 14.5, our analytical estimations show that the model behaves on average as a linear network model where, apart from the bursting dynamics, the MFR of each cortical area is highly proportional to the total input received (see Fig. 14.14).

Several open questions remain and a large set of possible implementations can be tried out:

First, we are aware of the arbitrary manner in which "natural behavior" was characterized: based exclusively on keeping the MFR at biologically reliable levels observed experimentally and on visual inspection of raster plots to avoid homogeneous dynamical responses. Thus, more convenient methodology founded on different measures would be desirable in order to characterize and classify the observed dynamics. Such measures could depend on, e.g., temporal correlations, frequency content, information transfer, etc.

Second, in our simulations, the small-world network topology following the Watts-Strogatz model was used for the internal neural connections within one cortical area. This topology has already been shown to enhance signal propagation and network synchronization which are so important for exchange of information.

It would be desirable, however, to introduce more realistic internal connectivities modeling finer cortical structures like layers, columns and if possible, the morphology of cortical neurons. Current sparse knowledge of detailed connectivity at the neuronal level makes such an implementation improbable in the nearest future. An intermediate solution might rely on taking just a small set of cat's cortical neurons, extracting their approximate local topology and randomly replicating it in order to mimic the internal structure of a cortical area. On the other hand, representing cortical layers and all the available experimental data about their interconnectivity offers an interesting opportunity to improve the internal architecture of the model. A first initial step should be the modeling of hierarchical organization of the cortex introducing hierarchical subnetworks for each cortical area rather than the small-worlds used here.

And finally, the linear behavior described by the model is not expected in real brains. As a complex system per excellence, the brain does not perform only such trivial behavior. Further modifications might include the introduction of delays and improved simulation strategies to limit the continuous spread of activity typical of pathological situations. For future work, we should remark that the dependence of the mean firing rates on the intra-areal connectivity among neurons is yet to be tested. Observing and characterizing brain activity under external stimuli is also of high interest.

# References

1. W. S. Armbruster, Evolution of floral form: electrostatic forces, pollination, and adaptive compromise, New Phytol., 152(2):181–183, 2001.
2. M. Golubitsky, I. Stewart, P.-L. Buono and J. J. Collins, A modular network for legged locomotion, Physica D, 115(1):56–72, 1998.
3. M. Golubitsky, M. Pivato and I. Stewart, Interior symmetry and local bifurcation in coupled cell networks, Dyn. Sys., 19(4):389–407, 2004.
4. H. Levine and E. Ben-Jacob, Physical schemata underlying biological pattern formation—examples, issues and strategies, Phys. Biol., 1:14–22, 2004.
5. S. Douady and Y. Couder, Phyllotaxis as a physical self-organized growth process, Phys. Rev. Lett., 68(13):2098–2101, 1992.
6. V. M. Eguíluz, M. Ospeck, Y. Choe, A. J. Hudspeth and M. O. Magnasco, Essential nonlinearities in hearing, Phys. Rev. Lett., 84(22):5232–5235, 2000.
7. P. Martin and A. J. Hudspeth, Compressive nonlinearity in the hair bundle's active response to mechanical stimulation, Proc. Natl. Acad. Sci. USA, 98(25):14386–14391, 2001.
8. PhysicsWeb, Best of physicsweb, Best of Physics in Biology, http://physicsweb. org/bestof/biology, 2006.
9. C. Koch and I. Segev, The role of single neurons in information processing, Nat. Neurosci., 3:1171–1177, 2000.
10. B. J. O'Brien, T. Isayama, R. Richardson and D. H. Berson, Intrinsic physiological properties of cat retinal ganglion cells, J. Physiol., 538(3):787–802, 2002.
11. R. H. Masland, The fundamental plan of the retina, Nat. Neurosci., 4(9): 877–886, 2001.
12. C. F. Stevens, Models are common; good theories are scarce, Nat. Neurosci., 3:1177, 2000.
13. L. C. Jia, M. Sano, P.-Y. Lai and C. K. Chan, Connectivities and synchronous firing in cortical neuronal networks, Phys. Rev. Lett., 93:088101, 2004.
14. J. van Pelt, I. Vajda, P. S. Wolters, M. A. Corner, W. L. C. Rutten and G. J. A. Ramakers, Dynamics and plasticity in developing neuronal networks in vitro, Prog. Brain Res., 147:173–188, 2005.
15. A. Van Ooyen, Competition in neurite outgrowth and the development of nerve connections, Prog. Brain Res., 147:81–99, 2005.
16. E. M. Izhikevich, Dynamical systems in neuroscience: the geometry of excitability and bursting, MIT Press, 2007.
17. E. M. Izhikevich, Which model to use for cortical spiking neurons, IEEE Trans. Neural Netw., 15(5):1063–1070, 2004.
18. R. C. Elson, A. I. Selverston, R. Huerta, N. F. Rulkov, M. I. Rabinovich and H. D. I. Abarbanel, Synchronous behaviour of two coupled biological neurons, Phys. Rev. Lett., 81(25):5692–5695, 1998.
19. M. Rabinovich, A. Volkovskii, P. Lecanda, R. Huerta, H. D. I. Abarbanel and G. Laurent, Dynamical encoding by networks of competing neuron groups: winnerless competition, Phys. Rev. Lett., 87(6):068102, 2001.
20. C. J. Rennie, P. A. Robinson and J. J. Wright, Unified neruophysical model of EEG spectra and evoked potentials, Biol. Cybern., 86:457–471, 2002.
21. J. J. Wright, C. J. Rennie, G. J. Lees, P. A. Robinson, P. D. Bourke, C. L. Chapman, E. Gordon and D. L. Rowe, Simulated electrocortical activity at microscopic, macroscopic and global scales, Neuropsychopharmacology, 28: 80–93, 2003.

22. P. A. Robinson, C. J. Rennie, D. L. Rowe, S. C. O'Connor, J. J. Wright, E. Gordon and R. W. Whitehouse, Neurophysical modeling of brain dynamics, Neuropsychopharmacology, 28:74–79, 2003.

23. H. R. Wilson and J. D. Cowan, A mathematical theory of the functional dynamics of cortical and thalamic neuron tissue, Kybernetik, 13:55–80, 1973.

24. M. Bazhenov, N. F. Rulkov, J.-M. Fellous and I. Timofeev, Role of network dynamics in shaping spike timing reliability, Phys. Rev. E, 72:041903, 2005.

25. G. Tanaka, B. Ibarz, M. A. F. Sanjuan and K. Aihara, Synchronization and propagation of bursts in networks of coupled map neurons, Chaos, 16:013113, 2006.

26. G. A. Ascoli, Progress and perspectives in computational neuroanatomy, Anat. Rec. (New Anat.), 257(6):195–207, 1999.

27. P. C. Bressloff, Resonantlike synchronization and bursting in a model of pulse-coupled neurons with active dendrites, J. Comput. Neurosci., 6:237–249, 1999.

28. S. M. Korogod, I. B. Kulagina, V. I. Kukushka, P. Gogan and S. Tyc-Dumont, Spatial reconfiguration of charge transfer effectiveness in active bistable dendritic arborizations, Eur. J. Neurosci; 16:2260–2270, 2002.

29. P. C. Bressloff and S. Coombes, Synchrony in an array of integrate-and-fire neurons with dendritic structure, Phys. Rev. Lett., 78(24):4665–4668, 1997.

30. L. da F. Costa, Morphological complex networks: can individual morphology determine the general connectivity and dynamics of networks?, oai:arXiv.org:q-bio/0503041, 2005.

31. L. F. Lago-Fernández, R. Huerta, F. Corbacho and J. A. Sigüenza, Fast response and temporal coherent oscillations in small-world networks, Phys. Rev. Lett., 84(12):2758–2761, 2000.

32. M. I. Rabinovich, P. Varona, A. I. Selverston and H. D. I. Abarbanel, Dynamical principles in neuroscience, Rev. Mod. Phys., 78:1213–1265, 2006.

33. S. Boccaletti, V. Latora, Y. Moreno, M. Chavez and D.-U. Hwang, Complex networks: structure and dynamics, Phys. Rep., 424:175–308, 2006.

34. H. Markram, The blue brain project, Nat. Rev. Neurosci., 7:153–160, 2006.

35. R. D. Traub and R. K. Wong, Cellular mechanism of neuronal synchronization in epilepsy, Science, 216:745–747, 1982.

36. A. Morrison, C. Mehring, T. Geisel, A. Aertsen and M. Diesmann, Advancing the boundaries of high-connectivity network simulation with distributed computing, Neural Comput., 17:1776–1801, 2005.

37. J. W. Scannell, G. A. P. C. Burns, C. C. Hilgetag, M. A. O'Neill and M. P. Young, The connectional organization of the cortico-thalamic system of the cat, Cereb. Cortex, 9:277–299, 1999.

38. J. Rinzel and G. B. Ermentrout, Analysis of neural excitability and oscillations, in Methods in neuronal modeling: from synapses to networks, ed. C. Koch and I. Segev, 135–169, MIT Press, Cambridge, MA, 1989.

39. P. Kudela, P. J. Franaszczuk and G. K. Bergey, Changing excitation and inhibition in simulated neural networks: effects on induced bursting behavior, Biol. Cybern., 88:276–285, 2003.

40. R. Kötter and F. T. Sommer, Global relationship between anatomical connectivity and activity propagation in the cerebral cortex, Phil. Trans. R. Soc. Lond. B, 355:127–134, 2000.

41. M. P. Young, C. C. Hilgetag and J. W. Scannell, On imputing function to structure from the behavioural effects of brain lesions, Phil. Trans. R. Soc. Lond. B, 355:147–161, 2000.

42. W. J. Freeman, Tutorial on neurobiology: From single neurons to brain chaos, Int. J. Bifurcation Chaos, 2(3):451–482, 1992.
43. P. beim Graben and J. Kurths, Simulating global properties of electroencephalograms with minimal random neural networks, Neurocomputing, doi: 10.1016/j.neucom.2007.02.007, 2007.

# Index

# Understanding Complex Systems

Jirsa, V.K.; Kelso, J.A.S. (Eds.)
Coordination Dynamics: Issues and Trends
XIV, 272 p. 2004 [978-3-540-20323-0]

Kerner, B.S.
The Physics of Traffic:
Empirical Freeway Pattern Features,
Engineering Applications, and Theory
XXIII, 682 p. 2004 [978-3-540-20716-0]

Kleidon, A.; Lorenz, R.D. (Eds.),
Non-equilibrium Thermodynamics
and the Production of Entropy
XIX, 260 p. 2005 [978-3-540-22495-2]

Kocarev, L.; Vattay, G. (Eds.)
Complex Dynamics in Communication
Networks
X, 361 p. 2005 [978-3-540-24305-2]

McDaniel, R.R.Jr.; Driebe, D.J. (Eds.)
Uncertainty and Surprise in Complex Systems:
Questions on Working with the Unexpected
X, 200 p. 2005 [978-3-540-23773-0]

Ausloos, M.; Dirickx, M. (Eds.)
The Logistic Map and the Route to Chaos –
From the Beginnings to Modern Applications
XX, 413 p. 2006 [978-3-540-28366-9]

Kaneko, K.
Life: An Introduction to Complex Systems
Biology
XIV, 369 p. 2006 [978-3-540-32666-3]

Braha, D.; Minai, A.A.; Bar-Yam, Y. (Eds.)
Complex Engineered Systems – Science Meets
Technology
X, 384 p. 2006 [978-3-540-32831-5]

Fradkov, A.L.
Cybernetical Physics – From Control of Chaos
to Quantum Control
XII, 241 p. 2007 [978-3-540-46275-0]

Aziz-Alaoui, M.A.; Bertelle, C. (Eds.)
Emergent Properties in Natural
and Artificial Dynamical Systems
X, 280 p. 2006 [978-3-540-34822-1]

Baglio, S.; Bulsara, A. (Eds.)
Device Applications of Nonlinear Dynamics
XI, 259 p. 2006 [978-3-540-33877-2]

Jirsa, V.K.; McIntosh, A.R. (Eds.)
Handbook of Brain Connectivity
X, 528 p. 2007 [978-3-540-71462-0]

Krauskopf, B.; Osinga, H.M.;
Galan-Vioque, J. (Eds.)
Numerical Continuation Methods
for Dynamical Systems
IV, 412 p. 2007 [978-1-4020-6355-8]

Perlovsky, L.I.; Kozma, R. (Eds.)
Neurodynamics of Cognition and Consciousness
XI, 366 p. 2007 [978-3-540-73266-2]

Qudrat-Ullah, H.; Spector, J.M.; Davidsen, P. (Eds.)
Complex Decision Making – Theory and Practice
XII, 337 p. 2008 [978-3-540-73664-6]

beim Graben, P.; Zhou, C.; Thiel, M.;
Kurths, J. (Eds.)
Lectures in Supercomputational Neuroscience –
Dynamics in Complex Brain Networks
X, 378 p. 2008 [978-3-540-73158-0]

Hunter, Mollie

COPY 1

You never knew her
as I did!

# You Never Knew Her As I Did!

# You Never Knew Her As I Did!

by Mollie Hunter

HARPER & ROW, PUBLISHERS

Library of Congress Cataloging in Publication Data
Hunter, Mollie, date
    You never knew her as I did!

    SUMMARY: Will Douglas, a seventeen-year-old page,
attempts to free Mary, Queen of Scots, from her island
prison.
    1. Mary, Queen of Scots, 1542–1587—Juvenile fiction.
[1. Mary, Queen of Scots, 1542–1587—Fiction.  2. Scotland—History—1057–1603—Fiction]  I. Title.
PZ7.H9176Yo 1981      [Fic]        81-47114
ISBN 0-06-022678-1                 AACR2
ISBN 0-06-022679-X (lib. bdg.)

*To my nephew and godson*
*Connal McIlwraith*
*A book of his own at last!*
*With love from*
  *Mollie Hunter*

## Foreword

This book tells of an episode in the life of Mary, Queen of Scots; and since she is one of the most controversial characters in all history, readers may wonder how much of the story is fact and how much is invention.

The answer here is that I have spent many years researching both the episode and its historical background, and in creating my novel I have kept faithfully to the findings of my research. Where there is any gap in the records of the period, the bridging inference I have drawn has invariably been the one I conceived as that most in line with known events and most true to the characters of those involved.

*M. H.*

*N*ews of her has come at last from England; and it is the worst I ever dreaded to hear. She has been executed—her head chopped off with an axe. "*. . . of the kind woodsmen use . . .*" the report says.

A voice far back in my mind is screaming denial of that report; and yet still I know it is true. I can trust the agent who wrote it. And the courier who brought it—the young fellow facing me now—was trained by me, personally. I know that he, too, is reliable. Besides, I have the proof of it all in the letters he also brought me. I must face it squarely, it seems. Mary, Queen of Scots—*my* Queen—is dead.

I wish the courier would stop staring at me—but perhaps he has never seen a man weep before this. He is so young—a boy, just turned sixteen. A boy can go safely where a man would be suspected. That was why I chose him for training, as I have chosen all my couriers for almost twenty years past now—oh God, twenty years! Has it been that long, the time she spent captive in England?

And for me? Twenty years in her service, scheming on her behalf in every royal court in Europe; even this raw young fellow knows it has been like that for me, does he not? And so he stares. Or does he wonder if I weep because I too may be one of the many who loved her simply as a man loves a woman?

"She is dead." I say the words aloud, suddenly not caring at all what he thinks of my choking voice. The years of service are at an end now. There is nothing more I can do for her. But that is hard to accept; so very hard. I wonder . . . is that why my memory is flooding now with the brave days past when all her fortunes rested on what I *could* do for her? Is it the helpless feeling I have now that makes me want so much to recall those days? The courier's presence is an intrusion on that desire.

"Go through to Minny," I tell him. "She will take care of your needs."

He looks blankly at me until I recollect he knows Minny only by the courtesy title I have given her. "Mistress Douglas," I say impatiently; and he turns to leave me.

The tide of memory begins to surge again; and now, thank God, I have the relief of letting myself drift with it. . . .

he seventeenth day of June
in the year of 1567; that was when they brought her, captive,
to our island Castle of Lochleven. I recall that very precisely.

We were waiting for her in a tight and silent group
on the landing stage there: Sir William Douglas, his brothers Robert and George, and myself Will Douglas, page
to Sir William—and also his bastard son, of course, although that relationship was never openly admitted.

At our backs, we had the grim and ancient walls of
the Castle. Facing us across the narrow strip of water between island and mainland, was the New House—the man-
sion Sir William had built to accommodate his mother
and seven sisters as well as his own growing brood of
children. It was five o'clock in the morning; dawn, with
nothing moving except the boat scudding out from the

mainland towards us. The eyes of the Douglas brothers were steady on the boat. Mine roved between it and their faces. How would they take the Queen's arrival here?

Sir William had honestly approved of the rebellion the Scottish lords had raised against her. More than that, he was hereditary Keeper of Lochleven Castle, and conscientious in his duties. Sir William, I thought, would have little trouble now in viewing the Queen as a Prisoner of State. And I could discount Robert Douglas. He was not part of the Castle garrison. Robert was there that morning only to act as witness to her actual landing. But what about George?

George was Commander of the Castle Guard. But George was still only twenty-four—the same age as the Queen; *and* handsome enough to have "Pretty Geordie" for his nickname. He was chivalrous too; a true romantic. And so how did *he* view the prospect of acting as jailer to a Queen who was also the most beautiful woman in Europe?

The boat came near enough to identify the two armoured men keeping guard over the Queen. George turned to his brothers. "The two we expected," he said. "Lord Ruthven, and our own dear brother-in-law Lord Lindsay."

"A cunning man, Ruthven," Robert Douglas remarked. "Watch out for him, William."

"While you depart from here," Sir William grumbled, "washing your hands of the trouble I might have."

"Lindsay will be the one to give trouble," George said. "There is no one half so ruthless as he is."

The other two nodded agreement to this; and, plaintively, Sir William added, "God knows why sister Euphemia chose to marry *that* man."

4

*God knows indeed*, I thought. Everyone was afraid of Lindsay—even his own young daughter, Margaret. Or so I had been told, at least, by Sir William's daughter, Ellen. And how could I doubt Ellen when I knew the close friendship there was between her and Lindsay's girl?

Suddenly, I found, I was sorry for the Queen; and the feeling surprised me. I had never been given to spending sympathy on anyone except myself, after all; and this Mary, Queen of Scots, was no part of my troubles. But even so . . . She had been kind to me, in the past. How was she bearing up now, I wondered, under the fact of being so completely at the mercy of men like Ruthven and Lindsay?

I had my answer to that within seconds of the boat's touching at the landing stage.

Ruthven was first ashore, the warrant for the Queen's committal outstretched in his hand. Lindsay was close on Ruthven's heels. The two young waiting ladies seated behind the Queen gathered up small pieces of baggage, while the Queen herself rose to step onto the landing stage. Her long, amber-coloured eyes flashed an upward glance at Ruthven and Lindsay. Her voice came with a cutting edge to the sweetness I remembered for it.

"I will have your heads for this!"

In a flare of black satin skirts and stockings of gold thread, she was ashore then unaided, in a leap as nimble as any I have ever seen. She looked down at the two waiting lords—as indeed from her slender height she could look down on most men. Ruthven's face had gone corpse white. Lindsay was dusky red with anger. He choked, on venom struggling for voice, and got the words of his retort out at last.

*5*

"You adulterous bitch, you! You French whore!"

She turned her back on him. The rabble soldiery had yelled the same words at her, only thirty-six hours before, when she had been forced to surrender to Lindsay himself and the others of her rebel lords. But this time she did not weep or tear her hair, as she had so despairingly done then. Calmly, instead, as Lindsay went on raving at her back, she faced Sir William and told him,

"I have been brought here by force, Sir William, and you know that. You know also that you have no right to detain me."

"Your Grace, I have the instructions of the Privy Council." Sir William was red with embarrassment over Lindsay's scene as he struggled to state his case to her. "And if they are agreed you are to be held in Lochleven Castle, I have no choice but to do my duty as its Keeper."

"Your first duty," she retorted, "is to me. And I defy the warrant of your instructions. The Privy Council—as you very well know—is ruled now by those same lords who are in rebellion against me."

"Because of your misgovernment, your Popish idolatry, your whoring with the Earl Bothwell!" Behind her back, Lindsay was still obscenely shouting, and now there were tears in her eyes. But now also, I realised, she had found one man there to champion her.

Another voice roared out suddenly, the voice of young George Douglas. In a parade-ground bellow that easily drowned the sound of Lindsay's snarl, he began a series of orders to the contingent of the Castle Guard lined up behind the landing stage. The stamp of feet, the clashing of arms, were added to the din of George's commands. And then, in the small silence that followed the end of this noisy display, the Queen made a move that abruptly snatched the initiative from us all.

She began walking alone towards the gate of the Castle. And short though the distance was between that gate and the landing stage, her action still stated her case better than any words could have done. Lochleven, it reminded everyone there, was a Royal castle. She, and she alone, was its owner. And every step she took now was both a denial of anyone's right to escort her there like a criminal, and an assertion of her own right to enter it at any time she pleased.

We stood like figures in a tableau watching her; the soldiers frozen in the last position George had given them, her own ladies fixed in the very postures they had taken on stepping ashore. I would have raised a cheer for her then, if I had not been as dumbstruck with surprise as all the others! And yet, with every second I watched her, I was aware also of another feeling creeping in on me.

She was walking, as always, with a willowy grace to all her movements, light and sure on her feet as a dancer. Yet even as I observed this, it seemed to me I could see her as she had been on the last occasion she had entered Lochleven.

Two years before that time, it had been, on one of her many justice tours around the country. And she had been dressed all in her favourite white, with rubies and pearls thick clustered on the satin gleam of her gown, a hawk on her wrist with jewelled jesses trailing from its legs. All around her too, as she moved with that dancer's step to the gate, there had been a chattering, shifting, laughing throng of courtiers competing for her favour.

*All around her* . . . That was the key, I thought, to the

feeling her gesture had stirred in me. It was not the difference in dress that mattered; not the contrast between her former splendour and the present plainness of her appearance. It was that sense of her utter aloneness; that oddly poignant sense . . .

George Douglas had something of the same feeling. I thought so, at least, from the way that *he* stared after her, and was more than ever certain that gallantry had been his motive for drowning out Lindsay's voice with the sound of his own. George, I concluded, was not taking at all kindly to his own share of putting her under lock and key!

Captain Drysdale of the Castle Guard was the one who broke the spell she seemed to have cast over us all. In a harsh voice that startled the silence, he ordered the soldiers to double forward. The wooden ranks broke, and became soldiers clumsily trotting in pursuit of the Queen. Their moving forms began to hide her from my view. Like a small beacon vanishing in mist, the moments of her aloneness vanished into the general stream of time. But still the impact of her action had not been lost. As I turned again to the group on the landing stage, Sir William was speaking to Lord Ruthven; and his voice was sharp with protest.

"I know all that, my lord. I do not need it read to me like a lesson that we rebelled because we could not—would not—accept the Earl of Bothwell as the Queen's husband. But Bothwell has now been routed in battle. And she *is* still the Queen—which means she still has powerful friends here at home, as well as among her kinsmen in France. And so, how long *can* I keep her against her will?"

"Put it like this," Ruthven suggested. "We have popular

feeling on our side so far as her marriage to Bothwell is concerned, and she herself sees now what a mistake it was. She must stay here until she agrees to divorce him."

"She'll not do that!" Sir William retorted. "Not now she has let it be known she is already seven weeks gone with Bothwell's child. If she divorces him now, it could be declared a bastard. And she'll not take *that* risk."

"You look too far ahead," Ruthven said smoothly. "There is no certainty she will carry the child to term. But even if she does—" He glanced around, smiling slightly; and then, with a movement of his hands that clearly showed his meaning, he added, "A pillow over its face, and a little downward pressure—"

"You will not make *me* party to killing a child!" Indignantly, Sir William interrupted. George Douglas chimed in on exactly the same note, only to be silenced by Lindsay shouting,

"Enough! We did not come here to argue!" Scowling, he turned then to Robert Douglas. "You, sir, were required only to witness the delivery of our warrant. Also, the fact of the Queen entering her prison. And your one final duty, therefore, is to make quick report to the lords with a statement of your witness."

"William?" Robert Douglas looked at Sir William, as if seeking his consent to leave; and was told irritably,

"Yes, yes. You have done all we asked of you."

"In that case . . ." Robert backed away from the group on the landing stage, then turned towards the waiting boatman. I guessed from his face, before he hurried into the boat, how relieved he was to have finished with his share in the proceedings, and did not blame him for that, considering how ugly they had been till then.

*9*

The silence following his departure lasted until the boat was well away from the landing stage; and then, abruptly, Lindsay told Sir William,

"Now we can talk privately as those who *do* have responsibility for this matter. And so we will have plainer speech than Lord Ruthven gave you."

"My lord—" Ruthven began warningly, but Lindsay brushed his words aside.

"The lords in Privy Council," he went on, "have debated further on the Queen since you were instructed to be her Keeper; and the result of their debate is that they are agreed she has shown herself quite unfit to rule this country. Their first intention, therefore—that of forcing her to separate herself from Bothwell—is now being published only as a blind to their final aim; which is this: She must give up the Throne entirely."

"My lord," Ruthven tried again to interrupt, "you were not authorised to tell Sir William—"

"My God, Ruthven!" Fiercely Lindsay rounded on the other man. "D'ye think that I—a man of the Reformed Faith—will tolerate that Catholic whore defiling my country's Throne a day longer than I can help? I have God's authority for speaking out now, I tell you—never mind the fact of common sense dictating that her jailers *must* know the truth of her situation now!"

Sir William and George were glancing most uneasily at one another while all this was going on. And long before the interchange was finished, I had discreetly interposed Sir William's bulk between Ruthven and myself. If the man objected so much to the others knowing the real facts, I reasoned, *my* presence there might suddenly become unwelcome to him. And I did not want to be disappointed of any further revelations!

Sir William coughed to bring attention back to himself, and then prompted, "You were about to say, Lindsay . . . ?"

Ruthven made a show of turning away as if he had shrugged off all further responsibility; and, still flushed with rage, Lindsay faced again to Sir William.

"I was about to say," he snapped, "that Mary Stuart must promise to abdicate the Throne in favour of the infant son of her earlier marriage. And that we—Ruthven and myself—have been instructed to stay here until we get her signature on a Deed that will confirm that promise."

Once again, George and Sir William exchanged glances; then, slowly, Sir William asked, "And suppose you succeed in that, who will be appointed Regent to rule on the child's behalf?"

Lindsay began to grin, a grin that grew into a leer as he told Sir William, "You *know* the answer to that!"

We all did. It was the Earl of Moray who would certainly be appointed Regent; Moray, the Queen's elder half-brother. But Moray was also the elder half-brother of Sir William, and of George. Moray was the bastard borne by their mother, Lady Margaret Douglas, to the Queen's father. And the full story of that long-ago affair was there now, in the leering grin on Lindsay's face.

"And so," he was continuing, "there is no need for me to tell you how very much to the interest of the whole Douglas family it will be, if she does abdicate. Or how very important it is for you to keep her secure until we have her signed promise on that."

"And then?" George asked suddenly. "Will she be set free then?"

"Oh, for heaven's sake, George!" Lindsay exclaimed scornfully. "She is no harmless cage bird to be loosed out of mere pity! Let her away from here at any time, and she will immediately gather a counterforce to the rebellion—if *that* is what you want!"

A flush leapt into George's handsome face. "I only asked," he said angrily, "in order to know what choice is open to her."

"What choice?" Lindsay echoed; and smiled his leering smile again. "That is easily answered too, my lad. She can abdicate and be imprisoned for life. Or she can die."

# 3

heard nothing further then
of the Queen's affairs. Lindsay strode off towards the Castle
gate as soon as he had spoken. The others followed him—
but not before Sir William had ordered me to fetch his
mother from the New House. I unmoored one of our boats
with my mind very much alive to all I had heard, and
the effect it could have on my own situation in the Castle;
and all the while I was rowing across to the New House,
I went on thinking of this.

I was already sixteen years old, after all, and yet my
bastard birth had given me no hope of fortune. What was
more, I had a passion for gambling, and I had just lost
every penny I once owned. I had also begun to find an
insufferable dullness in the only life offered by a small
castle tucked away on an island of two acres set in a loch
that was itself deep in the heart of the Scottish country-

side—all of which had brought me very near the point of deciding I would leave the island altogether for some more lively place. London, perhaps? My only talent—apart from skill at cards—was for playacting, and I had heard there was a lot of that going on these days, in London.

Yet here now, right under my nose, as it were, was a situation that would put Lochleven at the very centre of events. And so why go seeking abroad for the kind of excitement I was now more likely to find at home? The Castle could be assaulted in some attempt to rescue the Queen. Or she might make her own bid for freedom—and it was I who had charge of the Castle boats!

I had a certain advantage, also, in being the bastard of the house. I was not quite a gentleman, yet the family accepted me among them. I was not merely a servant, yet I still had some of a servant's duties. I was therefore free of both sides of the household; and that, in the game of intrigue that was bound to be played now, could make an important figure of me!

I moored the boat at the landing stage for the New House, ran past the stables behind the house itself, and sent a servant to rouse Lady Margaret Douglas. But the Old Lady—as we all called her—was already awake. And as I should have known, of course, she was expecting my summons. Sir William never moved a step without *her* advice; which meant, I thought wryly, that we would probably have to resign ourselves now to her almost constant presence in the Castle!

It took her only minutes to get the whole story of the Queen's arrival out of me; and when she stepped into my boat for the return journey, she was fairly crowing her satisfaction at the prospect of Moray becoming Regent.

"But *you* are not to speak of that yet," she reminded me sharply; "or you can expect no more favours from me, my lad."

I gave her my promise of silence and in a moment or two she was adding resentfully, "Although Moray should have been King, of course. And might have been too, if that young woman had never been born."

I said nothing to this. It was the Old Lady's favourite grievance that Queen Mary's birth had robbed her precious Moray of the Throne. And her favourite delusion too, of course; but it was not my place to remind the Old Lady that the law did not permit a bastard to be King.

"Remember," she warned me after we had landed and I was escorting her to the Great Hall of the Castle, "no gossip from you in the servants' quarters, my lad, or I will have Sir William whip you. But if you are good—" She fumbled inside her cloak and brought her bony hand out again with a gleam of silver in it. "Here!" She thrust the hand at me. "Here is some gambling money."

*As if she could buy my loyalty!* But that, of course, was the other side of my situation in the Castle. Simply because I had a foot in either camp, as it were, I was trusted by neither. I almost pushed the bribe back at her, but had not the heart to do so. She was smiling—the old hussy— and I knew how fond she was of me. I thanked her instead, as prettily as she expected, and then did what I had been itching for the past half hour to do.

I ran to tell Minny of my plans—Minny the laundry maid, the one person in the Castle who did trust me; Minny, who had got that pet name for "mother" long ago from me, because it was she who had clucked over and cared for me as a child. And Ellen Douglas too, of course. And Margaret Lindsay. Minny had so long been "Minny" to all three of us that no one ever called her anything else now.

But I was still her favourite, I knew; and there had been times, indeed, when I wondered if she actually *was* my mother. My build, after all, was slight—just like hers. Like her, too, I was dark haired, dark eyed, quick in all my movements. The Douglases, on the other hand, were all big and fair—like Ellen Douglas, who was as tall as I was although she was still only fourteen.

Like Ellen Douglas . . . I checked in the doorway of Minny's laundry, disappointment rising in me at the sight of Ellen there with her. I had been so set on speaking immediately and in private to Minny about my plans! She and Ellen both turned towards me. Minny paused from her task of counting pillow covers into Ellen's arms, and threw a mock challenge at me.

"So! I hear you have lost all your money again. But there is no use coming to me for more, my lad. *I* have nothing to spare for gambling."

I spanned her waist with my hands and swung her up— she was only a featherweight—to sit on her own ironing board. She looked comical, perched there with her feet dangling, and I laughed at her.

"Minny, Minny, Minny! Money, money, money! Is that all you can think of to say?"

She swung out her hand to give me a dunt on the ear, missed her mark, and then began scrambling down from the board the way a little girl scrambles down from too high a seat. Ellen and I laughed heartily at this, and Minny herself was grinning. I steadied her last step, and told her,

"No, sweet heart, I am not looking for a gambling stake.

Not yet, anyway. I came to give you news—to tell you that the Queen is here."

Minny nodded towards Ellen. "Your news is late. Ellen has just this moment told it to me."

"But the Queen," Ellen said primly, "is not here *as* the Queen. She is a Prisoner of State now; and that is how we have to look on her."

I was fond of Ellen, as a rule; and not just because she was my half-sister. She was lively, which had made for good games between us as children. Also, I liked to look at pretty girls, and Ellen had the kind of peach-blossom complexion and blond curls that made her very pretty. But I did not care for Ellen when she pattered the pious opinions of others, and behind her fair face now I thought I could see the plain, sallow one of Lord Lindsay's daughter, Margaret. Rudely, I demanded,

"Who told you then, Mistress Prim, what we must do or not do? Your little friend, Margaret?"

Ellen went red with annoyance. "Margaret Lindsay," she retorted, "is *not* little. She is a year older than me, and you know that! And Margaret has heard her father talking. He says the Queen has brought all her troubles on herself, and—"

"I know what Lord Lindsay says!" Sharply I interrupted Ellen. "I heard all his ravings at the Queen when she landed."

"Well?" Ellen challenged. "He was telling the truth, was he not? She *has* been very wicked, and—and wanton, too."

Minny laughed at this, and teased, "Now Ellen, what can you know of words like 'wanton'?"

Ellen looked even more annoyed then. "I know as much

as anyone else," she retorted. "I know why the Queen had Lord Darnley for a husband before she married Bothwell—because Darnley was tall and handsome, and—"

"And because she fell in love with him," Minny interrupted. "Is that so bad, Ellen? And just think what she got for her pains there!"

Ellen shrugged and looked away. I said loudly, "A poxy rogue of a man!" But Minny hushed me sternly for this, and herself went on telling Ellen the extent of the disaster that love had brought to the Queen. Darnley, she reminded her, had been proved a vicious fool, a corrupt creature unworthy of any woman. Everybody had hated Darnley. And so it was small wonder that the Queen—even before her child by him was born—had also come to hate him.

"It was no marriage at all then," Minny pointed out, "and in the end—"

"In the end," Ellen chimed in, "Darnley was murdered. By Bothwell. Everybody says it was Bothwell who murdered Darnley. And everybody says, too, that Bothwell was the Queen's lover *before* she took him for a husband. So what does that make of her, Minny!"

"You be quiet, Will," Minny commanded before I could get out the furious shout that rose in me with this. Then she turned to Ellen. "I am ashamed," she said coldly, "to hear you repeat such stories. You should have your mouth washed out for talking slander like that."

"How do *you* know it is slander?" Ellen demanded; and, sharply, Minny countered,

"I know it because nothing has ever been proved against the Queen. Nor has she ever had a chance to clear her name of the things charged to it. And to speak ill of anyone without proof *is* slander."

"And I told you, did I not, where she gets it?" I pushed Minny aside with this, determined now to add my share. "She listens to Margaret Lindsay parrotting what her father says. And Lord Lindsay hates the Queen. If she were a saint he would find something bad to say about her. As for Margaret Lindsay, she just tattletales out of jealousy, because she is plain and the Queen is beautiful."

"That is not true," Ellen retorted. "Margaret does not like to hear unkind things of anyone."

"She would be wise then," Minny remarked, "to keep her own mouth shut. And you had better learn that too, my girl, if you want to live a peaceful life."

Ellen's face changed suddenly with this. All the liveliness went out of it. Her mouth drooped at the corners. Her brows drew together in a frown. "Who wants a peaceful life?" she asked sullenly. She looked from Minny to me. "I could die sometimes, it is so dull here," she added. "You feel that too, do you not, Will?"

I nodded, finding I was beginning to forgive her, after all. Poor Ellen, having to divert herself with gossip, the same way as I had to find diversion with secret trips over to the Inn at the mainland village of Kinross!

"Pax, then?" I offered; and eagerly Ellen agreed,

"Yes, pax, Will. I—I do not believe those stories any more than you do. And I am sorry now I spoke harm of the Queen, because she was always kind to us—was she not? Both of us."

Memory came crowding back, the way it had done at the landing stage. I said, "She used to let you fairly cram yourself with gingerbread!" The smile spreading over Ellen's face then persuaded me to make up with her in style. I took out the Old Lady's silver coin. "And you can have this now," I told her. "It was to have been my gambling stake, but you can use it instead, to buy gingerbread for yourself and Margaret."

Ellen caught the coin as I flipped it to her, and I grinned to see in her face the thought of John Kemp's bakehouse in Kinross. Ellen always drooled at the mere smell of hot gingerbread! Minny watched indulgently as she thanked me, and told us,

"Now that's more like my two good bairns! And who knows. Having the Queen here could be just what is needed to liven things up for both of you."

Ellen's eyes gleamed afresh at this. "Yes," she said eagerly, "it could be exciting—for all of us!" She looked at me, waiting for my agreement on this, and for a moment I was tempted to take her, also, into my confidence. But some inner voice of caution told me otherwise. If the Queen did abdicate, the voice warned, and if Moray did become Regent, there would be soaring fortunes for the whole Douglas family—Ellen included. It would be foolish, therefore, to let her know that *my* share of the excitement would be found in helping the Queen to escape!

"No doubt at all of it," I said airily; and with a smile of delight at this, Ellen went off with her pillow covers, loudly protesting as she went that she would never quarrel with me again.

"Though I have heard *that* story before," Minny said, laughing as she closed the door on all this. "But now that you have finished with that particular squabble"—she turned to me, her eyes inviting —"you can tell me, can you not, just why you came rushing in the way you did?"

"Indeed I can!" I was suddenly full of my plans again, and rattling them off to her as quickly as I could. "And so," I finished, "you can see now what *I* was thinking when Ellen said it could be exciting here for all of us."

"I can see why you did not tell her what you have just told me," Minny said drily. "And I am warning you, Will. You may think the Old Lady is ambitious for the Earl of Moray, but I have lived long enough among the Douglas family to know he is even more ambitious for himself. I should be sorry for *anyone* he finds standing between himself and the prospect of being Regent."

It was not like Minny to be so serious with me. Minny had always been more inclined to laugh than to scold at any of my various escapades. "You are seeing me underground before I am even dead," I teased her. "But I have a lot of living to do yet, Minny. And that still includes some gambling!"

"But you have just given Ellen—" she began, and stopped herself to stare at me drawing my knife and then slicing off one of the silver buttons on my doublet.

"My next gambling stake—see!" I held the button up between thumb and forefinger, broadly grinning at her.

"But that set of buttons," she protested, "was the Old Lady's present to you on your last birthday. She will kill you for that!"

I shook my head. "What the eye does not see, the heart does not grieve for. And the Old Lady will not see me lacking one of my buttons because you, sweet heart, can easily find an exact copy to take its place."

Minny's eyes slid to the box where she kept all her replacement buttons. "I could, I suppose," she sighed; and within minutes of that, she had stitched on a button that was enough like the others to deceive even the sharp eyes of the Old Lady. I gave her a hearty kiss of thanks for her work, and she was still laughing at what she called my "impudence" when I ran off to find out whatever I could about the Queen's new circumstances.

*4*

*S*he had been locked up with her two young *femmes de chambre* in the small round tower built into the southeastern angle of the wall that circled all the Castle buildings. But that would at least give her privacy, since the small tower lay the whole width of the courtyard away from the square tower that was both our main defence structure and the family living quarters.

As for Ruthven and Lindsay, I was glad to learn from George, we would not have them all the time with us in the main tower. Lindsay had decided to stay with his wife in the New House, and Ruthven had also made the New House his preference.

"Comfort first for both of them, you see, although they are both so intent on making misery for the Queen," George added sourly. And sourness being so unusual in George, I felt I had been right in thinking that here was one Douglas, at least, whose sympathies would be with her. In everything else I had expected, however, I turned out to be completely wrong.

There were no great raids, no dramatic bids for escape, no excitement at all, in fact; or none, at least, of the kind I had hoped for. Sir William's shocked face, as he hurried back next morning from his first official visit to the round tower, gave notice of what we had to expect instead. The Queen had fallen ill. Breathlessly, he gave out the news. So violently ill that it seemed she would die!

We were a full company in Lochleven that morning. The Old Lady and Margaret Lindsay had come across with the lords Ruthven and Lindsay from the New House. The table for breakfast in the Great Hall was still littered with the remains of bread and ale. There was silence while Sir William leaned on the table, as if for support. I rushed to help him into his chair; and then, smoothly into the silence, came Lord Ruthven's voice.

"In that case, Sir William, we shall all be saved a lot of trouble. Shall we not?"

"Well spoken, Ruthven!" Lindsay exclaimed. "Oh, well spoken." He gave a great guffaw of laughter on the words, and I was not surprised to see Margaret Lindsay wince at all this. She had always been a nervous sort of girl; and she shrank even more then, when George banged the table and shouted angrily,

"Are you Christians, you two, or are you wolves!"

"*Geordie!*" Eyes brilliantly glaring in the ruined beauty of her face, the Old Lady commanded George to silence. Then, briskly as usual, the mixture of cunning and kindliness that was *her* nature took charge of the situation.

"William," she instructed. "You will send immediately

to Edinburgh for the Queen's own physician. And her own cook."

"But hold hard!" Lindsay interrupted. "If the Queen does recover, that means there will still be the Deed of Abdication standing, unsigned, between Moray and the Regency. Between *your son* and the Regency!"

"Y'are a fool, Lindsay," the Old Lady retorted. "You always were a fool."

"Eh?" Lindsay stared at her, mouth hanging open, rage beginning to take over from his astonishment; but still there was no weakening of the Old Lady's brilliant glare. Acidly she told Lindsay,

"I have more sons than Moray. And if the Queen dies so soon after entering William's custody, it is *he* who will be blamed. Poison—that is what will be suspected. And taking the Queen prisoner is one thing—no more, in fact, than the fortunes of war. But poisoning her is another matter altogether; one that would be certain to raise strong feelings against the whole Douglas family—Moray included. And where would his chance of the Regency be then?"

Lindsay backed down, muttering, before this tirade, but the Old Lady was relentless. "And besides all that," she added, "neither I nor any of my family are the monster of cruelty *you* are, my lord." The look that had pinned Lindsay shifted, and transfixed Ruthven instead. "*Or* a snake in the grass like you, my lord."

Ruthven tried, but failed to outface her; and had to content himself with a mutter that no one could distinguish. I wanted to laugh, then. There was no one so formidable as the Old Lady when she was in her stride. And yet, I thought, that was partly why I was as fond of her as she was of me. She, at least, was never dull when she was quenching people with that glare and flicking the acid of her tongue at them!

George was on his feet by that time, knowing well that he was the one who would be chosen to ride to Edinburgh; and within minutes of this, I was rowing him over to the New House stables. I got back to the island landing stage to find Ellen and Margaret waiting there for me. They helped me beach the boat, chattering all the time of the Queen, and George's mission to Edinburgh—but not a word, I noticed, of Lindsay's part in the scene before George had departed.

I chattered with them, aware of Ellen's tact in avoiding this—for Margaret's sake, of course—but without much patience, myself, for the girl. She was old enough now, after all, to take her father's failings for granted—just as I did, for instance, where Sir William was concerned.

"The Queen's physician," Ellen informed me, "is a Frenchman. That is because she prefers to have French servants. The Old Lady said so. And her cook is French too. He is a Provençal, by the name of Diderot. And the physician, the Old Lady said, is called Arnault—M'sieur Claude Arnault."

Ellen was good at picking up information! I went back to my duties, turning these scraps over in my mind and deciding it could very well be useful to strike up an acquaintance with the second of these men—this M'sieur Arnault, who would certainly be on close terms with the Queen. And to better that acquaintance, of course, if it seemed to be leading in the direction I wanted to go!

I was waiting by the gate when George returned that evening with both men in tow. The cook, I saw then, was a plump man, with a melancholy and rather handsome face; but the appearance of Arnault the physician was far from attractive. Short, swarthy skinned, with a great stomach and a warty nose—that was how he looked. But the little eyes above the warty nose were merry ones. I liked the knowing twinkle they held; and before long, I discovered, I had developed quite a liking for the man himself.

It was in the garden I first spoke to him, beside the herb bed at its far end. The garden itself lay outside the wall around the Castle courtyard, and stretched to take up all that part of the island not occupied by the buildings enclosed by the wall. The garden, also, was well planted with shrubs and trees, so that it was easy for me to stalk M'sieur Arnault on his herb-gathering expedition, and then to saunter casually from cover to talk to him.

The greeting I called brought him up from his stooped position at the herb bed. He had a fistful of greenery in his hand, and waving this towards the bed, he told me,

"This is shameful, the neglect of these good plants here."

The herb bed was certainly very weed-grown. I said defensively, "It was well looked after in the Old Lady's day."

"The Old Lady—? Ah yes, Lady Margaret Douglas."

"She used to be Keeper of the Castle," I explained. "After she was widowed, and Sir William was still a minor. That is why he depends on her so much. And *everything* was in good order then—or so they tell me, at least. But Lady Agnes— Sir William's wife, she is—well, she is very placid. She does not much care what happens or how things look, so long as life goes on somehow."

"Herbs," Arnault retorted, "can *be* life. Herbs are the raw stuff of medicines. And I need them, just as every physician does."

I jumped for the opening he had given me. With a nod to the herbs in his hand, I asked, "And your medicine from these—will that cure the Queen?"

Arnault cocked his head at me, little eyes suddenly aglitter with sardonic humour. "That," he told me, "could be called a strange question—coming from a page. Could it not?"

I shrugged. "I do not think so. You arrived only yesterday; and the Queen was very ill then. But you do not seem too concerned now, M'sieu, and so I wondered—"

"If that fat little Frenchman really could cure her," Arnault interrupted. "And so you lay in wait for me, you followed me, and now you make a casual chance to ask me. Is that not so?"

Again I shrugged, but in admission of his charge this time, and feeling not a little embarrassed by the way he had seen through my manoeuvres. Arnault chuckled, a rich fat sound, at the sight of the flush on my face. Then, as suddenly as if he had been struck, his grin vanished. His face became lowering, dark with menace. In a voice loaded with suspicion, he asked,

"But why are you so curious, my young friend—or should I say, *my young enemy?* Are you another of those who wish so much that the Queen should die?"

I backed from his look stammering, "No! M'sieu Arnault, believe me, no! I am on *her* side."

Arnault looked me up and down. "You are more than page to Sir William. You are also, I believe—ahem—a Douglas by blood. And so what have you to gain by being a friend to the Queen—here, in this nest of her enemies?"

"Why, nothing," I told him. And then, with laughter beginning suddenly to bubble up in me, I added, "And everything!"

Arnault was quick to understand. "Of course, of course." He nodded, with some of the menace clearing from his face. "You have here in this Castle a dull life. And so you look for excitement—but for its own sake only. You gain no reward; but to you, that is still everything—yes?"

"Yes. Exactly yes, M'sieu." The laughter was coming through my voice now, and I did not try to damp it down. "And if the Queen were to try to escape from here—"

"Hush!" Arnault's hand shot out to clamp over my mouth. "Do not say that word. Not yet, at least. First of all, she must be cured of this illness."

I pulled his hand from my face. "But can you do that, M'sieu? Can you?"

Arnault drew himself up as much as his fat stomach allowed. "I am a good physician," he said proudly. "With modesty, I may say, an exceedingly good physician. What is more, I recognize the nature of the Queen's illness. She has had it before; I cured her then, and—with God's help—I will do so again."

I said "Amen" to this, but not with any great certainty, and Arnault noticed my lack of fervour. "I see," he remarked drily, "that you are not much given to habits of devotion. But I am, my young friend. And if you truly wish for the Queen to be well again—"

"I do, M'sieu," I interrupted. "In all truth, I assure you of that."

"Then pray for her," Arnault retorted. "As I shall! Because a word in God's ear, I must tell you, will often do

more good than any medicine. And one more thing—"
A fat forefinger beckoning brought me closer to himself.
"That man who came here with me—Diderot, the Queen's
cook; he is blindly loyal to her, you understand?"

I was not sure I did understand, but I nodded all the
same, and Arnault went on, "He is also an expert with a
meat cleaver. Not that you have anything to fear from
that of course, unless"—the big warty face came even
closer to my own, and with warm breath hissing on my
cheek, Arnault finished—"unless I find you have played
me false today. And then, my young friend, one word
from me in Diderot's ear, and he will split you as neatly
as he would split a chicken."

I stepped nervously back from him, rubbing the moist
patch his breath had left on my cheek. Here was another,
it seemed, who was "blindly loyal" to the Queen. And
so he was still testing me out, using Diderot like a threat
to frighten me. And, I realised, he had come very close
to success in that! As coolly as I could then I told him,

"I am in no danger, M'sieu, from Diderot's cleaver."

Arnault's face changed swiftly again into a mask of sardonic humour; and that was when I decided I liked the
man. He was not unlike myself, after all—in some respects,
at least. That piece of playacting over Diderot, for instance;
that was the kind of thing *I* liked to do. There was something attractive, also, in his humour. And of course, he
was as conceited as I was myself! I thought of him saying,
"With modesty . . . an exceedingly good physician," and
grinned in reply as he asked,

"So I make myself clear now, do I?"

"Very clear, M'sieu."

"Good. And you will remember my advice on prayer?"

"Whenever I can, M'sieu."

Arnault was satisfied at last. "In that case," he told me, "you must also hope with me that God will not be asleep in Lochleven this summer. And meanwhile, you can bend your back to help pick these herbs, so that I can give Him what assistance I may."

## 5

*I* found it a nuisance, that promise I had given Arnault, and my prayers for the Queen were not so long or so frequent as they might have been. But I did pray. And she did get well; well enough at first to eat the delicate-flavoured concoctions Diderot prepared for her, then well enough finally to go walking in the garden with her ladies—Jane Kennedy and Maria de Courcelles—on either side of her.

We watched them from a little distance, Ellen and Margaret and myself. Kennedy was a tiny creature, no bigger than a child of twelve, but de Courcelles was nearly as tall as the Queen herself. They looked like two women walking with a little girl, Ellen remarked; then she exclaimed aloud as the three of them moved into a patch of sunlight.

It had been a miserable summer that year, and this was

one of the few fine days we had had. The sun's rays bathed the Queen's head, making the red-gold of her hair seem suddenly like a nimbus around it. Margaret, too, drew a gasping breath of admiration, and whispered to Ellen,

"I would like to see it all spread out for combing—would you not?"

I heard the conversation that followed, without really hearing it. The Queen was walking with some of her easy, dancing grace beginning to show again in her step, and that was a pleasant sight. It would have been a reassuring one too, I thought, if I had not known that Lindsay and Ruthven had already begun to plague her daily for the promise they wanted. And how could she get really well when that was happening?

She was holding them both at bay with argument. Arnault had told me so. He had also told me how much these arguments exhausted her. And in addition to all that now, of course, there was this new emissary from the lords—this man, Sir Robert Melville, who had begun to travel back and forth between them and the Queen.

I was curious about the same Sir Robert. He had a foppish appearance, certainly—all ringlets and ribbons and laces and high-heeled shoes. In his manner, also, he was more like some fool of a dancing master than the skilled politician he was supposed to be. But his face! . . . There now, we were into a different matter! The face, I was sure, was that of a clever man—thin featured, long lipped, with languid-drooping eyelids that most successfully veiled the shrewdness behind them.

I wondered whom I could ask about him. Minny was not acquainted with the circles in which men like Sir Robert moved. Nor was Arnault of any help, except to tell me that the Queen had held several private conversations with Sir Robert. There was nothing for it, I realised, except to be even more than usually observant when he visited; and was finally rewarded for my pains in this when I saw him one day in the garden, deep in conversation with George.

They were talking secretly. From the watchful looks they cast around them, I was sure of that. I began looking for a chance to coax George into telling me something of their conversation; but that chance had still to come my way when I discovered that God had been asleep in Lochleven, after all, that summer.

It was Arnault who proclaimed it so; Arnault with every line of his body drooping, the light gone from his eyes, face haggard, the step that carried him into the Great Hall no more than a shamble. He gave his news briefly, ignoring all the others there, and speaking directly to Sir William.

The Queen had fallen ill again; but not, this time, with the feverish malady that had previously attacked her. This time, Arnault said, the Queen's illness was due to the fact that she had miscarried of her child.

I watched him as he stood there, small and fat in his black gown, yet with the sorrow in his face making him somehow dignified. What reception would he find for the news that had moved *him* so deeply? My mind went back to the conversation I had overheard on the day the Queen was brought to the island, and I feared for the little man. Sir William spoke, awkwardly, his face embarrassed.

"You may—er—tell Her Grace that I—er—send her my commiserations."

"Commiserations?" That was Ruthven, slyly, a mocking smile flashing out with the word.

"Aye, be damned to such talk!" Lindsay exclaimed. "The brat was Bothwell's. And now that this has happened, it is just as Ruthven said it would be. One complication, at least, has been removed from the scene."

I looked at the Old Lady, waiting for her to blast them as she had done on the last occasion they talked so unfeelingly; but the Old Lady only shrugged, and said,

"It *is* a problem solved, of course. And we should be grateful for that."

I gaped at her words. Was this what power, and the craving for power, did to people? The Old Lady *loved* children—who knew that better than I?—yet here she was now, talking of the death of a child as casually as she would have spoken of a discarded pawn on a chessboard!

"I will inform Her Grace of your opinions." That was Arnault again, but speaking this time in a voice that carried a quiet and deadly hatred. "And meanwhile, if you will excuse me . . ." He turned to Sir William. "I must ask your permission to leave, sir. I—I find myself suddenly unwell."

"M'sieu, I regret—er—" Sir William was immediately all flustered concern. In a flurry of words he granted Arnault's request, and hurriedly added, "My page will escort you."

I saw George's eyes on Arnault as I armed the little man from the Hall, and was half expecting it when I heard him following swiftly behind the two of us. The tremble I had felt when I gripped Arnault's arm was still there; but it was rage that shook him, I guessed, rather than the illness he had claimed, and so I had not dared speak to him. Neither had he spoken to me by the time George caught up on our progress across the courtyard; but from George himself, the words came tumbling.

"M'sieu! M'sieu Arnault! Not all of us here are savages. I pray you, M'sieu, to believe that, and to tell Her Grace how heartily sorry I am for her misfortune."

Arnault straightened, shaking my hand from his arm. He stared at George. Then he nodded, a curt little nod, and said,

"I am your good servant, M'sieu Douglas."

The service had been softly spoken, I noticed, in spite of the curtness of the nod; and as Arnault turned and walked from us, I was glad of the comfort George's words must have given him. On impulse, I said,

"I wish *I* had thought of sending such a message."

"You?" George looked his disbelief at me. "You never cared for anyone in your life—except yourself!"

The thought of Minny flashed into my mind. "That is not true," I said resentfully. "Not entirely, anyway."

"I am sorry, Will." George looked away from me, biting his lip. "I had forgotten Minny. This whole business of the Queen, you see—it upsets my mind so that I cannot think properly."

Abruptly he turned towards the outbuilding that held the mews for the hawks he was so fond of training. My resentment slid from me. I caught his arm and quickly asked,

"Does it upset you enough to discuss it with Sir Robert Melville?"

"How do you—? You were *spying* on me!"

George had more than his share of the temper that ran in the Douglas family. I backed from him as he struck

*35*

at my clutching hand and furiously accused me.

"Wait, George, wait," I begged. "I did spy on you. I admit that. But only to see if I could find out what sort of man Melville is—whether he is friend to the Queen, or foe. Because *I* am on her side, George. And I will help her if I can."

George stared at me. With a jerk of his head then, he commanded me to follow him towards the mews; and there, between the fluttering lines of his perched hawks, he faced me again.

"Did you mean what you said about the Queen?" he asked. "Truly mean it?"

I nodded. "Yes. I am quite decided on that."

"Why? Why should *you* want to help her?"

"Because—" I stopped there, with my mouth open on the same answer I had given Arnault. What would a man of George's noble character think if I gave my true reason for wanting to take a hand in the Queen's affairs? My mind raced, working on all the possibilities.

I wanted George to trust me, as Minny did. I had always wanted that. But I had worked to earn Minny's trust; and if I lied to George now . . .

"You are very slow to answer." He was watching me, his face suspicious. I called up all my courage, and told him,

"For the excitement of it. Because it will make life here less dull for me."

George shrugged and turned away to gentle the bird perched nearest him. "That sounds like you," he remarked; and his tone was so disparaging that I blurted out,

"And also because she was kind to me. When I was only a small boy—" I stopped, feeling foolish over what

I had been about to say. George glanced at me over his shoulder, and quietly urged,

"Go on, Will."

"She—she disliked to hear those who jeered 'bastard' at me. Remember? She had her own name for me. 'My little orphan'—that was what she called me. You remember that too—do you not, George?"

"I remember," George said. And then, speaking low as if to himself, he added, "I remember too much about her, I think, for my own comfort."

"George—" I ventured a touch on his shoulder. "About Melville. *Is* he her friend?"

George turned full towards me and said, "Give me one good reason why I should tell you anything about Melville."

"I have given you one already," I protested. "I want to help the Queen. And—and you can trust me, George."

"Can I?" George came closer with this; close enough to peer into my face. Softly then, he added, "You are a liar, Will. You know you are."

"I admit that." My heart was hammering by this time. I could hear the unsteadiness in my own voice. Yet still I knew I had to convince George there and then, or never have his trust at all. And so I pushed myself on to tell him, "But I spoke the truth about myself to you just now, when I could have pretended to some nobler reason for wanting to help the Queen. I always will do that, I promise you. And I will never, never repeat anything you might tell me."

There was a long moment of silence between us; a very long moment. Then George said, "I believe you."

I found myself grinning—foolishly grinning, no doubt,

such a shock of pleasure went through me then; but I sobered quickly enough when George said,

"About Melville, then. Outwardly, he is working on behalf of the lords; but in secret, he *is* the Queen's friend. And, he says, she must be persuaded to sign that Deed of Abdication. There is no way of keeping her life safe otherwise. And so he has concocted a plan to defeat this situation. Also, he has smuggled in to her various tokens that the plan will succeed—pledges of support from some of the lords still faithful to her, along with a letter from the English Ambassador, written at the direct command of Queen Elizabeth of England."

I let out such a whistle of surprise at this that all the hawks were startled from their perches. George frowned annoyance at me, but once he had the birds quiet again, he went on,

"That letter told the Queen something which is perfectly true: A signature obtained under duress has no legal force; and it also reminded her that there is no power on earth that can take away her natural-born right to be Queen. Therefore, the English Queen urged, our Queen should sign the Deed. But only in order to safeguard her life. And afterwards, she will have every right to repudiate the signature."

"Afterwards?"

"After she escapes from here."

I stared at George. "Is Melville planning that too?"

George grinned—a very wry grin—and said, "Not he! Melville has sympathy enough for the Queen; but he is still the kind who always waits to see which way the cat will jump. And until he can tell whether it is to be the Queen or the lords who will finally triumph, he would not so far commit himself."

"But he thinks you will!"

"He thinks I might," George corrected, "and that was why he decided to sound me out on the matter. I have made no secret, after all, of my distress at the Queen's situation. And he knows perfectly well that she is more likely to sign the Deed if he can hold out to her even a hope of escaping afterwards."

"And will you, George—will you help her escape?"

George turned his back on me, making no answer. I gave him a moment, and then repeated, "Will you, George?"

"Oh, for pity's sake!" He whirled towards me again. "Lochleven is a State Prison and *my brother* is its hereditary Keeper! Or had you forgotten all that? It would be stark ruin for him if *I* helped the Queen escape."

I stood in silence, trying to imagine myself in George's shoes. Would I betray a brother for the sake of the Queen? And there was something else George had not mentioned—some*one* else, rather. Moray. His half-brother! If the Queen abdicated or died, and Moray became Regent, George's fortunes would soar along with those of the other Douglases. But if George tried to help the Queen escape, and was caught in this, he would face the powerful vengeance Moray could bring on him.

"You said you wanted to help." George broke into my thoughts. "And you can. Melville comes here again tomorrow, bringing the Deed with him; and he is convinced he can persuade the Queen to sign then—provided I am at hand to reassure her afterwards—"

"But you said—"

"I said nothing! And I said nothing to Melville either, except that I agreed he could tell the Queen I would stand her friend in any way I could. And so, when Melville arrives along with his two witnesses to the signing—two of the legal gentlemen they call Notaries Public—what I want *you* to do is to act as messenger for any meeting he might arrange between the Queen and myself. Do you agree to that?"

"With all my heart," I told him.

George smiled a little at the warmth of my reply, but was immediately serious again as he warned,

"And remember your promise. Not a word of all this to *anyone*."

I was about to give a very ordinary sort of assurance to this when I thought of a phrase that had a fine, heroic ring to it—the sort of phrase, it seemed to me, that could very well have been put into a play. I spoke my phrase aloud.

"I shall wear silence like a shield!"

I said this, I believe, in the grand manner suited to it; but all it drew from George was the kind of look I was more accustomed to receiving from the Old Lady.

"There are times," he told me, "when you talk just like a playactor."

I decided to ignore the look and treat the remark as the compliment it should have been. George and I, after all, were now fellow conspirators—were we not? My only reply, therefore, was to sweep him a bow, with a very elegant angle to my right leg and a fine flourish of my hands. But this had the effect of again startling the hawks; and, irritably, as he went once more to soothe them, George threw over his shoulder at me,

"And behave just like one too."

I bowed again, making such a comical caper of this second bow that George laughed in spite of himself.

"I know that," I told him. "But you must admit, George, that I have a considerable talent for it."

"Why d'ye call it a talent?" he demanded. "You make a fool of yourself, behaving so."

"I know that also." I had realised by then that I was about to reveal more of my secret self than I usually did; but by that time also, the exhilaration of having George's trust at last had gone to my head, and so I could not stop myself. "But one thing *you* do not know, George, is that clowning like this is a good way of hiding what is really in my mind. And that is something I have always found very useful!"

George looked at me in a puzzled sort of way. "So long as you remember," he said, "that clowning will not help the Queen."

"How can you tell that?" I asked. "The most dangerous of situations, George, can be disguised in laughter. And that means the Queen may yet have need of a Court Jester!"

*6*

---

*I* took my leave of George with this, still grinning at the way I had tied up his tongue with my arguments. I was not nearly so lighthearted, however, as I waited with him the next day at the landing stage to meet Sir Robert arriving with his Notaries Public.

They were like two draggled rooks flanking a peacock, I thought, looking at the contrast between his finery and their lawyers' gowns of shabby black. But their eyes were cold, their faces grim, and I did not care for the probing glances they gave me when George drew Sir Robert aside to speak privately to him.

I turned away from their looks, to see George walking off towards the garden and Sir Robert beckoning me towards himself. I closed with him, and he spoke low and rapidly to me.

"I have been assured I can trust you; and the plan is

this. I must have time alone with the Queen; and so take these Notaries to Sir William. Get from him the key to the Queen's tower. Remind him it *was* agreed I should see the Queen in private before he joins me here with the lords Ruthven and Lindsay. Do not let either of these two see you. And come quickly back with that key!"

A signal of his hand brought the Notaries to him. "Go with this page," he instructed; and immediately I began urging them across the courtyard, hurrying them along like a collie herding sheep. A few minutes of this was enough to exhaust my patience. I pointed ahead, and said,

"D'you see that flight of steps built against the Castle's outer wall? It leads to the level of the second floor; and the doorway at the top of the steps is the chief entrance to the Castle. It gives directly onto the Great Hall. And I will ask Sir William to meet you there."

I was off then, without giving either a chance to protest. I took the outer stair in a series of bounds, stepped through the Hall doorway, then ducked sharply on to all fours. Facing me, as I crouched like this, and about four feet inside the doorway, was the screen we called the "service screen"—solid wood in its lower half, latticework above that. I crawled forward and peered through the lattice. Lindsay was in the Hall, and Ruthven; but not Sir William. I ducked again, and considered my next hazard.

On my left, in the southeastern corner of the Hall, was the only way of reaching Sir William's apartments—the spiral stair that twisted downwards to the kitchen and upwards to all the floors above the level of the Hall. But the screen did not stretch the full breadth of the Hall. There was a gap between it and the south wall—the necessary gap that allowed entrance to the main body of the

Hall. I started towards the stair, crawling along behind the solid lower half of the screen. At the extremity of this cover, I paused to eye the space ahead of me.

If I moved fast enough, I decided, Lindsay and Ruthven might not be aware of me flashing towards the stair. I judged my distance, balanced myself, and rose in a catlike leap that carried me on to the fourth step. As silently as I had leapt, then, I raced upwards to the level of Sir William's apartments, and entered hurriedly on my own knock at his door.

Sir William was there, close in conversation with the Old Lady. The speed of my entrance took them by surprise, and they started apart with something that looked very like guilt on their faces. I rattled off my message, told Sir William where he would find the Notaries, and held out my hand for the key. Sir William began fumbling it off his belt, talking as he did so, in the flustered manner that always overcame him when he was embarrassed.

"Yes. Ah yes, Will. But I—um—I— You must—er—tell Sir Robert that I myself shall not be present when the—um—lords Lindsay and Ruthven go to speak to Her Grace."

"That a matter of importance to be attended to here will not allow of that." Quickly, smoothly, the Old Lady came in on Sir William's stumbled excuse. Much too quickly, I thought. And much too smoothly!

I thought again of the guilty looks while I repeated my manoeuvres for passing unseen by Lindsay and Ruthven; and raced back to Sir Robert feeling certain I knew what lay behind those looks. The pair of them had been planning that Sir William would deliberately absent himself from the lords' interview with the Queen. But why? My guess at the answer to this frightened me, and I ran all the faster.

I found Sir Robert striding back and forth at the Queen's tower, ribbons and laces all aflutter in the wind of his impatience. I handed over the key, blurting out as I did so,

"Sir William has excused himself from the interview between the Queen and the lords; and I can guess why. He has never cared to be mixed up in violence, and he does not want to witness what might happen then. Because you know how violent Lord Lindsay can be."

There was no veiling of Sir Robert's eyes now. They were wide and alert, bright with calculation. "There will be no call for anyone to use violence," he told me. "Not if I have this time alone with the Queen." Still speaking, he turned to unlock the door. "But I must still have you close at hand when the moment comes to fetch Mr. Douglas. And so come on, lad."

I hung back, not yet convinced. "But sir—what if the Queen will not agree to your plan? What if she will not sign the Deed? Should I not warn Mr. Douglas now that there could be danger to her?"

"The Queen will sign. She has a subtle nature, and it is the very subtlety of my plan that will appeal to her." Sir Robert had started up the stairs inside the tower, with this; and, calling back as he continued towards the Queen's chamber, he added, "So do as you are told, instead of wasting precious time here in argument."

I followed him up the stairs, still feeling uneasy, but reassured by one thing at least. Whatever else he had agreed to with Lindsay and Ruthven, Sir William would never countenance murder. And so the Queen's life was not really in danger. What was more, I argued, if I ran for

George now, I would be forcing him to a decision he had seemed unwilling to make. And he certainly would not thank me for that!

Sir Robert knocked on the door of the Queen's chamber, and a voice bade him enter. I slid in on his heels, louting low in her presence, as he did. Arnault was in the chamber with her, along with Kennedy and de Courcelles; and she herself was lying in bed, propped up on a great pile of pillows.

Everything about her was white—the pillows themselves, her bedgown, the cover and hangings of the bed; and the great mass of her hair was spread out against this whiteness in a wide and shining web of copper-gold. Her eyes travelled from Sir Robert to rest on my face.

"Will Douglas?" She spoke then, uncertainly, her voice questioning—because I had grown, of course, in the two years since she had seen me. I opened my mouth to answer; but before I could do so, she smiled and said, "Yes, it is. It *is* my little orphan!"

I bowed again, in a confusion of surprise and pleasure. She had remembered her name for me. Even in the midst of all this, she had remembered it!

Sir Robert had moved to the bed, and was quietly offering his sympathy on the loss of her child. But his face was anxious; and very soon, he was bending down to speak in an even quieter voice. The Queen answered him in French, the language always readiest to her tongue, and he continued in French also. I could make out little of what they said. My schooling had made me fluent enough in French, but those low tones defeated me. I stopped trying to listen, and watched their faces instead. Sir Robert's anxiety was still obvious; and who could blame him for that? He would be a famous figure, after all, if his plan succeeded; disgraced, if it was discovered. And if it failed altogether . . . The Queen, I noticed, kept interrupting his speech, her face vivid with anger sometimes, and sometimes downcast. She used her hands a lot as she talked, long white hands with round-tipped fingers that were as expressive as her face. Her voice rose in volume until at last I clearly heard her say,

"—and you take from me my only weapon, if I sign. Consider also, Melville, I am an anointed Queen. It goes against God, therefore, as well as against nature for me to strip myself of my Royal rights."

Melville's voice also rose. "God is all-seeing, Your Grace, and the signing is only a device to protect your life for the day you will regain your Throne. It will not be considered a sin in you to use any stratagem to that end."

The Queen sighed, a great deep sigh. It was her only answer to this last argument, and I thought, *She is wavering.* Abruptly then, she said,

"If it is true, what you have told me of George Douglas, I must see him immediately this business is over."

"Then you *will* sign, Your Grace?" Sir Robert straightened, relief beginning to smooth the worry from his face. Her hands fluttered in another of those expressive gestures.

"You are right. I was born a Queen. Nothing can ever make me less. And once I am free again . . ."

Sir Robert bent quickly in the gesture of kissing her hand, and in the very moment of his doing so, the door of her chamber was burst roughly open. Lindsay strode in, with Ruthven and the two Notaries at his heels. Lindsay's voice rang out, the gesture that went with the words taking in Arnault and the two *femmes de chambre.*

"Out!" he roared. "Out of here, all of you!" They backed in confusion from him, Arnault scowling, the two young women in a flutter of distress, and withdrew into the inner chamber of the Queen's apartments. Lindsay glared down at the Queen. "Now," he told her grimly, "we will have no interruptions with women weeping and that damned physician of yours speaking above his station."

He had not noticed me thus far. Either that, or he meant to ignore my presence. He was accustomed, after all, to see me stand waiting to be sent running with this or that message. I stayed where I was—in a far corner of the room with my back pressed to the wall. Lindsay turned to the Notaries.

"The Deed," he commanded. From inside his black gown, one of the Notaries produced a long roll of yellow parchment. From inside *his* gown, the other Notary produced a pen and an inkwell. Their gestures were so polished, so smooth with practice, that they were like a pair of conjurors performing sleight of hand. Lindsay snatched the Deed, and threw it down before the Queen.

"The law demands that you know the exact contents of the Deed," he told her; "and we mean to observe the law. You will read it aloud, therefore; every word of it."

The Queen had drawn herself up to sit very erect. "You will find yourself hard put to it to observe the law in this case, my lord," she said tartly; "because *I will not read that Deed*. And furthermore, I will not sign it."

Sir Robert began to exclaim in dismay, but the sound he made was overtopped by Lindsay roaring at him,

"You told us, Melville! You smooth-tongued viper, you! You told us you could persuade her to sign!"

"And I did, I did!" Reproachfully, Sir Robert turned

to the Queen. "Your Grace, what is this? You agreed to sign."

"Yes, indeed," the Queen retorted. "I did agree—but that was before the sight of Lord Lindsay reminded me of the kind of men to whom that would have given my kingdom."

Ruthven and Sir Robert both began speaking, at this; but Lindsay's uplifted hands silenced them. "I have the best answer to that," he told them. He stepped close to the bed, and bent towards the Queen. "Rise up," he ordered. "If you will not sign, I am ordered to take you to a place where—" He stopped, as if measuring the power of his next words, and then finished, *"Where I can give a good account of you!"*

There was no mistaking the menace of these last words. They were a death threat, if ever I heard one. But the Queen did not flinch from them. Bluntly, she told him,

"I am sick. I cannot rise."

"You mean you will not!"

"No. I cannot. But I would not if I could."

I have never seen a man look so baffled as Lindsay did then—nor a woman so determined as the Queen. She had won that round, I thought; because—short of dragging her forcibly from her bed—how could Lindsay meet such an answer? Ruthven came forward to pluck at Lindsay's sleeve, and then to whisper in his ear. Lindsay nodded. Ruthven went to whisper to one of the Notaries. The Notary nodded, and sidled towards the Queen's bed. The Deed lay where it had fallen in front of her. She had not deigned to lay so much as a finger on it.

Nervously, like a hard-mouthed dog bringing off a bad retrieve, the Notary snatched it up, unrolled it, and began to read aloud. The Queen closed her eyes. Her face smoothed of all expression, she lay back against her pillows. The droning voice of the Notary continued—interminably, it seemed to me. I did not understand a fraction of the legal jargon of it all, but the reading did end at last.

"There!" Delicately, Ruthven took the Deed from the Notary's hands, and laid it again before the Queen. Where the Notary's actions had been awkward, his were smooth, insinuating; like those of a snake gliding into a good position for a strike. "We have observed the law, Your Grace, in spite of you. And you now know the contents of the Deed."

"And which, in spite of all you have said," Lindsay added, "you will now sign."

The Queen's eyes flew open. Her breath had begun to come hard. A patch of feverish-bright colour had appeared high on either cheekbone. With one hand she swept the Deed off the bed, and at the same time cried,

"I will not sign. Before God, my lord, I swear that to you!"

"And before God, I swear to *you*, Madam! I am not willing to cut your throat; but if you do not sign, you compel me to it!"

Lindsay's hand travelled to his scabbard as he spoke. There was the whistling noise of steel against steel as his sword leapt free. The Queen screamed, a long shrill note of terror; and with arm extended towards her, the naked blade making a quivering line of light between them, Lindsay roared,

"You will sign. Or I will kill you."

"Never." She managed the word in the faintest of voices.

And then, on a stronger, rising note, in her accustomed French, *"Jamais! Jamais de ma vie!"*

The room was all confusion by then, with Sir Robert scrabbling on his knees for the Deed, Arnault and the Queen's two ladies rushing in from the inner apartment, the Notaries cowering back as far as they could from Lindsay; and with myself—all indecision now ended—sidling towards the door so that I could go dashing for George Douglas. In the midst of all this, also, I was aware of Ruthven throwing himself on his knees beside the bed. His face white as porridge, he babbled as he knelt,

"In the name of Christ, Madam, I beg you to sign. *I* do not wish you dead; but he lies—Lindsay lies when he says he is not willing to kill you."

The Queen paid no heed to Ruthven, or to anyone else there. Her gaze was fixed on the point of Lindsay's sword blade, and she was pressing as far back from it as she could. With every fraction her pillows yielded, the sword point travelled nearer. It was aimed at her throat. Lindsay's face above it was bloated and purple with rage, his eyes glittered supernaturally bright; but the hand gripping the sword hilt was rock steady.

At the back of my mind I heard a voice saying over and over, *He cannot kill her now—not in front of all these witnesses.* Yet Lindsay looked mad. At that moment, he *was* mad. He could, he *would* kill her. The shock of this realisation held me like one paralysed, and the others there seemed to be similarly held. We were like a roomful of statues—except for Lindsay. His right arm was still moving; slowly, surely bringing the sword point closer to the Queen's throat.

She could go no farther back from it. Motionless, eyes

wide open and glazed with terror, she pressed against her pillows. The sword point was two inches from her throat.

"Sign." The single word came from Lindsay with a rumbling, grating sound.

"No." Her lips scarcely moving, she answered it.

"Sign." The sword was an inch from her.

"No." Her answer this time was little more than a sigh.

The sword point touched her throat. She screamed again; and the sound was like a signal of release for the frozen horror in the room. All three of her attendants started towards her. Sir Robert thrust the Deed in front of her. With shaking hands that spattered ink all over it, he offered her a pen. Her right hand closed on the pen. Lindsay swung his sword up and away from her. With her left hand, then, she took the handkerchief Arnault offered her, and pressed it to the place where the sword had rested. Lindsay stood looking down at her, the congested blood draining from his face, his eyes taking on a more normal appearance.

"I will say this," he told her. "There is much courage in you."

Instantly she flung back at him, "But less in you, my lord, to threaten where it should be your duty to protect."

I thought Lindsay would explode then, into another bout of manic fury; but I was wrong. He looked at her, apparently dumbfounded by the way she had so swiftly gone back to the offensive; and almost querulously, he asked,

"Why do you parley with me? In God's name, why? You know we have you in a corner."

Her hand, the right hand holding the pen, went disdainfully wide of the Deed. "Would you not parley, my lord, if you had as much at stake at I?"

She was going to fight it out to the bitter end, I realised—perhaps even to a further threat on her life; and meanwhile, one of the Notaries was immovably blocking the doorway, which meant there was no way I could get out to fetch George. I edged nearer the door, hearing the continued debate going on between Lindsay and the Queen, but no longer listening to it. If there was only some way I could make that Notary move! . . . But the fellow seemed fixed as a rock.

I watched him as intently as he was watching the Queen and Lindsay, his eyes going back and forwards from one to the other, his whole face showing how fascinated he was by the argument between them. And the Queen seemed so endlessly ready to twist and turn in search of some way out of signing that Deed! I began to despair of the whole situation. But there was one factor in it that I had forgotten. The Queen was sick. And she was so near now to the end of her reserves that her strength could no longer match her will. Her head drooped. Her hand fell to the paper. Almost fainting, she scrawled her name.

"I sign this"—brokenly she spoke, as her hand went through its motions—"under protest." The pen dropped from her hand. "I shall repudiate this signature"—her eyes met those of Lindsay—"the moment I am free to do so."

"You will never be free." Lindsay snatched up the Deed and stood towering over her. "You swore to have my head. But I shall make sure the Privy Council acts swiftly on this Deed. And then we will see whose head will roll!"

I did not hear what followed this. In the few seconds since the Queen had signed, the Notary blocking the doorway had moved nearer the bed. As Lindsay spoke, I was easing the door open; and seconds later I was running like the wind, calling for George as I entered the garden, and not caring in the least who heard me do so.

George came running to meet me. I beckoned him on, then turned in my tracks and ran back to the tower. George's longer strides caught up with mine. We ran side by side, with me gasping out my tale as we ran. Then George drew ahead of me. I caught up with him again outside the door to the Queen's chamber, where he stood saying fiercely to Sir Robert,

"You have mismanaged this, Melville. You should have had my brother here to restrain Lindsay."

There was nothing suave or elegant about Sir Robert now. His figure sagged, as if something inside of him had broken. The lines of his face were twisted. With his laces and ribbons all drooping from that sagging body he looked, in fact, like a puppeteer's doll with its strings slackened.

"But Mr. Douglas," he faltered, "it was Sir William's own decision not to be present."

"Then you should have insisted otherwise," George retorted; and pushed brusquely past him into the Queen's chamber.

She was lying back against her pillows. Only Arnault and the waiting ladies were in the room with her now. Her face was as white as the white of her bed. Her eyes were closed. And she lay so still that—except for the gleaming mass of her hair—she could have been a white marble angel on a tomb of white marble. I checked my step, as transfixed by the sight as George was. And then the marble angel bled!

From the white, slender column of throat above the coverlet, a globule of blood welled darkly, broke, and trickled downwards in a thin crawl of scarlet. Arnault exclaimed in anger, and stepped forward with a piece of linen in his hand; but George was even quicker to move. On one knee beside the bed, he watched as Arnault's linen checked the bleeding. In a whisper, Arnault told him,

"Be easy, M'sieu. Her Grace bleeds freely from any wound. And Lindsay's sword did no more than scratch her skin."

George nodded, but his face was still anxious. Arnault stepped back at last, satisfied with his effort. George stayed where he was, his face on a level with that of the Queen. Her eyes opened. Their wide, golden-brown gaze was trancelike, at first. It rested on George, and recognition took over from the tranced look.

"Mr. Douglas." She spoke slowly, and very faintly.

"Your Grace." George's answering voice was firm, but I could see the tremble in the hand he had laid beside her own on the coverlet. "Tell me, Your Grace, how I can serve you."

"Lindsay—" she managed. "He said—he said he would take me from here to some secret place. He means to—to murder me."

"No, Your Grace. Not while I am alive to prevent that!"

Her eyes fixed full on George. "I must be set free," she said; "publicly set free, so that I may clear my name of the slanders on it. That is the only way to undo my situation now."

"Command me then, Your Grace," George answered gently. "Tell me your will."

The Queen stayed silent a moment, as if gathering her strength; and when she spoke again, her voice gained speed and urgency with every word.

"Moray—my brother Moray. He had no part in the rebellion. He was in France at the time. But the lords will make him Regent now; and he must not, must *not* agree to that. Will you go to him, Mr. Douglas? And will you beg him to come and see me? Will you tell him I am relying now for my freedom on *his* influence with these rebellious lords?"

I stood watching the dismay gather in George's face as she spoke. This was the test he had tried to avoid; the test between chivalry to her, and his own natural loyalty to the family cause. But how was it possible, I wondered, for her to be so deceived in Moray when even such as I knew the strength of the man's ambition? How *could* she place such hopes on him when his rise to power could be achieved only through her own downfall? George, I guessed, was struggling with the same questions.

"Your Grace," he pointed out, "if my lord Moray had wanted to help you, he could easily have been here by this time."

"He does not know how I am treated!" The Queen's answering voice was a cry of anguish. "He cannot know, or he would never have permitted it. He loves me, Mr. Douglas. My brother loves me!"

The Queen's hands had become as agitated with this as her voice. George reached out and took one of those restless hands in his own. He looked intently at her; and she, with tears in her eyes, looked back at him. She was beautiful always, but with those great tear-glittered eyes now, she was heartrendingly so. In a voice so low I barely caught the words, George told her,

"Majesty, sweet Majesty, I cannot see you so distressed! I will take your message to Moray. And if *he* fails you, I swear to you now, *I* will still stand your friend in any way that will once more make you my true and only Sovereign."

A flicker of some emotion I could not name ran across her carven-pale features. Still without taking her eyes off George's face, she raised her free hand to him. He bowed his head and kissed the tips of its long white fingers. With equal gentleness, then, he kissed the fingers of her other hand.

The pact was sealed. George had finally committed himself to be now and forever the Queen's man. I was moved to see this happen—with pleasure for her, a certain apprehension for him; and for myself . . . I was not sure of the feeling I had on my own behalf then, but it felt remarkably like envy of George.

# 7

*I* had news the next afternoon that both surprised and dismayed me. The Queen was to be transferred to the main tower of the Castle, to the suite of rooms on the floor directly above the Great Hall; and two soldiers were to be placed daily on guard outside the door of her new lodging. That was the surprising part of the news. Dismay came when I learned the names of the soldiers chosen for the guard—Kerr, and Newland, two of the most ill-natured fellows in the barracks, and both of them also sworn enemies of the Queen.

It was Ellen who told me of the new arrangement— an Ellen whose face was all alight with excitement and an air of smugness that warned me there was more yet to be said. Ellen, I knew, was a great believer in saving the icing on her cake for the last bite!

"You are lying," I told her calmly; and instantly she rose to the bait.

"I am *not* lying, Will Douglas! These are my father's

orders, and it was the Old Lady who told me so."

"Very well." I shrugged, affecting only a passing interest. "What else do you know?"

"A lot more than you think! I know that the Queen is to have all writing materials taken from her, and that she is to be watched night and day. And Will"—Ellen's rosy-fair face flushed an even deeper pink with excitement—"it is my mother who is to sit with her by day; but at night, it is Margaret Lindsay and I who will keep watch. We are to live in the new lodgings too, and to sleep there, *in the Queen's own bedchamber!*"

I stared in astonished disbelief. "Did the Old Lady—?"

"Yes, she did. 'Children also can be spies.' That was what she said to us." Ellen grinned triumphantly at me. "And that is what Margaret and I have to do now—to spy at night on the Queen's every move, so that we can tell the Old Lady whenever she does something that might mean she is planning an escape."

An escape . . . but what chance would there be of that now? I stared blankly at Ellen, remembering how calm Sir William and the Old Lady had remained when George had told them of his mission to Moray, and realising at last the reason for this. It had not needed George, after all, to point out that public opinion would sooner or later force Moray to visit the Queen of his own accord. Yet even so, the mere fact that George was carrying the Queen's message to him must have made them suspect his further intentions. And this was their answer, this new strictness with the Queen. . . .

"You have not congratulated me," Ellen challenged. She was watching me closely—entering already on her role of spy, I thought uneasily; and immediately began to cover myself with a pretence of jeering at her for a country girl who would not know how to behave before the Queen. But nothing, it seemed, could dent Ellen's complacency then.

"You are just jealous," she told me, "because *I* will be at the heart of all that happens now. And you are still where you were at the beginning—only on the fringe of things!"

That was a shrewd hit on Mistress Ellen's part; a very shrewd hit. But I would prove her wrong yet, I vowed. I would prove her very wrong. But first of all, I would have to make a chance to speak to George before he left with that message.

"I'll take you ashore, George," I offered, "and help you to saddle up at the New House."

The Old Lady looked up sharply. "*You* have become very helpful all of a sudden," she remarked. "And very anxious to hang around George's neck too, I notice."

"Leave him be, madam," George told her. "I shall be glad of his help."

I felt her eyes on my back as we went out together, and under my breath I said, "She misses nothing, that one!"

"She is afraid," George said. "The Queen is boxed up tight now, and yet still the great fear is that she will somehow escape and overturn that Deed of Abdication."

"If you have plans for that," I ventured, "if you *do* have plans, George, those soldiers at her door are going to be a problem for you."

George nodded. "I have thought of that. And I can see a way to counter it. But first of all, I must make contact with her supporters on the mainland. And even then, Will, my plan—*any* plan of escape—must wait until after Moray visits her. You heard how she spoke of him; how convinced she is that he will help her. And she herself will not consider escape until that hope is dead."

We had reached the landing stage by then. As he helped me push a boat out, George added, "And the closer the Queen is held meantime, of course, the less likely it is that harm can come to her—which is something to be grateful for, is it not?"

I pulled off from the shore, considering this, and was suddenly struck by the answer to Ellen's jeering that I was still only on the fringe of things. It would be quite easy for me to coax that dear silly little half-sister of mine to chatter about everything that went on in the Queen's apartments. I might even be able to persuade Margaret Lindsay to talk, nervous as she was. And Lady Agnes Douglas, of course, was such a kindhearted, placid sort of woman. I should not have much problem there in getting snippets of information to add to the watching brief I could hold while George was away. Then there was the pattern of the soldiers' duties, the timing of the meals sent up to the Queen's apartments—everything I learned in that line might eventually prove useful to George's plan for her escape. . . .

I found myself slipping as easily into the role of spy as Ellen seemed to have done, and was much amused at first to realise how much her smugness had been dented by the realities of her task.

"I am not sure now that we should be there at all," she confessed to me. "The Queen, you see, is still always so kind that it—well, it shames me sometimes to know I must report on all her doings."

"And Margaret," I asked, "does she also feel like that?"

"Margaret keeps herself to herself," Ellen said, "even more than she used to do. I do not know *what* she feels, Will."

I took to watching Margaret each time I escorted Sir William on his official visits to the Queen; but if Lindsay's daughter had now also been touched by the tenderness that had blossomed in Ellen, she was certainly keeping that well hidden. I seldom heard her speak, and there was never any expression at all in the thin, sallow face she kept bent down to her sewing.

Sir William began threatening to beat me for the way I seemed to be "forever idling about and staring," as he called it; but on the fifth day of George's absence from the Castle he had a letter with news that meant he could keep me busy enough. Lord Lindsay, it seemed, had made good his boast about hurrying the Council on over the Deed of Abdication. On that very day of the letter's arrival, the Queen's baby son was to be crowned King James the Sixth of Scotland! And Moray had already been appointed Regent.

I went with the rest of the family when Sir William summoned us to hear the letter read aloud. Lady Agnes listened to it with the vague smile that was usual with her when she felt matters were going well. The girls sat in awed silence. But the Old Lady, of course, was irrepressibly delighted.

"My son, the Regent." She tried the words over on her tongue, and crowed with pleasure at the sound of them. Sir William beamed at her with answering pleasure.

"My brother, the Regent!" he responded. And then, as if the words had been the spark for an explosion of energy in him, he jumped to his feet with the first of a long string of orders for the way he meant to celebrate the news.

There was to be a bonfire in the courtyard, fireworks to be set off at the height of the bonfire, salvoes of cannon to be fired from the Castle walls; and all of it had to be ready for that evening. I was sent running immediately with his first instruction for all this; and after that, he kept me so constantly employed on his preparations that I had hardly a moment to think my own thoughts about it all.

It did cross my mind, all the same, that it was most unfeeling of him to have such a celebration with the Queen so close at hand to witness it; and when he sent me that evening to fetch the girls to see the display, this thought came back even more strongly to me.

"Your father," I told Ellen, "has no more imagination than a louse. Either that, or he is just too unfeeling himself to realise how grieved the Queen must be at the reason for all this."

"She does not know the reason." Ellen spoke in a very small voice that was quite unlike her usual confident tones. "He has not told her yet."

"Not told her!" I stared around the courtyard at the leaping flames of the fire, the fireworks spurting red and gold. "But why? He knows she can see it all from her window. And she has a *right* to be told the reason for it."

Ellen looked away without answering. I gripped her shoulder, and said angrily, "*You* could have told her. Why did you keep quiet?"

Ellen looked uncertainly at Margaret Lindsay, but Margaret was no help to her. "I—I wanted to speak," she faltered. "I said to Margaret that we should."

"Well?" I swung to face Margaret Lindsay. She startled away from me, and said defensively, "We had no orders. And do you think she would have believed, anyway, even if we had told her?"

"Why should she not? *You* heard it from Sir William. She would have had to believe you."

"I am getting tired of you, Will Douglas." Margaret stared at me, her face suddenly hostile. "You seem to think you know this Queen. But you do not. She was *born* a Queen. She has always lived as a Queen. She cannot conceive of herself as anything less. And can you not see what that means? She simply cannot believe—even now, in spite of all that has happened—she cannot make herself believe that the lords would really act on that Deed. And so what was the point of *us* trying to tell her that they have?"

Ellen spoke again, still in that small ashamed voice. "She has bidden us tell my father to come to her. She has jumped to the conclusion, you see, that all this must be for someone's birthday, and—" Ellen paused, seeming almost to choke on her words, and then rushed on to a finish. "You know how she loves birthday games, and how good she is at giving presents. She wants to join in the celebration."

I let them both go then, I was so sick at heart to hear all this. Yet still, when Sir William called me to him a few minutes later, there was no way to avoid his summons. I had to trot faithfully as always at his heels when he made his way up to the Queen's apartments; and, standing just inside the doorway there, I saw how she received him. She was smiling as she stood with one hand laid on the back of the chair beside her, the other hand fingering the gold crucifix that hung always around her neck; and when she spoke to him, her tone was light.

"Well, Sir William. You are all very merry tonight. But

am I not to be allowed to join in the celebrations? It would please me, I must tell you, to add my share to the birthday scene."

And she would if she could, I thought despairingly. I still had the little gold comfit box she had given *me* on my tenth birthday. The taste of the sweets that had been in it filled my mouth suddenly again, as if I had just that moment eaten them, and my rage at Sir William was so strong that I wanted to strike him. And yet, I knew well, he was not a cruel man. For all my anger then, I knew it was not even true to call him an unfeeling one. It was simply a stupid man who stood there telling the Queen,

"But Your Grace, we have no birthday among us today. We are just celebrating the news."

"The news?"

"Why yes, Your Grace. The Deed you signed is law now. Your little son has been crowned King, and the Earl of Moray has been appointed Regent."

Now she *had* to believe! I thought she might collapse then, and anticipation of this made me feel sick again. But she took the blow without flinching, without even crying out. Her smile stiffened into a mask, then faded. The hand on the back of the chair tightened its grip. The fingers moving over the crucifix became still. And that was all. But not quite all, so far as Sir William was concerned!

He was waiting, still smiling with his own pleasure, for her to reply to his announcement. But it was no longer a lonely young woman, hungry for company, he had to deal with now. It was the Queen, with the full force of Royal dignity behind her continuing silence. I realised that as she gave an abrupt wave of dismissal, and then pointed this by turning haughtily from him.

He moved uncertainly to the door, looked at me as if asking what he had done to be treated so, then shrugged, and went out. I closed the door behind him. And no one pays attention to a page. As I drew the door shut, the Queen sank to her knees beside her chair, and laid her head on her outstretched arms. I stood outside the door and listened; and the sound that came to me then was that of bitter, bitter sobbing.

I could not get that sound out of my mind. Especially at night, before I slept, it was with me. And two days later, when George returned to Lochleven, it was with me again as I told him of that night's events. George listened to me, his face tight with the same anger I had felt; but when I had finished, he sighed and said,

"But it is still not easy, Will, to kill hope—especially in someone who clings as strongly as she does to her hope of Moray. And she is now buoyed up in that, you see, by the mere fact of my bringing back word that Moray *will* come to visit her."

"But his own credit now demands he must do so," I argued. "Even I know that, George. And you must surely have tried to get her to see that too."

George sighed again. "Of course. I put it as strongly as I could to her."

"And if she finds that Moray will not do anything for her?"

"I talked with her on that as well," George said. "I told her how I used my time away from here to get in touch with some of those still faithful to her cause—"

"With the Earl Bothwell?" I could not help eagerly interrupting then, it was so long since I had heard mention of Bothwell; but George shook his head to this.

"There can be no help from that quarter. Bothwell is now an outlaw, so ferociously pursued by Moray's forces that even the most loyal of the Queen's supporters dare not rally to him. And so I have informed her. Yet her spirit remains strong, in spite of that; and if Moray does indeed fail her, I have her permission to put my own plan of escape into practice."

I pushed the picture of the fugitive Bothwell to the back of my mind as it began to fill with visions of an assault force scaling the Castle walls; but George stopped me the moment I began to speak those new thoughts aloud.

"There will be no assault," he told me. "You know my brother. He would certainly think it proper to die in defence of his duty—and of his womenfolk, of course, because even they would be at risk in that kind of situation. And one thing I will *not* do is to put the lives of my family in danger."

"Then tell me what will happen," I begged. "And will I be allowed to take part in it?"

"Yes, I will need you," George admitted. "But as for the plan itself, I have already told you that the Queen's permission for that depends on whether or not Moray destroys the hope she rests on him. And so there is no point in speaking further of it until after he has been to see her."

I was tantalised out of my mind by this, but George was stubborn. I could not make him yield to tell me more; and so—like everyone else in the Castle after that—there was nothing I could do except to summon patience and wait for the promised day of Moray's visit to the Queen.

# 8

e was dressed all in black
when he did appear, nearly two weeks from that time—
no other points of colour about him except the silver that
cased his sword, the silver buckles on his shoes, the white
ruff around the neck of his black doublet.

Tall and thickset, his dark hair and beard looking even
darker against this one splash of white, he came treading
slowly into the Great Hall. I was impressed, in spite of
myself, at the air of sombre authority in his appearance;
but every lion has its running dog, I suppose, and so I
was not surprised to see Lord Lindsay following first
among those he had brought with him.

I noticed, also, that George was scowling at Lindsay's
presence. Sir William did not look well pleased, either,
at this. There was a low and hurried conversation between
him and Moray; and the result, it seemed, was that Moray
thought better of once more imposing Lindsay on the

Queen. With a glance around his party, he beckoned the Earl of Morton to him, then the Earl of Athole; and it was these two, finally, who escorted him to the Queen's apartments.

We were a very subdued gathering once they had gone. Lindsay, and the rest of Moray's escort, stood apart from those in Sir William's household, their conversation held down to the lowest of whispers. Lindsay himself kept total silence; but even in silence his heavy features gave out menace, and this added to the uneasiness of that subdued atmosphere.

It was all setting Sir William very much on edge, I realised; and as the hour for supper drew near, he became quite markedly so. Would Moray sup with the Queen, he wondered; or with us? He was like a broody hen fussing over her nest, I told myself contemptuously. And then was suddenly ashamed of the thought.

Sir William, after all, had always been content to leave the politics of the Queen's situation to the Old Lady. Yet it was still he who carried the responsibility for the Queen's imprisonment; and so it was not a trifling point, this question of where Moray would sup that night. If he showed himself friendly enough to the Queen to sit at table with her, I realised, it could be a sign that she was managing to charm him into a promise of her freedom. And how then would Sir William be rewarded for his strictness to her?

The smell of hot food wafted suddenly into the Hall. I saw, beyond the service screen, the figure of the Provençal cook Diderot, holding a silver dish at arms' length before him. Some maids bearing other dishes clustered behind him, all of them waiting for Sir William to lead their suppertime procession up to the Queen's apartments. He went

69

towards them, face creased with the worry of the moment. There was a further short wait while he tasted the food and the wine, then Diderot and the maids formed into procession to follow him upstairs.

The clattering of their feet on the stone steps echoed back to the Hall, faded, and died into nothing. For long moments of silence then, everyone in the Hall looked ceilingwards, as if that could somehow give a view of what was happening in the apartments above. We heard a distant murmur of voices, as if the door to these apartments had been opened. One of the distant voices became suddenly clear, rising in pitch above the others; the Queen's voice crying out in protest,

*"My lord Moray! Brother! You used not to think it beneath you to give me the napkin at supper!"*

Moray had refused to sup with her. The silence in the Great Hall was rippled over with voices as this realisation came to everyone there. I looked towards Lindsay, and then towards George. Lindsay was grinning. George, when our eyes met, gave the smallest of shrugs; and plainer than words, it seemed to me, the gesture said,

*"Did I not tell you so!"*

The stairs echoed again to the tread of feet. Then Moray and the other two were once more among us, with Moray's cold gaze slowly taking in each face there before Sir William led him ceremoniously to the family supper table.

I felt a twinge of fear at the renewed experience of that gaze. The warning Minny had once given me flashed back into my mind, and when I leaned over Moray's shoulder to pour his wine, my twinge became a fear so strong that I found myself breaking into a sweat. I must be mad, I thought, to make *this* man my enemy—because that was exactly what I would be doing if George's plan went forward, and if I took part in it. Nor would my lack of years save me if I were to be caught out in that—not from one so merciless as Moray!

I took the night to sleep on these fears, and the morning brought with it the reassurance that I could easily put distance between myself and Moray's vengeance. It was not so difficult, after all, to get a ship's passage to France; and many young Scotsmen had made good careers in that country. And there was England too, of course. Moray's writ did not run there either, and the thought of playacting in London still attracted me.

I went straight in search of George, once Moray and his retinue had departed, and found him checking weapons in the barracks armoury. I made sure there was no one else within earshot, then quickly challenged,

"The plan, George—does it go on?"

George nodded. "More than ever now. There is no hope of Moray ever arranging the Queen's release. He made that crystal clear to her last night; and again to Sir William and myself before he left this morning. And so listen carefully, Will." He paused to check, as I had done, against being overheard. "We are going to take the Castle *from the inside*, with the help of Captain Drysdale, and another eight men of the Guard. I have sounded out all of these, and all are willing."

"And my part in this?" I spoke so eagerly that George smiled as he told me,

"Your part will be the next step in the arrangements I have already made. The members of the force I have gathered are all discreetly quartered on the mainland, and I want you to take a boat over there now. At the Kinross

Inn, then, you will seek out a gentleman by the name of John Beaton; and you will give him this message from me: *The barge is ready for loading.*"

"The barge is ready—" I checked myself there. "Do you mean *our* barge? The one we send every week to the mainland for our supplies?"

George nodded. "The very same. Beaton and his men will seize it when it touches tomorrow as usual, at the mainland landing stage. When I see them bringing it back here I will give the order to make prisoners of the Queen's guards, and to overpower Sir William—but without hurting him, of course. The rest of the Guard will then be leaderless, we will have Beaton's force to assist us—not to mention the joint authority of Captain Drysdale and myself; and so, in the end, they are not likely to offer much resistance."

I did not wait to press George for any elaboration of this; nor did he need to tell me anything further of my part. George knew as well as I did that too long an absence from my usual duties would earn me a beating from Sir William. And he was well aware also of my method of slipping away from the island with the least risk of being noticed.

I ran straight for our east landing stage—the one we used when a wind from the west made the loch too rough to beach our boats at the more convenient stage facing west to the mainland. The east stage was where I always kept my own little boat. I set my usual course in this, and twelve minutes after speaking to George I was tying up at the public landing stage for the ferryboats plying out of Kinross. To my right then, I could see the pepper-pot shape of the roof on the tower that buttressed the Kinross Inn; and another minute or so later, I was strolling casually into the Inn's taproom.

There were only a few people there. I noted Jock Matheson, the innkeeper; the fat and talkative woman who was Mistress Matheson; two men playing dice with another man watching them; a group of four men sipping slow and silent at their ale; and one man sitting alone over a glass of wine. The solitary man was dressed like a gentleman, while the others were clearly not of that standing. I studied him covertly while Matheson drew the ale I ordered, and decided to take a chance on this being the mysterious John Beaton.

There was an empty stool beside the one he occupied. I sauntered towards it, and politely asked, "Sir, if I may share your corner—?"

His glance at me, I thought, was sharper than any gentleman had a right to give a stranger. On the other hand, if he was expecting a messenger such as myself . . . Without any answer except a wave and a nod, he granted my request. I set my ale on the trestle in front of the stool, seated myself, and spoke low to him.

"I thank you, Mr.—er—Beaton, is it not?"

This time there was no doubt of the sharpness in his glance at me, nor of the sudden wariness in his long, thin features. "Who told *you* my name is Beaton? Who are you?"

"I am Will Douglas, from Lochleven." I sipped my ale, counselling myself to go a little more carefully now. "And if you will tell me *your* full name, sir, I may find you are indeed the same Mr. Beaton who should have the message I carry."

"Go on drinking your ale," he told me, "and talk about sport, the weather—anything—before you say more on that score."

I drank my ale; and, quiet as ever, he continued, "My

full name is John Beaton. I am brother to Archbishop Beaton, the Queen's Ambassador to France. And the message *I* expect is from Mr. George Douglas."

I talked about the good sport there was in flying hawks from Lochleven, about fishing for trout in the loch, about the miserably wet summer we were having. Then at last I slipped into our conversation George's message—*"The barge is ready for loading."* Our murmuring voices went on from there.

*"And the day?"*

*"Tomorrow."*

*"The time?"*

*"It touches the mainland at nine o'clock in the morning."*

*"All other procedures will be as agreed?"*

*"That is what I have been told."*

*"Then you may let Mr. Douglas know he need not fear for my share."*

John Beaton rose with leisurely ease. His voice louder now, he spoke in the same vein he had earlier advised for me. "Well, my lad, I am much obliged to know the fishing is so good this year. But I have a few matters calling my attention now; and so . . ."

With a brief bow to myself, a nod, and a pleasant "Good day" all around, he was gone. I spun out the business of finishing my ale so as to give myself a decent interval before following his example; and by the time I left the Inn I was fairly hugging myself over the success of my mission.

The elation that bore me up then persisted long after I had got back to the Castle, long after I had given Beaton's message to George. And lying awake that night with my mind painting bright pictures on darkness, I rehearsed my own possible share in the next day's rescue scene. I would be on hand, I vowed, when George went with the soldiers to overpower the guards on the Queen's door. Not that I had ever used a sword before, of course, except in fencing lessons; but given the chance, I still saw no reason why I should not be able to put in at least a few bold and useful strokes.

I slept uneasily that night, perhaps because of my excitement, and in the dreams that plagued my sleep it seemed to me I could hear faintly the sound of hammers striking on metal. I woke to an odd sense of foreboding I could not name until memory of the previous day came rushing back; and the one thing I wanted to do then was to watch the barge being pushed out for its trip to the mainland.

As soon as I could after Sir William had unlocked the gate in the courtyard wall, I ran down to the west landing stage. The usual mooring place for the barge was empty—and yet it was still only eight o'clock in the morning, which meant I could not possibly have been too late to see it go out. I looked hurriedly right and left. And then I saw it—dragged right up on land, and chained with thick steel chains to the outer wall of the Castle.

The dream sound of hammers striking on metal came rushing back to me; and this time, I knew, I had not dreamed it. More than that, I realised, the barge was not intended to sail that day—or any other day, for that matter of it. There was no way of slipping those chains free. They were, as they had been intended to be, a fixture.

"A good piece of work, eh?"

The question sounding sudden and harsh at my elbow startled me out of my trance of dismay. I turned swiftly to see Captain Drysdale looming over me—Drysdale, who was to have been our accomplice, and who now sounded so complacent over the ruin of our plan! Years of dislike for the man's surly nature gathered instantly to back up the suspicion in my mind. Warily, I said,

"It was Sir William ordered it done, I suppose."

"Who else?" Drysdale grinned, the thin-lipped grin that always made him look like a fox, and smugly added, "But on my advice, of course."

I wanted desperately to ask why he had been so treacherous, but could not, for fear of letting him know my own involvement. And what about George? Did he know yet that his plan had been foiled? I made an excuse to leave Drysdale and hurried back through the courtyard gate. But it was another hour after that—well after the barge's usual departure time—before I could find George; and one look exchanged between us then was enough to make each of us aware that the other knew what had happened.

"I have warned Beaton," George said quietly. "I rowed out as soon as I saw the barge had been chained."

"And you know that we have Drysdale to blame?"

"I know. And I was a fool to trust him. There is no need," George assured me grimly, "to rub salt into *that* wound."

I bit back the reproach I had been about to make, and asked instead, "But why? Why did Drysdale pretend to be with you and then go straight to betray the plan to Sir William?"

"Because—" George hesitated, and then said all in a rush, "Because he thinks the same of the Queen as Lindsay does. He believes all the vile stories about her. And so he hates her. As do the two men he has put on guard at her door. He chose them specially for that reason—which is something else we have to put to our account with him."

George was flushed by then, and almost inarticulate with anger. I waited for him to calm down; and, more quietly at last, he added, "Or so I learned, at least, from my brother when he challenged me about the barge."

I gasped in dismay. I had forgotten George would have to reckon with Sir William's anger over the escape attempt! Hurriedly I asked,

"But what did he say about you? Will you be banished the Castle?"

George shook his head. "That would be as good—or as bad—as admitting that Lochleven is not so secure as my brother ought to have made it, and thus to draw the lords' censure down on his own head. I am to be barred from the Queen's presence, that is all—for the time being, at least."

"And so what do we do now?"

"I try, if I can, to discover the Queen's wishes before we decide on that. But as for how I am to do so . . ."

George paused, frowning over the question this posed. But *I* had the answer to it, all ready on the tip of my tongue. Arnault! Fat, cunning, and "blindly loyal" M'sieur Arnault had daily access to the Queen; and where could we find a better go-between than that little man?

*9*

$\mathcal{G}$ive her time" was Arnault's
first advice. "Her Grace is still melancholy over her failure
with Moray—too much so to be ready yet for active planning."

We bore patiently for a week, and by the end of that
time Arnault was reporting back to us,

"Her Grace is utterly determined on one thing, Mes-
sieurs. She will not let her cause go by default. She will
not cease, therefore, to demand both her freedom and the
right to appear before the lords to defend her name. She
enquires, therefore, M'sieu Douglas, if you are still willing
to be her messenger to the Earl of Moray."

"If that is her wish," George agreed. "But what of escape,
M'sieu Arnault. Has she spoken any further of that?"

Arnault was enjoying his role of intermediary, it seemed
to me. His tone became measured and rather pompous
as he told us then,

"Indeed she has. But she is now convinced that escape in itself is not enough. It will be of no use to her, in fact, to take advantage of any plan to free her until she first knows she can rely on having an army strong enough to defeat the one that Moray is certain to bring against her. To that end, therefore, she wishes to establish a courier service, with myself to smuggle out the letters she writes, and with you to carry them ashore to M'sieur John Beaton—who will then pass them further along the chain."

George said thoughtfully, "That makes sense, I agree. But M'sieu Arnault—"

Arnault checked him with one pudgy hand upheld. "And since the Queen's supporters in Scotland are in such disarray," he continued, "she intends to write also to France, to her kinfolk there, asking them to supply troops to back up her Scottish force, and thus be sure of final victory."

"But all that will take weeks to arrange," George exclaimed. "Months, perhaps, if the French are slow in their agreement. And time is short. Winter will soon be on us; and you know that armies can neither march nor invade in winter weather."

"You asked to know Her Grace's mind," Arnault pointed out. "And I have spoken it, exactly as she instructed me."

"I realise that." George was looking agitated now. "But delay could be dangerous for her. I have told you already how dangerous it could be."

"But she does not accept that," Arnault retorted. "And I am not the one to bring it home to her. What is more, I doubt if anyone could—yourself included. Her faith in her own innocence of any crime, M'sieu, is too strong for that."

Neither of them had been paying the least attention to me in the course of all this. Nor could I make head or tail of that remark about the Queen's "innocence"; and when George gave no answer to it, I grew impatient.

"Writing materials," I reminded the other two. "If the Queen is to have a courier service, she must first have pens, ink, and paper."

Arnault raised a grin for my benefit. "You have such common sense," he told me, "that I could almost believe you have French blood in you."

I glanced aside at George and said, "All I know of my blood is the name I bear." This was enough to raise some sort of grin from him too; and after that, we turned to the practical side of arranging the courier service.

Paper was the real problem; because paper, of course, was far too precious to be easily available. But Arnault had a small stock that was enough to begin with, and when this ran out, I robbed Sir William's desk for more. Until, that is, I discovered Lady Agnes at the same game; and since Lady Agnes knew I was well aware she had never been able to write more than her own name, she was immediately startled into confessing the reason for her theft.

" 'Tis for a—a journal. A private journal the Queen wishes to keep of her days here," she stuttered, and peered over the desk at me with big, cowlike eyes nervously blinking. "But you are not to tell, Will, because you know Sir William has ordered she is not to have any writing materials. And you would not want to get both Her Grace and me into trouble, would you?"

I promised faithfully I would not tell; and gave silent thanks, meanwhile, to the providence that had made Lady

Agnes so stupid, as well as so kindhearted. I also took great care, thereafter, to shuffle around the contents of Sir William's desk so that he would not notice when Lady Agnes thieved paper from it. But that was tame work. And it was always George, of course, who took the Queen's letters to the mainland.

I longed for diversion—any kind of diversion to while away the waiting time; and so I took to gambling again, with some of the Castle Guard for my opponents and another of my silver buttons for a stake.

"But this is the last time," Minny warned as she sewed a replacement on for me. "I'll not protect your hide again from the Old Lady, if you lose this one like you lost the other."

Minny always meant what she said; and so, after I had won enough to buy back my first button and have a little money over, I bowed out of the game. That was when I started slipping across to the mainland to see a girl I knew there—the daughter of the fat and talkative Mistress Matheson, who was landlady of the Kinross Inn. Muriel, they called this daughter. She was slender, with dark hair, and dark eyes that sent out signals telling me I could make progress with her. But the fat mother was watching us too closely for that; and so my chance of diversion at the Inn also came to nothing.

"I envy you," I told George. I had glimpsed him now and then at the Inn talking low and earnest to John Beaton, and others who were strangers to me. But George's eyes had avoided mine on these occasions; and now he told me,

"The less you know at this point, my lad, the better for all of us."

"Why?" I was feeling resentful by then, and the sharpness of George's answer showed how well he realised this.

"Because ignorance will be your only shield, of course, if anyone questions you about me."

I tried another tack. "Have you thought any more about escape, George? Because we should be ready with some kind of plan, should we not? Even if the Queen herself wants to delay the matter."

"I have thought—of course I have," George admitted. "But every possibility that comes to me finishes up by seeming wilder than the one before."

"To take the Castle by storm—" I began; but fiercely George cut across me.

"I told you my reasons against that! And I have made those clear too, to Beaton and the others. I will *not* risk the lives of my family."

I shrugged, and gave in. Our talk began to centre on other prospects for escape, but always we came up against the same obstacles. Even if we could get the Queen past the guard on her door, the gate in the courtyard wall was still the only exit from the Castle. By day, also, when the gate stood open, there were soldiers on guard there too. At night, at seven o'clock precisely, Sir William locked the gate. And the key to the gate was never out of his possession.

"But we must keep worrying at the problem," George insisted. "Because I *will* be banished from Lochleven if I am discovered to be still active on her behalf. And that will leave only you to keep in touch with me then; you, and only you, to carry out any possible rescue plan."

I liked the sound of that! Not because I wished George any harm, of course; but the prospect of rescuing the Queen entirely on my own *was* an attractive one. "I will keep on thinking," I assured him. "And if we can find a plan and you are banished, I promise I will carry it out as faithfully as you would yourself."

The glow of my conversation with George did not last. It was still only a far-off possibility, after all, this prospect he had held out; and the immediate prospect was only that of yet another dull winter in Lochleven. My envy began to stretch to cover the situation of Ellen and Margaret, so well placed to enjoy themselves now that the Queen had recovered from her fit of melancholy. And I was not alone in this.

The Queen still wept sometimes in the night, the girls said, and she prayed as much as she had always done. But even so, she was making good use now of the lute, the cards, the chessmen, and all the various other things she had asked for to lighten the burden of her imprisonment; and every day in her apartments now, there was music, and singing, games of skill and chance, lively conversation, and romps that were even livelier.

"And dancing," Ellen reported enthusiastically. "Margaret and I have learned some beautiful dances from her."

"Have you indeed!" Sir William exclaimed; and the slight edge to his voice reminded me that Ellen was the favourite among his children, the only one he had not been content to leave under the Old Lady's care in the New House. Sir William, I thought, was missing the company he was accustomed to have from Ellen; and when I went with him each day on his official visits to the Queen, it seemed to me that these visits were becoming longer.

"I must say," he told the Old Lady, "that Her Grace

seems to have become most resigned to her situation."

The Old Lady shrugged. "Her religion is a comfort to her, I suppose. And she does have a sweetness of nature which could also account for that. Besides, her wings have been well and truly clipped, now that the matter of the Regency has been so firmly settled."

"Then why should we not be more friendly to her?" Lady Agnes asked. "*I* find it pleasant enough to spend the day sitting with her."

The Old Lady gave Lady Agnes a glance that said, *You would!* But Lady Agnes was at the start of yet another addition to her family, and was far too content with herself to notice this. "The Queen," she went on, "is making such a beautiful tapestry, it is a joy to behold. And there always seems to be so much to talk about, too, with her and her ladies."

That drew the Old Lady! She could never bear to be left out of anything; and quite apart from that, she still craved the kind of Court gossip she had enjoyed in her youth. The Old Lady also began to visit the Queen's apartments; and sharp as her tongue was, it could also be a very amusing one. The length of Sir William's own visits increased when he found both her and Lady Agnes in the Queen's apartments. The visits themselves became less and less official, the traffic in and out of the Queen's apartments more and more casual; until finally, out of all this, it began to seem the most natural thing in the world for the whole family to be gathered there for at least some time each day. And if the rest of us could be there, then why not George also?

Casually one day, the Queen put this question to the Old Lady. There was no one, after all, who had a neater foot in the dance than young Mr. Douglas—or so *she* had been told, at least. And so it was a pity, was it not, that her ladies were being deprived of the opportunity to admire his skill?

Little Jane Kennedy and tall Maria de Courcelles looked modestly downwards. The Queen smiled, and teasingly reminded them how handsome Mr. Douglas was. Oh yes, Kennedy agreed demurely, she had noticed that. De Courcelles lifted her own very handsome head to acknowledge it would indeed be so much more pleasant to have the company of young Mr. Douglas also.

I watched the faces of Sir William and the Old Lady as they talked. Sir William had the air of a man who pretends not to hear the conversation around him, but the Old Lady was fairly preening with pleasure in all this talk about George. "Pretty Geordie"—it was she, I remembered, who had so fondly invented that nickname for him; and as my eyes came to rest on her then, I wondered how I could have been so blind to the strategy of all the seemingly harmless diversions that had brought us to this moment.

"You could have told me," I reproached George, "instead of leaving me to guess."

The arguments bred from that conversation about him were over by then. Sir William had finally agreed with the Old Lady that George had learned his lesson; and now George had the same freedom of the Queen's apartments as the rest of us had. He grinned, to begin with, at the peevish tone of my voice, and teased me,

"But maybe you are not so good an actor as you think you are!"

"And what has that to do with it?"

"A great deal, my lad. The Queen was well aware I would never be allowed into her presence until she had disarmed all suspicion of herself. And if there had been one unguarded look or word from you, my brother might just have guessed where all her entertainments were leading him."

"I would still have preferred the chance to act the fool," I retorted, "rather than to know you have made one of me."

"Oh come," George coaxed. "Think of what has been accomplished. There is no longer any need for Arnault to act as go-between. And supposing there should be a situation where I have to see the Queen alone? Her guards will let me pass without question now. And is all that not worth a little sacrifice of pride on your part?"

I could not think of any argument against this; and so went sulkily off instead to the game of cards the Queen had commanded I should play with her. She was a worthy opponent, I had discovered; but she had a gambling streak that ran as deep and wide as my own, and I happened that day to be holding the better cards. At the end of the game, she paid up handsomely; and I was not only feeling in better mood by then, but in better shape also to play this other game she had so skillfully devised for us.

"I told you—remember?" Ellen challenged me. "I said it could be exciting for all of us to have her here. And it *is* now. You have to admit that, Will. We are like people at Court now, are we not? And all of us really enjoying ourselves!"

"*All* of us?"

The smile Ellen had turned on me began to falter. "I know what you mean by that, Will Douglas; but you are wrong. Margaret does not laugh and sing the way I do because it is her nature to be quiet. And *she* cannot help the way her father behaved to the Queen."

"You spoke of disarming suspicion," I told George. "But I do not think we can yet be certain of Margaret Lindsay where that is concerned."

Jane Kennedy took up the lute for one of the galliards she played so well, and it was George who won the Queen's hand for the dance. The eyes of the Old Lady and Lady Agnes followed its movement, and Lady Agnes said wistfully,

"I wish I could dance like that."

The Old Lady glanced from the light, floating form of the Queen to the plump one of Lady Agnes; and with bright eyes sparking malice, she asked, "How could you when you do not look like that, even when you are not with child?"

I caught Ellen's eye and grinned to see her so embarrassed at this. The Old Lady and Lady Agnes went on bickering. Ellen moved to give me a pinch in the ribs for my impudence. Under cover of the music we started our own argument, then were silent again as Kennedy stopped playing, and George led the Queen back to her chair.

"But of course, Your Grace," he was telling her, "my

sisters would be deeply honoured to meet you." The Queen leaned forwards in smiling anticipation as he turned to announce to the rest of us, "Her Grace has expressed a desire to meet the Seven Lochleven Porches. And she is curious to know why they are called so."

We glanced from one to the other, smiling at this, before the Old Lady said, "When you meet my daughters, Your Grace, you will understand. They look like their nickname—all of them as tall, and slender, and straight as the pillars of a porch."

"And swan necked too," George added, "which gives a beautiful curve to the summit of each Porch!"

Our smiles broke into laughter then, but hard on this laughter came a voice, the small and clear voice of Margaret Lindsay saying,

"But Your Grace, my mother is one of them. And surely you cannot wish to meet now with *her*."

Embarrassed silence gripped the room. But the Queen's smile did not falter, and when she spoke, her voice was gentle. "Why not, child? Lady Lindsay has not harmed me, any more than you have."

A flush leapt into Margaret's sallow face. Impulsively she knelt to kiss the Queen's hand. And it was Margaret herself, the next day, who shyly led her mother as the first of the Seven Lochleven Porches to make their curtsies to the Queen.

Genteelly then, they sipped the wine I poured for them, congratulated Lady Agnes on her condition, clustered around to hear the Queen explain the intricate pattern of her tapestry work. They scolded me, too, for having spent so long without visiting them. I teased them in reply, for taking so long over finding husbands for themselves;

and thought how right Ellen had been. It really was most pleasant, this little Court the Queen had created around her own presence.

I looked again towards where she sat in a high-backed chair, with Sir William and George and the Porches gathered about her, Ellen and Margaret crouched at her feet. The polished wood of the chair made a dark frame for her head and face. The red-gold of her hair, the paleness of her skin, glowed from the frame with their own soft and mysterious light that moved as she moved when she glanced from one to the other of the surrounding group. And they were all so fascinated by her!

This little Court, I thought, was a sort of dream she had spun for them, with herself as the bright centre that drew their eyes inward from the world beyond the walls of Lochleven—the real world of power and war over kingdoms. And—except for George—not one of them could see how skillfully she was charming their attention away from her purpose of reentering that real world.

Sir William, I noticed, was bending a little, the better to hear her speaking. Ellen and Margaret were looking up into her face, and Margaret's eyes were as adoring now as those of Ellen. The Porches were all listening intently to her, and all standing very deferentially in the correct Court positions. She began speaking directly to the seven of them, telling them first that she had been accustomed always to walk or ride for some part of every day, and then wistfully confessing how much and how often this made her long to be allowed again into the fresh air of the garden.

And there was Sir William now, flinching back from the accusation in seven sisterly pairs of eyes, the reproach in seven sisterly voices! There was Sir William being forced to agree,

"Why yes, if the day is good enough, Her Grace may take a turn in the garden. I should be pleased to escort her."

So he was taking no risks there, even although he had been cornered into this further permission. But he had at least granted it. And it did at least mean that the Queen could now be seen leaving her apartments without an instant alarm being raised. One more small victory, I thought, for the Court of Dreams!

The Lochleven Porches departed, there was space again for dancing; and when Kennedy struck a chord on the lute, Sir William bowed Maria de Courcelles to the floor. She was nearly as tall as he was, and the glossiness of her dark hair looked well against his ruddy fairness. But Sir William's features were coarsened with middle age. Moreover, he was nowhere near so slim as George, who had now taken the Queen on his arm; and so it was George and the Queen who made by far the better couple. The Old Lady called Ellen to her, and said,

"You see there, Ellen, how tall the Queen is. And yet George is an even height with her. Now let that be a lesson to you because you are going to be tall, like your aunts, the Porches. And half the reason that six of *them* are still unmarried is that they make themselves look ridiculous by consenting to dance with men who are not even their equals in height."

Ellen bent to whisper something to the Old Lady. I thought, from their faces, it might be a question. The Old Lady drew back from Ellen and looked towards George and the Queen. I looked also, and found I could not take my eyes from them.

The Queen was dancing with her head back tilted, so that her smile at George seemed to slant a little. The amber gleam of her eyes reflected her smile. The light in the room made a glisten on the delicate skin of her throat, like the glisten of the pearls fringing the white of the heart-shaped coif on her hair. And George was gazing at all this with so deep a longing in his eyes that there was no mistaking the fact. *George was in love with her!*

It was the sound of the Old Lady calling me to bring her a footstool that startled me back to the rest of the room. I hurried to obey the call; and as I knelt to lift her feet onto the stool, she said softly,

"You are not to get mixed up in any plans that George may have. D'ye hear me?"

I was too surprised to make an instant recovery from the effect of this. I tried to rise, mumbling some sort of denial of any intent whatsoever, and found myself pinned there by one bony hand pressing into my shoulder. Under her breath, the Old Lady hissed at me,

"Stop lying, Will. I always know when you are lying. You were very thick with George for a while back there. But you will let him take his own risks from now on, or answer to me for it."

I nodded, not daring to trust my voice; and she released me, to sit back and resume her watch on the dancers. To my further surprise then, I saw that her gaze at George and the Queen had an air of grim satisfaction about it— like that of an old tigress, I thought, watching her cub at play!

I rose towards the tray of wine, and while I pretended to put the cups on it in order, I tried to bring some order also into my thoughts on the scene that had just taken place. But there was one thing about it that defeated me.

I simply could not see why the Old Lady—of all people—should seem pleased that George was in love with the Queen. And, I realised, I would never keep abreast of events now until I had the answer to *that* riddle!

## 10

Minny went off at a tangent with *her* answer.

" 'Tis not easy," she said wistfully, "being in love. It can be a sore thing, Will. A very sore thing."

*Harking back to her own young days*, I thought, and turned swiftly from the uncomfortable feeling that Minny's sore experience of love might have been the reason for my own existence.

"That was not my point, Minny," I reminded her. "It was the Old Lady I asked about—why she should be so content to see George in love with the Queen."

Minny stared blankly at me. "Why should she not? Mr. George is handsome. His blood is most noble, and he is of an age to give the Queen many children. It would be a good match—and the Old Lady would certainly delight to see a King's crown on the head of her pretty Geordie!"

"But Minny, the Queen is already married. To Bothwell."

Minny's stare became one of pity. "You *are* green, Will," she told me. "Forget Bothwell. He is a fugitive now. An exile in Norway, they say. The Queen could get a divorce from him for any one of half a dozen reasons. Or else she could have the marriage to him annulled. And the Old Lady knows that as well as I do."

"But even so," I protested, "you are going too fast. The Queen cannot make a King of George until she has won back her Throne. And she cannot do that until she has got out of here and fought a battle with Moray. But even if she does succeed in all these things, and even if she owes her success to George's help, it does not mean to say she would marry him."

"No," Minny admitted. "But the chance is there, is it not? And trust the Old Lady to be wide awake to that!"

"And Moray?" I demanded. "What about him and her pride in him as Regent? There would be no more power for Moray if the Queen were to win back her Throne."

"Of course," Minny agreed. "But to have George for King would give the Old Lady even more cause for pride, would it not? And besides, laddie, she knows very well there is nothing she can do now to stop him in his tracks— not if he is as much in love with the Queen as you say he is."

"She could warn Moray."

"No, no." Decisively Minny shook her head. "She might be content for Mr. George to 'take his own risks,' as you called it. But he is still her youngest, her precious ewe lamb. And so *she'll* not be the one to put his very life at stake."

Suddenly then, I remembered what the Old Lady herself had said, months before, when she had insisted on Arnault being brought to treat the Queen's illness. *I have more sons than Moray.* Minny was right, I realised. The Old Lady would indeed let matters take their course now. And if Moray *was* destined to be toppled eventually from his high position as Regent, why should she not console herself with the hope of George being raised even higher?

But supposing the Queen did not want to divorce Bothwell—or to have her marriage to him annulled? I would need someone other than Minny to answer that for me, I decided. Minny was shrewd enough in reading the Old Lady's mind; but the someone I needed now would have to be equally shrewd in reading the mind of the Queen!

Arnault, at first, was very decisive. "You can be certain of one thing," he told me. "As a faithful daughter of the Roman Church, the Queen would never consider divorce."

"And annullment?"

"Well, yes," he allowed. "She might agree to that—but only if she saw a need for it."

"A political need—is that what you mean?"

"Precisely. But why do you ask all this?"

"Because George is in love with the Queen. And the Old Lady thinks she might marry him."

To my surprise then, Arnault chuckled and said, "I knew that would happen. I knew it!"

"Oh, so you knew, did you?" I challenged. "Because that was also part of the Queen's strategy?"

"Not at all, my young friend." Arnault continued to grin at me. "But what happens when you put a honeypot down among bees? They swarm towards it, they cannot resist its attractions. And that is the measure of Her Grace's charm. That is how she has succeeded in drawing you all into this—what did you call it?—this Court of Dreams. As for young M'sieur Douglas, has she shown him any more favour there than she has to the rest of you?"

I thought back to the moment I had realised George was drowning deep in love with the Queen, and compared her expression then with the inviting looks sent out by the innkeeper's black-eyed daughter. "No," I admitted. "She has never played the coquette, or given him any sign of love. And would she marry him, M'sieu Arnault, without love?"

Arnault's grin turned sour at this; and grimly, he said, "She did without love in her marriage to Bothwell."

"But why?" I was asking this, I realised, with all sorts of confused thoughts about the Queen rushing suddenly into my mind. Arnault must have sensed this, because he answered me with a question of his own. "Why do *you* think she married him?"

The memories lurking behind my thoughts began to surface—Lord Lindsay shouting "Slut" and "Whore" at the Queen, myself and Ellen quarrelling over the way Ellen had parrotted Lindsay's opinions of her character. "How can I tell," I hedged, "when all I know is what people say of her?"

"And people say she is a light woman, a wanton!" Arnault spoke angrily now, his dark face beginning to flush. "They say that Bothwell was her lover, even before Darnley was murdered. And do you know how much truth there is in that?"

I shook my head. "I told you. All I do know is rumour."

"But now you are troubled for young M'sieur Douglas," Arnault jeered. "And so now also, you will let rumour condemn her."

I looked at him then in even greater confusion. Those memories and the pictures that had surfaced along with them were so much at variance with my recent recollections of a Queen who could be one moment a laughing girl, and the next a mysterious presence of majesty—or a kindly young woman teaching the steps of a dance; the Queen who was high-spirited, yet always repelled by anything gross or coarse, who wept in the night, and spent long hours in private prayer . . . Awkwardly I told Arnault,

"You do me wrong, M'sieu. It is just that—well, I do not believe the rumours about her, and I do not want to believe them. But I would like to understand."

"Understand what?" Arnault looked with hostility at me. "The truth? Because that was what I offered you."

I nodded, too overawed by his expression to do more. Arnault gave a grunt of grim satisfaction. "Then let me instruct you a little," he said. "There are certain matters a woman cannot hide from her physician, any more than she can hide these matters from her confessor. It is therefore as Her Majesty's physician I speak when I say that the stories you have heard of her affair with Bothwell are false—all false."

He was glaring at me, daring me to speak further. I took my courage in both hands, and told him, "I need to hear more than that, M'sieu Arnault, if I am truly to understand. Because she did marry Bothwell after Darnley was murdered. Only a little more than three months afterwards! And everyone says it was Bothwell who did the murder."

"And for once," Arnault jeered, "everyone is right! Which proves?"

I almost failed then, before the sardonic note in his voice; but having come thus far, I had to nerve myself to go on. "I do not know what it proves," I told him. "But people also say it was she who incited Bothwell to the murder. And to buy his silence over it, she had to marry him."

"Then people lie! Once again I tell you they lie, they lie!" Arnault was exclaiming loudly now, the flush on his face growing deeper with every word. "It was not till the day *after* the Queen married Bothwell that she knew he was the murderer. And if you do not believe me on that, ask the Spanish Ambassador to her Court. The English Ambassador too. *They* heard her calling in anguish for a knife to kill herself!"

Arnault was shaking with anger by this time. I shrank from him, hardly able to believe that a man of his sardonic humour should also have in him so much of passion. "But believe me you shall," he was finally insisting, "for the very good reason that I was there when she seized the knife; and it was I—Arnault—who wrenched it from her hand."

I gave the little man a moment to calm himself; and then, as quietly as I could, I told him, "You are so sure of your facts, M'sieu Arnault, that I must believe you. But none of this answers the very first question I asked, because I still do not know why the Queen should have chosen to marry Bothwell."

Arnault said wearily, "He had one great virtue in her eyes. She was surrounded by treacherous, self-seeking men, but he had always been utterly loyal to her. She knew she could count on him to support her in the terrible confusion following on Darnley's murder. And you have seen Bothwell in her retinue in the days when she used to visit this place. You must know the nature of the man."

Did I? I recalled Bothwell as I had seen him in the Queen's entourage—a fiercely energetic man, swarthy, almost ugly, yet with something compelling in the glitter of his dark eyes. . . .

"Ambitious," Arnault said, "a clever soldier, and a man who knows only one way with women—the Queen included. Shrewd enough, too, to make sure the chance of marriage to her did not slip by him then. And so he simply arranged matters so that she would have no honourable course except to marry him. He took her by force. And that, my young friend, was the extent of the 'choice' she had."

I said nothing to this. I was too appalled, then, to speak. Arnault was silent for a while also, his gaze faraway and mournful; but finally he roused himself enough to say, "It is the tragedy of this poor young Queen, you see, that she has always needed to lean on someone stronger than herself. And for the short time that mattered, alas, it was her mistake to lean on Bothwell."

"She is leaning on no one now!" I said this impulsively, and was surprised myself at the force of conviction behind it. "It was not George who planned her courier service or the way she has disarmed suspicion of herself. Nor was it you, M'sieu Arnault."

Arnault rubbed thoughtfully at the stubble that passed, with him, for a beard. "True," he agreed. "And it is also true there have been other times when she has surprised everyone by acting for herself—with skill too, and great courage. But that was always in situations she could defeat with one bold stroke; and this present one is different from those others. This one has forced her into a waiting game, just when delay spells danger to her. And so, you see, all the strength she has brought to it could count for nothing in the end."

That made twice I had heard this mysterious reference to "delay" and "danger." I would not let it pass this time, I decided; and bluntly I told Arnault, "No, I do not see. And I have been kept long enough in the dark over it. I think it is high time you told me, M'sieu Arnault, what *is* this danger you speak of?"

Arnault did not answer straightaway. Instead, he turned from me to sit at his desk, elbows propped, chin resting on clasped hands. From this position he gave me a long, considering look, and only then did he say,

"It is not I who should rattle the skeleton in the Douglas family cupboard. Ask your friend—your blood relative M'sieur George, and see if *he* is willing to tell you."

I waited by the west landing stage when I knew George had a letter to take to the mainland, making my duty of looking after the Castle boats an excuse to linger there. It was a cold day, real November weather, with sleet in the wind driving over the loch. George came hurrying down to the boats with the fur collar of his cloak pulled well up around his ears; and immediately I tackled him:

"I have been speaking to Arnault, George."

"And—?" George was so impatient to be off he was barely listening.

"He told me about the Queen and Bothwell—the truth about the marriage between them."

That startled him into attention. He shot me a puzzled look, and asked, "Why did you want to know?"

What would he say, I wondered, if I told him the real reason? I decided to hedge, and said defensively, "Why should I not, when I find her so different from the picture that rumour has painted of her?"

George began to smile a little. "Good lad," he said softly. "I always knew you were a good lad behind that careless face you wear."

That was the conversation finished so far as he was concerned, but I caught his arm as he slipped the mooring rope of the boat, and urgently told him,

"There was something else, George. Arnault spoke of the Queen playing a waiting game, and the danger this could be to her—just as you did once. But when I asked him what the danger was, he would not tell me. He said I must ask you."

George looked sharply at me. "Was that all he said?"

"No. He said—" I hesitated, and then awkwardly quoted, " 'It is not I who should rattle the skeleton in the Douglas family cupboard.' "

"Arnault," said George grimly, "has quite a sense of humour!" He gave a savage little jerk of the mooring rope, and then asked, "What else did he tell you?"

"About Darnley's murder. He said it was true that Bothwell did it, but that the Queen did not know that till after she married him."

"Exactly!" George exclaimed. "And so the skeleton, you see, is beginning to rattle!"

I had my own share of the Douglas temper, and that

was when I felt it beginning to rise. I said, "Look, George, you said the day might come when I would have to act alone for the Queen. That means I will be running the same risks as you do now—perhaps more, because you are Moray's half-brother, and I am not. And so I think it only fair that you stop being so mysterious with me."

George nodded. "Yes. You are right. But 'tis the mere fact of being Moray's half-brother that makes it difficult to tell you. Because he has always been good to me, Will. More than good . . ."

I swallowed my rising anger and waited while George looked out into the driving sleet.

"Bothwell killed Darnley," he said at last; "but Bothwell was only the cat's-paw in that. There were others who wanted Darnley dead. They encouraged him to the murder; and the bait they held out was that—once the deed was done—they would support his ambition to marry the Queen."

"Others?"

"Men of Moray's party. The very ones who made the marriage an excuse to raise rebellion against her."

"But George!" I caught George's arm and shook it in an attempt to make him face me. "Moray was not even in the country when the lords rebelled. You know that! He was in France."

"Of course." Now George did turn, of his own accord, his face pinched by the cold and working with an emotion I could only guess at. "Moray always has excelled at making bullets for others to fire!"

I dropped my hold on George's arm and took a pace or two back and forwards as I tried to put order into all this question and answer. Then, when I thought I had it clear in my mind, I said,

"So what you are telling me is this. The murder of Darnley was just the first step in a plot to bring the Queen down. Bothwell, and Bothwell's ambition to marry the Queen, was the tool in that. And it was Moray who was the real force behind it all. Is that it?"

George turned from me again without making any answer; but his silence was answer enough. I let the silence endure for a moment before I asked, "And the danger to the Queen now?"

George's head whipped round to me as if it had been pulled on a string. "Can you not guess? The whole country accepts now that Bothwell is guilty of Darnley's murder, even although he has never been legally declared so. But once Moray manages to turn all the talk about him into an official declaration of guilt, how long do you think the Queen can escape being charged with the same crime?"

I stared in horror at him. He was shaking. The face that normally looked so smooth and handsome was transformed by his distress to an ugly mask. "And I love her!" The voice that came out of the mask had a strangled sound. "Oh my God, how I love her!"

I gave him a moment or two to recover himself before I asked, "How will Moray go about this—making a legal charge against Bothwell, I mean?"

"He will convene a Parliament—bring witnesses to testify before it. And," George added bitterly, "you can depend on it he would have done so already if it had not been for the need to prepare the statements of these witnesses. But torture, you see, takes time."

I found myself inwardly flinching from the picture conjured up by those final words. And George, it seemed, had had enough of our conversation. He began clambering into the boat. I leaned on the gunwale to steady it; and as he picked up the oars, I asked,

"How long have you known all this about Moray? Since the Queen was first brought here?"

"Known!" George gave a little laugh. "That is the wrong word, Will. No one *knows* anything about Moray. That is his strength. He orders his lieutenants—men like Lindsay—in hints, and half hints. And as for those outside that charmed circle, they just have to guess and piece small things together—slowly, as I have done—before they can finally arrive at the truth."

"But George"— I kept my hold on the gunwale, unwilling to let George go until I had the last piece of the puzzle about Moray in place—"why does he hate the Queen so? He must hate her, if all you say of him is true."

George shook his head. "You do not understand," he said. "Moray does not hate her. The Queen is right, in fact, to claim that he loves her. But what she has not recognised is that he loves power even more. And he had that, you see, so long as she leaned on *him* for advice—which she did, always. Until she married Darnley, that is; and then he lost all influence over her. But that was also when he discovered it was no use to fight back openly against the marriage—you remember?"

I nodded. I had been only fourteen then, of course; but even so, I could not have failed to be impressed by the events of that time. Moray in the field with one army, the Queen at the head of her own troops opposing him— the excitement in the Castle had been tremendous then!

"Then you will remember too," George said, "how brilliantly the Queen defeated *that* bid for power. But this at least you can say for Moray. He learns from his mistakes—learns well too; because these more devious methods that got him the Regency will bring him more now than power. He will have money from them too—much money; all the rich pickings to be had from administering the Royal estates of the Queen's little son. And it is to safeguard both kinds of gain, now, that the Queen must be branded with Bothwell's guilt. Not only dethroned, you see, but ultimately destroyed as well."

I had heard enough. I let the boat go then—indeed, George gave such a shove of the oars on his final word that I could not have held it longer. I stood watching him pull across to the mainland, feeling a great pity for the Queen walking so blindly into the trap that had been set for her. And still too trusting to see the danger it held. Because what was it Arnault had said? "Her faith in her own innocence of any crime . . . is too strong for that."

The loch grew wilder yet with the sleet battering on it. I turned away from its heaving grey water and walked towards the Castle, feeling cold now from more than the winter wind driving on my back.

That cold November was followed by a brief spell of most peculiar weather—certain days of a humid calm quite unnatural to the season. It was just before the middle of December that this spell began; the very same time as Sir William had word that Moray was once more coming to see the Queen. And for the three days between that advance notice and Moray's actual arrival, I could feel the effect of this odd weather adding to the general air of uneasiness in the Castle.

The Queen herself, in fact, was the only one who seemed untouched by this uneasiness. With an appearance of total composure she listened to Sir William telling her the nature of Moray's visit, stressing all the while how formal this was intended to be; and calmly, when the explanation was finished, she said,

"Very well, Sir William. You may make it known to

Moray and his party of lords that I will grant them an audience."

I gawked at the form of this announcement, as much as Sir William did; but there was still another surprise in store for us.

"In the Presence Chamber," the Queen added. "You may bring them to me there."

Sir William looked helplessly at her. It was she herself, certainly, who had ordered a Presence Chamber to be built in Lochleven. But that had been when she was a reigning Monarch, entitled to such dignities. But now she knew— had he not just been at pains to impress it on her?—the reason for Moray's visit. A Parliament had at last been held. And the proceedings Moray had brought before that Parliament had confirmed the very fears George had voiced for her. Yet here she was, after months of seemingly tame acceptance of Moray's Regency, suddenly reasserting her Royal rights over him!

There was no denying, either, that she meant to have these. Coldly she listened to Sir William's protests. As coldly, she dismissed them with a swift reminder of the way she had been forced to sign the Deed of Abdication; and suddenly then, I thought I understood what lay behind her attitude.

*She had seen the danger!* She had accepted at last that the net cast for Bothwell was closing on her also. And she was determined to give Moray open challenge on that!

"Did you convince her of it?" I asked George. "Or did she come to it of her own accord?"

"Half and half," George admitted. "That day I talked

to you about it, Will—I pulled off from the landing stage then, knowing that I *had* to make her see it. And praise God, she became gradually more and more amenable to argument."

"So she'll wait no longer for aid from France?"

"No—praise God again! She is ripe and ready for escape, with as little delay now as possible."

I felt my heart give a most tremendous leap of excitement. In a rush of words I exclaimed, "Then I can get her out! I have been thinking and thinking ever since the last time we spoke of it, and I am certain now that there is only one way to get over all the difficulties. *She must walk openly out of the Castle.*"

George stared in disbelief. "Have you lost your wits?"

"No, I have found them! I shall need some help in the plan, of course; but only from Minny, and only at the beginning of it. The rest will be up to the Queen herself, then. And so listen—"

I went on speaking as rapidly as I could in my efforts to convince George. But I was still not making sufficient impression. I could see that all the time I talked; and disappointment rose sharply in me when he said finally,

"It could succeed, I suppose. But there are so many ifs to it . . ."

"So you will not let me try it—is that what you mean?"

"No, no, you are leaping ahead," George reproved. "It does have some merits. And if we are forced eventually to rely on it, then I suppose we must. But that will only happen, remember, if I am not here myself to think of a better plan."

The lump of disappointment in my throat suddenly vanished. With cold alarm creeping in to replace it, I asked,

"Why do you talk like that? Is it—is it Moray? Has he begun to suspect you?"

George shrugged. "How could he fail to suspect? I have continued to carry the Queen's letters to him, after all; and so he knows my sympathy for her. Besides, there is gossip and guesswork all over the country about our life here in Lochleven. And Moray has spies everywhere—including the one in our very midst. Or had you forgotten about Drysdale?"

Captain Drysdale—I *had* forgotten about him! "But you said yourself," I protested, "that Sir William would keep quiet over the business with the barge. You said he would have to, in case it had a bad effect on his own position here."

"True," George agreed. "But what had Drysdale to lose by telling Moray of it? And you know what a grasping creature Drysdale has always been where money is concerned. What did he stand to gain by way of a bribe for information supplied?"

And Drysdale also hated the Queen! The cold feeling of alarm began to reach right through me. I said quickly, "You should get away—clear out altogether while Moray is here. There would be nothing then, to remind him of his suspicions."

"Maybe so," George agreed. "On the other hand, to run might be to admit guilt. But I *will* lie low, I promise you. And Moray, when he comes here tomorrow, will not find anything that draws his eye to me."

I wanted to take George's word for that; wanted most anxiously to do so. But George, I remembered gloomily, was Commander of the Castle Guard. And that would bring him very much under Moray's eye—would it not?—

when his soldiers had to escort the Queen to the Presence Chamber.

It stood apart from the Castle's main tower—a single-storied building with its doorway in the gable end facing the courtyard, a large window high up in the opposite gable built into the courtyard wall. The one room inside this building was a spacious, vaulted chamber, unfurnished except for a thronelike chair of red velvet placed on the dais beneath the large window. Sir William inspected it on the morning of the day Moray was due—yet another of those oddly calm days; and then fretted away the next few hours with complaints that Moray would blame *him* for the Queen's insistence on using it.

At two o'clock the soldier on lookout duty came hurrying to tell him that Moray's party had been sighted approaching the mainland landing stage. George was given his orders to escort the Queen to the Presence Chamber, and Sir William went down to our own west landing stage to greet Moray. There he stood with one hand on my shoulder as he watched the approaching boats; and even if I had not known how nervous he was by that time, the strength of his grip would have told me so.

"She will ruin me yet, Will," he muttered. "Do you know that? She will ruin me!"

It came to me with some force then that it was *I* who could be the instrument of his ruin, and I was more dismayed than I could have believed possible. Sir William had never been less than kind to me, after all, despite the occasional—and usually justified—beatings I had suffered from him. I found myself reaching up my own hand to cover the one holding my shoulder. Sir William looked down at this, then drew me close to him; and I was still gripped by the feeling this roused in me when the boats with Moray's party in them touched at the landing stage.

Moray was sombre in black again, but the numerous lords who were with him this time were peacocked out in reds and blues and silvers, glittered with jewelled rings and the gold of neck chains, plumed with feathered hats that matched the glowing colours of their velvet doublets. I was swamped in a sea of colour, deafened by a rumble of voices, pushed and jostled from Sir William's side as they all crowded around him and Moray.

The rumble became shot through with individual cries of protest as the crowding men caught the drift of the conversation between these two. I made out Lord Lindsay's voice among those who cried loudest; and once more, heard the ugly epithets he had used of the Queen at her own landing there. And that decided me. I knew for certain now how false these were. And as for my feelings towards Sir William, *he* had never acknowledged me as his son. It was the Queen who had drawn the sting that word "bastard" had so often dealt me. I would not weaken towards Sir William again, I vowed; not so long as she had only George and myself to stand between her and the fate *this* mob had planned!

The party fell into place behind Moray leading them towards the Castle gate, and I regained my accustomed place at Sir William's heels. From halfway across the courtyard, I saw that the door to the Presence Chamber stood open. Lined up outside it were the soldiers of the Queen's escort, with George Douglas commanding them to attention. The voices of Moray's party died away as we neared the door, until the only sound breaking the odd stillness of that day was the tramp of our own feet. We crowded into the Presence Chamber, Moray still in the lead, Sir William behind him, myself behind Sir William.

The Queen was standing beside the red velvet chair in the pose that was usual for her—one hand on the back of the chair, the other fingering her gold crucifix hanging on its slender golden neck chain. She was in black, slashed with purple, and her head was crowned with light from the window behind her. She made no move at the sight of the crowding lords. As the last man entered, the voice of George Douglas came from beyond the door, with an order to the soldiers to "Stand easy." And then, as suddenly as a stage effect that had been carefully planned and practised, there was a moment I would not have believed if I had not been there myself to see it happen.

From the moveless air outside—the air that had not held a breath of wind for all of three days—came a blast so powerful that the lattice of the window behind the Queen was burst wide open. She swayed to the force of the blast. The gaudy group assembled before her swayed also, and opened mouths of superstitious terror. The Queen's grasp on her chair brought her quickly upright again; and with eyes sweeping contemptuously over all the gaping faces, she cried out,

"The heavens speak, my lords, and send a wind for traitors!"

Sir William rushed to the window and brought the wild flapping of the casement under control. "Yes, bolt it again," the Queen told him, "but treachery has already entered here!"

Moray stepped forward, clearing his throat. "Your Grace—" he began. "Sister—"

Fiercely the Queen rounded on him. "You call me 'sister' and I always have been a sister to you, heaping love and favours on you, pardoning your former rebellion, forgiving every act that was not the act of a brother. But what have your actions been to me, my lord?—for I will never call you 'brother' again. How have you treated me?"

"Your Majesty—" Moray tried again, but still the Queen would not let him speak.

"Yes, my Majesty! And what have you tried to do against that, Moray? I have asked you often enough for the freedom I should justly have. I have asked even more often for something I value far more highly than freedom—the chance to appear before the Privy Council to defend myself against the vile innuendoes cast on me. But you, unjust and base as you are, have acted in every way towards me as an enemy. Because do not think, my lord, that I am now as ignorant as I once was of the meaning behind all your manoeuvres—your sly off-going to France, leaving behind you a rebellion already plotted; your delay in first coming to visit me here in my prison. *And* the reason also for your even longer delay in convening the Parliament that has now confirmed the outlawry of the Earl Bothwell, stripped him of all his offices and dignities, laid him under pain of treason, and declared him guilty of murder.

"You needed time to bring me down with him; time to manufacture the 'evidence' that would link *my* name to murder also, and so justify my continued imprisonment here. And time also, my lord, to persuade Parliament to confirm that you and these other lords should have indemnity for all your treacherous actions."

It was magnificent, I thought. *She* was magnificent, poised there like some young storm goddess in the trail of that violent wind. And all her accusations were hitting home to Moray. He was staring at her, clearly aghast at the degree of truth in her tirade—but was it wise to humiliate him so completely? And in front of all that assembly of lords, too! One of them—the Earl of Morton—had stepped to Moray's side now, and was feebly trying to pacify her.

"Your Grace," he pleaded, "my lord Moray was appointed Regent at the instance of the Council. And it *is* as the Council that we come here now to inform you of Parliament's decisions. As to your freedom, I personally will do everything in my power—"

"You *have* no power!" Sharp as a scythe cutting through corn, the Queen cut through Morton's words. "You are Moray's creature. You are all his creatures. And I would rather wear out my life in perpetual imprisonment than accept freedom now at his hands."

She paused, looked directly at Moray, and then let her eyes travel over the other faces there. Moray did not return her look, and he did not try to take advantage of the pause to speak again. I wondered why. Was it rage that tied his tongue? Or astonishment that so gentle a nature could suddenly show such force? Or was he simply too taken aback, still, by her knowledge, to even try some denial of its truth?

"My lords"—the Queen's eyes had returned to Moray, and she was speaking more quietly now—"I vow this to you. I shall never again speak face to face with the Earl of Moray. *But I shall regain my freedom!* And as there is a just God in heaven, that freedom will be to Moray's disgrace, damage, and ruin."

Proudly then she stepped down from her place on the dais and walked straight towards Moray. Still without looking at her, he moved a step to let her pass. But the Queen did not pass. Instead, she stopped and looked intently at his averted face, at the hands that seemed to hang nerveless at his sides. Then, in one swift movement, she had hold of one of these hands in a gesture that forced his eyes to hers.

"You will repent your perfidy to me!" Her voice was breaking now, with tears barely restrained; but she mastered the break, and finished, "Sooner or later, and cost what it might, I tell you that you shall repent!"

The lords gave way in silence before her as she continued her proud exit from the Presence Chamber. She had been right—they *were* all Moray's creatures. And where she had so thoroughly asserted herself as still having Sovereign right over him, there was none of them—not even one so brash as Lord Lindsay—who would dare to challenge her. . . .

"She has some informant here, I tell you!" Moray's voice, as he faced Sir William in the Great Hall, was nearer to a snarl than to human speech. "She knew too much about me *not* to have an informant. And I know who it is, d'ye hear? *I know!*"

He glared all around him—at Sir William, at Lady Agnes, the Old Lady; even at myself. There was no one else there. The lords had been dismissed, back to their boats; for this, it seemed, was to be a family inquisition.

"Now, James—" the Old Lady began soothingly; but Moray ignored her. Head lowered like that of a bull about to charge, his very beard quivering with anger, he fixed his glare finally on Sir William, and roared,

"But you will banish him from this place! I command it. Banish George Douglas from Lochleven—or I will have him *hanged!*"

Lady Agnes screamed. The Old Lady gasped. Even Sir William flinched from the impact of Moray's final word. Moray ignored the women. With his gaze still on Sir William, he went on,

"As for you, you have been much too lenient with her, or she could not have written the letters George carried to me. But that must end. You will seize all her writing materials, or answer to me for it. And mark this, brother; mark it well. If she does escape, as she claims she will do, you will answer to me for that too!"

Abruptly, on this, he turned on his heel and stalked from the room. The quivering silence he left behind was broken by a sob from Lady Agnes. The Old Lady shook herself out of what seemed an almost trancelike state and began wildly to babble that her Geordie was innocent of all harm; that nothing and nobody was going to take him away from her. Sir William turned to me.

"Fetch George here," he ordered nervously. "And for God's sake, boy, be quick about it."

With a hand pushing in the small of my back, he thrust me to the door. I took the flight of steps outside it in a series of bounds, then raced for the likeliest place to find George. He was where I had guessed—in the barracks; and as fast as I could, I told him my story. But George was not so shaken by it as I had expected.

"I was fairly sure it would come to this," he told me, "after the way the Queen spoke in the Presence Chamber."

"You heard it all?"

"Every word. The door was left open—remember? And her voice carried well. But I still cannot leave here. Not now. Not with matters at this stage."

I looked in bewilderment at him. "But you have no choice! I told you, George. Sir William has been *ordered* to banish you."

"Ah, yes." George gave the beginnings of a smile. "Sir William has been ordered. But who really orders Sir William? And what if I plead so prettily with the Old Lady that she cannot find it in her heart to let my brother send me away from her?"

"Even if she knows the danger you would be in then?"

"I could persuade her otherwise, I think." George was speaking half to himself now, working out his ideas as he did so. "It will be very hard for the Old Lady to accept that any of her sons could deliberately kill his brother. . . ."

"But supposing you did manage to persuade her of that," I argued. "And supposing *she* managed to persuade Sir William to let you stay. The danger would be no less real, would it not?"

The smile slid away from George's face. "I will have to be taken before I am hanged," he told me; "which means I shall just have to keep one step ahead of Moray's men. And so, come on, Will. You were told to hurry back with me."

I fell into step with him across the courtyard. Together we mounted the steps to the Great Hall, then I stood aside to let him enter ahead of me. The Old Lady and Sir William both rose at the sight of him, and one glance at their faces told me of the argument they must have had while I was away. Sir William looked sullen now, rather than agitated;

and although it took a lot to make the Old Lady weep, she *had* been weeping.

"Geordie!" she exclaimed, and held out her arms to him. George went swiftly to her, his own arms held out in reply; and as they came together in embrace, I had the feeling in my bones that there was only one way the argument could go now. George would be allowed to stay— for a time, at least. Yet it could not be all that long, I realised, before Moray discovered that his command had not been carried out. And so it looked very much as if George *would* be forced to rely on the escape plan I had put to him!

Carefully I began laying the groundwork my plan needed.

I would have little problem, I knew, in persuading Minny to give the help I wanted from her. If she would not, for the Queen's sake, do as I asked, she would do it for mine. But Minny, I soon discovered, was more than willing to be active in the Queen's cause. Minny had already heard the soldiers of the escort tell of the dressing down she had given Moray. All the servants in the Castle, indeed, had heard their story. And Minny was delighted by it.

"Because," said she, "it is not only high time that Moray learned what it feels like to have his nose put out of joint. It goes against the grain, also, to see such continued injustice against the Queen."

I was intrigued by this. What did Minny know, after

all, about affairs of State? "*I* know," I said, "that the stories about the Queen are false. I have good warrant for knowing. But how can you be so sure?"

Minny grinned at me, the impish grin that always made her look half her true age. "You have a poor memory, Will," she remarked. "Do you not recall how I checked Ellen for speaking slander about the Queen?"

I remembered that very well. "But," I added, "I remember also thinking then that you were doing no more than giving some of your usual common-sense advice."

"There was a bit more to it than that," Minny informed me. "I have noticed something about men, Will. Whenever they are jealous of any power a woman may have, they always go the same way about trying to drag her down—tarnishing her good name first, branding her with words like 'whore,' in the certain knowledge that some at least of the mud they fling will stick. And that, I think, has been the fate of this poor Queen."

We were sobered by this, myself as much as Minny. I settled down immediately to telling her of the part I meant her to play; and once I had got her agreement to it, I went off to wait with what patience I could, for the result of this.

Minny had no news for me the next day; but that was because—for once—the Old Lady had not brought her maid across with her on her daily visit from the New House. And it was through Kirsty, the maid, that Minny would have to reach the Old Lady's ear. On the day following, however, we had better luck. Kirsty came with the Old Lady as usual, and Minny had the chance to send her message.

"And then," she reported gleefully, "I did just what

you told me, Will. When the Old Lady came across to see me, I told her, I just told her straight, 'Well, your Ladyship, it has come to this. I cannot go on any longer with the extra burden of work that washing for the Queen and her ladies has put on me. My back is near to breaking with it all. And besides, there is the ironing of it—all those delicate laces and frills, it takes too long, altogether too long. So there it is, your Ladyship. I have made up my mind to take service with some other family that can be more kind to me—unless, of course, you can farm out all that extra work to someone else. . . .' "

"And what did she say to that?" I was laughing by then, the way Minny herself was laughing, but still thinking ahead to the next move. "What did she say?" I repeated, because Minny was not looking quite so triumphant now.

"Once a week," Minny said. "She would not agree to bringing in help more than once a week. But the arrangement is to start right away, Will. 'You can send a message over to some respectable woman in Kinross,' she told me, 'and as soon as I have approved of her, she can start coming once a week to collect the linen from the Queen's apartments.' And that is better than nothing, is it not?"

Once a week would suit my plan nicely, I thought; and Minny had done well, to fool someone as sharp as the Old Lady. I told her so, in the warmest of terms, and straight away started to get from her a note of all the washerwomen in Kinross.

I was surprised, then, to find that such a small place could produce so many of these; but it was not just any woman who would do, of course, and so the list narrowed finally to the five that Minny reckoned might be worth my while to visit. But, I decided, I could not do all that in one trip to the mainland, since that would mean being too long absent from my duties with Sir William.

I would have to allow myself two days to make my choice among them. And meanwhile, there was George. He would need to be told that my plan was now at least under way; but I would have to be very careful how I went about that. Sir William had yielded very reluctantly to the Old Lady's insistence that George should be allowed to stay on in the Castle; and I might come under suspicion too, if he and I were now seen with our heads together.

I waited for my chance to speak privately to him, and found this the next morning when I saw him on his way to the north tip of the island—the place he most favoured for the flying of his hawks. I stalked him through the garden to that point, and had all the talk I wanted before I used the cover of the garden again to reach the east landing stage and my own boat waiting there for me.

I was feeling very pleased with myself, accordingly, as I oared the boat swiftly towards the mainland and my search among the washerwomen of Kinross. George had been more amused than annoyed to know what I had done thus far. He had even admitted to a change of mind over the prospect of success for my plan. And not only that. He had actually gone so far as to say he would give official notice of it to John Beaton!

I pulled lustily for the mainland's public landing stage, tied up my boat among the ferryboats waiting there for hire, and set off to find the houses of the five women on my list.

The luck that began that day for me did not last. Of the three women I managed to see before I had to go back to the Castle, two already had all the work they could manage. The third—who had just the right height and build for my purpose—could not accept my offer because she had a sick child to look after. That brought me to the end of the time I could spare from my duties without the risk of Sir William noticing my absence; and so I had to resign myself to starting all over again on the following day.

I had no chance to speak again with George, either, before I did so. George, it seemed, had taken off early that morning for the mainland; and from the sour look of Sir William as he remarked on this, I was left in no doubt that he suspected the purpose of George's absence. I kept discreetly silent through all his grumbles on the subject, and tried to increase my own protection against suspicion by being more than ever attentive to him. His sour mood slowly lost its edge until he reached the point where he was ruffling my hair and telling me,

"You, at least, are a good lad, Will. *You* have the sense to know which side your bread is buttered."

"Yes, sir," I said humbly. "Thank you, sir." And obediently, when he ordered me to do so, I entertained him with card tricks and played the fool long enough to turn his humour completely sweet again. I did all this with a reasonably good grace too, knowing very well how his relations with the Queen had cooled since her scene with Moray, and—in spite of myself—feeling some compassion for the way he was now denying himself the pleasures of her apartments.

So the greater part of that day vanished from beneath my grasp, until it was five o'clock in the afternoon, and almost dark. That still gave me two hours, however, before I had to serve Sir William's supper, and so I did not hesitate when the chance came to slip away from him. Within a quarter of an hour from that time I was tying up my boat again at the mainland stage, and hurrying off to search for the first of the two women still on my list.

Mistress Jean Lawson, she was called. She lived only a few minutes away from the landing stage, and the moment she answered my knock on her door I realised that luck could once more be with me. Mistress Lawson, I saw then, was an elderly woman, but she was still tall enough and thin enough for my purpose. Not only that, she was willing also to come straight away with me to the Castle, to be approved by the Old Lady.

"If you will just give me time to get my shawl," she said, "and to let my goodman know where I have gone at such a darkling hour."

"With pleasure, mistress," I told her. "And I will wait for you at the landing stage."

I strolled the hundred yards from her house to where my little boat bobbed beside the larger craft still waiting for hire. The streets of Kinross were dark and deserted now, and there were only two men left waiting beside their ferry boats. I talked idly with them until Mistress Lawson appeared beside me. They made some joking remarks to her then, but I cut this short; and seconds later, she was seated in my little boat while I rowed most cheerfully with her to the island.

I had covered half the distance there when we both became aware of some commotion on the landing stage we had left so quiet behind us. I rested the oars momentarily, to look back; and dark as it was by then, I thought I could see a horse at the water's edge, with its rider in the act of dismounting.

"Someone in a hurry to find a boat hire," Mistress Lawson remarked; and sure enough then, we both saw one of the ferryboats pushing off from the landing stage.

*Could it be George?* I bent to the oars again, and rowed at a quicker pace than before. The larger boat, with two men at the oars, began to pull up on me. I touched at the landing stage only half a minute before they did, and leapt ashore with the mooring rope in my hand. The passenger from the other boat came hard on my heels—and it was George. I called out to him. I heard him order his boatmen to wait for his return to the mainland. Then he closed with me; and low in my ear he said urgently,

"Moray has sent men to make sure the ban on me is enforced. And they are only half an hour behind me now. But I must see the Queen before they see Sir William. I *must*, Will."

"I can clear the coast for you." I spoke rapidly in reply, all the speculation that had been racing through my mind crystallising instantly into a plan of action. "This woman with me—we will use her as the decoy in that."

I turned quickly to my own boat, helped Mistress Lawson ashore, and told her, "Now, mistress, if you will just follow this gentleman and myself . . ."

George and I set off at a pace that might have been a little too brisk for her liking, but she came close enough behind us, for all that. "Mistress Jean Lawson, to see Lady Margaret Douglas," I told the sentries at the gate, nodding towards her. They let her pass, and as we walked across the courtyard, I told George the details of my intentions.

At the foot of the outside stair leading up to the doorway

of the Great Hall, George stood aside. I beckoned the woman to follow me up to the door; and George—as I had warned him he must do—fell in close behind her. Once through the door, I led her along behind the service screen, with George all the time following and keeping low behind the screen's solid lower half. I reached the entrance gap at the screen's southeastern limit. There I stopped, pulling the woman to a halt with me so that our presence side by side in the gap would entirely block any view beyond it.

The Old Lady was in the Hall along with the two girls, all three of them with their heads together over the nightly report the girls had to make. I paused in the gap for the moment George would need to cross it behind cover of Mistress Lawson and myself, and thus gain the inner stair leading up to the Queen's apartments. Then, to the Old Lady I announced,

"This woman, your ladyship, is Mistress Jean Lawson. Minny sent me to fetch her to you."

Firmly I thrust Mistress Lawson forward; and myself turned towards the stairs. It was Lady Agnes who would be in the Queen's apartments at that hour. That was the likelihood behind all my calculations. But supposing Sir William was there too? The excuse that would get rid of Lady Agnes would certainly not do for him! I sped upwards, trying hard to figure some alternative device for getting him out of George's way.

Kerr and Newland, the two surly fellows guarding the Queen's door, stared stonily at me as I knocked there. Then their eyes went indifferently past me—as they must earlier have gone past the accustomed sight of George, mounting still higher up the stairs as if climbing towards the level of his own quarters. The voice of Lady Agnes answered my knock; and I went in to find her seated in a group with the Queen and the Queen's ladies, all four of them busy with their sewing. I made my bow to the Queen, and raised my head to say,

"If it please Your Grace . . ." Deliberately I held her gaze for seconds longer than was proper to this formality. I saw her eyes become suddenly alert before I turned to tell Lady Agnes,

"If Her Grace permits, your ladyship, there is a Mistress Lawson just arrived to seek employment as a washerwoman. Minny sent me to Kinross to fetch her."

"But surely—" Lady Agnes looked at me in some puzzlement. "There must be some mistake, Will. *I* know nothing of any Mistress Lawson."

"Then perhaps," the Queen suggested gently, "you had better find out."

"Er—well, yes. With Your Grace's permission." Lady Agnes rose in a flustered sort of way, and made her curtsy to the Queen. I sprang to open the door for her, and followed her to the stairs, calling loudly after her the "all-clear" signal that would be heard by George waiting in the half dark of the landing one flight above me:

*"In the Hall, your ladyship. The woman is waiting in the Great Hall."*

"Come here to me, Will." The Queen's voice summoned me back into the room, and I hurried to her. She and her ladies had thrust aside their sewing, and I could see from their faces that she *had* guessed there was something in the wind.

"Pray, Your Grace—" I blurted the words out over her questions. "Waste no time. Mr. George is closely pursued by Moray's men, and he must see you. This is his last, his very last chance—"

"Your Grace!" George's entrance, George's exclamation, interrupted me. In swift strides he was across the room and going down on one knee before the Queen. "I have gathered all the latest intelligence from John Beaton," he told her; "and here is the burden of it. Moray has over-reached himself with his Parliament, and opinion in the country is now veering in your favour. The common people, Beaton reports, are content to see Bothwell pursued with all the rigours of the law, but they do not want to see *you* further hounded. They want their Queen returned to them, Your Grace. They have become uneasy at your continued imprisonment—"

"And the lords?" Her face in a glow of delight, the Queen broke in. "What support can I expect there?"

"All that you formerly had. Huntly, Arbroath, Galloway, Ross, Fleming, Herries—" George went on reeling off a list of names that ended with "—and chiefest of all, of course, your old friend Lord Seton. But besides that, some of the lords who *were* on Moray's side are now talking of supporting you instead—even some of those who were here with him, less than a week ago. They have become uneasy, it seems, at the power Moray now has. And so Lord Seton has sent me to ask your permission to assure all these of your renewed favour, if they do return to their proper allegiance."

"Lord Seton shall have it, Mr. Douglas. I will write to him now!" The Queen jumped to her feet, rushed to her desk—and then remembered! There, at least, Moray had been obeyed; and Sir William had been thorough in raiding her supply of writing materials. The only remnant of these was a pen—an old one with its quill all broken—that had been left on her desk as being too battered for further use. With some vague thought of lending a hand in this crisis, I drew my knife to try sharpening its point. But the Queen was already far ahead of me. Snatching up the empty inkwell from her desk, she thrust it into Jane Kennedy's hands.

"Soot from the fireplace," she commanded. "And you, Maria, fetch water."

Kennedy rushed to scrape soot into the inkwell. De Courcelles ran to the bedchamber. I kept on sharpening the point of the quill, stealing glances all the while. Now Kennedy was coming back to the desk to place the inkwell on it. De Courcelles was running back from the bedchamber, a little jug of water in her hand. The Queen had picked up the largest of her tapestry needles, and was standing ready for them both.

De Courcelles handed the water to her. She tipped some of it in among the soot, mixed the result swiftly and thoroughly with her tapestry needle, then took her handkerchief of fine white lawn from the cuff of her gown, and spread it out on the desk before her.

"Listen at the door," she commanded Kennedy and de Courcelles. They ran to crouch down with heads pressed against the door, while she seated herself and said, "Now the pen. And keep firm hold of this while I write."

I handed her the pen; and from my side of the desk, I steadied the top and bottom right-hand corners of the handkerchief. George did the same for the left-hand side, so that the white square of stuff lay taut and flat on the desk before her. She wrote her message—a mere six lines; and then, in her bold Italian hand, the Royal signature

*Marie R.* Our hands released the white square, and the Queen laid down her pen.

"I will signal to you, Your Grace," George told her, "to let you know this has been delivered. Watch for me when you walk in the garden. I shall ride my horse out into the water of the mainland shore, and come as close as I dare."

"And my escape?" she asked. "Is there a plan for that?"

George nodded towards me. "Will has something in hand. He will tell you of it himself. And trust him, I beg you. He is loyal."

Surprise—or was it doubt?—showed briefly on the Queen's face; but she did not voice her feelings. Instead, she reached up quickly to take off one of her earrings—one of the pear-shaped pearls that were her favourite jewels. With her fingers working on it, she told George,

"Then take my letter to Lord Seton with the hope that I shall soon be leading my forces in person." The pearl was loose now. She held it out to George with one hand, gave him the handkerchief with the other, and added, "And when the moment comes to receive me on the mainland, send this pearl back to me as a signal that all is ready there."

George stowed the pearl and the handkerchief inside his doublet, speaking as he did so. "I will do more than that, Your Grace. The boy's plan may fail. In which case, we must think of another, and perhaps yet another. But when I send the pearl back it will be not only the sign that we are ready for you. It will tell you also that matters have reached the point for an escape plan that *must not* fail."

Swiftly then he knelt to add, "And so, Your Grace, adieu.

And, as God hears me, I shall be your true and faithful servant until death."

The Queen opened her mouth to reply; and in the same moment came a cry from the listeners at the door, "Hush! Oh, hush!"

George was on his feet on the instant, his head turned in a listening attitude. Everyone listened, breath held, bodies motionless. Footsteps! We could all hear them now—faint, but rapidly growing louder; the footsteps of a number of men coming quickly up the stairs. George moved to the door. I whispered urgently,

"You'll be killed!"

"Not I! I'll fight my way through!" He was at the door, his hand on the latch. Over his shoulder he threw a last word to the Queen. *"Trust the boy!"*

The noise of footsteps had a voice added to it by then, Sir William's voice loudly roaring George's name. I made to follow him as he plunged through the doorway, but the Queen's hand restrained me. Her eyes were big with fright. For her letter? For George's life? I was frightened only for George at that moment.

We could hear the muffled sound of two men's voices now—George and Sir William, each trying to outshout the other; but it was Sir William's voice that finally overtopped the general din of accusation and counteraccusation.

"—and think yourself lucky I have not shot you with my own hand, instead of standing between you and Moray's men. But I *will* shoot you, if you do not take yourself instantly out of Lochleven. I will have the Castle cannon turned on you, by God, if you come within half a mile of here again!"

A sound of footsteps descending the stairs, and voices that faded as the footsteps faded; there was nothing more to be heard after that. George was finally banished—but he had at least escaped with his life. I let out a long, shivering breath of relief. The Queen and her ladies did so too. The Queen's eyes met mine; and with a little smile that seemed oddly sad to me, she said,

"So, Will Douglas, I am in your hands now."

"An it please Your Grace." It was the force of realising the truth of this, rather than good manners, that brought me down on one knee before her then. "I will be your saviour, Your Grace, if you will let me."

"My little orphan," she asked, "why should I not? I *have* no help now, except you."

"And no one, Your Grace, who is more loyal to you."

She held out her hand to me as I said this. I dared to take the hand in my own, and to kiss it—just as George had done when he had first sworn allegiance to her. And just as had happened with me then, I had a feeling I could not place. I looked up into the Queen's face. She was still wearing that oddly sad little smile; and suddenly I knew the nature of my feeling. I had lost something, I realised. I had lost my envy of George.

# 13

The Queen was as charmed by my plan as I was myself.

I had been careful, after all, to make sure that Mistress Lawson resembled her in height and build; and the guards would soon become used to seeing the woman come weekly to the Castle. Kerr and Newland would have their instructions to let her enter the Queen's apartments to collect and deliver the linen for washing; and once that pattern had been well established, the rest would follow quickly.

A sufficient bribe to make the favour worth her while, and Mistress Lawson could be persuaded to lend the Queen her dress and shawl. With these for disguise, then, the shawl pulled well forward to hide her features and a bundle of linen tucked under her arm, there would be nothing to stop the Queen walking calmly out of her apartments and down to the boat awaiting the return of the "washerwoman" to the mainland.

That was more or less the sum of it; and the moment I learned from Minny that Mistress Lawson's employment had been confirmed, the Queen was anxious to write letters that would arrange the events to follow her escape. Lady Agnes, however, was now too far gone with child to be nimble in thieving further supplies for writing the supposed journal, and so it was I who did so. The Queen wrote her letters during her supper hour—the only time she was allowed to sit in private; and then, with these hidden alongside a further letter that would be my credential as her courier, I went ashore to see if I could find John Beaton.

I was nervous, at first, about this—perhaps unreasonably so; but it was still only on the previous day that George had kept his promise to signal the safe delivery of the handkerchief letter and Sir William had acted on his threat to turn the Castle cannon on him. I kept low in my boat, thinking of the panic this had caused—especially in the Old Lady. That had upset me—but still not nearly so much as the sight of the cannon shot whistling towards George. I tried to console myself by remembering that Drysdale's eagerness to train the guns had been the cause of the shots' going wide of their mark. And *I* could not become the target either for Sir William's anger or Drysdale's spite, I argued, until I too became suspect. I had an uncomfortable journey all the same, and was still shaking a little from the effect of it by the time I had tied up my boat and was heading for the Kinross Inn.

It was Muriel, the black-eyed daughter of the house, who brought back my self-esteem. I deliberately sought her out there as the person least likely to report my doings, and discovered from her that Beaton had taken up almost permanent residence at the Inn.

"Although he keeps himself very much to himself," she added. And she smiled as she spoke, that demurely enticing smile she knew I never could resist. I kissed her soundly then, without any noticeable resistance on her part; and with my faith in myself once more high in consequence of this, I followed her along the passage that led to Beaton's room.

No voice answered my knock on its door; and the door itself, I found, was locked. I bent to the keyhole and softly called my name. A key rattled in the lock, and Beaton opened to me. He looked just as I remembered him—lean, elegant, and wary eyed as ever. Without a word of greeting he relocked the door, then took the letters I handed him and began to scan the unsealed one that was my credential from the Queen. I glanced around the room as he read.

It was small, and sparsely furnished. One narrow pallet bed, two chairs, a small writing desk—that was all it held. Its situation was on the ground floor. The lattice of its window stood open, in spite of the freezing weather outside, and opposite the open window was the doorway of the Inn's stables. It flashed through my mind that Beaton had made very sure he would not be trapped there by any unfriendly visitor. Then I became aware that his gaze had lifted from the letter to fix on me.

"So!" He tapped the letter with the fingers of his free hand. "You are the Queen's new courier, eh? And she has accepted the plan of escape you have worked out for her."

"That," I said boldly, "is because it will succeed. It is too simple, sir, *not* to succeed."

"Let me be the judge of that," Beaton advised. "And since the whole process of mustering an army will depend on it, I must have more detail than Mr. Douglas could give when he told me of it."

I explained my plan, then, as fully as I could. It would be no problem, I pointed out, for the Queen to be secret in making the exchange of clothes with Mistress Lawson. The hour of the woman's visit had been fixed for ten o'clock in the morning, and it was well known to everyone that the Queen always stayed late abed. Ellen and Margaret would therefore have been up for hours while she was still lying in her bedchamber; and that meant she would be alone there when Mistress Lawson arrived to collect the linen. But Maria de Courcelles would follow, when the woman went into the bedchamber. Maria would help to speed the exchange of dress between them. And when the "washerwoman" finally departed again with her bundle, Maria would still be heard from the bedchamber in apparent conversation with the "Queen."

As for how long it would be before all this could happen, five or six visits from Mistress Lawson, I had reckoned, would be sufficient for everyone—and not least the guards—to cease paying any attention to her entering and leaving the Queen's apartments. "And that," I finished, "would bring us to the fourth of February. Or, to be completely safe, we could say the eleventh."

Beaton had begun to smile as I spoke; and with a rather caustic note of humour then, he said, "You have the makings of a spymaster in you, I see. And the plan is certainly feasible. But the timing of the escape must still relate to the weather conditions that will follow; and also to the question of our being able to bring superior numbers to the field."

I was dismayed. I had been so determined there would be no further shilly-shallying over freeing the Queen. Not now that *I* was in charge. Firmly I said, "If you are thinking of the pleas the Queen has made for aid from France—"

"I am," Beaton interrupted, "but only in the sense that France's failure to help makes it all the more important for us to be able to muster superior numbers from our own resources. And we cannot be sure of doing that until we are sure also of yet another of the lords now wavering in his allegiance to Moray—no less a person, in fact, than the Earl of Argyll."

"*Argyll!*" I stared in disbelief. "But he is Moray's own brother-in-law!"

"Exactly. And that should show you how the opposition is now crumbling! What is more, Argyll can bring a greater number to the field than any of the other lords. He could bring us up to an army of six thousand, in fact—more than Moray could ever hope to raise. But he is cunning, a most cunning man. We need more time yet to bring him to a firm promise of support. And so what do you think now, my lad, of that escape date you gave me?"

I was not anxious to say what I thought, but could see no help for it. "I suppose," I admitted, "that I will have to be guided by your choice."

Beaton sketched me a bow. "In that case," he said, "we will forward the date of the Queen's escape to the twenty-fourth of March—thus giving ourselves not only the time we need to work on Argyll, but also the advantage of spring weather for the opening of our campaign. Agreed?"

That last question, it seemed, was only a formality before he sat down at his desk and began to write. I waited in silence till he had finished. He sealed his letter, took a packet of other letters from his desk, then looked up at me.

"Just one more point then, Will. About this Mistress

Lawson on whom so much depends. Are you quite, quite sure she will fall in with your plan?"

Beaton had been too long away from the Queen's company, I thought. He seemed to have forgotten how her charm worked on everyone—witness the Court of Dreams! I smiled at my own recollection of this, and told him,

"I am quite, quite sure that the Queen can persuade Mistress Lawson to do so. And besides, Her Grace is generous in her rewards."

I gave him just a glimpse then, of the gold the Queen had insisted I should have for *my* services; but he frowned at this, and said sharply,

"Put it away!"

I slid the gold out of sight. He handed his letters to me, and went on, "Hide these too, until you can give them to the Queen. This newly written one also, where I have explained the reasoning behind the choice of date for her escape."

I put the letters inside my doublet; but I had no intention of ending our interview on that note. I had a question to ask him; one that had been burning in my mind, and I meant at least to try for an answer to it.

"Er—and can you tell me, sir," I ventured, "if George Douglas is still unharmed? And also, perhaps, where he has hidden himself?"

Beaton gave me the wariest look I had yet had from him. Then, to my utter surprise, he said, "Will, I have learned from George that you feel tenderly for the Old— for Lady Margaret Douglas. But you do understand, do you not, that you could not give her any information now about George without betraying your own involvement in the Queen's affairs?"

He had caught me out so completely in this that I could only stammer. "I am sorry. She has been so concerned about him, you see. And I, too—"

"Have been just as concerned," Beaton finished for me. "But there is no need to worry, either on her behalf or on your own. I do assure you of that. George *is* in a safe hiding place. And this very day, as it happens, he has sent a message to his mother to tell her so."

I let out a sigh of relief at this. It had not been comfortable, living with the Old Lady's distress over George; and if it had not been for Beaton's cautious attitude, I realised, I could have allowed that to lead me into a considerable blunder. I saw him smiling a little at my expression, and was encouraged then, to say,

"I take your meaning completely, sir; but— Is there some chance, perhaps, that I might see George again before the date of the escape?"

Beaton shrugged. "It is possible, I suppose—*if* one of his visits to me here happens to coincide with your further courier work for the Queen."

"And if the Queen herself asks when she will see him again?"

"Tell her," Beaton said, "that the day she sets foot on the mainland, she will be met by an escort of fifty young gentlemen under the command of Lord Seton. And that George Douglas will be one of the fifty."

I left him with this and hurried from the Inn, resisting the temptation to see Muriel on my way out. The number "fifty" was echoing in my mind, and I rowed back to the Castle with my thoughts making it up to fifty-one. Because, I vowed to myself, I would be a member of that company! As soon as I saw the boat with the Queen in it touch at the mainland landing stage, I would set out in my own little boat to join them. But to make sure I *could* do that, I would first of all have to create an excuse for being at our own landing stage at the time she left from there. . . .

Sir William was pleased when I began a regular weekly check of all our boats. "You are improving, Will," he told me. "You used not to be so conscientious in your duties."

"It saves time in the end," I said modestly; and continued to go down to the west landing stage every Wednesday morning just before the time Mistress Lawson was due to arrive there. I stayed until she left again, around half an hour later; and made a point, meanwhile, of noting small things that might later be of use to the Queen. Finally, when I thought I had the whole situation well surveyed and there was a chance to talk without being overheard, I relayed my information to her.

"The woman's boat," I reported, "comes in always at the most convenient point—our west landing stage. And she has no regular boatman. She just takes whichever one happens to be free for hire. Also, before she embarks on her return journey, she makes a habit of throwing her bundle ahead of her into the boat."

"She is not the talkative kind either," the Queen murmured. "And so—thank God—I shall not have to speak with any of those hired boatmen."

"But she is talking to you, is she not? I mean, you have managed to—" I stopped, as the Queen began smiling at this.

"But of course, Will. Gold is a great loosener of tongues!"

"And she *has* agreed to do as you ask?"

"Most certainly. She would have done so, I think, even without the bribe!"

So here, I thought, was at least some proof that George had not been mistaken in the intelligence he had risked so much to bring. First Minny, and now Mistress Lawson—the signs indeed were that the Queen was regaining the sympathy of the common people. . . . I sensed eyes on me at that point, and looked up to see Margaret Lindsay's gaze fixed on myself and the Queen. Immediately then, I felt uneasy.

Margaret Lindsay seemed always to be watching the Queen, these days—and not furtively, either, in the way that was called for by her role as a spy. There had been a change in that; a change I could date from the moment she had led her mother as the first of the Porches to make her curtsy to the Queen. And now, it seemed to me, Margaret's constantly following gaze had a doglike devotion in it. She had got even thinner, too, since that time of the Porches' visit; and her sallow face now bore a look of strain on it. I had begun to fancy, too, that the eyes in that face turned hostile whenever they rested on me alone; and I did not like the feeling this gave.

I made an excuse to rise and get away from the uncomfortable effect she had on me. But Margaret spoke suddenly, in a voice that seemed high and thin, compared to her usual tones.

"I dreamed. Last night I dreamed such a dream of you, Will Douglas!" Her eyes had shifted from the Queen's face, to mine. There was a sort of fixed stare to them; and both her look and her words gave me a chill that was like fear. But what had I to fear from Margaret Lindsay?

"You brought a bird into the Castle, in my dream."

Still in those high tones, she went on; and still she stared. "A big, black bird it was, like a raven. But it was bigger than a raven. It was enormous. It followed you, flying after you. And then it swooped down to seize the Queen, and flew off with her. With my Queen!"

Her hands released the sewing they held. It slid off her lap, to the floor. The hands hovered, empty, for a moment. Then one of them jerked upwards, to cover her mouth. From behind the covering hand came in a muffled scream,

"With *my Queen!*"

The sound had the room instantly in a stir, with Ellen shrinking back from Margaret, the Queen and her ladies starting towards her. She was rocking and moaning now, as if in some sort of fit; but when the Queen knelt to soothe her, the moaning turned to a quiet weeping that was quite pitiful to witness. The Queen grasped both of the girl's hands in her own and said gently,

"You are not well, child. You cannot be well, or you would not have such dreams."

Words began to come through the sound of weeping, broken words, fitfully uttered. I thought I heard "escape" among them, and "leaving me." The Queen's hearing was sharper than mine—or else it was her closeness to Margaret that allowed her to hear more distinctly. She clasped the girl's hands more tightly, and told her,

"But of course I shall escape. I must! And so must Kennedy and de Courcelles. We shall all leave here. But listen to me now. This is what we shall do—"

Briefly, over Margaret's shoulder, her eyes warned me to make myself scarce; then she bent closer yet to the girl and began whispering in her ear. I edged to the door, and saw Kennedy mouthing at me as I went,

"Arnault . . ." Silently her lips formed the word. "Fetch Arnault!"

I ran quickly up the stairs to Arnault's chamber, and blurted out, "In the Queen's apartments—Margaret Lindsay in a kind of seizure. Babbling about me, and of escape for the Queen!"

"Ah, Dieu!" With a sigh of resignation, Arnault heaved his bulk upright, grabbed a box that held some potions and pills, then hurried from the room. I paced about after he had gone, my mind a blank except for one thought. Margaret Lindsay's clinging devotion to the Queen threatened more of danger than her spying had ever done!

Arnault came back at last, puffing from the exertion of climbing the stairs, and collapsed thankfully into his chair. "I have given her a soothing powder," he announced. "She will sleep now."

I faced him, nerves all ajangle, and exploded, "Damn your soothing powders. You should have found out if she knows anything about me!"

"I did." Arnault gave me his usual sardonic grin. "The girl is given to humours and vapours of all kinds. Therefore, she has nightmares. Also, she has become abject in her love for the Queen. And she does not particularly like you. Therefore her nightmares take the form of your doing something dreadful that will steal the Queen away from her. That was all there was to her outburst, my young friend. I questioned her closely—as closely as any physician could; and I do assure you that was the whole reason for it. She knows nothing of your plan. Absolutely nothing."

Yet still, I thought, she had some *sense* of what I intended. And it was this sense—this sort of animal feeling of suspi-

cion about me—that had surfaced so strangely in her
dreaming mind. It was an eerie feeling to have about her.
And then there was Ellen. How far had *she* been influenced
by Margaret's ravings? And what danger might that spell,
if the girl made some further outburst? I turned this over
in my mind for a day or two, and then—very obliquely—
I began probing Ellen on the subject. But there, of course,
I had reckoned without Ellen's usual bluntness of speech.

"Just stop beating around the bush," she told me; "and
answer directly. *Are* you mixed up in some escape plan
for the Queen?"

I put on my best air of injured innocence, and countered,
"What on earth makes you ask that? You are not having
nightmares too, are you?"

Ellen laughed. "I sleep soundly. I always have."

"And Margaret Lindsay is like her father—not violent,
as he is; but crazy all the same. And so why ask such a
question?"

"Because," Ellen said blandly, "I saw you once with a
packet of letters. It was in the Queen's apartments. And
you did not know I was there, in the inner chamber. I
knew it was letters you had, too, because you dropped
the packet in your haste to hide it inside your doublet.
And that was when I had a good look at it."

My mind was racing as she spoke, darting here and there
in search of a way out. I managed a laugh when she finished, and told her,

"You have caught me, Ellen—or rather, you have caught
George. They were his letters—ones he had written to
the Queen. I promised I would steal them back for him.
It was just before he was banished, and he was afraid they
would be found and—and compromise her."

Ellen took the bait in that word "compromise." Rising

swiftly to it as a trout to a hatch of mayfly, she challenged, "Why? Were they love letters?"

"I am sorry, Ellen." Regretfully I shook my head. "I promised I would not say."

"They were! They *were* love letters!" Ellen's voice rose to a squeal of delight. The she sighed a long, quivering sigh. "Oh-h-h, Will, is it not romantic? To be in love—and in love with the Queen!"

"I never said he was," I reminded her.

"No—but I guessed that for myself long before he *was* banished. And I must have been right, if he thinks his letters would compromise her. And listen, Will—" Ellen drew closer, a light of conspiracy in her eyes. Her delight in my story, I guessed, was about to lead her into some indiscretion of her own. And I was right! Her voice dropping to a whisper, she told me, "I can give you a secret, Will, in return for yours. The Queen *will* escape, just as she said to Moray she would. And she is planning it now. She told Margaret and me so, the day Margaret wept over her nightmare about you. But there will be no more of these nightmares now, because *she is going to take us both with her!*"

"Are you sure? But how could she manage that?" I was only playing for time with my questions; time to realise the Queen had taken the only way possible of making sure that poor, unhinged girl would no longer be a danger to her. But Ellen took me seriously enough—and also seemed to realise, then, just how indiscreet she had been.

"Oh, there are ways," she said airily. "We talked about them. But I am not stupid enough to repeat *that* part of the conversation."

And I would not be stupid enough to spoil the Queen's work by pressing her to say more! As for the alarm over the letters, that was safely over too, now that Ellen had so completely swallowed my story; and the more I allowed her to think she had some secret of her own to hide, the less likely she was to have fresh suspicions about mine. I made pretence of being crestfallen, and said wistfully,

"As you please, Ellen. But I shall miss you when you *are* gone."

I had struck the right note there, it seemed—the one that pandered best to Ellen's vanity. She smiled kindly at me, vowed she would not forget me however far we might be separated by fortune; and left me, finally, looking the very picture of innocent smugness. I watched her go, thinking that Mistress Ellen was not nearly so shrewd as she imagined herself to be; and felt, like a tide rising in me, the exhilaration that comes from the knowledge of danger met head-on and thoroughly routed.

I was a different person then, it seemed to me, from everything I had been before. I was no longer just Will Douglas the page, the disregarded bastard of the house. I was Will Douglas, courier to the Queen of Scotland. I was the spymaster on whom she and all her friends depended, the one link between her and her gathering army. They could do nothing without me now; not one thing. What was more, the Queen had continued to show her gratitude with the gold she pressed on me every time I took her letters ashore to Beaton; so that suddenly also, I was rich! For the first time in my life I had money—real money, to save, to spend, just as I pleased. *Or* to gamble . . .

Now that, I told myself, would be a pleasure I had never imagined could be mine. To gamble with *gold* as my stake! The attraction of this thought began to pull me more strongly than anything I had ever felt before. And, I remembered, there were still two weeks to go before the twenty-fourth of March—plenty of time to have a few tastes of that pleasure. But cautious ones, of course.

I could not, for example, afford to draw attention on myself by lingering in the Kinross Inn. In the Castle guardhouse, on the other hand, I had a few card-playing cronies who were used to seeing me produce such odd stakes as the silver buttons I had lost and won back again. A late-night game with them would do no harm. . . .

I had more than one late-night game with my guardhouse cronies—several, in fact; but I was still at my self-appointed post as usual, and still as alert as ever, when the day of the escape came at last. Mistress Lawson came ashore, dragging her right foot a little as she walked, the way she always did. But that was another of those small things I had noted and been careful to pass on. The Queen had practised that limping walk, and I was sure she would not forget to use it.

Mistress Lawson passed me by with her usual morning greeting. When she came back, I knew, she would nod to me before she tossed her bundle into the waiting boat. And the Queen would not forget, either, about that nod. I settled down to my pretence of inspecting all the Castle boats, and then turned in earnest to the inspection of my own little craft. The rowlocks needed some greasing. The boatman who had brought Mistress Lawson across watched me at this task, and called out eventually,

"You're busy, then."

I recognised him as one of the two men who had been

waiting for hire on the night I had first brought Mistress Lawson to the Castle—one of the two who had joked with her on the Kinross side of the water. I gave him only a grunt in reply, not wanting to enter any conversation that would set his chaffing tongue at work on me. He got out of his boat and stamped around, calling out every now and then to the guards on the gate, the servant women who came to dip buckets of water out of the loch. His restlessness made *me* nervous, almost to the point of deserting my post to see if any hitch had occurred in my plan.

Then I saw her, the old washerwoman coming out through the Castle gate, head bent and shawl pulled forward, her bundle under her arm. The guards on the gate had not so much as glanced at her; and, I realised then, I could not myself tell whether it was the Queen or Mistress Lawson I was seeing. She limped the short distance from gate to landing stage, gave her usual parting nod in my direction, and tossed her bundle into the boat. With the stiff movements natural to an old woman, she stepped aboard. The boatman pushed off from the landing stage; and—if everything before that had gone according to plan—the Queen was on her way to freedom!

I straightened up from my own boat. From shore to shore the rowing time was between ten and twelve minutes, and I had Beaton's promise that she would be met on the mainland shore. In twelve minutes at the most, then, she would be riding towards her company of fifty— and I would be pushing out my own small boat to join them. As for the mount I would need, I could easily steal one from the stables of the New House!

Carried faintly across the water, I heard the sound of her boatman's voice. There was a laugh in the sound. Was he trying to joke with her, the way he had tried before to joke with Mistress Lawson? I saw him rest on his oars and lean towards the figure opposite him. One of his hands rose quickly to the shawl around her head. She jerked her head aside. Her arms flew up in the gesture of clutching the shawl more tightly around her. There was a swift flurry of water around the boat as the boatman began oaring it in a circle to turn it back towards our landing stage.

I felt nothing at that moment—not despair, or disappointment, or anger; just nothing. The boatman had begun shouting, loud and hoarse, an alarm call to the guards on the gate. They came running to the landing stage. The boatman looked back over his shoulder as he rowed. He was within thirty feet of the landing stage now, and his call to the guards came with every word in it distinct.

"I have her! I have your prisoner!"

His boat bumped against the landing stage. The shawled figure in it did not move; but the boatman was ashore on the instant, his voice coming out in a scared but triumphant babble.

"I teased her about keeping her shawl so close round her face. 'Come, mistress,' I said, 'will you not let me see how pretty you are?' Then I put up a hand to draw her shawl aside. She held it closer to her, and I saw her hands— so white, so soft. 'Those are a lady's hands!' I said; and I pulled the shawl aside. And there she was—the Queen! *The Queen!*"

I moved towards the boat, forestalling the action of the guards. The shawled figure raised her head to look up at me. Her face was deathly pale. Her eyes held a despair so profound that no words could have spoken it. She took the hand I held out to her. I helped her ashore; and then, with the guards following at our heels, I led her back into captivity.

*14*

*I* had a dog once of the kind known as a "licker." If he had a wound or sore of any kind, he would not only lick it clean, he would continue to lick and lick until the place was irritated beyond any chance of natural healing. And so it was with me after Sir William learned of the escape attempt.

The investigation he made then was thorough. His decisions were swift. Mistress Lawson, he declared, had been simply a dupe, and would be adequately punished by her loss of employment. But my case was different! *My* punishment would be a whipping, followed by banishment from the Castle—and that immediately, with no more than the clothes I wore and not even the chance of saying good-bye to Minny.

I had not the spirit left even to try seeking Minny out when the whipping was finished. I shrugged my doublet over my smarting shoulders, turned my back on the only home I had ever known, and rowed off in my little boat—but not to the mainland, where I might have had to face John Beaton or George Douglas. I made, instead, for another of the islands on the loch; the very small one known as Scart Island, where there was nothing except the remains of what had once been a fisherman's cottage. And there, among the fallen and crumbled stone of this ruin, I lay and licked at my wounds.

It was not that I blamed myself for the failure of the escape, because that, after all, had been due to nothing more than bad luck. But in spite of that, I knew, I still had to bear the blame of two serious faults committed *before* the escape. I had underestimated my half-sister, Ellen. And when I had had my fling at gambling with gold, I had not taken sufficient account of Drysdale's hatred of the Queen.

My innocent-seeming Ellen had not been so innocent after all—as I learned to my cost when she hurried to tell Sir William of the letters she had seen me handling, and when she boasted then, also, that she had not believed the lies I had told about them! As for Drysdale, he had heard whispers of my gambling in the guardhouse. And, as he had argued in the testimony he also had hastened to give, where could such as I have got gold *except* from the Queen?

Then there had been mad Margaret's story of her dream about the great black bird. That had made a great impression—on the Old Lady, even more than on Sir William. And I had not been to blame, either, for Margaret's wild imaginings. But what did that matter now? Her story had most effectively put the final nail in my coffin—and oh, my God, how I wished they *had* killed me; put me altogether out of my misery instead of just banishing me to writhe in lonely self-contempt on that deserted island!

I had had my chance to help the Queen, and I had let myself be found out in that. Now *she* was alone, the contact with the mainland broken, no other ally in the Castle able to renew it. And all through *my* carelessness, *my* folly. That was the wound I licked at, and licked again, lying with my face in the wet grass between the stones, clawing at the stones themselves till my nails broke and my fingertips bled, and I sat up at last wondering aloud just why I should be suffering such pain.

The reason came slowly to me. I was just over a month short of my seventeenth birthday, and I had never before felt the pangs of love. But gradually as I sat there despising myself, remembering her, and wondering at the root of a pain so intense, I began to realise why I could not pick myself up from this experience and go on as jauntily as I always had before.

I was in love with her. I, Will Douglas, fortuneless, homeless, disgraced, was in love with the Queen; so in love, I would have died for her. And I wanted to die. Indeed, if shame could have killed, I would have died there and then, on Scart Island.

I began to weep. I had not wept since I was twelve years old, on the day I had fought a lad much bigger than myself to a standstill for jeering "bastard" at me; but I wept then till I was exhausted, and fell asleep where I lay. Cold woke me. I had left the Castle with no more than the clothes on my back, and they were rimed now with the frost of late March. I got to my feet and wandered aimlessly around the small confines of the island.

I could see the Castle from there. And so long as *she* was in the Castle, I thought, I could not bear to be out of sight of it. Without any idea beyond this I began to set up a sort of camp for myself, rebuilding the stone of the ruined cottage till it gave me a certain amount of shelter, gathering twigs to start a fire. I had the fishing tackle I kept always in my boat; and with that, I reckoned, I could supply myself with trout from the loch. I could go into Kinross, too, to buy oatmeal and salt—but not till I could depend on that gallant company of fifty being scattered back to their homes. Not till I thought Beaton and George would have ceased to look out for me as the rascal who had brought disaster to all their plans for the Queen.

There was only one way, I had found, to deal with the dog that was a "licker." That had been to cut the bottom out of a bucket, and then to invert the thing over his head so that it sat around his neck like a funnel-shaped collar projecting well beyond the tip of his muzzle. He could sleep, eat, and drink, while he wore this collar. He could do anything he normally did, in fact—except for licking his wound. Every time he tried that, he found the rigid frame of the bucket making a barrier between his tongue and the sore place. And the life—if it could be called a life—I had begun to invent for myself on Scart Island would serve the same function for me, I thought, as the bucket had done for the dog.

I kept my fire going with driftwood, and sat for hours beside it fashioning rabbit snares from strong sticks and bits of my fishing line. I fished for whole days at a time, sometimes from the shore of the island, sometimes drifting in my boat. I made catapults from wood and rabbit skin, and shot at sea gulls bobbing on the waves close inshore. I raked over the ruins of the cottage for bits and pieces of pottery, and found also a rusty cooking pot that I scoured with sand until this, too, became of use to me. I let a full week pass like this, however, before I dared to go into Kinross for oatmeal and salt. But no one, it seemed, noticed me there; and so every now and then after that I made further trips to renew my supplies, going quickly into the village, buying furtively, and returning as quickly to my hiding hole on Scart Island.

My life was firmly set then, in its new pattern. But there was still one terrible flaw in the reasoning that had led me to create it. I could keep my hands always busy. But I could not control my eyes. They were still always on the Castle, and on my distant view of Kinross; still always watching for the kind of movement that might be a sign of rescue or escape for her. And then there were the nights. I could not sleep, and yet neither could I work at night. The nights were when I was free to lick my wound till it bled again.

I began going into Kinross after dark, instead of in the daytime; heading always for some drinking den or other, and always choosing those lying well away from the Kinross Inn. I still had some of the Queen's gold left, and so I had every chance of getting drunk in these places. And that was all I wanted then—to get so drunk that my mind would stop working, and the pain would go away. And some of the loneliness too, perhaps. But there was one night when I did go to the Kinross Inn. I was looking for Muriel; for her black eyes, the enticement of her smile. That was what I told myself, but in my heart I knew it was not these attractions that had drawn me there. It was simply that damnable loneliness, and the knowledge that there was no one I *could* speak to now, except Muriel.

She was not there. I waited in the shadows outside the place till I saw the potboy come sauntering out, and asked him to take her a message for me. He looked me up and down, at my wrinkled clothes, my unkempt hair, and sneered,

"She'll not want to hear from the likes of you. Not where she is now, at any rate."

"Where? Where is she? Has she left the Inn?"

"More than three weeks ago. Gone to work at the New House, she has. Kirsty Thomson—her that used to be maid to Lady Margaret Douglas—she left to get married." A snicker of laughter, and then, "Not before time, if you ask me, seeing how round her belly was! And Mistress Muriel has got her place."

So there was not even a hope of seeing her again! I turned blindly away towards the outskirts of the village and one of the drinking dens I knew I would find there. All the things I might have said to Muriel went through my mind as I headed to it. And Muriel, I told myself, would have understood. Girls always did understand, did they not, about being in love? But who was there to talk to now?

I reached the place—a dark, dirty, malodorous hole, it was—and started drinking. I spoke to no one, and gradually all the faces around me began to blur in my sight. I was reaching the stage where I could stop thinking, stop feeling. My head began to droop. I felt a great desire to let it sink down altogether and to let myself sprawl with arms outstretched over the table. I swayed forward, and was in the very act of collapsing into this position, when a hand gripped the back of my neck and roughly jerked me back to the upright.

Mouth slack, eyes wildly squinting, I tried to see who had hold of me. The stool I sat on was kicked away. The hand on my neck took all my weight for a moment, then was joined by another hand gripping my shoulder. I mumbled protests, but the owner of the hands was merciless. He pushed me ahead of him to the door, then out into the yard. A man stood in my path there, a bucket of water in his hand. The man behind me said sharply,

"Douse him!"

The man with the bucket tipped its contents over me. The man behind me did the same with another bucket of water. I staggered upright from the second impact, wiped water from my face and eyes, and saw George Douglas and John Beaton standing watching me. Beaton put down his bucket the moment he saw I could stand of my own accord, and walked rapidly away. George held his ground, and asked,

"Can you walk?"

I shook more water from me, and questioned in my turn. "How did you find me?"

"The potboy. He is one of a number we paid to report any sight of you. Now, can you walk?"

"What's that to you?"

"Look," George told me grimly, "I risk my neck standing here talking to you. But I do it because I need you. *We* need you."

"I failed." Low and sullen, the words came from me. "I am of no more use to you."

"Yes, you are. Because we *must* have an ally inside the

Castle. *And I can get you back there.* Now, if you are willing to take this last chance to help the Queen, will you walk with me to the New House?"

I found myself walking by his side, without any decision made, my mind still fuzzy with all I had drunk; and on our way to the New House, I listened as best I could to what he had to say.

"I have had secret meetings with my sisters," he told me; "and you know the tender feelings they have always had for you. Also, from the occasional sightings there have been of you in Kinross, I knew you still to be in the vicinity. And so I put it to the Porches that remorse for your involvement in the escape attempt was the feeling that had held you here. They, in their turn, discovered for me that Sir William—believe it or not—has been pining for you. He is fonder of you than anyone supposed, it seems! And of course, he feels sorely bereft of all the services you used to give."

Guilt stabbed suddenly through the dullness of my mind, but George was still talking, telling me how he had persuaded the Porches to work on the affection Sir William still seemed to have for me.

"And the upshot of all this manoeuvring," George finished, "is that my brother is now willing to have you back in the Castle with him."

I could not speak for a time, I was so overcome by this news; and George let me be until I finally managed to ask, "Is it—is all this because you have some other form of escape planned?"

"We will talk of that," George said, "once the Porches have seen that you are bathed and fed and given some decent clothes."

We were at the stables of the New House by then. Just behind the stables lay the back entrance to the house itself. George stopped in the shadow of the stables, and asked, "But meanwhile, Will, do you want to go back?"

Again I found I could not speak at first, but I did at last bring out a mutter of assent. "Then come to Beaton's room at the Inn tomorrow night at nine," he told me. "You will learn all you need to know then. And, Will—" He hesitated a moment, as if wondering how to put what he had to say next, then reached out a hand to me and finished, "Welcome back, also, to our ranks!"

Our hands met in a long grip, but when he withdrew his at last, I still hung back from going to the door of the New House. I was not only soaked to the skin, after all. I had lived for five weeks in solitude and half starvation, wearing the same clothes all that time, and drunk for part of it. I was ragged, filthy, more of a stinking scarecrow than a human being. What would those gentle ladies in the New House *say* if I walked in there? George read my mind.

"Go in, Will," he urged. "I have prepared them for what they will see. And they still want you to come to them."

I nodded, took my courage in both hands, and went up to the door of the New House. It yielded under my hand, and I stepped into the light and warmth the Porches had waiting for me.

## 15

*I* was prompt, the following night, in keeping my rendezvous at the Kinross Inn; but entering Beaton's room, I found, was no longer a simple matter of calling my name through the keyhole. Two burly young men stood outside his door, both armed cap-a-pie; and they held me there at sword point until Beaton appeared to identify me in person.

I followed him back into his room, feeling shaken by this experience; and found George there also, along with two other men. One of these was a youngish fellow, with a great air of worldly experience about him. The other— a soldierly-looking man with curly grey hair—I guessed to be in his fifties. George made the introductions. The younger of the two was John Sempil, husband to one of the Queen's former ladies-in-waiting. The older man, I was somewhat awed to learn, was no less a person than her old friend Lord Seton. Both of them looked me up and down with some wonder in their eyes.

The Porches had done their best, of course, to make me presentable again; but even a bath, a good meal, and decent clothing had not been enough to give me back my normal appearance. For a start, I knew, I had lost so much flesh that my clothes were loose on me and my face had become gaunt. But what I did not realise was how my expression had altered. John Sempil looked from me to George, and said,

"I thought you said this was a merry fellow!"

"So I did," George agreed. "And he was—as I have always known him."

"Then, by God, he has changed," Sempil exclaimed. "I would not care to meet *him* in a dark alley, if he had some grudge against me!"

Lord Seton said brusquely, "Enough! If he can act the part of his former self, that is all we need."

I looked at George to see if he knew what was meant by that. George said quietly, "Will, you asked if we had another escape attempt in mind. And we have. One, this time, that *must* succeed. One that *must* have you there in the Castle."

I looked away and said in a flat voice, "I told you. I have had my chance."

"Listen, Will." George took me by the shoulders, so that I was forced to meet his eyes again. "The time has never been more ripe for the Queen's escape. All her supporters are ready and waiting for her. We are still managing to hold that fox, Argyll, to the promise that will give us superior numbers. And if we do not take advantage of that *now*, the chance of victory for her may be gone forever."

They were all intently watching me. I looked aside again, biting my lip, saying nothing. George gave my shoulders a little shake. "If I explain to these others what I have in mind," he persisted, "will you tell us then whether or not you will do as I ask?"

Still I had nothing to say; but George had apparently decided now to take my silence for consent to proceed. He turned from me, took pen and paper from Beaton's desk, and rapidly sketched the main features of the Castle. The others looked over his shoulder as he worked, with both Sempil and Lord Seton occasionally throwing curious glances at me.

"There!" George sat back at last, and began demonstrating from his sketch. "There we have the gate in the courtyard wall. There also, on the second floor of the main tower, is the Great Hall with the screen that runs across the greater part of its width. This spiral stair in the Hall's southeastern corner leads up to the Queen's apartments on the floor above. There is no way she can leave the tower, therefore, except by coming down that stair, slipping past this gap between the end of the screen and the south wall of the Hall, then using the protection of the screen to gain the doorway of the Hall. From that point she must proceed down the outer stair that leads to the courtyard. And there is no way she can leave the courtyard itself, except by that gate in the wall."

Sempil leaned forward to put his finger on the sketch. "That spiral stair," he said, "must go down to the area below the Hall as well as leading to the apartments above it. What lies below?"

"The kitchen," George told him. "And of course, the kitchen also has a doorway into the courtyard. But to leave the tower by *that* door would mean passing through a whole horde of servants there—not to mention the guards who come to the kitchen for meals and gossip in their off-duty times."

Lord Seton said, "And I suppose you have ruled out the possibility of escape through a window of her apartments?"

"Yes, indeed, my lord." George tapped the plan decisively. "All her windows are under observation from one angle or another. Besides which, she would risk injury in the descent from so high a point; and the boy could not carry her out."

I had heard enough by then. George was doing no more than repeating all the points we had so often discussed before. And he had still said nothing to solve the final problem of getting her through that gate in the courtyard wall. Curtly I told him, "You are wasting your time with all this. I cannot get her out at all."

"Oh yes, you can!" George looked up at me, smiling a little. "If you do as I say, she can walk out of the Castle any time you choose."

"Past the guards on her door?" Resentfully I returned his look. "Past those on the gate? I tried that—remember? And you know how long it took to arrange."

"Ah yes. And time is of the essence now. But, Will—" George's smile was growing wider. "Just think of the *pattern* of the guards' duties. You are as familiar with that— are you not?—as I am."

I was, of course. And the pattern never varied. At seven o'clock every evening when Sir William locked the gate in the courtyard wall, the guards who had been on duty there stood down for the night. And so did those on the Queen's door. Why should they not, after all? The doorway of the Great Hall *was* the only way the Queen could get out of the tower and into the courtyard. The gate in the wall *was* the only way out of the courtyard itself. And the moment Sir William sat down to supper after he had locked the gate, he put the key—

"*The key . . . !*" I did no more than breathe the words, but George heard them.

"Yes, Will, the key to the gate!" He nodded to me, his smile broadening to a grin of conspiracy that took in the others also. "Because, you see," he explained to them, "the guards on the Queen's door and those on the courtyard gate are dismissed from duty once that gate is locked for the night. And so, if the boy waits till after it *is* locked, and then steals the key, he rids himself of the whole problem in smuggling the Queen past those guards."

"Of course, of course!" Eagerly Lord Seton agreed. "He simply unlocks the gate again, lets the Queen through—"

"And rows her ashore," George finished triumphantly, "in his own little boat."

I began to laugh, weakly, foolishly. It was such a simple idea; so brilliantly simple! How had George come to think of it? And why had *I* never thought of it before? I could visualise exactly, too, the situation that would make it possible for me to steal the key; and already, in my mind's eye, I could see that happening. George rose and came towards me, asking,

"Can you do it? *Will* you do it?"

I nodded, still laughing. Lord Seton frowned at this, and said sharply, "Steady yourself, boy, and give us a proper answer."

Suddenly then, through the foolish sound I was making, I heard again my own agonised weeping on Scart Island. Behind all the faces watching me, I saw *her* face. And it was not the sharpness of Lord Seton's voice that steadied me enough to answer,

"Yes, my lord. I will do it."

Seton eyed me in silence for a moment; then, abruptly, he asked, "And you do understand the risk you will be running this time?"

Did he think I was afraid? I faced squarely up to him, and said with all the force I could muster, "I told you, my lord, I would do it. And I will, supposing I were to be hanged tomorrow for it."

A look of grim triumph spread over Lord Seton's features. Quietly, from behind me, John Beaton said, "I told you once you had the makings of a spymaster in you. But I think now—" He paused, and I turned to see him looking around the other three. "I think now," he went on, "I can speak for us all when I say that you have also the makings of a brave man."

There was a general murmur of assent, to which Lord Seton added, "But you do realise how quickly you will have to work?"

"Yes," I told him. "But it is still up to you, my lord, to tell me exactly what time I do have."

"My lord," George interposed, "the boy can return tomorrow—Friday—to the Castle. That will bring us to April thirtieth."

"And we could be completely ready three days from now—by May second." Lord Seton muttered this half to himself. "If we could have her out by that date . . ."

May second—that was my birthday. George began speaking, and I brushed this thought aside to listen to him. "If you will permit us to explain to the boy, my lord," he suggested, and then turned from Lord Seton to me.

"Once you *are* back in the Castle," he told me, "what we plan is this. I will divert suspicion of our intentions by sending word to the Old Lady that I mean to leave the country for good, to seek my fortune in France. And on the day before the escape, I shall appear openly at the New House under pretext of saying good-bye to her."

He paused then to look at Beaton, who immediately added to this, "And on the day of the escape itself, I shall also appear openly in Kinross with a bodyguard of ten men to escort Mr. Douglas on his supposed journey to take ship for France. But those ten men, of course, will be the Queen's immediate escort when she touches at the mainland."

"Of which escort," Sempil chimed in, "I shall be one. And the moment we have her among us, we will ride to rendezvous with Lord Seton's company."

"Fifty of them," Lord Seton himself finished, "waiting in a hollow on Benarty Hill, from whence we can directly observe Her Grace's journey ashore."

They had it all so well planned this time! And yet, once I *did* have that key in my possession, I still could not get her out of the tower without the risk of Sir William seeing her as she passed by the gap at the limit of the Great Hall's service screen. How could I plan for that? And in so short a time, too? Friday, Saturday, Sunday; and they wanted me to get her out *on* the Sunday.

Sunday, May the second . . . The date rang again in my head, with the emphasis this time on the word "Sunday," and the feeling also that there was something I ought to have remembered, apart from the second of May being my birthday. *Sunday*, May the— I had it! The first Sunday in May was always a holiday. And I could make good use of that holiday coinciding, this year, with my birthday.

Very good use! I turned to Lord Seton, and told him,

"I can have her out on Sunday, my lord."

There was a moment's silence, and then a loud sigh of relief from all four men. I addressed them generally. "But you will have to bear in mind that there is only one circumstance in which I can steal the key. And because of that, I must be sure the escort will meet my boat no later than a quarter after seven o'clock on that evening."

"At the landing stage for the New House," George said quickly. "Can you bring her in there?"

I hesitated to answer this. From our own west landing stage to the landing stage of the New House would certainly be my shortest crossing. But Sir William had that particular crossing directly under observation from the west window of the Great Hall—and at the very time, too, when I would have to make it. Cautiously, I asked,

"Why the New House?"

"Because I will have to be quickly informed of the moment to bring up my reinforcements," Lord Seton told me. "And the very fact that my hiding place overlooks the crossing to the New House means that the Queen herself could wave the signal I will need."

"And a further good reason," George added, "is that the horses in the New House stables are the *only* ones immediately available for pursuit. But we mean to foil that pursuit by stealing those horses to make mounts for our own men."

I thought of the time when *I* had planned to steal a mount for myself from the New House stables. But stealing all of Sir William's horses was a bold as well as a cunning stroke! As for the signal to Lord Seton, there was no doubt that one given by the Queen herself would be the speediest way to send news of her approach. I would think of that west window, I decided; and I would find some method of dealing with Sir William's view from it. I glanced from George to Lord Seton, and said,

"Very well, my lord. I will bring Her Grace in at the New House. And I will ask her to wave a scarf or some other such thing as soon as I think we are far enough from the island for her to do so safely."

"Good!" Lord Seton rose briskly on the word, and glanced around the other three. "So is it all settled now?"

Beaton said quickly, "Except for one thing, my lord." His glance flickered between George and myself; and the glance was followed by a silence—a silence, I began to realise, that held a certain awkwardness. George spoke, without looking directly at me.

"Er—Will, I—um—I said enough earlier to let you understand that this is the one attempt that *must not* fail. And you remember the pearl the Queen gave me to be returned as the signal for that? Well, she must have it back now. But not by your hand, because—" He paused, flushing in embarrassment; and it was then that the reason for the feeling of that silent moment dawned on me.

*She* had to be as much convinced as they were of the urgency of this attempt. But I was "the merry fellow," the fool who had let myself be outwitted by a little girl like Ellen, the boaster who had swaggered under the spying eyes of Captain Drysdale. If the signal of the pearl came now from me, she might not—probably would not—take it seriously enough.

"—because of the reputation you have got for yourself," George continued awkwardly. "And so I will have to find some other way of sending it to her."

I had swallowed my medicine by then, bitter though the taste of it had been. I could even see the answer to

George's dilemma; and, ironically enough, I thought, it was my own scapegrace ways that had provided it.

"*Muriel!*" I uttered the one word of my answer aloud; and then, as George frowned in puzzlement at me, I repeated, "Muriel—Muriel Matheson, daughter to the innkeeper here. She is maid to the Old Lady now. And Muriel will do anything you ask of her, if she knows it is to be done for me. Give *her* the pearl when you go on Saturday to say good-bye to the Old Lady. And then, when she attends the Old Lady on the daily visit to the Castle, she can give it to the Queen."

"H'mm." George considered this, and then said, "This girl—Muriel—would have to spin some tale to account for the Queen's jewel being in *her* possession."

He looked expectantly at me; and so did the other three. *Waiting for the accomplished liar to speak* . . . I said impatiently,

"Well, that is easy enough. Tell her to say she got it from the boatman who discovered the Queen in her guise as the washerwoman. The earring fell off that day when he so roughly pulled her shawl aside. But he has only just found it, and given it to Muriel to be returned to the Queen."

Lord Seton asked, "Could the girl keep her face straight, telling such a tale?"

"Muriel?" I glanced at him, and laughed. "I know her, my lord; and she is a born deceiver! Besides, I can brief her on what she has to do. I will find some chance for that before I leave the New House tomorrow."

"In that case . . ." Lord Seton sketched a bow in the direction of George, and John Beaton, motioned Sempil to come with him, and then held out his hand to me. "The next time I see you," he said, "I hope it will be in the presence of Her Grace. And with God's help, boy, I think you will make that possible."

The grasp of his hand in mine was firm, dry, and warm. The thought, *I could trust such a hand*, flashed across my mind. Then he was gone, with Sempil following, and George telling me,

"And now, Will, I had better see *you* on your way."

We walked together out of the Inn, and there was nothing then to stop me taking my leave of him; yet still I lingered. My head was full of the plan that had come to me in outline, back there in Beaton's room; and the plan linked up so neatly with something I had once said to George that I could not help asking him,

"George, do you remember how you reproached me, once, for playing the fool? And do you remember what I said to you when you told me, then, that clowning would not help the Queen?"

Frowningly, George strove to recollect our one-time conversation, and managed at last, "You said something like—'the most dangerous of situations can be disguised in laughter.'"

"That was exactly what I said." I grinned at him, with the impulse that had prompted the remark bubbling up again inside me. "But I finished then by telling you also that the Queen might yet have need of a Court Jester."

"Ye-e-e-s," George agreed cautiously. "I believe you did. And so?"

"And so," I told him, "I plan for Sunday to prove me right in that. And believe me, George, if only I can jest enough on that day—and jest in sufficient earnest, too— then there is no doubt we *will* have our Queen again."

## 16

*I* took a loving leave of the Porches next morning—but not before I had spoken secretly to Muriel and got her consent to do what George would ask of her on the day following. Being a lady's maid suited her, I thought. She looked more charming than ever in the frilled cap and velvet-bodiced gown the Old Lady had provided. She flashed her coquettish smile at me when I told her so; and for old times' sake, then, I sealed our bargain with the kiss she expected of me.

The Old Lady saw me before I was too far away from her room and made a guess at where I had been. She knew what Muriel was like by then, I suppose, and she certainly thought she knew me! She had to hide a smile all the same, I noticed, before she said,

"So you are already up to your old tricks, are you?"

I grinned at her with all my former impudence, and

said, "Maybe, your ladyship. But you are too beautiful still, are you not, to be jealous of a little lady's maid?"

The smile came out of hiding then, and finished in a shout of laughter. "We've missed you, you fool," she told me, gave me a token box on the ear, and went on her way still chuckling.

And that was how it would be for the next three days, I vowed. I would show a properly humble gratitude for my reinstatement, of course, but I would still go on being the Will Douglas they were all expecting to meet again— as impudent as before, as much of a fool as I had ever been. I would play a role, in effect; the role of my former self. And if I could not succeed in that role, then I was not the actor I had always supposed myself to be!

I made straight for Sir William's apartments when I got to the Castle, and was brought up short there by the sight of Lady Agnes lying in bed with a new-born infant in her arms.

"Will!" She smiled her usual vague smile up at me. "You're back. Then you can take the baby from me a moment, if you please."

I took a gingerly hold of the baby while she prepared to heave herself out of bed; and then stood noting the accumulation of Sir William's personal gear lying around. They needed a nurse for him too, not a page, I thought; and the moment Lady Agnes took the child back from me, I set about putting things back to the order I had always kept for them. Lady Agnes prattled on while I worked, proud of being so quickly on her feet again after the birth, proud that this would enable her to resume her duties with the Queen.

But that will not stop *me*, your ladyship, I told her silently; and let her chatter flow over me while I thought ahead to the details of my plan. My time was so short. And there were so many of these details to be worked out! A nurse came bustling in to relieve Lady Agnes of the baby while she retired to dress herself, and minutes after the nurse came Sir William. He looked me up and down, unsmiling, his face closed to any welcome. I knelt to him, kissed his hand, and said humbly,

"I thank you, sir, for your forgiveness, which is more than I deserve of you."

"Er—well, I thought— Since you are young, and foolish by nature . . . And—er—if you promise never to repeat your fault . . ."

Sir William was stuttering as he always did when he was embarrassed; and that, I thought, was the chance to have the ball in my court. I looked up at him and said with pretended fervour, "Of course I promise, sir. And sir, you *are* glad to have me back—are you not?"

"Certainly. Certainly I am." Sir William flushed a little, and at last began to smile. "I—er—I have missed you badly, boy."

"Then, sir, we must celebrate!" I jumped to my feet with the words, and rushed cheerfully on, "You are glad to have me back, I am glad to be here, and—have you forgotten, sir?—Sunday is my birthday. And Sunday is also a holiday. We could have a whole day's celebration then!"

"Oh, Will!" Lady Agnes had reappeared in time to hear this, and she stood looking now in dismay at me. "But I am just out of my confinement, and a whole day's celebration would mean all sorts of preparations. The food, the wine—there would be so much to do!"

"But *you* would not need to do it." Pleading, I held out my hands to her. "We could hire Mistress Matheson to bring over all we need from the Kinross Inn."

"And who would pay for that?" Sir William enquired—but not too sternly, I thought. I flashed a grin at him, and said, "*You* might, sir. As a birthday present for me, perhaps?"

Sir William chuckled then, as I had hoped he would. "You have not changed," he said. "My God, you have not changed one little bit!"

"Oh, but I will, I will," I told him, and cut a clown's caper that made him chuckle all the more. "Once I am seventeen on Sunday, I promise I will be the most sober fellow alive. If *you* will promise to let me have this one last fling with my celebration."

"I will think about it," Sir William conceded. "But meanwhile, my lad, it is back to duty for you. And the first of your duties, remember, is to attend on *me*."

Briefly then, he kissed Lady Agnes, made a casual show of admiring the baby, and beckoned me to follow him from the apartment. I kept at his heels as he went downstairs, and stopped when he did, outside the Queen's apartments. He turned to look back at me then, his face once more severe.

"I wish to let Her Grace also see the extent of my forgiveness," he told me. "Therefore you will attend me, even in here. But remember your promise, and—I warn you—do not try again to take advantage of me."

Now was the time for another touch of humble gratitude. "I would not dream of doing so, sir," I protested, "unless it were to tell Her Grace, even before you do, how gracious you have been to me."

Sir William gave a satisfied nod, then knocked at the door. I entered the room a pace or two behind him, and once more found myself in the presence of the Queen. She was newly dressed, sitting before her mirror with her hair still loose about her shoulders. Her eyes met mine through a fine, floating veil of its gleaming strands, and I felt the tingling pleasure of that encounter in a flush that seemed to run from my hairline to my heels. Sir William's bow and his opening words gave me a few seconds to recover myself; but I was still feeling the effect of that flush when she answered,

"Yes, indeed, Sir William. I do see how forgiving you have been. And how kind to me too, to let me enjoy again the company of my little orphan."

Sir William laughed. "Not so little now, Your Grace. He will be seventeen on Sunday."

I seized on the chance he had given me, and said quickly, "And I am so pleased to be back, Your Grace, that I mean my birthday celebration to make Sunday a holiday to remember. If—" I hesitated, glancing at Sir William. "If I am allowed to, that is."

"But of course you must be allowed!" Reproach in her voice, the Queen followed my glance. "Why should he not, Sir William?"

Sir William shrugged, and did everything he could to avoid meeting the Queen's gaze. But I kept my eyes hard on her, and had the satisfaction of seeing her become aware of the intensity of my look. "Lady Agnes," Sir William was saying lamely. "She is just out of her confinement, you understand. And all the preparations that would be involved—she could not undertake that. It would mean hiring from the Kinross Inn—a costly business; and—" His eyes finally settled on me. "And I have not yet decided whether he is worth all that."

"Then I shall decide for you." Smiling, the Queen rose and went to her desk, lifted its lid, and called, "Will, come here to me."

I went towards her and stood by the desk, with one hand resting on its edge. She saw the tiny wedge of paper I let fall then, from my fingers into the desk itself, but did not blink an eye to betray her knowledge while she told me,

"I will give you this to pay for your celebration, Will, on condition I am invited to it. Now, does that please you?"

I looked from the roll of gold coins in her hand to Sir William, and said, "If it also pleases Sir William, Your Grace."

"Well, Sir William?" Smiling, she turned to him. "Does the boy have his celebration, or does he not?"

"Your Grace—" Sir William was flushed with embarrassment. "You are too generous! But if you insist on paying, then by all means I agree that the boy should have his way."

And now was the time to revert again to impudence! Quickly I chimed in, "In that case, sir, I will be as kind to you now as the Queen has been to me. I will choose from you a birthday present that will cost you nothing, because it is a gift that money cannot buy. I choose . . . I choose . . ." I stopped, holding him in suspense till he finally demanded,

"What?"

"I can guess!" the Queen exclaimed. "Sunday is the day for May games; and because it is your birthday, you want the right to preside over them. You want Sir William to let you be the Lord of Misrule!"

"For that one whole day—yes, Your Grace, that *is* what I want." I turned from her to smile at Sir William. "If you will grant me that favour for my gift, sir?"

I had him in a position now where he knew he would sound churlish if he refused—especially so since he was already aware he had been made to seem mean over paying for the celebration itself. I wondered if the Queen would guess that my manoeuvring towards this point was connected with the folded-small note I had dropped into her desk. I saw her fingers stray towards it as Sir William hesitated over his consent, and prayed she would wait till we were gone before allowing those impatient fingers to make any further move.

Sir William's doubting look began to turn into a smile. He liked May games as much as anyone, after all; and he had already admitted his pleasure in my return. I stifled the long breath of relief I wanted to draw when he finally voiced his consent; and gave him, instead, the kind of fulsome thanks that flattered him into a mood in which he would have granted almost anything I asked. But that, for the moment, was all I wanted from him; and my next target—to be reached as quickly as I could—was some method of disabling every boat on the island, except my own.

"You spoke of duties," I reminded him once we had left the Queen's apartments; "but who has been looking after the boats while I was gone?"

"Captain Drysdale," he told me. "Not the most suitable person, I agree. But who else was there?"

"Sir William—" I stopped in my tracks, so that he had to halt too. "I was seduced from all duty to you, I know. But I have repented, and these boats are important to us. I *must* make sure they are all still sound."

"Then make sure, make sure." Amiably he waved me off, and I sped down to the boats to work out the best way of putting them temporarily out of action. The loss of the key, I was well aware, could be discovered within seconds of my stealing it. But that loss would have to be connected first to the fact that I, too, was missing; and so I would have perhaps three minutes in hand for the exercise of getting the Queen from her apartments to my boat. Any time after that, however, could see the pursuit started; and so I *had* to delay that pursuit by at least the further ten or twelve minutes I needed to row her to the mainland.

A hole in the bottom of each boat would have served my purpose, of course. But the boat timbers were too sturdy to be easily and quickly holed. I could not damage them too soon before the actual moment of escape, either, since that would be to risk too early a discovery of their condition. Something more subtle was called for. . . .

I stood staring at the boats, and thinking. There was a rough westerly wind blowing—the kind we sometimes had to suffer for days at a time; and according to our usual practice then, the boats had been drawn up on shore at the sheltered east landing stage. As usual, also, they were all held there by one chain slotted through an iron ring fixed to the prow of each boat. And that chain, I realised, was my answer. One end of it was firmly stapled to a wooden bollard. The other end was coiled around a second bollard. And so long as that chain could not run freely through the rings on the prows of the boats, *none of them could be released.*

There was one way I could make sure of that! A wooden peg driven through each link in the loose end of the chain would be enough to prevent it running through those iron rings. And that exercise, I reckoned, was something I could complete in the very last quarter of an hour before the escape. But what were the chances of my being observed in this? There would be no danger from the guards on the gate, since that was set in the north wall of the court-yard. But what about windows? There was a big east win-dow in the Queen's sitting room; and Sir William would be there, serving supper to her, at the very time I would be working on the chain. Could I risk him seeing me then, from that east window?

I would have to, I decided. There was no way, after all, of avoiding that risk, any more than I could avoid all the others still to be faced. But I did at least have a plan for dealing with them as they arose! With a certain wry satisfaction in the way that plan was taking shape, I turned from the boats and went hurrying off to see Minny.

She had been expecting me, I realised, when I saw the expression on her face. But she had still not been expecting the changed expression *I* wore. The gladness of her look changed suddenly to pity; and without a word beyond her first cry of greeting, she rose to fold me in her arms. We stood thus together for a long moment before she let me go, and said quietly,

"You are going to try again. Is that it?"

"Yes. And this time, Minny, I am going to succeed. It is life or death now."

"Your death?"

"Certainly, if my plan goes wrong."

"Oh, Will—" Her lips quivered. "Whatever happens then, you will be going away again. But for good, this time."

I reached for her hands and pressed them between my own. "Only at first, Minny, because I must escape with her. But I will let you know where we have gone, and then you will join me. To live with me. To look after me. Will you do that?"

The tears that had threatened came freely then. "God knows you need looking after!"

"Mi-i-i-nny!" I gave her hands a gentle shake. "There is no time to cry if you want to save my life. I want you to work for me—to work harder and quicker than you have ever done."

"Tell me, then." Minny wiped the tears away with the back of her hand, and faced me bravely. "But let me tell you one thing first, Will. You are—"

The brave front vanished. Her head dropped. The hands between my own trembled. I had to make it easier for her. I said gently, "I'm your son. No need to tell me. I have often felt it so—Minny."

She looked up at me, slowly. "And you—you do not blame me?"

"Now," I said, "you are just being foolish. And at this moment, Mistress Douglas, a foolish woman is no use to *me*."

She tried to answer the smile I was giving her, and—being Minny—she succeeded. "Now you are more like yourself," I told her; and as quickly as I could then, I outlined what she would have to do.

I wanted her, I told her, to make me two of the costumes worn by such of the soldiers' wives as were local women. Full skirts of black, tight bodices of scarlet, and high-crowned black hats—that was the kind of dress that distinguished these women. And any kind of distinctive dress, of course, is the best for disguise.

"Because then," I explained to Minny, "the onlooker assumes from that dress that he is seeing only the kind of person he *expects* to see wearing it."

"But why two?" Minny asked. "Why two, for God's sake?"

I hushed her down to launch into the next part of my tale—a vivid recollection coming to me as I spoke of the Queen walking in the garden with her maids, and Ellen remarking that Kennedy's lack of inches made the three of them look like "two women walking with a little girl." But if the Queen were to try escaping alone, I reminded Minny, her unusual tallness would be the very feature that could betray her disguise. It was a common thing, on the other hand, to see one of the local women with a young daughter in a dress identical to her own. And so, if the Queen took Kennedy with her, their differing heights would be just what was needed to mislead people into the impression I wanted to create—that of a soldier's wife and child, walking hand in hand across the courtyard.

Minny began looking to the chests and cupboards that held all her spare bolts of cloth. "I have red flannel," she muttered. "I can manage the black, too. And I think I already have one of those hats around here somewhere." She rose towards one of the cupboards, and then turned to ask, "How long do I have to do all this?"

"Till Sunday afternoon."

"Your birthday—!"

"Yes. I have persuaded Sir William to let me celebrate it with May revels that will take up the whole of that day. The Queen will be at the revels too; and in the middle of the afternoon, when her apartments are empty and therefore unguarded, you can slip into them to hide the costumes in the chest that lies to the left of her writing desk."

"I had better get to work!" Minny threw open the cupboard door and began to rummage along the shelves, then turned to ask, "But Will, how will you let the Queen know what is expected of her?"

My mind slid back to the previous night in the New House and the writing of the note I had so recently dropped inside the Queen's desk. "No need to worry about that," I answered Minny. "I have already found a way of telling her."

"But will she do as you say? I mean"—Minny faced again to the shelves to hide her embarrassment—"after the last time?"

I got to my feet and asked, "Can you promise to do your share, Minny?"

She threw a hurt look at me. "Do you need to ask that, Will?"

"Then trust me for the rest," I told her. "And come tomorrow, Minny, when the Queen receives a certain signal that will be sent to her then, you can rely on it that she will trust me also."

I do not think Minny would have argued any further on this but I could not have stayed, in any case, to find out. The morning was flying away from me. It would soon be time for me to serve Sir William's dinner, and I still had two other people to see. I decided on Arnault before dinner, and Diderot afterwards; and hurried off to Arnault's room. He looked up at me as I came in and said sarcastically,

"*Eh bien!* So we are once again honoured!" Sharply he closed the book he had been reading, and rushed on without giving me a chance for any word in reply. "And have you learned yet of how *we* persisted in our efforts to arrange the Queen's escape once you had paid the penalty of your folly? Do you know anything at all of the shifts and stratagems we had to contrive to keep in touch with the good young M'sieur Douglas and his helpers? But no! How could you be aware of that? And what do you care anyway, steeped in self-love as you are! But let me tell you this, young sir, let me warn you now—"

He was wound up to a high pitch, I realised; and so there was nothing for it but to let him ramble on till he had let out all the spleen and bitterness that seemed to have built up against me. I deserved it, after all, and he could not say anything I had not already said to myself. I waited with bent head until his voice tailed away into mutterings of disgust and despair. He had turned away from me by then, to slump forward over his desk. I touched his shoulder, ignoring the way he flinched from the feel of my hand, and quickly told him,

"I am not here to defend myself from the past, M'sieu Arnault, but to tell you of the future. At seven o'clock on Sunday evening, the Queen will be as free as a bird. That is a promise. But it is not one I can carry out unless *you* promise to help me at a crucial stage of her escape."

His head jerked up. His face turned in astonishment towards me. Rapidly and briefly I told him of the meeting in the Kinross Inn, the plan that had sprung then to my mind, and the part he would have to play in it. He listened without once taking his eyes off me, the surprise on his face changing to wonder, and then to delighted approval. To my great embarrassment then, he jumped to his feet, seized me by the shoulders, kissed me resoundingly on either cheek, and showered as many blessings on me as he had previously flung curses.

As quickly as I could, I freed myself from his embrace. It was past time, I reckoned, for the Queen's dinner to be served, which meant Sir William would soon be calling me to serve his own. And I could not so soon risk my welcome home by being late for that! With Arnault's voice still following me, I hurried out and dashed downstairs to the kitchen.

Diderot was there. I caught a glimpse of his big, handsome face in the corner where he usually sat after he had finished his work on the Queen's dinner. I stood getting my breath back while the trays for the Great Hall were arranged, then placed myself as usual at the head of the service procession, and walked upstairs to the Hall to find a very jovial Sir William waiting for me to put his dish in front of him.

"Ah ha, Will!" He rubbed his hands. "Like old times, eh?"

"Just like old times, Sir William." But my hand was trembling as I poured his wine; and as the meal went on, I could have shouted with vexation at the way Lady Agnes toyed slowly with her food.

"Tomorrow morning, Will," Sir William said between mouthfuls, "you can go over to the Inn and arrange what you want with Mistress Matheson. Eh?"

I was swallowing down my own dinner as fast as I could. I said, "Yes, sir. Thank you, sir." And turning to Lady Agnes he remarked complacently,

"You see, my dear—a reformed character!"

I rose quickly in my role of reformed character to supervise the clearing of the table, went soberly out of the Hall and back down to the kitchen. Diderot spoke only French, I knew; and so, with every step I took, I was rehearsing what I would say to him. *"Be brief,"* I warned myself. That would be the best way to make my meaning absolutely clear to him.

Diderot was at the chopping block, butchering the carcass of a sheep. He had a long and sharp boning knife in his right hand. His cleaver—the cleaver Arnault had threatened he might use on *me*—lay within his reach. I had a moment's panic when I thought I could not remember the French word for "cleaver." Then I had it—*fendoir; le fendoir.*

I spoke low to him, ignoring the suspicion that looked out at first from his face. "Diderot, you know why I was banished. I am back for the same purpose. *But you must help.* At seven o'clock on Sunday, the Queen and one of her maids will come down the steps from the Great Hall. They must reach the gate without being stopped by any servant, or by any of the guards. And I want you to make sure of that. You understand?"

Diderot nodded. His eyes went to his cleaver. I said, "Yes, my friend. Come out from the kitchen at seven o'clock on Sunday night. I will meet you. Stand so that you can watch what happens in the courtyard while the Queen walks to the gate. And if anyone tries to lay hands on her—"

Diderot's face lifted quickly to the kitchen's smoky ceiling. Then he looked down again, the question about Sir William plain in his eyes. I shook my head, and said,

"No, not him, Diderot. Not with the cleaver, at least. And no woman with the cleaver, either. But with anyone else, my friend, you have a free hand."

Diderot reached out to exchange the knife for the cleaver, and with pudgy fingers caressing its heavy blade, he promised quietly, *"Personne ne me passera pas."*

No, I thought; no one *would* get past him. And now, in all the rest that remained to be done, the responsibility would be mine alone.

# 17

*S*unday the second of May dawned bright and clear, with the west wind still blowing but much moderated in force. I was out early, supervising the servants at setting up a long line of tables in the garden, and then secretly visiting Minny to see if she had the two costumes ready. She was still hard at work on them; but, she promised, she would have them finished on time. Ellen passed me on the stairs as I went in from hearing this, and said,

"*You* look cock-a-hoop today."

"And why not," I asked, "on my birthday?"

Ellen surveyed me again and pronounced her next verdict. "You have grown thinner."

I had not spoken directly to her since my return; and now, I guessed, she wanted to prolong the conversation. But I was too wary to be trapped again by Ellen! "And taller," I retorted; "*and* older. Which means I have no time now for little girls like you!"

I chucked her under the chin as I spoke, and carried on upstairs, laughing at the disgusted face she made. In my own room—a cupboard-sized hole in the wall two steps down from Sir William's door—I changed into my best clothes. Today, we would all be wearing our best. Either that, or those who felt like it would wear a masquerade costume. But the important thing for me in wearing my best was that it meant I could also wear my sword. I belted this on and went down to look out for the boats coming over with the wine and food from the Kinross Inn.

Mistress Matheson came with the first of these, talkative as always. I gave her instructions on laying out the provisions, and then went off to cut myself a straight stick of hazel to use as my wand when Sir William appointed me Lord of Misrule for that day. The tables were laid out with flagons of wine all along their length, when I got back with my stick. I groped in the bushes for another flagon I had hidden there—one I had chosen for its distinctive shape, and then filled with coloured water—and placed it also on the table. The Lord of Misrule, I intended to inform everyone, would not condescend to drink any but his own wine—and that a strong one, fit for such an occasion!

Ellen came dancing out into the garden, dressed in a gown of pale-blue satin that gave her a look of almost ethereal beauty. Solid, sensible Ellen. Sly little Ellen! But she was so pretty now that I forgave her. Margaret Lindsay trailed behind her, looking wan in some sort of dull green. But Margaret, I thought, would have looked wan, whatever colour she wore. I gave them both some wine, and drank so heartily myself that Ellen exclaimed,

"You will be drunk before dinnertime!"

"Certainly," I told her. "And I shall be drunk after dinnertime too."

"Then you will be a fit Lord of Misrule," Sir William's voice said from behind me.

I looked around and saw that the garden had begun to fill up with family and servants. Muriel was there, trailing behind the Old Lady with a cloak and a pile of cushions in her arms. I raised my eyebrows in question to her, and she answered with a long, slow wink that I hoped was intended to assure me of the safe delivery of the pearl. The Porches came swirling round me, each of them with her attendant admirer—except for Lady Lindsay, who was now arm in arm with Margaret. Diderot strode up to the tables, proudly carrying a huge dish piled with fruit made out of marzipan. His personal gift to the Queen, I guessed. Diderot prided himself on his skill in such confections. But where was the Queen herself? The sun was high now, yet we still could not start the May games without her.

I kept my eyes on the garden entrance and finally saw Jane Kennedy and Maria de Courcelles come in, side by side. But still no sign of the Queen. My eye was caught by a sudden flash of violet and white. There was a tall young man in particoloured costume of violet and white coming into the garden, closely followed by a contingent of the guard; and for a wild moment I thought it was George I could see.

That was the very style of George's holiday dress—velvet cap, breeches and doublet of satin, silk stockings—all in those colours of violet and white, with sword scabbard and shoe buckles of silver. But this young man was slimmer than George had ever been. And younger, perhaps, since he had no beard . . .

He came striding straight towards me, laughing, face tilted up to the day's brightness. The sun gleamed off the tendrils of hair escaping from under the violet-and-white cap; and now I could see they were not the golden fair of George's hair. They were golden-red. And for each step of that laughing progress, there were ladies curtsying to the ground, men doffing their caps and sinking down on one knee, voices that murmured, *Your Grace. Your Majesty!* And I thought I had already known this Queen of mine in all her moods!

I got hurriedly down myself just before she reached me, wondering as I knelt how I had come to forget her love of masquerade. And this, of course, was her favourite one; the one most flattering to her height and her long, slim legs. . . . But how had she managed to get hold of George's clothes?

She told me herself, standing in front of me, hands on hips, long legs straddled, the very picture of a debonair and handsome young man. "You see me dressed now in these borrowed plumes by courtesy of their owner's brother!" She was laughing again as she spoke, enjoying the impact she was making. "But get up, Will, get up. It is you, and not I, we must all kneel to this day."

I still had my hazel wand in my hand. I turned from her, holding it out to Sir William. He touched me on the shoulder with it, and loudly cried,

"William Douglas, I name you now for this whole day and over all the souls in Lochleven, as—*Lord of Misrule!*"

In the roar of acclaim that followed I rose swiftly and seized the wand of my office from him. I could cut any madcap caper I pleased now. And so at last I had got from Sir William himself the one excuse he would *have* to believe if he observed me in some unusual action. From all that company, too, it was now also my privilege to choose one I could command at any time to follow me and imitate whatever I myself might do. . . .

With my wand raised high and commandingly, I swung towards the Queen and ordered, "Kneel, then! Kneel, young Sir Particolour, and be dubbed, in your turn, my servant."

She went down on both knees before me, arms spread out, head bent, like someone kneeling for execution, her laughing face almost touching the ground. I brought the tip of the wand down on her velvet cap, and cried,

"I name you now, servant to Will Douglas, Lord of Misrule in Lochleven. And for this whole day, I bind you to follow me and obey all my commands."

She scrambled to her feet, calling out for wine to sustain her in the task, and the call was a release that sent everyone surging to the tables. I grabbed my own flask of "wine," rapidly poured a cup for her, then tilted the flask to my lips and drank. Our eyes met as she tasted the contents of her cup. I flourished my flask at her and announced,

"I have drunk a fair amount of this already, young sir, and I intend to drink much more before Lochleven is shut up for the night."

"Then I must also drink more," she answered, "if I am to keep pace with you."

I poured again for her, shouting, "You hear that my lords, ladies, and gentlemen? You must *all* drink up, to keep pace with the Lord of Misrule."

They were all only too willing. The wine went round

and round till I judged they were merry enough to start the games; and, whooping like a madman then, I led them all off through the garden to get branches of birch for the traditional procession of bringing in summer. The Queen leaned on my shoulder, pretending breathlessness, as I sent them careering into a thicket of young birch trees. It was our first chance of private talk since my return on Friday, and we made the most of it.

"*I have got the pearl back from Mr. Douglas. Yesterday, by the hand of the Old Lady's maid.*"

"*So you will trust me now for all the details in my note?*"

"*I must. The pearl was agreed on as a sign.*"

"*Then I can promise the two costumes will be in the chest, waiting for you.*"

"*And Maria can keep the girls occupied in the inner chamber while Jane and I change!*"

"*Yes—but remember the dangerous point of your exit from the Castle. Sir William sits at supper with the gap at the end of the service screen plainly visible on his left. And you will have to pass over that gap.*"

"*I have my excuse if he sees me. I am following the commands of the Lord of Misrule!*"

She had been quick to understand! I glanced over my shoulder at her, smiling, all ready to tell her of the part Arnault would play at that point of the escape. She forestalled me, her voice suddenly anxious.

"*The pursuit, Will—have you thought of that?*"

"*Of course. We have delayed that as much as we can. You will have a clear start on them.*"

Ellen burst from the bushes crying, "I can hear the musicians. They are coming!"

"Then follow me. Take your branches and follow!" I

waved my wand, shouting to her and the others to form up in procession. They came streaming from all directions with their branches of birch held aloft, and fell in behind me as I led their triumphal progress back to the stretch of grass left clear for the dances. The musicians hired from Kinross were already there, banging and blowing away. I shouted to them to change their tune to "The Dance of Robin Hood," and hastily selected the characters. Arnault, I announced, could be Friar Tuck. He was fat enough. My servant Queen would be Robin. Diderot was big enough to be Little John, Ellen pretty enough to be Maid Marian. Sir William could have the plum role of Sheriff. And I myself would be Point of Arrow, slaying anyone I chose.

Off we went, with myself making sure it was the maddest "Robin Hood" ever danced, and with only those "slain" early in the game having any breathing space at all. I kept hold of my flagon, all the same, throughout it all, and made great play of continuing to drink from it. As I did also in "Blind Man's Buff," "The Farmer in the Dell," "Poor Roger," and all the other games that followed.

"There is some devil in you today," the Old Lady told me at last. "I have never seen you quite as mad as this before."

Matters were quieter by then, mostly because I had eventually given them all a chance to sit down and eat. But the Old Lady, I noticed, was only pecking at her food, and she had taken very little of the wine the others were drinking so freely.

"And I always thought you were fond of George," she went on. "But you do not seem to care one bit now that he is going so far from us all."

I leaned towards her, sorry that this part in my masquerade had to be played, but still determined to play it convincingly. "George," I said thickly, "has made only one mistake. He has saddened you by saying good-bye the day *before* my celebration, instead of waiting till the day after it."

The Old Lady pushed me roughly back from herself. "You are heartless! I always said that of you; and now I *know* I was right."

"Not at all, your ladyship. Tomorrow I shall care. But today I shall make even you laugh—and that in spite of yourself!"

I was on my feet with the words, and then leaping onto the table. Shouts, yells of laughter, greeted my pose there. I shouted to the musicians to play. They struck up in loud discord, and I went dancing down the length of the tables, teetering from foot to foot among all the cups and dishes littered there, apparently managing to avoid each obstacle only by that miracle of balance accorded to the supremely drunk. Halfway down the tables I struck another pose. The musicians rewarded me with a long drumroll, and with a flourish of my wand then, I shouted,

"The Lord of Misrule commands! Let Adam now be Eve. Let Eve be Adam!"

We had been merry before, but this was now the signal for licence, and they seized on it as swiftly as they had earlier seized on the wine. Within seconds they had scattered among the shrubbery and were exchanging dress—man with woman, and woman with man. Sir William reappeared wearing the gown that belonged to his child's nurse, and roaring with laughter at himself. The Porches exchanged with their swains. A girl I took to be Muriel ran past me, and turned out to be a kitchen boy. Diderot came prancing out of the bushes in a dress that met only halfway about his middle, and was followed by a young serving girl with his doublet billowing amply around her. I myself seized Ellen and traded my holiday clothes for her ethereal blue satin.

I spread the skirts of this in a clumsy curtsy to the Queen, and then led them all off in a wild game of "Follow My Leader." They came pell-mell after me, all of them exactly imitating me, prancing through the shrubbery, skipping along the top of a terrace wall, dancing around and around the flower beds, giving tongue the while with the same sounds as I made: crowing like a cock, neighing like a horse . . . *How long could I keep this up? How long did I need to keep it up? All of them were convinced beyond doubt now that anyone as drunk as I could not have a single serious thought in his head. . . .*

Still wildly dancing I stole a look at the Queen following immediately behind me. Her face had a look of white exhaustion. There were drops of sweat standing out on her skin. It was time to let her rest before the final exertion of the escape. I brought my cavalcade to a halt, and—much to Ellen's disgust—made a further change of clothes with her. If I could divert the company with my own efforts for another hour, I reckoned, that would bring us to five o'clock. The Queen could retire with good grace from the revels then, and her retirement would be the signal for the rest to depart also. With my doublet and sword belt dangling from one hand, I ran back to the site of our earlier games, all the others following in a ragged tail.

I had always been able to do handstands and other acrobatic tricks. Or since I was seven, at least, I had been able to perform like this. I had been solemnly trying to teach myself to do so one day, all alone on the mainland shore of the loch, when my bumbling efforts had been seen by one of a group of strolling players visiting Kinross. He was an old man, this player, one of the clowns of his group. And he had been scornfully amused at first, by my failures. But then he had started to teach me all his own secrets of balance. And that old man had taught me well!

I threw my doublet and belt to Sir William to hold. He guessed my intention—as indeed he should have done, considering the number of times I had performed for *him*. A shout from him took the place of any announcement I might have needed to make. Some of the guests collapsed to the benches by the tables. The rest formed a wide circle around me. I grinned at them, and called,

"I will perform better drunk than sober, I promise you!"

There was a roar of applause. I judged the space I would have, balanced on the balls of my feet as the old man had taught me, then ran forward to let a double somersault carry me into the last hour of my role as the Queen's Jester.

# 18

I had spent a good part of Saturday cutting and shaping the pegs I meant to drive through the links in the boat chain. From this, I had gone on to stowing them in my fishing bag, along with a small iron mallet, and then hiding the bag under the shoreward end of the east landing stage. There was no one about there when I went down at the time I had set for myself that evening, and my bag had not been disturbed. Quickly I drew it to me, tipped out the contents, unwrapped the loose end of the boat chain from the bollard, and set methodically to work.

A peg through every fourth link would be enough to begin with, I had reckoned, and then a peg through every second link. I hammered each one firmly in, rewound the chain onto the bollard, and then stowed my emptied bag and the mallet back into hiding.

I had worked fast, as well as methodically; but I had one more move to make before I was finished. I lifted a pair of oars from one of the boats, and hurried with these to the southeast corner of the courtyard wall. Beside me then, I had the outer aspect of the little round tower where the Queen had first been imprisoned. And there was a window in this outer wall; the one window through which she *could* escape if I failed to get the key to the gate. Always providing, of course, that I could lead her safely across the stretch of courtyard between the main tower and this little one!

The window was fully eight feet from the ground. I upended the oars, leaned them against the wall beneath the window, jammed their blades firmly into the ground, then stood back to give a critical look at my work. With the oars to help her descent, I decided, and with myself waiting underneath to break her fall, she should be able now to get safely down from that window. And if we *were* unlucky enough to be seen crossing the courtyard together, we could still give the excuse that would have served if I had been caught working on the boat chain. I could protest I was only carrying out another jest in my role as Lord of Misrule. She could plead she was still the servant who was bound to copy all my antics.

I turned and ran back on my tracks, knowing then it would be a matter of seconds only until Sir William finished serving the Queen's supper and came to lock the gate.

The guards there were used to seeing me rush in at the last moment. They let me pass with no more than a glance. Almost immediately on my right then was the retaining wall of the stair leading down from the entrance to the Great Hall. I sidled quickly along its length. The last three steps of the flight projected beyond the southeast angle of the tower. I took a flying leap over the projection; and, safely hidden then from the view of anyone descending the stairs, I ran for the tower's kitchen entrance.

My timing had been exact. I heard Sir William's voice within seconds of gaining concealment in the kitchen doorway. He was talking to Kerr and Newland, both coming off duty as usual at the same time as he went to lock the gate. Sir William sounded as if he had sobered up to the point of being now in a sour sort of mood. Then I heard another voice—Drysdale's. I pressed back into further hiding. He and Sir William were talking about *me!* I caught something from Drysdale about "that boy," and "fooling around with the boats." Had I been seen, after all, from that big east window in the Queen's sitting room?

The voices of Sir William and Drysdale faded in the direction of the gate. I looked obliquely from my hiding place and saw Kerr and Newland still standing together, still talking. Were they also discussing some sight of me at the boats? I waited with held breath for the return of Sir William and Drysdale. If there was any delay in dismissing the guards on the gate, I realised, it would almost certainly be because of some suspicion about the boats. I began to pray that Sir William's sour mood meant he would be unwilling to take the trouble of going down to check on these.

Drysdale came back into my view, followed by the guards relieved of duty at the gate. Kerr and Newland turned to greet these two, then the whole group walked on towards the kitchen door. My prayer—if it had been needed—was answered. I moved swiftly into the kitchen, and called out,

"Sir William is ready to be served."

There was a flurry of maids and stewards reaching for the dishes that would be served to the others at table in the Hall. I picked up Sir William's own dish, took a damask napkin from the neatly folded pile standing ready, then led the procession of servants up the kitchen stairs, past the end of the service screen, and into the Hall.

Sir William was already in his seat at the head of the table, Lady Agnes on his right, the Old Lady on his left, and Lady Lindsay beside the Old Lady. Arnault, too, was seated as usual, in the place opposite my own—one that was not too near the family company, yet still not too near either, to the servants' position below the salt. I placed the dish I carried in front of Sir William, then went to the window immediately to the right of where he sat—the one that gave him a view of Kinross, and that would therefore also let him see any boat heading directly westwards from the island. Quietly I began to close the shutters over this west window.

"Leave those!" Sir William bellowed. "What d'ye think you are at, boy—shutting out *my* view without *my* permission!"

"But sir—" I looked innocently around. "It was to protect you. The wind is still from the west. And it has become such a cold one again."

"I fear your page is right, Sir William." Arnault spoke up with a great air of professional gravity. "May is a treacherous month; the time when men—and women too—are subject to chills and rheums of various kinds. And that is the very thing likely to happen to you, if you expose yourself to a cold wind after the exertions *you* have had today."

Sir William eyed him without much favour. "What d'ye

mean to imply by that, M'sieu? That I am not entitled to disport myself occasionally, as other men do?"

"Sir William, I am merely saying—" Arnault's voice flowed smoothly on while I continued my quiet closing of the shutters. Sir William interjected occasional grumbles into the flow, but I could see he was heeding Arnault's advice. Sir William was a fit man, but still always careful of his health! I wondered if I could dare to close the shutters of the north window also, but decided against that risk. The north window would certainly allow Sir William an oblique view of the crossing I had to make; but I still had one card left to play in that part of the game.

The key to the gate was my next objective. It lay where Sir William always placed it when he supped—on the table, close to his right hand. I came back to the table, busily shaking out the folds of my napkin, my eyes fixed on the key's heavy iron shape.

"Your wine, sir?" I leaned towards the flagon of wine in front of Sir William, and grasped it with my right hand. The cup for the wine stood beside the flagon. I let my napkin drop carelessly to the table before I reached for the cup with my left hand. I poured the wine and put the flagon back in place. The napkin had fallen in loose folds that completely covered the key. I put the wine cup handy to Sir William's grasp, swept up napkin and key together in one servile movement, then bowed myself back from the table.

Arnault had been watching me, and Arnault's tongue became busy again the moment the napkin dropped over the key. "As I was saying, Sir William," he began, "it is not the quantity of wine a man drinks so much as the question of whether he has the stomach for it. Also, I have noted in you, sir, a certain choleric tendency which means you may find an overindulgence difficult to digest. And if you will pardon me for saying so—"

"Good God, man," Sir William interrupted, "are you daring to suggest I cannot hold my wine?"

The long and hard outline of the key under the napkin was firmly in my grasp by that point. With Arnault launching into a vehement defence against Sir William's charge, I moved unobtrusively from the table and towards the screen. I could count on Arnault to go on with the distraction of his argument. Arnault, I knew, would continue to keep Sir William's attention diverted towards his place on the right of Sir William's own, and therefore away from the sight of the Queen and Kennedy passing the gap at the end of the screen on the left of Sir William's position.

The three-minute time margin I had counted on was already being eaten into; yet still nothing done in the course of those three minutes could be so hurried as to call attention to myself. I schooled myself to a walking pace as I left the Hall, went down the outer stair, and around the southeast corner of the tower. Diderot was standing at the entrance to the kitchen, lounging there like a man taking the air after a hard day's work indoors. His cleaver hung from the broad leather belt around his waist, attached there by an inch or so of chain snapped onto a steel ring. He moved forward at the sight of me. I closed with him, and looked up to the south window of the Queen's apartments.

The note I had smuggled in to the Queen had warned her to have Jane Kennedy on watch there. I saw Jane's head peering from the window, and waved upwards. On the instant, Jane withdrew her head. I turned to Diderot and told him,

"Watch out. You will see them in fifteen seconds from now."

Diderot nodded and moved to lean back against the south wall of the tower. I walked soberly away from him towards the gate, counting the seconds as I went. I stood by the gate with the key—still under the napkin—in my right hand, my face turned towards the stair. If I did not see the Queen and Kennedy by the time I had finished counting, *something* would have gone wrong.

The longest fifteen seconds in my life came to an end with the appearance of two figures in red and black at the head of the stairs. They started walking hand in hand down the steps—the tall one and the tiny one, the "soldier's wife" and her "daughter." Some members of the Guard came out of the barracks, noisily arguing with one another as they moved towards the kitchen. The Queen could not see them from her position on the steps. But she had heard them. And she had stopped where she was, on the fourth step from the bottom. One more downward step, she had realised, would bring her into the soldiers' view. And she did not know of Diderot waiting there to protect her passage to the gate! There had been no chance for me to warn her of that precaution.

Someone had to decide what to do. The Queen was motionless still; and Diderot himself would not be able to see her unless she moved the one further step that would also let her be seen by the soldiers. I turned to the gate, unlocked it, then vigorously beckoned to the two figures on the stairs.

The tall one began moving down again, taking the small

one with her. Hand in hand at the foot of the stairs, they turned to walk towards the gate. I saw Diderot sauntering into view at their rear, as they walked; but his cleaver still hung at his belt, and there was no outcry from the soldiers still arguing behind the angle of the wall.

I was holding the gate open just far enough to let the two figures pass beyond it. They slipped through the space, with myself crowding on their heels. I closed the gate, relocked it, then got rid of the key by dropping it down the barrel of the cannon beside the gate. My action brought a nervous giggle from Kennedy; but the Queen made no sound. I whispered to her,

"Stay as near as you can to the wall, Your Grace. There may be soldiers on the parapet above us. And keep close behind me."

Rapidly then, I led the way from the gate to the north-eastern corner of the courtyard wall. The east landing stage lay diagonally ahead from there, and only twenty yards away. I could see my own little boat bobbing at the edge of the shrubbery to one side of the stage. Over my shoulder, I said,

"Walk hand in hand, now, as you did before."

Immediately then I stepped from the sheltering shadow of the wall, and walked quickly to the landing stage. A swift glance showed me the other two following at the same brisk pace. The landing stage was clear—no, God damn it! There was someone there—a woman, just rising from the cover of the shrubbery. Minny! It was Minny waiting there. I became aware of the footsteps behind me faltering as she spoke.

"Will . . . I had to come. To say—"

I turned to reassure the Queen just at the moment I

realised that Minny—from sheer force of habit, no doubt—was sinking down in a curtsy to her. "No! For God's sake, Minny!" I made a lunge to grasp her and bring her upright again. "You could have us all caught if anyone saw you at that!"

She was trembling, and I had not meant to be so rough. I turned to the Queen and told her, "Get quickly into the boat, Your Grace. And keep well down till I tell you otherwise."

Minny had recovered herself enough to reach for the painter of the boat. I let her continue unwinding it from its bollard while I steadied the boat to let the Queen and Kennedy climb aboard. We came face to face in the moment before I myself leapt aboard; and bravely, Minny finished,

"—to say good-bye, Will."

"Not good-bye"—I bent and kissed her cheek—"but 'till we meet again.' "

"God keep you, son." I was thrusting with an oar to push the boat out as I caught Minny's last words; but she was out of sight before I had the chance to look up again. I spoke to the Queen and Kennedy, crouching low in the boat.

"I am rowing north just now, to bring us to that end of the island. The trees in the garden will give us cover till we reach it. Once I have cleared the island's north end, I will row northwest to the small island called Alice's Bower. That will keep us just out of sight of the north window in the Great Hall. And Alice's Bower will also give us cover when I turn west there, to head for the New House landing stage."

I concentrated all my strength of mind and body, then,

on clearing the north tip of the island and reaching Alice's Bower faster than I had ever done before. There was no movement from the other two, as I rowed. I thought briefly of Minny allowing herself to be locked out for the night so that she could say good-bye to me. Minny would be discovered when the alarm went up. But I *would* send for her. And soon. The Queen spoke softly into my thoughts,

"Sir William saw you tonight at the landing stage. From my east window."

My stroke grew rough for a moment before I said, "I was holding up the launch of the Castle boats—delaying pursuit, as I said I would."

"I guessed that. I distracted his attention by pretending to faint. And made such a show of it, he was not sure if he *had* seen you, after all."

So it was her quick-wittedness I had to thank for the answer to that prayer I had put up! I steadied my stroke and saw the north tip of the island beginning to drop away from me. A minute later, Alice's Bower loomed up on my left. I oared around for my final change in direction. The island slid past, and we were four minutes from the mainland shore. Gaspingly, I said,

"You may sit up now, Your Grace, and wave your signal to Lord Seton."

The Queen and Kennedy both struggled up to sit opposite me. The Queen swept the countrywoman's black hat from her head; and I saw that, underneath this, she had bound her hair up in a gauzy veil of red and white. Quickly she unwound this, and with her hair tumbling free then, she rose in the boat to wave the veil back and forwards, back and forwards. Kennedy steadied her as she waved, and it was Kennedy who said,

"There is someone waiting on the landing stage—an armoured man."

I looked over my shoulder, and immediately stopped rowing. It was not George waiting there. Neither was it Beaton, or Sempil. I hailed the waiting figure loudly, roughly.

"Who are you?"

The answer came back indistinctly, and I thought I caught the name "Wardlaw." But I knew no Wardlaw! I looked up at the Queen and told her,

"This man is a stranger to me."

Quietly she answered, "And to me."

So we could be running into a trap! I opened my mouth to speak the words, but she checked me with one uplifted hand. "Go on rowing, Will," she told me; and calmly seated herself again.

"But, Your Grace—"

"Yes!" she interrupted. "It could be a trap. But what can I do now, except to dare it?" Her eyes flashed suddenly with the light of that first challenge I had seen her give to Lord Lindsay. Straight-backed, with the ring of ultimate authority in her voice, she commanded, *"Go on rowing!"*

I was no longer in charge of the escape. I bent to the oars and pulled the last hundred yards to the landing stage. The voice of the man there sounded again as I began closing with it.

"Did you not hear me? I am Wardlaw—James Wardlaw, your guide to the rendezvous with Lord Seton."

I looked back over my shoulder. The man, Wardlaw, had come right to the edge of the stage. He turned, as I looked, and beckoned forcefully in the direction of the boat shed. From beyond the angle of the boat shed wall a band of horsemen came clattering, George Douglas in the lead, Beaton and Sempil riding just behind him.

Wardlaw seized the painter I threw to him, and pulled the boat close to the landing stage. I leapt ashore, and would have turned then to hand the Queen out of the boat—except that there was no need for me to do so. George had flung himself off his horse and come rushing to the edge of the landing stage. The others of his band had all followed suit, and there was a score of hands already waiting to help her ashore.

I stood back from the first flurry of greetings, and it was only when they brought the Queen in triumph to the waiting horses that I realised these had only two spare mounts among them. George turned from helping the Queen into the saddle of the first of these spare mounts. His eyes met mine, then went beyond me to where Jane Kennedy stood. I stood watching his indecision and miserably regretting my own failure to warn him she would be part of the escape plan.

The rest of the troop was mounted now, reins gathered, all ready to go. They too were looking towards myself and Kennedy. And so was the Queen. She urged her horse towards us, bent down, and said quietly to Kennedy,

"Jane, my gentle little Jane, I need no one now to play the lute or to dress my hair for me. But I do need soldiers."

Kennedy smiled up at her. "And I, Your Grace, am glad to be spared the battle that faces you. But I will find myself an outfit, and follow after you, to dress your hair for victory, *after* the battle."

Standing on tiptoe, she kissed the face bent down to her. The Queen sat straight again in the saddle, and turned to her escort.

"My lords," she cried. "Gentlemen! I owe my freedom to the boy, Will Douglas. And I will not leave this place till I see him mounted and ready to go with me."

George was already bringing the spare horse up to me, leading it along with his own. Laughing, he told me, "Mount and ride, Will! Mount and ride!"

I swung myself into the saddle at the same time as he did, laughing along with him. "Beside me, Will!" the Queen cried, and urged her horse towards the waiting figure of Wardlaw the guide.

"Through the village of Lochleven," he called to her, "and then another mile is all you have to the rendezvous."

*Through the village of Lochleven* . . . Did Wardlaw know what he was doing? It was not likely we would encounter opposition in so small a village; but there was no mistaking the identity of the Queen now—not with that mass of red-gold hair tumbled about her shoulders. What would the people of the village do when they saw her; the ordinary people who were neither soldiers nor spies, but simply her subjects? What would *their* attitude be to her now?

It was the last-but-one risk she would have to face before the final risk of the battle with Moray's men, I realised; the risk of being jeered at again, of having such names as Lindsay had used thrown once more in her face. And, I vowed, I would kill any man who dared use these names to her now.

I saw the first of the villagers within seconds of our start—a young man who stared in amazement at first, then smiled, tugged off his hat, and bowed low as she passed him. The Queen smiled back, and waved to him. And all through the village it was the same. With each little group of people we passed I saw how right—how gloriously right—Beaton had been in the intelligence sent through George. From each of these little groups I saw that first start of amazement, then smiles, waves, and a bending of the knee to her. And smiling, graciously acknowledging the people's pleasure in her return, the Queen led her troop through them towards the rendezvous.

Once out of the village, however, she called to Wardlaw to quicken the pace; and he did so—with a vengeance, too. A good horse can run a mile in little more than two minutes; and these horses stolen from Sir William were good ones that did not balk at the rough terrain they had to cover before we saw Lord Seton and his troop. They too were riding hard, sweeping down the shoulder of Benarty Hill towards the meeting point, and there was no slowing in the pace of either troop when they did join us there. Seton simply pointed to the track ahead and shouted,

"Farther southwards yet, Your Grace. To Queensferry, and your first resting place in freedom!"

And then the battle, I thought; the battle that would decide all. The Queen turned a laughing face towards me before she urged her horse to the head of our joint force. And she was so beautiful! The glow of colour on her skin was like the opalescent glow of a seashell. The amber of her eyes gleamed with dark, contrasting brilliance. In the wind of our passage her long hair was like a mass of golden-red light swirling around her head.

She was free; restored to subjects who loved her again, and so she was—she *must* be—riding to victory. And she owed that to *me*. That was the final message I saw then in her beautiful, laughing face; and in the pride and glory of that moment, I could have wept. . . .

I did weep, I did weep! With my head leaning on my hands, and the tears slowly dripping onto the courier's report of her execution. I wept for the victory that should have been hers; the victory that was snatched away when the fox, Argyll, failed at the last moment to bring up the support we needed. I wept for her flight from the battlefield, for the mistaken decision to seek a refuge in England, and for all the long years of captivity she had endured there.

Why had she not fled to her own kinfolk in France instead of trusting to the English Queen to help her? Why had she not realised how insecure that woman felt in her own Throne—and how much she feared that this beautiful and popular young Scottish Queen would become a focus of rebellion for *her* subjects? I wept for that too; and for the folly of those Englishmen who had finally justified the fears of *their* Queen.

My Mary, my Queen, would have been alive still if the plots of those Englishmen had not given their Queen an excuse to get rid of the "threat" to her throne. *My* Queen had never plotted against the life of that other one. My Queen had only ever wanted them to free her from her English prison. And she had not listened to me. For almost twenty years I had been master of all the spying and scheming on her behalf; yet still she had not listened when I warned her that those Englishmen's plots were only a trap set to kill her.

I reached a hand blindly to the papers on my desk. Her

letter to me lay there, the letter she had written a few hours before her execution. She had made her last will and testament then too. And she had remembered me in that. The wording of the bequest was in her letter—*to my little orphan, Will Douglas* . . . I could not bear to read the words again. The letter drifted from my hand.

Now it was Jane Kennedy's letter that lay on top of the other papers; "gentle little Jane" who had got her outfit and gamely followed behind us on that first day's ride to freedom. Jane had been there with her to the last moment of her life. Jane had attended her throughout the whole scene of the execution. . . .

. . . *I took the scarf that was to bind her eyes and stood in front of her. She bent to kiss me. Her cheek was cold where it touched mine. I was weeping so that I could scarcely see to tie the scarf. She knelt down, spreading her arms wide on either side of her.* . . .

I had seen her kneel like that—in the garden, that May day she had acted the part of my servant. But then she had been laughing as she knelt, laughing with pleasure in the masquerade she was playing, and in anticipation of the freedom soon to follow. . . . My eyes blurred so that I could not read any more.

I dropped Jane's letter also, and sat with my head bowed over it. The door to my study opened and Minny came in, the young courier following behind her. Minny's hair was white now. Minny was old. And my Queen was dead. I heard my own voice reaching out from a daze of anguish,

"All that beautiful hair, Minny. Do you remember that hair? She cut it off so that she could ride as fast as she had to, in her flight from that battlefield. And as far. Sixty miles, Minny. Sixty miles in the saddle on one day with never a complaint out of her, and her wonderful hair cropped as short as my own. . . ."

"Will—" Minny took me by the shoulder and gently shook me. "The courier wants to speak to you."

I looked beyond her to the courier, and Minny moved aside to let him speak directly to me. He was awkward to begin with; his round, innocent face flushing as he told me,

"It was to comfort your distress, sir, that I asked to speak to you. To tell you something about *her*."

I said roughly, "What can *you* tell me? You never knew her as I did."

"Sir, you are wrong! You knew her only as the restless Queen in captivity. I knew her in death. And she was changed then—utterly changed; no longer struggling against her situation, but totally accepting it instead."

The words had burst from him with all the fervour *I* had ever been capable of at his age; and when he saw I did not mean to stop him, he rushed on in a torrent of speech,

"I determined to be there, sir, as soon as I knew that some of the common people would be permitted to view the execution. I saw it all, sir. I saw her enter the hall with none of the pomp or privilege of a Queen allowed to her, but still so calm and dignified that she *looked* royal. And she died firm in her faith, sir. She told her execution- ers so. She told them also that she trusted in God to receive her spirit. And all this she said with a face so serene that it seemed to us she was almost happy to die."

*My Queen, my Mary, my love* . . . It had taken George Douglas more than eight years of hopeless devotion to her before his longing for children had led him to marry another. But I had never married. The mist in front of my eyes would not let me see the boy, but still I heard him continue,

"And from first to last, sir, she never faltered in anything she had to do. Her speech was clear, her bearing as regal as ever. And her courage, sir—" The boy paused; and then, in a voice that choked with the strength of his feeling, he finished, "The courage of this Queen, sir, was matchless!"

I blinked to clear my sight. The boy was looking beyond me, his eyes filled now with the light of some inward vision. He spoke again, a strangely yearning note in his voice,

"I have heard, sir, how beautiful she was in her young days. But I have still to see a woman more beautiful than she was then. And I swear to you, I do not think I ever shall. Yet it was not her features alone that made me think so. It was something within herself, something—"

Again he paused, groping for words. I heard a ghost whisper in my ear; the sardonic ghost of long-dead, "blindly devoted" Arnault telling once more his analogy of the honeypot and the bees. The boy, I thought, did not yet realise what had happened to him. But I knew. I could see how, in the last few moments of her life, the boy had sensed the strange attraction that had never failed to draw men to her. I rose and went towards him, no longer resenting the way he had contradicted when I said, "You never knew her as I did." Nor could I grudge him this one last memory that was all he had to set against the host of memories that were mine to cherish. I let my hand fall on his shoulder, and told him,

"You will find the words yet, for what you wanted to say."

I felt the quiver that ran through him then. I drew Minny to me with my other hand, while the voice that had so often before cried out in my heart, cried out again, *"My Queen, my Mary, my love. . . ."* But I owed that boy something for the way he had given of his own small share of experience in the attempt to comfort me. I owed him, I thought, some acknowledgment of *his* feeling. Some right to mourn with me. Quietly I said,

"God will take care of her soul. We can all be sure of that. But we—you and I together, lad—we will remember *her.*"

*"Our* Queen!" I sensed the words proudly repeating in the boy's mind as I walked him and Minny from the room with me. And silently again my heart's cry responded, *"My* Mary. *My* love. . . ."

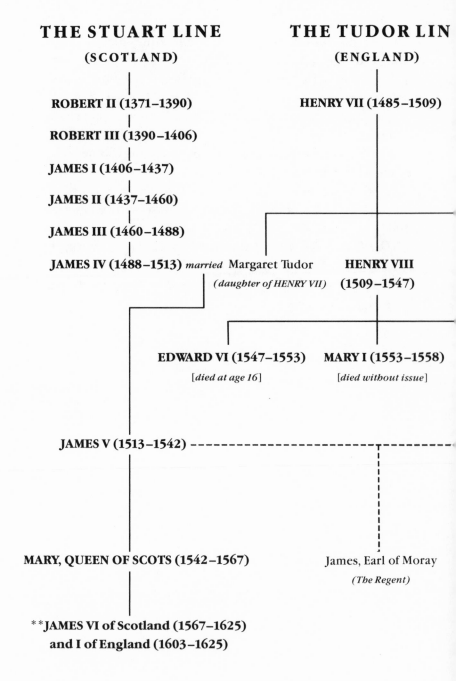

**THE STUART LINE**
(SCOTLAND)

**THE TUDOR LIN**
(ENGLAND)

**ROBERT II (1371–1390)**

**HENRY VII (1485–1509)**

**ROBERT III (1390–1406)**

**JAMES I (1406–1437)**

**JAMES II (1437–1460)**

**JAMES III (1460–1488)**

**JAMES IV (1488–1513)** *married* Margaret Tudor  **HENRY VIII**
*(daughter of HENRY VII)*  **(1509–1547)**

**EDWARD VI (1547–1553)**   **MARY I (1553–1558)**
[*died at age 16*]   [*died without issue*]

**JAMES V (1513–1542)** - - - - - - - - - - - - - - - - - - - - - - - - -

**MARY, QUEEN OF SCOTS (1542–1567)**   James, Earl of Moray
*(The Regent)*

**\*\*JAMES VI of Scotland (1567–1625)
and I of England (1603–1625)**

# THE STUARTS AND TUDORS*

\* Names in capitals indicate crowned sovereigns; dates in
parentheses indicate period of reign; solid lines indicate
legitimacy; dotted lines indicate illegitimacy.

\*\* Became James I of Great Britain at the Union of England and
Scotland in 1603. Tudor ancestry traced back through James
IV's marriage to Margaret Tudor.

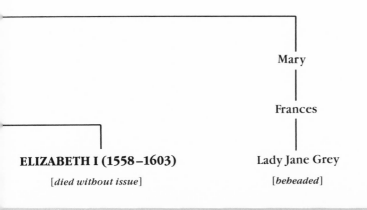

Mary

Frances

**ELIZABETH I (1558–1603)**

*[died without issue]*

Lady Jane Grey

*[beheaded]*

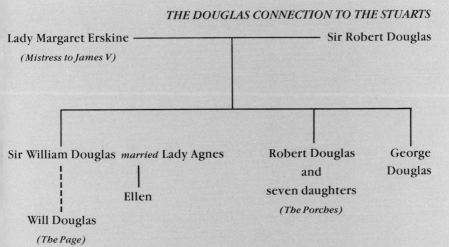

*THE DOUGLAS CONNECTION TO THE STUARTS*

Lady Margaret Erskine ———————————— Sir Robert Douglas

*(Mistress to James V)*

Sir William Douglas *married* Lady Agnes

Ellen

Will Douglas

*(The Page)*

Robert Douglas
and
seven daughters

*(The Porches)*

George
Douglas

## ABOUT THE AUTHOR

Mollie Hunter enjoys writing about her native Scotland, from the prehistory of her Carnegie Medal winner *The Stronghold* to the folklore that inspires fantasies such as *The Kelpie's Pearls* and *The Wicked One*. The Highland glen where she lives with her husband, Michael, is rich in history, and its atmosphere combines with years of research to give authentic flavour to her historical novels.

But Mollie Hunter is not all romantic Celt. Early years in a Lowland village near Edinburgh have given her the setting for *A Sound of Chariots* and *The Third Eye*, two powerful novels that speak realistically of the problems of young-adult life.

In the words of one eminent critic, she is "Scotland's most gifted storyteller"; and this, of all the awards her writing has won, is the accolade she most values.